Lecture Notes in Mathematics 1663

Springer
Berlin
Heidelberg
New York
Barcelona
Budapest
Hong Kong
London
Milan
Paris
Santa Clara
Singapore
Tokyo

Yulia E. Karpeshina

Perturbation Theory for the Schrödinger Operator with a Periodic Potential

Springer

Author

Yulia E. Karpeshina
Department of Mathematics
University of Alabama
Birmingham, AL 35294-1170
USA
E-Mail: Karpeshi@vorteb.math.uab.edu

Cataloging-in-Publication Data applied for

Die Deutsche Bibliothek - CIP-Einheitsaufnahme

Karpešina, Julija E.:
Perturbation theory for the Schrödinger operator with a periodic potential / Yulia E. Karpeshina. - Berlin ; Heidelberg ; New York ; Barcelona ; Budapest ; Hong Kong ; London ; Milan ; Paris ; Santa Clara ; Singapore ; Tokyo : Springer, 1997
(Lecture notes in mathematics ; 1663)
ISBN 3-540-63136-4

Mathematics Subject Classification (1991):
35A35, 35B20, 35B27, 35B30, 35B40, 35C10, 35C20, 35J10, 35P15, 35P20, 81C05, 81H20, 81Q05, 81Q10, 81Q15

ISSN 0075-8434
ISBN 3-540-63136-4 Springer-Verlag Berlin Heidelberg New York

Typesetting: Camera-ready TEX output by the author
SPIN: 10553275 46/3142-543210 - Printed on acid-free paper

Table of Contents

1. Introduction.

The Schrödinger operator with a periodic potential describes the motion of a particle in bulk matter. Therefore, it is interesting to have a detailed analysis of the spectral properties of this operator. Both physicists and mathematicians have been studying the periodic Schrödinger operator for a long time [1]. The most significant progress has been achieved in the one-dimensional case [2] The two and three dimensional cases are still of great challenge.

Initially, physicists observed that the spectrum of the periodic Schrödinger operator has a band structure and is semibounded below (see f.e.[BS, Ki, Mad, Zi]). Moreover, according to the famous Bethe-Sommerfeld conjecture [BS] there exist only a finite number of gaps in the spectrum. The eigenfunctions of each band can be described as "Bloch functions", which satisfy quasiperiodic conditions in the elementary cell [Bl]. This means that they can be parameterized by the number of the band and the quasimomentum, which is a parameter of the quasiperiodic conditions. The eigenvalues of Bloch eigenfunctions with a fixed quasimomentum form a discrete set.

For physical applications it is important to have a perturbation theory of the Scrödinger operator with a periodic potential. In one-dimensional situation the perturbation theory was constructed by Carvey D. Mc. [C1] – [C3]. However, in many dimensional situations its construction turns out to be rather difficult, because the denseness of Bloch eigenvalues of the free operator increases infinitely with increasing energy. Under perturbation, the eigenvalues influence each other strongly and the regular perturbation theory does not work. The main aim of this book is to construct perturbation formulae for Bloch eigenvalues and their spectral projections in a high energy region on a rich set of quasimomenta. The construction of these formulae is connected with the investigation of a complicated picture of the crystal diffraction.

Another problem, considered here, is a semi-bounded crystal problem, i.e., the Schrödinger operator which has the zero potential in a half space and a periodic potential in the other half space. The interaction of a plane wave with

---

[1] see f.e. [A], [Ag], [Ar] – [DavSi], [Di] – [DyPe], [Ea1] – [GiKnTr2], [GorKapp1] – [HøHoMa], [KargKor] – [K17], [Ki] – [Le], [Mad] – [Out], [Pav], [PavSm1] – [Rai], [ReSi4] – [SheShu], [Si2] – [Zi].

[2] see f.e. [Av1], [BelBovChe], [Bent1], [Bu1] – [BuDm3] , [C1] – [C3], [Ea2], [Fir1] – [FirKor], [FroPav2], [GaTr1, GaTr2], [Ha], [KargKor], [Kohn] – [LaPan], [MagWin] – [McKTr2], [Ol], [PavSm1], [PavSm2], [ReSi4], [SheShu], [Ti], [WeKe1], [WeKe2].

a semicrystal will be studied. First, the asymptotic expansion of the reflection coefficients in a high energy region will be obtained, this expansion is valid for a rich set of momenta of the incident plane wave. Second, the connection of the asymptotic coefficients with the potential will be established. Based upon these, the inverse problem will be solved, this problem is to determine the potential from the asymptotics of the reflection coefficients in a high energy region (a crystallography problem).

Let us describe briefly some previous results.

I.M. Gelfand began the rigorous study of the periodic Schrödinger operator [Gelf]. He proved the Parseval relation for Bloch waves in $L_2(R^n)$. The expansion theorem was proved by E.Ch Titchmarsh [Ti] in the one-dimensional situation and by F.Odeh, J.B. Keller [OdKe] in the many-dimensional case. V.L. Lidskiy applied E.Ch Titchmarsh's method to prove the Parseval formula in manydimensional situation [Ea2]. The first rigorous proof of the fact, that the spectrum of the periodic Schrödinger operator is the union of all the Bloch eigenvalues, corresponding to different quasimomenta, was given by F.Odeh, J.B. Keller [OdKe]. M.S.P. Eastam gave another proof of this fact [Ea1, Ea2]. L.E. Thomas showed that the spectrum of the operator is absolutely continuous [Th]. Wilsox C. studied analytical properties of eigenvalues as functions of quasimomenta [Wil].

The detailed investigation of the band structure is still a challenging problem. A first step in this direction was made by M. M. Skriganov [Sk1]–[Sk7]. He gave the proof of the Bethe- Sommerfeld conjecture. He considered the operator

$$H = (-\Delta)^l + V \tag{1.0.1}$$

in $L_2(R^n)$, $n > 1$, where V is the operation of multiplication by a smooth potential. M. M. Skriganov has proved the conjecture for certain n, l, including the physically interesting cases $n = 2, 3, l = 1$ (the Schrödinger operator). He developed the subtle methods of arithmetical and geometrical theory of lattices. This makes proofs sometimes different for rational and non-rational lattices. For example, in the case $4l > n + 1$, only a proof for rational lattices is given.

Another beautiful proof of the Bethe-Sommerfeld conjecture in the dimension two, using an asymptotic of a Bessel function, was found by B.E.J. Dahlberg and E. Trubowitz [DahTr].

However, one can suppose that Bethe and Sommerfeld were guided by the ideas of perturbation theory for the many-dimensional case. The different approaches to the construction of this theory one can find in [FeKnTr1, FeKnTr2], [Fri], [K4] – [K15], [Ve1] – [Ve7]. As it was mentioned before, its mathematical foundation is a complicated matter, because the denseness of Bloch eigenvalues of a free operator ($V = 0$) increases infinitely with increasing energy. The Bloch eigenvalues of the free operator are situated very close to each other in a high energy region. Therefore, when perturbation disturbs them, they strongly influence each other. Thus, to describe the perturbation of one of the eigenvalues, we must study not only that eigenvalue, but also the surrounding ones. This causes analytical difficulties, in particular, "the small denominators problem".

The first asymptotic formula in the high energy region for a stable under perturbation Bloch eigenvalue has been constructed for $l = 1$ by O.A. Veliev

[Ve1] – [Ve7]. The stable case corresponds to nonsignificant diffraction inside the crystal. The validity of the Bethe-Sommerfeld conjecture for $n = 2$ and $n = 3$ is a consequence of this formula. The formula in [Ve1] – [Ve7] reproduces the first terms of the asymptotic behavior of the eigenvalue in the case of a smooth potential.

The first results about the unstable case, which corresponds to a significant diffraction inside the crystal, were obtained by J. Feldman, H. Knörrer and E. Trubowitz [FeKnTr2] (more precisely about these results see page 16).

In our consideration the perturbation series both for an eigenvalue and a spectral projection are constructed. The method is based on the expansion of the resolvent in a perturbation series. The series converge for a rich set of quasimomenta and have an asymptotic character in the high energy region. They are differentiable with respect to the quasimomentum and preserve their asymptotic character. The particular terms in the series are simple and can be calculated directly. This is the first method which works for a general class of potentials, including potentials with Coulomb and even stronger singularities. This perturbation theory is valid not only for the proof of the Bethe-Sommerfeld conjecture, but, moreover, for the description of the isoenergetic surface. Many other physical values can be determined using these formulae. In the unstable case the perturbation series are constructed with respect to an auxiliary operator, which roughly describes the diffraction inside the crystal.

One of the main difficulties is to construct the nonsingular set, that is, the set of quasimomenta for which the perturbation series converge. This difficulty is certainly of a physical nature. Convergence of the perturbation series for an eigenfunction shows the perturbed eigenfunction to be close to the unperturbed one (the plane wave). This means that this plane wave goes through the crystal almost without diffraction. But it is well known that, in fact, the plane wave $\exp i(\mathbf{k}, x)$ is refracted by the crystal, if $\mathbf{k}$ satisfies the von Laue diffraction condition (see f.e.[BS], [Ki], [Mad]):[3]

$$|\mathbf{k}| = |\mathbf{k} + 2\pi\mathbf{q}|, \tag{1.0.2}$$

for some $\mathbf{q} \in Z^3 \setminus \{0\}$. The refracted wave is known to be $a \exp(i(\mathbf{k} + 2\pi\mathbf{q}, x))$, $a \in C$. This wave interferes with the initial one $\exp(i(\mathbf{k}, x))$ and distorts it strongly. This means that the perturbation series diverges if $\mathbf{k}$ is not far from the planes (1.0.2). Here the question arises: does the series converge when $\mathbf{k}$ is not in the vicinity of (1.0.2)? It turns out that it does, when $2l > n$. However, it is not enough for the control of the convergence of series when $2l \leq n$. There are some additional diffraction conditions arising in this more complicated case. Another problem is that when eliminating the singular set (where the series can diverge), we must take care that this set does not become "too extensive". This means that it must not include the whole set of quasimomenta which correspond to a given energy. We shall show that the nonsingular set is rather rich – it has an asymptotically full measure on the isoenergetic surface of the free operator. Geometric considerations are made in the explicit form for a smooth potential.

[3] In this equation we suppose a cell of periods to be unit.

In the case of a nonsmooth potential the formulae for the nonsingular set are less explicit, but, nevertheless, the set can be determined by a simple computer program.

To construct the nonsingular set in the simplest case $2l > n$ and $V \in L_\infty$, we delete from the isoenergetic surface S_k of the free operator (S_k is a sphere of radius k centered at the origin of R^n, $0 < \delta \ll 1$) the momenta belonging to the $(k^{-n+1-2\delta})$-neighborhood of the planes $|\mathbf{k}| = |\mathbf{k} + 2\pi\mathbf{q}|$, $\mathbf{q} \in Z^n \setminus \{0\}$. In the rest of S_k the perturbation series converge and the perturbation of the isoenergetic surface is asymptotically small in a high energy region ($k \to \infty$).

The situation becomes more complicated as soon as we lift the restriction $2l > n$. When $2l \leq n$, but $4l > n + 1$ and V is smooth, we have to delete from S_k some vicinities of the planes

$$|\mathbf{k} + 2\pi\mathbf{m}| = |\mathbf{k} + 2\pi\mathbf{m} + 2\pi\mathbf{q}|, \quad \mathbf{m} \in Z^n, \ \mathbf{q} \in Z^n \setminus \{0\}, \tag{1.0.3}$$

$|\mathbf{k} + 2\pi\mathbf{m}| \approx k$. We call relations (1.0.3) *the Generalized Laue Diffraction Conditions.* The size of the vicinity to delete depends on k and $\mathbf{m}, \mathbf{q}$. Thus, the nonsingular set becomes less extensive than in the case $2l > n$, but nevertheless, it has an asymptotically full measure on S_k. From equation (1.0.3) one can see that the formulae for the nonsingular set depend only on the periods of the potential.

Special considerations are needed for a non-smooth potential. We introduce the concept of the "number" of states and consider its geometrical aspects. The nonsingular set is described in the terms of the number of states.

The situation is most complicated in the case of the **Schrödinger operator**. The singular set has a part which depends essentially on the potential, even when it is smooth. To construct the nonsingular set one has to delete a neighborhood of the surfaces:

$$|\mathbf{k} + 2\pi\mathbf{m}|^2 = |\mathbf{k} + 2\pi(\mathbf{m} + \mathbf{q})|^2 + \Delta\lambda_{\mathbf{mq}}(\mathbf{k}). \tag{1.0.4}$$

Here, as before, $\mathbf{m}, \mathbf{q} \in Z^3$, $|\mathbf{k} + 2\pi\mathbf{m}| \approx k$. The new terms $\Delta\lambda_{\mathbf{m}}(\mathbf{k})$ are smooth functions of $\mathbf{k}$ determined by the potential. For many $\mathbf{m}$, $\Delta\lambda_{\mathbf{mq}}(\mathbf{k}) = 0$, but for a number of $\mathbf{m}$ the functions $\Delta\lambda_{\mathbf{mq}}(\mathbf{k})$ essentially differ from zero; they are the perturbations of Bloch eigenvalues of the free operator in the one-dimensional situation by some periodic potential V_q. We call equations (1.0.4) the *Modified Laue Diffraction Conditions.*

The case of the Schrödinger operator with a nonsmooth potential accumulates all described restrictions on the nonsingular set.

In the case when t is at the diffraction surface (1.0.2), the refracted wave arises in the crystal and there exists a splitting of the degenerated eigenvalue. Suppose that $\mathbf{k}$ satisfies the von Laue condition $|\mathbf{k}| = |\mathbf{k} + 2\pi\mathbf{q}|$ for a unique $\mathbf{q}$: it is generally known that the plane wave $\exp i(\mathbf{k}, x)$ is refracted by the crystal for such k. Physicists consider the refracted wave to be $a \exp i(\mathbf{k} + 2\pi\mathbf{q}, x)$, $a \in C$ (see f.e. [Ki, Mad, Zi]). The resulting wave is a linear combination of the initial and refracted waves. The mathematical study of this problem (Chapter 2) shows that this is a good approximation for the case $2l > n$. Taking a model operator

H_q – which roughly accounts the refraction and splitting – as the initial operator instead of H_0, we construct the perturbation series for t near a diffraction plane. The rigorous study of the diffraction at the Laue diffraction planes (1.0.2), in the case of the Schrödinger operator ($n = 3, l = 1$), shows that such simple approximation is not sufficient any longer. We have to represent a refracted wave as a linear combination of the waves $a_n \exp i(\mathbf{k}+2\pi n\mathbf{q}, x)$, $n \in Z$. This will be an approximate refracted wave. It is constructed by using a model operator, roughly describing the refraction inside the crystal. This operator has a more complicated form than that in the case $2l > n$. Taking the model operator as the initial operator instead of H_0, we construct the perturbation series for t near the diffraction surface.

The perturbation series near the nonsingular set and the planes of diffraction make it possible to describe an essential part of the perturbed isoenergetic surface.

In the case of a semicrystal we consider its interaction with an incident plane wave $\exp(i(\mathbf{k}, x))$. Let $\mathbf{k}$ belong to the nonsingular set for the whole crystal. Therefore, a wave close to $\exp i(\mathbf{k}, x)$ can propagate inside the crystal. For the wave $\exp i(\mathbf{k}, x)$ to "penetrate" actually inside the crystal, we have to eliminate the interaction of the incident wave $\exp i(\mathbf{k}, x)$ with the surface. To do this, we have to impose more restrictions on the nonsingular set. Nevertheless, this new nonsingular set has an asymptotically full measure on S_k. Under some new restrictions on the nonsingular set we construct a high-order asymptotic expansions of the reflected and refracted waves for $\mathbf{k}$ belonging to this nonsingular set. Furthermore, we show that the relations between the asymptotic coefficients and the potential are not very complicated. That is why one can determine the potential from the asymptotic expansion of the reflected wave (only if the potential is known to be a trigonometric polynomial).

Now we want to describe the results more concretely. We study the operator (1.0.1) in three cases, as geometric and analytical difficulties increase:

1. $2l > n$;
2. $4l > n + 1$, $(2l \leq n)$;
3. $n = 3, l = 1$.

The case $2l > n$ is considered in Chapter 2, the case $4l > n + 1$ – in the Chapter 3, and the case $n = 3$, $l = 1$ (the Schrödinger operator) is studied in Chapter 4.

Let us write potential $V(x)$ in the form :

$$V(x) = \sum_{m \in Z^n} v_m \exp i(\mathbf{p}_m(0), x), \tag{1.0.5}$$

where $(\cdot, \cdot)$ is the scalar product in R^n and $\mathbf{p}_m(0)$ is a vector of the dual lattice:

$$\mathbf{p}_m(0) = 2\pi(m_1 a_1^{-1}, ..., m_n a_n^{-1}).$$

The potential V is real by assumption, so $v_m = \bar{v}_{-m}$. We suppose $v_0 = 0$; this assumption does not restrict the generality of our considerations.

We consider a potential, which satisfies the condition

$$\sum_{m \in Z^n \setminus \{0\}} | v_m |^2 | m |^{-4l+n+\beta} < \infty, \tag{1.0.6}$$

$$| m | = (m_1^2 + ... + m_n^2)^{1/2},$$

for some β obeying the following inequalities:

$$\begin{array}{ll} \beta > 0 & \text{if } 2l \leq n, \quad n \neq 2 \text{ (in particular, } n = 3, l = 1) \\ \beta \geq 2l - n & \text{if } 2l > n, \quad n \neq 2 \text{ or } \ 2l > 3, n = 2; \\ \beta > 1 & \text{if } \ l < 3 \ n = 2. \end{array}$$

The potential does not need to be smooth to satisfy this condition. For example, in the case $n = 3$, a function, which behaves in a neighborhood of some point x_0 as $| x - x_0 |^{-\zeta}$, $\zeta < 2l$, in particular, a Coulomb potential, satisfies this condition.

For the sake of simplicity we assume that the potential has orthogonal periods $a_1, ..., a_n$, however all the results are valid also for non-orthogonal periods.

It was shown [Gelf, OdKe, Ea1, Ea2, Th] that the spectral analysis of H can be reduced to studying the operators $H(t)$, $t \in K$, where K is the unit cell of the dual lattice,

$$K = [0, 2\pi a_1^{-1}) \times ... \times [0, 2\pi a_n^{-1}).$$

The vector t is called quasimomentum. The operator $H(t)$, $t \in K$, acts in $L_2(Q), Q = [0, a_1) \times ...[0, a_n)$. Its action is described by formula (1.0.1) together with the quasiperiodic conditions:

$$u(x_1, ..., x_{j-1}, a_j, x_{j+1}, ..., x_n) = \exp(it_j a_j) u(x_1, ..., x_{j-1}, 0, x_{j+1}, ..., x_n), \tag{1.0.7}$$

$$j = 1, ..., n.$$

The derivatives with respect to x_j, $j = 1, ..., n,$, must also satisfy the similar conditions.

The operator $H(t)$ has a discrete semi-bounded spectrum $\Lambda(t)$:

$$\Lambda(t) = \cup_{n=1}^{\infty} \lambda_n(t), \ \lambda_n(t) \to_{n \to \infty} \infty.$$

The spectrum Λ of operator H is the union of the spectra $\Lambda(t)$,

$$\Lambda = \cup_{t \in K} \Lambda(t) = \cup_{n \in N, t \in K} \lambda_n(t).$$

The functions $\lambda_n(t)$ are continuous, so Λ has a band structure:

$$\Lambda = \cup_{n=1}^{\infty} [q_n, Q_n], \quad q_n = \min_{t \in K} \lambda_n(t), \quad Q_n = \max_{t \in K} \lambda_n(t).$$

The eigenfunctions of $H(t)$ and H are simply related. If we extend the eigenfunctions of all the operators $H(t)$ quasiperiodically (see (1.0.7)) to R^n, we obtain a complete system of eigenfunctions of the operator H.

Let $H_0(t)$ be the operator corresponding to the zero potential. Its eigenfunctions are the plane waves:

$$\exp(i(\mathbf{p}_j(t), x)), \; j \in Z^n, \; \mathbf{p}_j(t) = \mathbf{p}_j(0) + t. \tag{1.0.8}$$

The eigenfunction (1.0.8) corresponds to the eigenvalue $p_j^{2l}(t) = \mid \mathbf{p}_j(t) \mid^{2l}$. Thus, the spectrum of H_0 is equal to

$$\Lambda_0(t) = \{p_j^{2l}(t)\}_{j \in Z^n}.$$

Using the basis of the eigenfunctions of $H_0(t)$ one can write the matrix $H(t)$ in the form

$$H(t)_{mj} = p_m^{2l}(t)\delta_{mj} + v_{m-j}, \tag{1.0.9}$$

where δ_{mj} is the Kronecker symbol. Of course, the free operator is diagonal in this basis.

Note that any $\mathbf{k} \in R^n$ can be uniquely represented in the form:

$$\mathbf{k} = \mathbf{p}_j(t), \; j \in Z^n, \; t \in K. \tag{1.0.10}$$

Thus, any plane wave $\exp i(\mathbf{k}, x)$ can be written in the form (1.0.8). Naturally, we can rewrite the von Laue diffraction conditions (1.0.2) for (1.0.8) as follows:

$$p_j^{2l}(t) = p_{j+q}^{2l}(t) = k^{2l}, q \neq 0. \tag{1.0.11}$$

Similarly, the Generalized von Laue diffraction conditions and the Modified von Laue diffraction conditions can be represented as follows:

$$p_{j+m}^{2l}(t) = p_{j+m+q}^{2l}(t), \; q \neq 0, \; p_j^{2l}(t) = k^{2l} \tag{1.0.12}$$

$$p_{j+m}^{2}(t) = p_{j+m+q}^{2}(t) + \Delta\lambda_{mq}(\mathbf{p}_j(t)), \; q \neq 0, \; p_j^2(t) = k^2. \tag{1.0.13}$$

We will use formula (1.0.8) for plane waves and formulae (1.0.11) – (1.0.13) for the diffractions conditions.

In physical literature, the important concept of the isoenergetic surface of the free operator is used (see f.e. [Ki, Mad, Zi]). It is said that a point t belongs to an isoenergetic surface $S_0(k)$ of the free operator H_0, if and only if, the operator $H_0(t)$ has an eigenvalue equal to k^{2l}, i.e., there exists $m \in Z^n$, such that $p_m^{2l}(t) = k^{2l}$. This surface can be obtained as follows: the sphere of radius k centered at the origin of R^n is divided into pieces by the dual lattice $\{\mathbf{p}_m(t)\}_{m \in Z^n}$, and then all these pieces are transmitted into the cell K of the dual lattice. Thus, we obtain the sphere "packed into the bag" K (Fig.1).
Note that the selfintersections are described by the von Laue diffraction conditions (1.0.11).

Let $S_1(k) \subseteq S_0(k)$. We say that $S_1(k)$ has an asymptotically full measure on $S_0(k)$ if the relation

$$\frac{s(S_1(k))}{s(S_0(k))} \to_{k \to \infty} 1 \tag{1.0.14}$$

holds, where $s(\cdot)$ is the area of a surface.

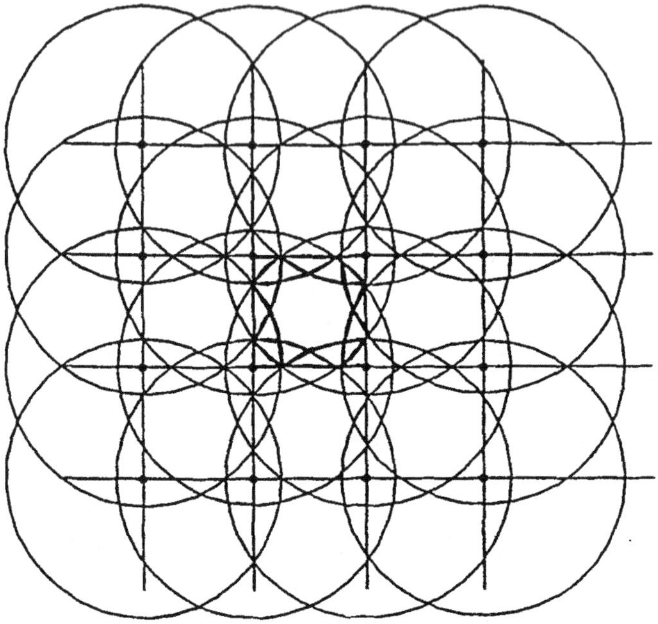

Fig.1 The isoenergetic surface of the

free operator for n=2

In Chapter 2 we consider the case **2l > n**, where V is a trigonometric polynomial. This simplest situation is described in order to clear up the basic method of our considerations – the formal construction of perturbation series and the description of the nonsingular set for which these series converge. In this chapter we introduce the factor α, $-1 \leq \alpha \leq 1$ in front of the potential, and consider the operator:

$$H_\alpha = (-\Delta)^l + \alpha V, \tag{1.0.15}$$

$$V(x) = \sum_{j \in Z^n, |j| < R_0} v_j \exp(i(\mathbf{p}_j(0), x)), \quad R_0 < \infty \tag{1.0.16}$$

We describe the nonsingular set $\chi_0(k,\delta)$ for this case as $S_0(k) \setminus A_0(k,\delta)$, where $A_0(k,\delta)$ is the $(k^{-n+1-\delta})$-neighborhood of the selfintersections of $S_0(k)$. If $t \in \chi_0(k,\delta)$, then (1.0.11) does not hold, i.e., there is a unique j such that $p_j^{2l}(t) = k^{2l}$. Moreover, for all $m \neq j$:

$$\left|p_m^{2l}(t) - p_j^{2l}(t)\right| > 2k^{2l-n-\delta}.$$

This inequality means that the free operator has a unique eigenvalue k^{2l} in the interval $(k^{2l} - 2k^{2l-n-\delta}, k^{2l} + 2k^{2l-n-\delta})$. We will prove that the nonsingular set $\chi_0(k,\delta)$ has an asymptotically full measure on $S_0(k)$.

We construct perturbation formulae as follows. For t of $\chi_0(k,\delta)$, we prove the convergence of the standard perturbation series for the resolvent:

$$(H_\alpha(t)-z)^{-1}=(H_0(t)-z)^{-1}\sum_{r=0}^{\infty}(-V(H_0(t)-z)^{-1})^r, \qquad (1.0.17)$$

when z belongs to the circle C_0 of the radius $k^{2l-n-\delta}$ centered at the unperturbed eigenvalue $p_j^{2l}(t)=k^{2l}$. Let $\varepsilon(k,\delta)$ be the interval cut out by the circle C_0 on the real axis: $\varepsilon(k,\delta)=[k^{2l}-k^{2l-n-\delta},k^{2l}+k^{2l-n-\delta}]$. Using the well-known formulae for the number n_α of perturbed eigenvalues in $\varepsilon(k,\delta)$ (see f.e. [Kato, ReSi4])

$$n_\alpha=-\frac{1}{2\pi i}\mathrm{Tr}\oint_{C_1}(H_\alpha(t)-z)^{-1}dz, \qquad (1.0.18)$$

and the convergence of the series for the resolvent, we prove that the perturbed operator $H_\alpha(t)$ has also only one eigenvalue in $\varepsilon(k,\delta)$. Then, using the formulae for the perturbed eigenvalue $\lambda(\alpha,t)$ and its spectral projection $E(\alpha,t)$

$$\frac{\partial\lambda(\alpha,t)}{\partial\alpha}=-\frac{1}{2\pi i}\mathrm{Tr}\oint_{C_0}V(H_\alpha(t)-z)^{-1}dz, \qquad (1.0.19)$$

$$E(\alpha,t)=-\frac{1}{2\pi i}\oint_{C_0}(H_\alpha(t)-z)^{-1}dz, \qquad (1.0.20)$$

we obtain series for the perturbed eigenvalue and its spectral projection. To describe these formulae more precisely, we introduce functions $g_r(k,t)$ and operator-valued functions $G_r(k,t)$, $r=0,\ldots,$ $t\in K$, which are the results of applying the integral formulae (1.0.19) and (1.0.20) to r-th term of the series (1.0.17):

$$g_r(k,t)=\frac{(-1)^r}{2\pi i}\mathrm{Tr}\oint_{C_0}((H_0(t)-z)^{-1}V)^r dz, \qquad (1.0.21)$$

$$G_r(k,t)=\frac{(-1)^{r+1}}{2\pi i}\oint_{C_0}((H_0(t)-z)^{-1}V)^r(H_0(t)-z)^{-1}dz. \qquad (1.0.22)$$

Let E_j be the spectral projection of the free operator, corresponding to the eigenvalue p_j^{2l} : $(E_j)_{rm}=\delta_{jr}\delta_{jm}$.[4]

As it was mentioned before, the free operator has a unique eigenvalue in the interval $[k^{2l}-2k^{2l-n-\delta},k^{2l}+k^{2l-n-\delta}]$. Thus, the functions $g_r(k,t)$ and $G_r(k,t)$ are correctly defined when $t\in\chi_0(k,\delta)$, because both integrands have a unique pole at the point $z=k^{2l}$. Each of the contour integrals equals the residue of the corresponding rational function at this point. For example,

$$g_1(k,t)=0 \qquad (1.0.23)$$

$$g_2(k,t)=\sum_{q\in Z^n,q\neq 0}|v_q|^2(p_j^{2l}-p_{j+q}^{2l})^{-1}= \qquad (1.0.24)$$

[4] δ_{jr},δ_{jm} are the Kronecker symbols.

$$= \sum_{q\in Z^n, q\neq 0} \frac{|v_q|^2 (2p_j^{2l} - p_{j+q}^{2l} - p_{j-q}^{2l})}{2(p_j^{2l} - p_{j+q}^{2l})(p_j^{2l} - p_{j-q}^{2l})},$$

$$G_1(k,t)_{rm} = v_{j-m}(p_j^{2l} - p_m^{2l})^{-1}\delta_{rj} + v_{r-j}(p_j^{2l} - p_r^{2l})^{-1}\delta_{mj}. \tag{1.0.25}$$

All the constructions are stable with respect to small perturbations of t, more precisely, with respect to perturbations of order $k^{-n+1-2\delta}$. Thus, the perturbation series for eigenvalues and spectral projections are constructed in the $(k^{-n+1-2\delta})$-neighborhood of $\chi_0(k,\delta)$. The following theorem holds.

Theorem 1.1 . *Suppose t belongs to the $(k^{-n+1-2\delta})$-neighborhood in K of the nonsingular set $\chi_0(k,\delta)$, $0 < 2\delta < 2l - n$. Then, for sufficiently large k, $k > k_0(V,\delta)$, and for all α, $-1 \leq \alpha \leq 1$, there exists a single eigenvalue of the operator H in the interval $\varepsilon(k,\delta) \equiv [k^{2l} - k^{2l-n-\delta}, k^{2l} + k^{2l-n-\delta}]$. It is given by the series:*

$$\lambda(\alpha,t) = p_j^{2l}(t) + \sum_{r=2}^{\infty} \alpha^r g_r(k,t), \tag{1.0.26}$$

converging absolutely in the disk $|\alpha| \leq 1$, where the index j is uniquely determined from the relation $p_j^{2l}(t) \in \varepsilon(k,\delta)$. The spectral projection, corresponding to $\lambda(\alpha,t)$, is given by the series:

$$E(\alpha,t) = E_j + \sum_{r=1}^{\infty} \alpha^r G_r(k,t), \tag{1.0.27}$$

which converges in the class $\mathbf{S}_1$ uniformly with respect to α in the disk $|\alpha| \leq 1$.

For the coefficients $g_r(k,t)$ and $G_r(k,t)$ the following estimates hold: [5].

$$| g_r(k,t) | < k^{-\gamma_0(r-1)}, \tag{1.0.28}$$

$$\| G_r(k,t) \|_1 < k^{-\gamma_0 r}, \tag{1.0.29}$$

$$\gamma_0 = 2l - n - 2\delta.$$

The operator G_m is nonzero only on the finite-dimensional subspace $(\sum_{i:|i-j|<mR_0} E_i) l_2^n$.

Let us introduce the notations:

$$T(m) \equiv \frac{\partial^{|m|}}{\partial t_1^{m_1} \partial t_2^{m_2} \dots \partial t_n^{m_n}}, \tag{1.0.30}$$

$$| m | \equiv m_1 + m_2 + \dots + m_n, \ m! \equiv m_1! m_2! \dots m_n!,$$

$$0 \leq | m | < \infty, \ T(0)f \equiv f.$$

[5] Here and below $\| \cdot \|_1$ is the norm in the trace class $\mathbf{S}_1$ (see f.e. [ReSi1])

Theorem 1.2 . *Under the conditions of Theorem 1.1, the series (1.0.26) and (1.0.27), can be differentiated termwise with respect to t any number of times, and they retain their asymptotic character. For the coefficients $g_r(k,t)$ and $G_r(k,t)$, the following estimates hold, which are uniform with respect to t in the $(k^{-n+1-2\delta})$-neighborhood in C^n of the nonsingular set:*

$$| T(m)g_r(k,t) |< m!k^{-\gamma_0(r-1)+|m|(n-1+\delta)}, \tag{1.0.31}$$

$$||T(m)G_r(k,t)|| < m!k^{-\gamma_0 r+|m|(n-1+\delta)}. \tag{1.0.32}$$

Next, we study the case, when there exists a real refraction inside the crystal, i.e., the perturbation is significant. Namely, we construct perturbation formulae for quasimomenta in a vicinity of the von Laue diffraction planes. We describe the Bloch eigenvalues and their spectral projections of H, which are close to those of a model operator, roughly taking into account the refraction inside the crystal.

The nonsingular set has an asymptotically full measure on $S_0(k)$. However, besides the nonsingular set there exists also the singular part of $S_0(k)$, where perturbed and unperturbed eigenvalues and eigenfunctions are, generally speaking, not close. For example, let us consider the plane wave $\exp i(\mathbf{p}_j(t), x)$, such that its momentum $\mathbf{p}_j(t)$ satisfies the von Laue diffraction conditions, i.e., $p_j(t) = p_{j+q}(t)$ for some $q \neq 0$. It is well-known that this wave is refracted by the crystal, the refracted wave being $a \exp i(\mathbf{p}_{j+q}(t), x)$, $a \in R$. The latter wave interferes with the initial one and distorts it strongly. Hence, the perturbed and unperturbed eigenfunctions obviously are not close when t satisfies the von Laue diffraction conditions. Thus, the existence of the singular set is connected with diffraction conditions, in most simple situation with the von Laue diffraction conditions.

We suppose t does not belong to the nonsingular set $\chi_0(k,\delta)$. This means that t is in a vicinity of a selfintersection of the isoenergetic surface. More precisely, t is not in the nonsingular set, if at least for one $j : p_j^{2l} = k^{2l}$ and one $q \neq 0$ the next inequality holds:

$$|p_j^{2l}(t) - p_{j+q}^{2l}(t)| \leq k^{-n+1-\delta}. \tag{1.0.33}$$

The set of t, which satisfies (1.0.33) for a given q, we denote by $\mathcal{K}S_q$. This set is obtained by partitioning a spherical layer $S_q = \{x : |x| = k, |(x, \mathbf{p}_q(0))| < k^{-n+1-\delta}\}$ into pieces by the dual lattice; after this, the pieces being translated in a parallel manner into K. We consider the subset χ_q of $\mathcal{K}S_q$, such that inequality (1.0.33) holds only for a single q. It will be proved that this subset has an asymptotically full measure on $\mathcal{K}S_q$. If t is in the $(k^{-2-\delta})$-neighborhood of χ_q, than the eigenvalue and spectral projection of $H(t)$ turns out to be close to the corresponding eigenvalue and spectral projection of the operator $\hat{H}_q(t)$, which roughly accounts the splitting and refraction:

$$\hat{H}_q(t) = H_0(t) + P_q V P_q, \tag{1.0.34}$$

P_q being the diagonal projection:

$$(P_q)_{ii} = \delta_{ij} + \delta_{ij+q}. \tag{1.0.35}$$

The matrix of $\hat{H}_q(t)$ has only two non-diagonal elements:

$$\hat{H}_q(t)_{j,j+q} = \overline{\hat{H}_q(t)_{j+q,j}} = v_q. \tag{1.0.36}$$

Thus, for the spectral analysis of $\hat{H}_q(t)$, it suffices to consider the matrix:

$$\begin{pmatrix} p_j^{2l}(t) & v_{-q} \\ v_q & p_{j-q}^{2l}(t) \end{pmatrix}. \tag{1.0.37}$$

We construct the perturbation series for the eigenvalue and eigenfunction of $H(t)$ with respect to $\hat{H}_q(t)$ (see Theorems 2.3 and 2.4 in Section 2.4).

In the Chapter 3 we suppose $\mathbf{4l > n + 1}$. On the basis of new geometrical considerations, a nonsingular set is constructed, first for a trigonometric polynomial, then for a smooth and, finally, for a non-smooth potential. In the case of a smooth potential these new considerations consist not only in deleting the selfintersections of $S_0(k)$ (which are described by the von Laue diffraction conditions $p_j^{2l}(t) = p_{j+q}^{2l}(t) = k^{2l}$, $q \neq 0$), but, moreover, in deleting some vicinities of the planes, which are described by the generalized Laue diffraction conditions $p_{j+m}^{2l}(t) = p_{j+q+m}^{2l}(t)$, $q \neq 0$, $p_j^{2l}(t) = k^{2l}$. The problem is to chose properly the sizes of the vicinities. On the one hand, they should not be too small, otherwise it is not possible to prove the convergence of the series. On the other hand, they should not be too large, because one has to prove that the nonsingular set $\chi_1(k,\delta)$ (which is $S_0(k)$ without the vicinities of (1.0.12)) has an asymptotically full measure on $S_0(k)$. We chose the size of vicinities, depending on q and m, more precisely, on $|q|$ and the distances between $p_{j+m}^{2l}(t)$ and k^2.

The principal theorems of the perturbation theory similar to Theorems 1.1, 1.2 are proved for the nonsingular set $\chi_1(k,\delta)$ $(\gamma = 4l - n - 1 - 8l\delta)$.

Let us note that we need some new geometrical considerations for the construction of the nonsingular set in the case of a non-smooth potential. We introduce a concept of the number of states. Let the number of states $N(\Omega, t)$ for the closed bounded set Ω in R^n and quasimomentum t, be the number of the lattice points $\mathbf{p}_m(t), m \in Z^n$, belonging to Ω. It is easy to see that

$$\int_K N(\Omega, t)dt = V(\Omega), \tag{1.0.38}$$

where $V(\Omega)$ is the volume of Ω.

In the case of a non-smooth potential, we consider a set of the regions Ω_m (Ω_m is a thin spherical layer near S_k) and choose the quasimomenta satisfying the relation: $N(\Omega_m, t) < V(\Omega_m)k^\delta$, $0 < \delta \ll 1$. We impose similar restrictions on the functions:

$$M(\{\Omega_q\}, V, t) = \sum_{q \in Z^n} | v_q |^2 N(\Omega_q, t), \tag{1.0.39}$$

where $\{\Omega_q\}_{q\in Z^n}$ are some regions, chosen in a special way, which will be described in Section 3.6. Using the concept of the number of states, we construct the nonsingular set of an asymptotically full measure on $S_0(k)$. The principal theorems of the perturbation theory similar to Theorems 1.1 and 1.2 hold. is on the isoenergetic surface.

In Chapter 4 we consider the Schrödinger operator ($n = 3$, $l = 1$). This operator is most interesting both from physical and mathematical points of view. In this case the series (1.0.26) and (1.0.27) do not converge, because $\gamma_0 = -8\delta$, i.e., γ_0 is negative. To construct the convergent series we shall take the auxiliary operator $\hat{H}(t)$, described below, as the initial one, instead of $H_0(t)$. Thus, we shall get the modified perturbation series. For these series the assertions similar to Theorems 1.1 and 1.2 hold. However, the operator $\hat{H}(t)$ depends on V, and therefore these theorems are not convenient to use. Fortunately, it turns out, that, with good accuracy, the modified series may be replaced by a sufficiently long part of the series (1.0.26) and (1.0.27) for $H_0(t)$. Furthermore, the estimates similar to (1.0.28) and (1.0.29) for these terms are valid. Thus we obtain simplified formulae of the perturbation theory. However, to justify them we need first to construct the modified series. Moreover, the latter give the exact values of $\lambda(t)$ and $E(t)$. Here, for the first time, the Bethe-Sommerfeld conjecture is proved for non-smooth potentials.

Let us construct the operator $\hat{H}(t)$ for a trigonometric polynomial $V(x)$:

$$V(x) = \sum_{j\in Z^3, |j|<R_0} v_j \exp(i(\mathbf{p}_j(0), x)),$$

To define $\hat{H}$ we start with the definition of a set, which we denote by $\Gamma(R_0)$. Let us consider $j : j \in R^3, 0 < |j| < R_0$. In this set some of the j are scalar multipliers of others. Let us keep from every family of the scalar multipliers only the minimal representative, i.e., the representative having the minimal length (see Fig.2). We denote by $\Gamma(R_0)$ the union of these minimal representatives. In other words, each $j : j \in R^3, 0 < |j| < R_0$ can be uniquely represented in the form $j = mj_0$, where $m \in Z \setminus \{0\}$, $j_0 \in \Gamma(R_0)$.

It is easy to see that

$$V = \sum_{q\in\Gamma(R_0)} V_q, \tag{1.0.40}$$

where V_q depends only on one variable $(x, \mathbf{p}_q(0))$.

Let us consider the following sets in Z^3:

$$\Pi(k^{1/5}) = \{j :| (\mathbf{p}_j(0), q) |< k^{1/5}\}, q \in \Gamma(R_0), \tag{1.0.41}$$

$$T(k^{1/5}) = \tag{1.0.42}$$
$$\{j : \exists q, q' \in \Gamma(R_0), q \neq q' :| (\mathbf{p}_j(0), q) |< k^{1/5}\}, | (\mathbf{p}_j(0), q') |< k^{3/5}\}.$$

We construct the diagonal projection P_q as follows:

$$(P_q)_{jj} = \begin{cases} 1 & \text{if } j \in \Pi_q(k^{1/5}) \setminus T(k^{1/5}), \\ 0 & \text{otherwise.} \end{cases} \tag{1.0.43}$$

Let $\hat{H}(t)$ be defined by the formula

$$\hat{H}(t) = H_0(t) + \sum_{q \in \Gamma(R_0)} P_q V_q P_q. \tag{1.0.44}$$

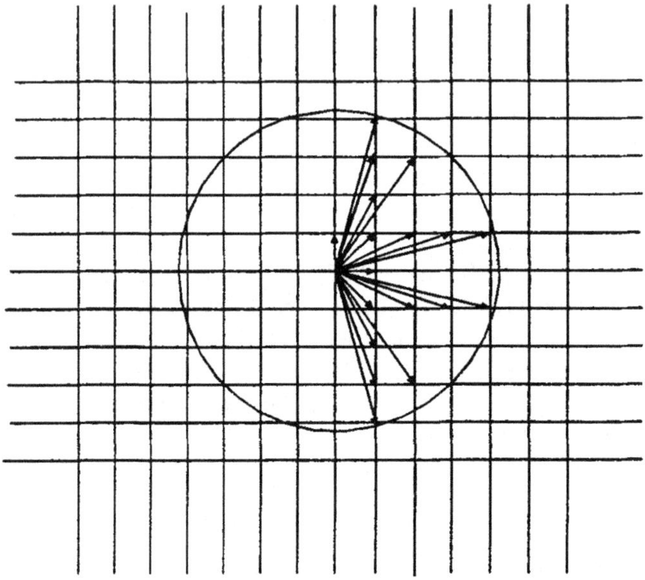

Fig.2 The set $\Gamma(R_0)$ in Z^2

We introduce the notation:

$$W = V - \sum_{q \in \Gamma(R_0)} P_q V_q P_q. \tag{1.0.45}$$

Let the functions $\hat{g}_r(k,t)$ and the operator-valued function $\hat{G}_r(k,t)$ be constructed for $\hat{H}(t)$ in the same way as $g_r(k,t)$ and $G_r(k,t)$ have been constructed for $H_0(t)$:

$$\hat{g}_r(k,t) = \frac{(-1)^r}{2\pi i r} \mathrm{Tr} \oint_{C_0} ((\hat{H}(t) - z)^{-1} W)^r dz \tag{1.0.46}$$

$$\hat{G}_r(k,t) = \frac{(-1)^{r+1}}{2\pi i} \oint_{C_0} ((\hat{H}(t) - z)^{-1} W)^r (\hat{H}(t) - z)^{-1} dz. \tag{1.0.47}$$

They are correctly defined in the $(k^{-2-2\delta})$-neighborhood of the nonsingular set $S_3(k,\delta,V)$. This set of an asymptotically full measure is constructed in Section 2.3. The theorem of expansion into modified perturbation series holds.

Theorem 1.3 . *Suppose t belongs to the* $(k^{-2-2\delta})$*-neighborhood of a nonsingular set* $S_3(k,V,\delta)$, $0 < 2\delta < 1/100$. *Then for sufficiently large k,* $k > k_0(V,\delta)$, *there exists a single eigenvalue of the operator H in the interval* $\varepsilon(k,\delta) \equiv [k^2 - k^{-1-\delta}, k^2 + k^{-1-\delta}]$. *It is given by the series:*

$$\lambda(t) = p_j^{2l}(t) + \sum_{r=2}^{\infty} \hat{g}_r(k,t), \tag{1.0.48}$$

where the index j is uniquely determined from the relation $p_j^{2l}(t) \in \varepsilon(k,\delta)$. *The spectral projection, corresponding to* $\lambda(t)$, *is given by the series*

$$E(t) = E_j + \sum_{r=2}^{\infty} \hat{G}_r(k,t), \tag{1.0.49}$$

which converges in the class $\mathbf{S}_1$.

For the coefficients $\hat{g}_r(k,t)$ *and* $\hat{G}_r(k,t)$ *the following estimates hold:*

$$| T(m)\hat{g}_r(k,t) | < m! k^{-1-\delta-r/20+2|m|(1+\delta)}, \tag{1.0.50}$$

$$\|T(m)\hat{G}_r(k,t)\| < m! k^{-r/20+2|m|(1+\delta)}. \tag{1.0.51}$$

Furthermore, we show that $\hat{G}_r(k,t) = G_r(k,t)$, if $r \le M(k)$, $M(k) = k^{1-\delta}R_0^{-1}$ and $\hat{g}_r(k,t) = g_r(k,t)$, if $r \le 2M(k)$.

Theorem 1.4 . *Under the assumptions of Theorem 1.3, the eigenvalue* $\lambda(t)$ *and its spectral projection* $E(t)$ *can be represented in the form:*

$$\lambda(t) = p_j^{2l}(t) + \sum_{r=2}^{2M(k)} g_r(k,t) + \varphi(k,t), \tag{1.0.52}$$

$$E(t) = E_j + \sum_{r=2}^{M(k)} G_r(k,t) + \psi(k,t), \tag{1.0.53}$$

$$M(k) = k^{1-\delta}R_0^{-1}.$$

For the function g_r *and the operator-value functions* G_r, *the estimates (1.0.39) and (1.0.43) are fulfilled, while for* $\varphi(k,t)$ *and* $\psi(k,t)$ *the following ones hold:*

$$| T(m)\varphi(k,t) | < m! k^{-k^{1-2\delta}} k^{2(1+\delta)|m|}, \tag{1.0.54}$$

$$\|T(m)\psi(k,t)\|_1 < m! k^{-k^{1-2\delta}} k^{2(1+\delta)|m|}. \tag{1.0.55}$$

We show in Section 3.5 that formulae (1.0.48), (1.0.49), (1.0.52), and (1.0.53) are differentiable with respect to quasimomentum t.

There is an interesting connection between the results of Thomas, L.E. and Wassell, S.R. [ThWal] and this result. These authors consider a semiclassical approximation ($h \to 0$) of eigenvalues and eigenfunctions of a periodic operator

$-\frac{\hbar^2}{2}\Delta+V$, $x \in R^n$, $n > 1$ at high energy. They show that most of the eigenvalues are stable with respect to perturbation V in a high energy region and constructed an asymptotic expansion for their eigenfunctions. The proofs are based on their results on the stability of Hamiltonian systems at high energy [ThWa2].

In sections 4.8 - 4.10 we consider the unstable case, i.e., the case when the perturbations of an eigenvalue and its eigenfunction are significant. Namely, we construct perturbation formulae when a quasimomentum is in a vicinity of the von Laue diffraction planes:

$$|p_j^2(t) - p_{j+q}^2(t)| \leq k^{\delta}. \tag{1.0.56}$$

In this case there is essential diffraction inside the crystal. The perturbed eigenvalues and eigenfunctions are not close to unperturbed ones. The simple operator (1.0.34), which was used to describe diffraction in the case $2l > n$ is not valid any more. This means that the picture of diffraction is, in fact, more complicated.

The unstable case was studied by J. Feldman, H. Knörrer, E. Trubowitz [FeKnTr2] in the two and three dimensional situations. In the three-dimensional case they study the eigenvalues of H, which are not close to the unperturbed ones, but can be approximated by eigenvalues of the operator $-\Delta + V_\gamma$, where γ some vector of the dual lattice and V_γ is independent of x in the direction γ i.e.,

$$V_\gamma(x) = V_\gamma(x - \gamma(x,\gamma)|\gamma|^{-2}) = \sum_{j:(\mathbf{p}_j(0),\gamma)=0} v_j \exp i(x, \mathbf{p}_j(0)).$$

It was proved that for arbitrary γ of the dual lattice and any eigenvalue of $H_\gamma(t)$ corresponding to a sufficiently large momentum in the direction γ, there exists a close eigenvalue of the operator $H(t)$ with the same quasimomentum, multiplicity being taken into account. The same result was proved for $n = 2$. Moreover, in the two-dimensional case it was shown that on the rich set of t the corresponding eigenfunctions are close too. O.A. Veliev discussed this problem [Ve6].

The approach developing here, has its own peculiarities. It provides formulae not only for unstable eigenvalues, but also for their spectral projections in three dimensional situation. The converging perturbation series with respect to the model operator $\hat{H}$ (see (1.0.44)), roughly describing also the refraction, are constructed. The series have an asymptotic character in a high energy region. They can be differentiated any number of times with respect to the quasimomentum. Thus, the eigenvalue and the spectral projection of $H(t)$ are close to those of the operator $\hat{H}(t)$ on the essential part of the singular set. Note, that the operator $\hat{H}(t)$ has a block structure. The diagonal part of $\hat{H}$ coincides with the corresponding part of the free operator: $(I - \sum_{q\in\Gamma(R_0)} P_q)\hat{H}(t) = (I - \sum_{q\in\Gamma(R_0)} P_q)H_0(t)$. The blocks of $\hat{H}(t)$ are determined by the orthogonal projections P_q, $P_q\hat{H}(t) = \hat{H}(t)P_q = P_q\hat{H}(t)P_q$. Each block is a "piece" of the matrix of the Schrödinger operator with the potential V_q, i.e., $P_q\hat{H}(t)P_q = P_q(H_0(t) + V_q)P_q$. Thus, each block is simply connected with the matrix of a periodic Schrödinger operator in the one-dimensional space, because V_q depends only on $(x, \mathbf{p}_q(0))$.

In the case of the nonsingular set we constructed perturbation series with respect to the operator $\hat{H}(t)$. The perturbed eigenvalue $\lambda(t)$ is asymptotically close ($k \to \infty$) to an eigenvalue $p_j^{2l}(t)$ of the diagonal part of $\hat{H}(t)$; i.e, $\lambda(t)$ is close to the eigenvalue $p_j^{2l}(t)$ of the free operator $H_0(t)$. Accordingly, the spectral projector of $H(t)$ corresponding to $\lambda(t)$ is close to that of $H_0(t)$, corresponding to $p_j^{2l}(t)$. We prove that in the case of the singular set (more precisely of its essential part) the eigenvalues and its spectral projection of $H(t)$ are close to those, corresponding to the block part of $\hat{H}(t)$. Thus, the blocks $P_q\hat{H}P_q = P_q(H_0(t)+V_q)P_q$, which are simply connected with one dimensional Schrödinger operators with the potentials V_q, roughly describe the refraction inside crystal for t of the essential part of the singular set.

From the geometrical point of view we consider the singular set not only as a part of the whole cell K, but, more precisely, we consider its "trace" on the isoenergetic surface $S_0(k)$ in K of the free operator for a fixed energy k^2. Namely, we study the perturbations of the eigenvalues $p_j^2(t)$ and its spectral projections, which satisfy the conditions $k^2 = p_j^2(t)$ and (1.0.56) for some given q and k. It is proved that for a fixed q and sufficiently large k there exists a rich set of t (an essential part of the singular set) for which perturbation series with respect to the model operator converge. Although the main geometrical problems arises because of fixed k (one has to prove that perturbation series converge on a rich subset of the surface in K), the consideration of the singular set with a fixed energy turns out to be very useful, because it enable us to describe the perturbed isoenergetic surface near the selfintersections of the isoenergetic surface of the free operator (they obviously are formed by the points satisfying the von Laue diffraction conditions). We can observe how this intersections are transformed into quasiintersections under the perturbation. This approach also turns out to be effective in solving a semicrystal problem (see Chapter 5).

In Sections 4.9 - 4.11 we consider the Schrödinger operator with a potential satisfying condition (1.0.6). In this case, the nonsingular set is, without going into details, the intersection of the nonsingular set for a non-smooth potential in the case of $4l > n+1$ (Section 2.6) and the nonsingular set for a smooth potential in the case of Schrödinger operator (Section 4.4). The main difficulties here are in the analytical part, that is in the proof of the convergence of the series.

We represent the potential V in the form $V = V_1 + V_2$:

$$V_1 = \sum_{|j|<k^\rho} v_j \exp(i(\mathbf{p}_j(0), x)),$$

$$V_2 = \sum_{|j|\geq k^\rho} v_j \exp(i(\mathbf{p}_j(0), x)), \quad 0 < \rho < 1.$$

We shall consider V_2 as a small perturbation of V_1, using

$$\sum_{|j|\geq k^\rho} |\, v_j \,|^2 |\, j \,|^{-1+\beta} \to_{k\to\infty} 0. \qquad (1.0.57)$$

We construct $\hat{H}(t) = H_0(t) + \hat{V}_1$, for $\hat{V}_1$, just as we do in the case of a trigonometric polynomial in Section 4.4. Let us define once more $\hat{g}_r$ and $\hat{G}_r$ by formulae (1.0.46), (1.0.47), where $W = V - \hat{V}_1$. The principal theorem of the perturbation theory, similar to Theorem 1.3, holds. In this case we can also simplify the modified series and obtain formulae similar to (1.0.44) and (1.0.46). But their accuracy decreases together with a decrease in the smoothness of the potential, that is when β decreases.

For physical applications it is very important to understand how the isoenergetic surface of H looks like. This surface is the set of t in K, such that there is an eigenvalue $\lambda_n(t)$ of operator $H(t)$, which is equal to a given λ:

$$S_H(\lambda) = \{t, t \in K : \exists \lambda_n(t) = \lambda\}.$$

In physics, this surface, considered for some special energy $\lambda = \lambda_F$, is called the Fermi surface. The forms of the Fermi surfaces for different crystals explain many of their properties. However, their description is mainly phenomenological (see f.e. [BS, Bl, CrWo, Ki, Mad, Zi]). The rigorous proof of the fact that the isoenergetic surface has zero Lebesgue measure in K one can find in L.E. Thomas paper [Th]. It is important to study the isoenergetic surface not only for quasimomenta in K, but in the whole complex space C^n, i.e.,

$$S(\lambda) = \{t, t \in C^n : \exists \lambda_m(t) = \lambda\}, \quad \lambda \in R,$$

here $\lambda_m(t)$ are analytic continuations of the band functions. L.E. Thomas proved [Th] that $\lambda_m(t)$ are analytic functions of each component of the quasimomentum in a neighborhood of the real axis, and they are not constants. Analytic properties of band functions as functions of many variables were studied by J.E. Avron and B. Simon [AvSi1]. It is important that the set $S(\lambda)$ can be described as the zero set of a regularized determinant $\Delta(\lambda, t)$ (see [DS, GoKr, Si1, Kuch1, Kuch2, KnTr]). The set of zeros of this determinant in the space $C \times C^n$ is called a Bloch variety. The Bloch variety contains all points which can be reached by analytic continuation of any band function, i.e.,

$$B(V) = \{(\lambda, t), \lambda \in C, t \in C^n : \exists \lambda_m(t) = \lambda\}.$$

P.E. Kuchment considered Bloch varieties for a wide class of periodic elliptic and hyperelliptic operators and showed that a Bloch variety either a proper analytic subvariety or the whole complex space (for the Schrödinger operator the last case is excluded by Thomas' Theorem). The close study of isoenergetic surfaces and Bloch varieties was developed in [Bä1] – [BäKnTr], [DyPe], [EsRaTr1, EsRaTr2], [GiKnTr1, GiKnTr2], [KnTr], [No]. H. Knörrer and E. Trubowitz [KnTr] directionally compactified a Bloch variety and, using this, showed that the Bloch variety is irreducible up to a translation. This means that the analytic continuation of any given band function is $B(V)$, from which all other band functions can be found. It is proved also that V is constant if $B(V)$ contains a component which is the graph of an entire function. The case of an isoenergetic (Fermi) surface was considered by D. Bätig, H. Knörrer and E. Trubowitz in [BäKnTr]. The directional compactification of the complex Fermi surface is made. One of the main theorems is that, if for a single real λ one of the components of the isoenergetic surface is a sphere, then V is a constant. D. Gieseker, H. Knörrer and E. Trubowitz studied Bloch varieties for the discretized periodic Schrödinger operators in two dimensions. They showed [GiKnTr1, GiKnTr2] that a Bloch variety is generically determines the corresponding potential, the Bloch variety itself being defined by a physical notion of a density of states. A toroidal compactification of the Bloch variety and the Fermi surfaces was made by D. Bätig [Bä1, Bä2].

We discuss here only the real isoenergetic surface in a high energy region, i.e. the case when λ is big enough and $t \in K$. The perturbation formulae constructed here enable us to describe the behavior of the isoenergetic surface near nonsingular set and a significant part of the singular set (see Sections 2.6, 3.8, 4.12 for different n and l). It is proved that there exists a unique piece of the isoenergetic surface in the $(k^{-n+1-\delta})$- neighborhood of each piece of the nonsingular set, the corresponding normals being close, too. It is shown that in a neighborhood of the essential part of the singular set, there exist pieces of the isoenergetic surface of the model operator, roughly describing diffraction. Then, we show that in the $(k^{-n+1+\delta})$-neighborhood of the each of these pieces, is situated a unique piece of the isoenergetic surface of H, corresponding normals being close. The proof of the Bethe-Sommerfeld conjecture is a simple consequence of the fact that the isoenergetic surface exists for all λ large enough. It is a challenging problem to describe the behavior of the isoenergetic surface when parameters λ and t are complex. The papers of Bättig, D., Gieseker, D., Knörrer H., and Trubowitz, E., are devoted to this problem [Bä1, Bä2, BäKnTr, GiKnTr1, GiKnTr2, KnTr]. In particular, in [BäKnTr] it is proved that if for a single λ Fermi surface is a sphere than the potential is a constant.

The perturbation series (1.0.27), (1.0.49) can be used for the calculation of the spectral projection of the operator H for a given quasimomentum t. However, this formula is not applicable directly, when one knows the value of the perturbed eigenvalue $\lambda_j(k,t)$, $\lambda_j(k,t) = k_0^{2l}$, and $n-1$ coordinates, say, $t_2, ..., t_n$, of point t, i.e., when t is on the perturbed isoenergetic surface corresponding to a given energy k_0^{2l}.[6] To use formula (1.0.27) or (1.0.53) in this case, we first have to determine t_1 from the equation

$$\lambda_j(k,t) = k_0^{2l}. \tag{1.0.58}$$

Practically, we can do this only approximately, taking $t_1 \approx t_{01}$, where t_{01} is the solution of the unperturbed equation $p_j^{2l}(t) = k_0^{2l}$. Substituting $t_1 \approx t_{01}$ in the series (1.0.27) or (1.0.53), we obtain rather crude asymptotic formula for the spectral projection. Similarly, we obtain an approximate formula for the normal to the isoenergetic surface of H. However, there exists another way to construct the eigenfunction for t on the isoenergetic surface. In this way we do not need to solve equation (1.0.58). We can construct the perturbation series for the eigenfunction with a given energy k_0^{2l}, and $n-1$ coordinates of quasimomentum, say, $t_2, ..., t_n$. Each term of this series explicitly depends on $k_0^{2l}, t_2, ..., t_n$. The

[6] This is the case we have in the semicrystal problem (Chapter V).

method consists of integrating the resolvent $(H(t) - k_0^{2l})^{-1}$ with respect to t_1 over the contour around the point $t_1 = t_{01}$, where k_0^{2l} is fixed. Expanding the resolvent in the perturbation series and proving its convergence, we obtain a new formula for the eigenfunction (see Sections 2.7, 3.9 and 4.13 for different n and l).

In the Chapter 5 we consider the operator

$$H_+ = -\Delta + V_+ \tag{1.0.59}$$

in $L_2(R^n)$, $n = 2, 3$, where V_+ is the operation of multiplication by the potential:

$$V_+(x) = \begin{cases} V(x) & \text{if } x_1 \geq 0; \\ 0 & \text{if } x_1 < 0; \end{cases} \tag{1.0.60}$$

V being a trigonometric polynomial. We suppose one of its periods a_1 lies on axis x_1 and is orthogonal to the others.

We denote the projection of a vector x on the plane $x_1 = 0$ by $x_{||}$.

Suppose the free wave $\exp(i(\mathbf{k}, x))$, $k_1 > 0$ is incident upon the crystal from the semispace $x_1 < 0$. We represent it in the form:

$$\exp(i(\mathbf{k}, x)) = \exp\left(i(k_{||}, x_{||})) + i(k^2 - |k_{||}|^2)^{1/2} x_1\right) = \tag{1.0.61}$$

$$= \exp(i(\mathbf{p}_m(t), x)) = \exp(i(\mathbf{p}_{m_{||}}(t_{||}), x_{||}) + i(p_{m_1}(t_1), x_1)),$$

where $\mathbf{p}_m(t) = \mathbf{k}$, $\mathbf{p}_m(t)$ is a vector of the dual lattice, corresponding to the potential V; $t_{||}$, $m_{||}$, $k_{||}$, $\mathbf{p}_{m||}(t_{||})$ are the projections of t, m, $\mathbf{k}$, $\mathbf{p}_m(t)$, correspondingly, on the plane $x_1 = 0$, and

$$p_{m_1}(t_1) = (k^2 - p_m^2(t_{||}))^{1/2} = (2\pi m_1 + t_1) a_1^{-1},$$

$$m_1 \geq 0, \quad 0 \leq t < 2\pi.$$

Since the potential V_+ is a periodic function with respect to $x_{||}$, the incident and reflected waves have the same quasimomentum $t_{||}$. Therefore, the reflected wave can be written as a linear combination of the following plane waves:

$$\Psi_-^0(k, k_{||} + p_{q_{||}}(0), x) = \exp(i(k_{||} + p_{q_{||}}(0), x_{||}) - i(k^2 - \left|k_{||} + p_{q_{||}}(0)\right|)^{1/2} x_1).$$

That is,

$$\Psi_{refl} = \sum_{q \in Z^{n-1}} \beta_{q_{||}} \Psi_-^0(k, k_{||} + p_{q_{||}}(0), x), \tag{1.0.62}$$

where $\beta_{q_{||}}$ are the reflection coefficients. One has to choose the reflection and refraction waves in such a way that the continuity conditions at the plane $x_1 = 0$, for the eigenfunction and its derivative with respect to x_1, are fulfilled.

Suppose t belongs to the nonsingular set for the periodic potential V. Then the wave $\psi(\mathbf{k}, x)$ close to $\exp(i(\mathbf{k}, x))$ can propagate in the crystal (see Theorem 1.4). Therefore, the continuity conditions can be satisfied with an accuracy to $o(1)$, the reflected wave being equal to zero. We prove that for a rich set of $\mathbf{k}$

there exists not only a wave close to $\exp(i(\mathbf{k},x)) \equiv \Psi_0^+(k,k_{||},x)$, but also there exist waves close to $\Psi_0^+(k,k_{||}+p_{q_{||}}(0),x)$,

$$\Psi_0^+(k,k_{||}+p_{q_{||}}(0),x) = \exp(i(k_{||}+p_{q_{||}}(0),x_{||}) + i(k^2 - |k_{||}+p_{q_{||}}(0)|)^{1/2}x_1),$$

$$|q_{||}| < k^{3\delta}.$$

In this case, we can satisfy the continuity conditions with a high accuracy $O(k^{-k^{2\delta}})$. Thus, we get an approximate solution. The question arises: is this approximate solution close to the accurate one, satisfying the continuity conditions precisely? To answer this question, first of all let us see whether reflected and refracted waves are always defined uniquely. Suppose the equation $H_+\Psi = k^2\Psi$ has a smooth solution Ψ_{surf}, which is a reflected wave for $x_1 \leq 0$ and a refracted wave for $x_1 \geq 0$. This solution is called a surface state. The surface states were discovered by Rayleigh [Ray] at the end of the last century and are still of great importance in modern physics [7]. Obviously, in the case of a surface state the reflected and refracted waves are not uniquely determined. However, the nondecaying component of the reflected and refracted waves are uniquely determined, because it can be shown that surface states exponentially decay as $x_1 \to \pm\infty$.

It turns out that the surface of the crystal can essentially influence the nondecaying part of the reflected wave too. This happens when there exists a solution of the equation $H_+\psi = k^2\psi$, which can be approximated with good accuracy by a reflected and refracted wave in the sense that the error in the continuity conditions on the surface is small. We call such a solution a quasisurface state. The quasisurface state can have strong influence on the asymptotics of the reflected and refracted waves, even far from the surface. Unlike the surface state, it can also influence the nondecaying component of the reflected wave.

In Sections 5.3 - 5.4 we formulate, in the analytic form, the conditions on $k_{||}$, under which the surface and quasisurface states are absent. In Section 5.5 we make the geometrical construction of the nonsingular set $\chi_5(k,V,\delta)$, on which these conditions are satisfied, and therefore, surface and quasisurface states are eliminated. This enable us to describe the reflection coefficients with a high accuracy. Namely, we represent them in the form:

$$\beta_{q_{||}} = \beta^0_{q_{||}} + O(k^{-k^{2\delta}}),$$

where $\beta^0_{q_{||}}$ are the reflection coefficients of the approximate solution. We also prove that the refracted wave is close to the approximate one. Then (Section5.6) we obtain a formula for the reflection coefficients, which will expose their connection with the potential in the most explicit form.

To describe the formulae for the reflection coefficients, let us introduce some new notations. Let $v_{m_{||}}(x_1)$ be the Fourier coefficients of V with respect to $x_{||}$:

$$v_{m_{||}} = w \int V(x) \exp(i(p_{m||}(0), x_{||}))dx_{||},$$

[7] see f.e. [Ae, AgMi, DaLe, DavSi, GroHøMe, JaMolPas, K1, K3, Ki, KurPa2, Mad, Lo, Pas, PavPol, Zi].

$$w = a_2^{-1} \text{ when } n = 2, \quad w = a_2^{-1} a_3^{-1} \text{ when } n = 3.$$

Let $V^{(r)}(0)$ be the vector, whose elements are the derivatives of the functions $v_{m_\|}(x_1)$ of r order with respect to x_1, at the point $x_1 = 0$:

$$V^{(r)}(0)_{m_\|} = (\partial^r v_{m_\|}/\partial x_1)(0),$$

and

$$k_1(m_\|) = (k^2 - |k_\| + p_{m_\|}(0)|)^{1/2}.$$

We obtain that on a rich subset of S_k, the approximate reflection coefficients $\beta^0_{q_\|}$ admit the following asymptotic expansion:

$$\beta^0_{m_\|} = -\sum_{r=0}^{[k^\delta]} \frac{(\Phi_r)_{m_\|}}{(-2ik_1(m_\|))^{r+2}} + O(k^{-k^{2\delta}}), \tag{1.0.63}$$

where Φ_r is a vector-valued function, depending on $V^{(r)}(0)$, $V^{(r-1)}(0)$, ..., $V(0)$. These functions depend also on $\mathbf{k}$. We describe them more precisely in Section 5.6. Here we note only that Φ_r depends on the highest derivative $V^{(r)}(0)$ in a simple way:

$$\Phi_r = V^{(r)}(0) + \varphi_r(V^{(r-1)}(0), ..., V(0)), \tag{1.0.64}$$

where φ_r is the polynomial vector-valued function of $V^{(r-1)}(0), ..., V(0)$ and $\varphi_0 = 0$. One can easily see that $\varphi_r = \tilde{\varphi}_r(\Phi_0, ..., \Phi_{r-1})$. This enable us to obtain the vectors $V^r(0)$ by the recursive procedure:

$$V(0)_{m_\|} = \lim_{k\to\infty} \beta_{m_\|} 4k_1^2(m_\|) \tag{1.0.65}$$

$$V^{(1)}(0)_{m_\|} = \tag{1.0.66}$$

$$\lim_{k\to\infty} \left((-2ik_1^2(m_\|))^3(\beta_{m_\|} - V(0)_{m_\|}/4k_1^2(m_\|)) - \Phi_1(V(0))_{m_\|}\right),$$

and so on. In Section 5.7 we use these relations for the solution of the inverse problem.

The author expresses her deep gratitude to B.S. Pavlov for many fruitful discussions, to M.Sh. Birman and V.S. Buslaev for their continued attention, to H. Knörrer, E. Trubowitz and J.K. Moser for the support and the hospitality at ETH, where an important part of the job was done, to J.R. McLaughlin for the support and the fruitful long-lasting hospitality at Rensselaer Polytechnic Institute, to L.D. Faddeev for his support of this work.

2. Perturbation Theory for a Polyharmonic Operator in the Case of $2l > n$.

2.1 Introduction. Isoenergetic Surface of the Free operator. Laue Diffraction Conditions.

In the study of the polyharmonic operator we begin with the case of $2l > n$. We use this relatively simple case to describe the main features of the perturbation procedure for eigenvalues and spectral projections, clearing out its connection with diffraction conditions. In fact, we based our consideration on physicists idea (see f.e. [Ki, Mad, Zi], that perturbation theory formulae are valid on the isoenergetic surface of the corresponding unperturbed operator, except in a vicinity of the points satisfying the von Laue diffraction conditions. In this chapter the perturbation series for Bloch eigenvalues and spectral projections on the rich set of quasimomenta are constructed. The proof of the convergence is based on a general idea of perturbation theory. However, the formulae arising in this chapter are valid in principle, for even more complicated situations (see f.e Chapter 3), where new diffraction conditions arise and new geometrical and analytical considerations are developed.

To begin with, we consider the free operator $H_0(t)$. Its eigenvalues are $p_m^{2l}(t)$, $m \in Z^n$. Let $N(k)$ be the number of points $p_m^{2l}(t)$ in the interval $(k^{2l}, (2k)^{2l})$. It is clear that $N(k)$ is the number of points $\mathbf{p}_m(t)$, which lie in the layer between the spheres S_k, S_{2k} of the radii k and $2k$, centered at the origin. It is obvious that $N(k) \approx k^n$. Thus, the average distance between the points $p_m^{2l}(t)$ in the interval $(k^{2l}, (2k)^{2l})$ is of order k^{2l-n}. Since $2l > n$, the eigenvalues of the free operator used are situated far from each other for large energies ($k \to \infty$). Thus, the usual perturbation theory can be applied to describe their perturbations. In particular, the general formulae for perturbations of eigenvalue $p_j^{2l}(t) = k^{2l}$ and the corresponding eigenfunction are valid for k large enough, if

$$\min_{q \neq 0} \mid p_j^{2l}(t) - p_{j+q}^{2l}(t) \mid > k^{2l-n-\delta}, \tag{2.1.1}$$

$0 < \delta < 2l - n$, $\delta \neq \delta(k)$, because $k^{2l-n-\delta} > \| V \|$ for sufficiently large k. Under the condition (2.1.1), the perturbed eigenvalue and the eigenfunction are asymptotically close to the unperturbed ones. We show in Section 2.3 that inequality (2.1.1) is valid on a rich set $\chi_0(k,\delta)$ of quasimomenta, belonging to the isoenergetic surface of the free operator, $\chi_0(k,\delta) \subset S_0(k)$. The set $\chi_0(k,\delta)$ has an asymptotically full measure on $S_0(k)$, when $k \to \infty$.

Next, we consider the case when estimate (2.1.1) does not hold. Suppose, that

$$p_j^{2l}(t) = p_{j-q}^{2l}(t) \tag{2.1.2}$$

for a unique $q, q \neq 0, q \in Z^n$. Using the notation $\mathbf{k} = \mathbf{p}_j(t)$, we rewrite equation (2.1.2) in the form of the von Laue diffraction condition (see f.e.[Ki, Mad, Zi]):

$$|\mathbf{k}| = |\mathbf{k} - \mathbf{p}_q(0)|. \tag{2.1.3}$$

It is generally known that the plane wave $\exp i(\mathbf{k}, x)$ is refracted by a crystal if $\mathbf{k}$ satisfies this condition. The refracted wave is $a \exp i(\mathbf{k} - \mathbf{p}_q(0), x)$, $a \in C$. This wave interferes with the initial one and distorts it. The resulting wave is

$$\nu_1 \exp i(\mathbf{k}, x) + \nu_2 \exp i(\mathbf{k} - \mathbf{p}_q(0), x), \tag{2.1.4}$$

$$\nu_1^2 + \nu_2^2 = 1.$$

The convergence of perturbation series with respect to H_0 shows that the perturbed eigenfunction is close to the unperturbed one, i.e., to the plane wave $\exp i(\mathbf{k}, x)$. We can see from formula (2.1.4), that it is not the case when equality (2.1.3) holds. Thus, we infer that perturbation series diverge, if $\mathbf{k}$ is in a vicinity of the planes (2.1.3).

Let $p_j^{2l}(t) = k^{2l}$ and the relation

$$|p_j^{2l}(t) - p_{j-q}^{2l}(t)| < k^{2l-n-\delta} \tag{2.1.5}$$

hold for a unique q, $q \neq 0$, $q \in Z^n$. We suppose that the distance from the points $p_j^{2l}(t)$, $p_{j-q}^{2l}(t)$ to other points $p_m^{2l}(t)$, $m \neq j, j-q$ is more then $k^{2l-n-6\delta}$:

$$\min_{i \neq j, j-q} |p_j^{2l}(t) - p_i^{2l}(t)| > k^{2l-n-6\delta}. \tag{2.1.6}$$

When the eigenvalue $p_j^{2l}(t)$ is perturbed by V, it "interacts" strongly with $p_{j-q}^{2}(t)$ and weakly with other $p_m^{2l}(t)$, $m \neq j, j-q$.

To describe approximately the interaction between $p_j^{2l}(t)$ and $p_{j-q}^{2l}(t)$, we consider the operator:

$$\hat{H}_q(t) = H_0(t) + P_q V P_q, \tag{2.1.7}$$

P_q being the diagonal projection

$$(P)_{mm} = \delta_{jm} + \delta_{j-q,m}. \tag{2.1.8}$$

The matrix of $\hat{H}_q(t)$ has only two non-diagonal elements:

$$\hat{H}_q(t)_{j,j-q} = \overline{\hat{H}_q(t)_{j-q,j}} = v_q. \tag{2.1.9}$$

Thus, for the spectral analysis of $\hat{H}_q(t)$, it suffices to consider the matrix:

$$\begin{pmatrix} p_j^{2l}(t) & v_q \\ v_{-q} & p_{j-q}^{2l}(t) \end{pmatrix}. \tag{2.1.10}$$

It is not hard to show that eigenvalues $\hat{\lambda}^+(t)$ and $\hat{\lambda}^-(t)$ of this matrix are rather far from each other:

$$| \hat{\lambda}^+(t) - \hat{\lambda}^-(t) | > |v_q|. \tag{2.1.11}$$

We shall construct (Section 2.4) perturbation series for eigenvalues and spectral projections of operator $H(t)$, considering $\hat{H}_q(t)$ as the initial one. These series converge and have an asymptotic character in a high energy region. They are infinitely differentiable with respect to t, and retain their asymptotic character. Hence, operator $H(t)$ has an eigenvalue which is asymptotically close to an eigenvalue of $\hat{H}_q(t)$. Their spectral projections are also close. The eigenfunction of H in the x-representation is close to the corresponding eigenfunction of $\hat{H}_q(t)$, which is given by formula (2.1.4), (ν_1, ν_2) being the eigenvectors of the matrix (2.1.10). Let $\chi_q(k,\delta)$, $\chi_q(k,\delta) \subset S_0(k)$, be the set of quasimomenta, satisfying conditions (2.1.5), (2.1.6). We shall show (Section 2.5) that set $\chi_q(k,\delta)$ is rather rich: it has an asymptotically full measure on $\mathcal{K}S_q$, $\mathcal{K}S_q$ being the part of the singular set, determined by inequality (2.1.5).

To investigate the case when inequality (2.1.5) is valid for more then one q, one should take operator $\hat{H}_q(t)$ with a more extended non-diagonal part.

Further, in Section 2.6, the proof of the Bethe-Sommerfeld conjecture and the description of the isoenergetic surface of operator H are given. According to the Bethe-Sommerfeld conjecture, there exists only a finite number of gaps in the spectrum of H. Furthermore, considering the isoenergetic surface, we prove that there exists a unique piece of this isoenergetic surface near each piece of the nonsingular set, and the functions describing this pair of pieces differ by the value of order $k^{-4l+2n+\delta}$. We shall show that there exist two parts of the perturbed isoenergetic surface in a vicinity of $\chi_q \cup \chi_{-q}$. They are asymptotically close to pieces of the isoenergetic surface of $\hat{H}_q$ and can be parameterized in an analogous way.

The perturbation series (2.2.15) yields the spectral projection of $H(t)$ for a given quasimomentum t. However, this formula is not applicable directly when it is given the value of the perturbed eigenvalue $\lambda(\alpha,t) = k_0^{2l}$ and $n-1$ coordinates of t, say $t_2, ..., t_n$, of quasimomentum t, i.e., when t is on the perturbed isoenergetic surface, corresponding to a given energy k_0^{2l}. To use formula (2.2.15) in this case, we have previously to determine t_1 from the equation

$$\lambda(\alpha,t) = k_0^{2l}. \tag{2.1.12}$$

Practically, we can do it only approximately, taking $t_1 \approx t_{01}$, where t_{01} is the solution of the unperturbed equation $p_j^{2l}(t) = k^{2l}$. Substituting $t_1 \approx t_{01}$ in series (2.2.15), we obtain a rather rough asymptotic formula for the spectral projection (Section 2.7) and for $\nabla\lambda(\alpha,t)$.

However, there exists another way to construct an eigenfunction for t being on the isoenergetic surface of H. In this way we don't need to solve equation (2.1.12). We can directly construct the perturbation theory series for the eigenfunction with fixed $k_0^{2l}, t_2, ..., t_n$. Each term of this series depends on $k_0^{2l}, t_2, ..., t_n$ in an explicit form. This formula yields the eigenfunction for a given $k_0^{2l}, t_2, ..., t_n$ to any accuracy. It will be used in Chapter 5 for studying the semicrystal. The case of a non-smooth potential is considered in Chapter 3 together with the case of $4l > n + 1$.

2.2 Analytic Perturbation Theory for the Nonsingular Set.

We consider the operator

$$H_\alpha = (-\Delta)^l + \alpha V \tag{2.2.1}$$

in $L_2(R^n), n \geq 2, 2l > n$, where V is the operator of multiplication by a real, periodic potential, a trigonometric polynomial:

$$V(x) = \sum_{q \in R^n, |q| < R_0} v_q \exp i(\mathbf{p}_q(0), x). \tag{2.2.2}$$

The parameter α, introduced for convenience, varies over the interval $[-1, 1]$. I. M. Gel'fand showed that the investigation of the spectral properties of H_α reduces to the study of a family of regular operators $H_\alpha(t)$ (see Chapter I) with the matrices

$$H_\alpha(t)_{mq} = p_m^{2l}\delta_{mq} + \alpha v_{m-q}. \tag{2.2.3}$$

Before proving the main result, we formulate the Geometric Lemma:

Lemma 2.1 . *For an arbitrarily small positive δ, $2\delta < 2l - n$, and sufficiently large k, $k > k_0(\delta, a_1, ...a_n)$, there exists a nonsingular set $\chi_0(k, \delta)$, belonging to the isoenergetic surface $S_0(k)$ of the free operator H_0, such that, for any point t of it, the following conditions hold:*

1. *There exists a unique $j \in Z^n$, such that $p_j(t) = k$.*

2. $$\min_{i \neq j} | p_j^2(t) - p_i^2(t) | > 2k^{-n+2-\delta}. \tag{2.2.4}$$

Moreover, for any t in the $(2k^{-n+1-2\delta})$-neighborhood of the nonsingular set in C^n, there exists a unique $j \in Z^n$, such that $| p_j^2 - k^2 | < 5k^{-n+2-2\delta}$ and inequality (2.2.4) holds.

The nonsingular set has an asymptotically full measure on $S_0(k)$. Moreover,

$$\frac{s(S_0(k) \setminus \chi_0(k, \delta))}{s(S_0(k))} =_{k \to \infty} O(k^{-\delta/2}). \tag{2.2.5}$$

The proof of the lemma will be presented in Section 2.3.

Corollary 2.1 . *If t belongs to the $(2k^{-n+1-2\delta})$-neighborhood of the nonsingular set in C^n, then, for all z lying on the circle $C_0 = \{z :\mid z - k^{2l} \mid= k^{2l-n-\delta}\}$ and any i of Z^n, the inequality*

$$2 \mid p_i^{2l} - z \mid > k^{2l-n-\delta} \tag{2.2.6}$$

holds.

Proof of the corollary. Let $t \in \chi_0(k,\delta)$. Taking into account the relation $p_j^{2l}(t) = k^{2l}$ and the definition of C_0, we see that

$$\mid p_j^{2l} - k^{2l} \mid= k^{2l-n-\delta}. \tag{2.2.7}$$

Thus, the estimate (2.2.6) is valid for $i = j$. From (2.2.4) it follows that

$$\min_{i \neq j} \mid p_j^{2l}(t) - p_i^{2l}(t) \mid > 2k^{2l-n-\delta} \tag{2.2.8}$$

Using relation (2.2.7) we obtain inequality (2.2.6) for $i \neq j$. It is easy to see that all the estimates are stable under a perturbation of t of order $2k^{-n+1-2\delta}$; therefore (2.2.6) holds not only on the nonsingular set, but also in the $(2k^{-n+1-2\delta})$-neighborhood of it. The corollary is proved.

As denoted in Chapter 1, E_j is the spectral projection of the free operator, corresponding to the eigenvalue p_j^{2l} : $(E_j)_{rm} = \delta_{jr}\delta_{jm}$[1] We define functions $g_r(k,t)$ and operator-valued functions $G_r(k,t)$, $r = 0, ..., t \in \chi_0(k,\delta)$ as in Chapter 1:

$$g_r(k,t) = \frac{(-1)^r}{2\pi i r} \mathrm{Tr} \oint_{C_0} ((H_0(t) - z)^{-1}V)^r dz \tag{2.2.9}$$

$$G_r(k,t) = \frac{(-1)^{r+1}}{2\pi i} \oint_{C_0} ((H_0(t) - z)^{-1}V)^r (H_0(t) - z)^{-1} dz. \tag{2.2.10}$$

We remark that to find $g_r(k,t)$ and $G_r(k,t)$ it is necessary to compute the residues of a rational function of a simple structure, whose numerator does not depend on z, while the denominator is a product of factors of the type $(p_i^{2l}(t) - z)$. For all t in the nonsingular set within C_0 the integrand has a single pole at the point $z = k^{2l} = p_i^{2l}(t)$. In the sums, arising in the construction of the integrand operators and their traces there only a finite number of terms, having a singularity at the point $z = k^{2l}$. By computing the residue at this point, we obtain an explicit expressions for $g_r(k,t)$ and $G_r(k,t)$. For example,

$$g_1(k,t) = 0, \tag{2.2.11}$$

$$g_2(k,t) = \sum_{q \in Z^n, q \neq 0} \mid v_q \mid^2 (p_j^{2l}(t) - p_{j+q}^{2l}(t))^{-1} =$$

$$\sum_{q \in Z^n, q \neq 0} \frac{\mid v_q \mid^2 (2p_j^{2l}(t) - p_{j+q}^{2l}(t) - p_{j-q}^{2l}(t))}{2(p_j^{2l}(t) - p_{j+q}^{2l}(t))(p_j^{2l}(t) - p_{j-q}^{2l}(t))}, \tag{2.2.12}$$

$$G_1(k,t)_{rm} = v_{j-m}(p_j^{2l}(t) - p_m^{2l}(t))^{-1}\delta_{rj} + v_{r-j}(p_j^{2l}(t) - p_r^{2l}(t))^{-1}\delta_{mj}, \quad (G_{jj} = 0). \tag{2.2.13}$$

[1] δ_{jr}, δ_{jm} are Kronecker symbols.

Theorem 2.1 . *Suppose t belongs to the $(k^{-n+1-2\delta})$-neighborhood in K of the nonsingular set $\chi_0(k,\delta)$, $0 < 2\delta < 2l-n$. Then for sufficiently large k, $k > k_0(\|V\|,\delta)$, for all α, $-1 \le \alpha \le 1$, there exists a single eigenvalue of the operator $H_\alpha(t)$ in the interval $\varepsilon(k,\delta) \equiv (k^{2l} - k^{2l-n-\delta}, k^{2l} + k^{2l-n-\delta})$. It is given by the series:*

$$\lambda(\alpha,t) = p_j^{2l}(t) + \sum_{r=2}^{\infty} \alpha^r g_r(k,t), \tag{2.2.14}$$

converging absolutely in the disk $|\alpha| \le 1$, where the index j is uniquely determined from the relation $p_j^{2l}(t) \in \varepsilon(k,\delta)$. The spectral projection, corresponding to $\lambda(\alpha,t)$ is given by the series:

$$E(\alpha,t) = E_j + \sum_{r=1}^{\infty} \alpha^r G_r(k,t), \tag{2.2.15}$$

which converges in the class $\mathbf{S}_1$ uniformly with respect to α in the disk $|\alpha| \le 1$.

For coefficients $g_r(k,t)$, $G_r(k,t)$ the following estimates hold:

$$|g_r(k,t)| < k^{2l-n-\delta-\gamma_0 r} \tag{2.2.16}$$

$$\|G_r(k,t)\|_1 < k^{-\gamma_0 r} \tag{2.2.17}$$

where $\gamma_0 = 2l-n-2\delta$. The operator G_r is nonzero only on the finite-dimensional subspace $(\sum_{i:|i-j|<rR_0} E_i)l_2^n$.

Corollary 2.2 . *For the perturbed eigenvalue and its spectral projection the following estimates are valid:*

$$|\lambda(\alpha,t) - p_j^{2l}(t)| \le 2\alpha^2 k^{2l-n-\delta-2\gamma_0}, \tag{2.2.18}$$

$$\|E(\alpha,t) - E_j\|_1 \le 2\alpha k^{-\gamma_0}. \tag{2.2.19}$$

Remark 2.1. Here we discuss only the selfadjoint case, and hence all arguments are carried out only for real $\alpha, -1 \le \alpha \le 1$. Actually, the majority of the formulas are preserved also for complex α, although completely new spectral phenomena arise here which we shall discuss in detail elsewhere.

Proof of the theorem. The proof of the theorem is based on expanding the resolvent $(H(t)-z)^{-1}$ in a perturbation series for z, belonging to the contour C_0 about the unperturbed eigenvalue $p_j^{2l}(t)$. Then, integrating the resolvent yields the formulae for the perturbed eigenvalue and its spectral projection.

It is obvious that

$$(H_\alpha(t) - z)^{-1} = (H_0(t) - z)^{-1/2}(I - \alpha A)^{-1}(H_0(t) - z)^{-1/2}, \tag{2.2.20}$$

$$A \equiv -(H_0(t) - z)^{-1/2} V (H_0(t) - z)^{-1/2},$$

(choice of the branch of the square root is arbitrary). Suppose $z \in C_0$. Using (2.2.6), we obtain:

$$\|(H_0(t)-z)^{-1/2}\| \le k^{-l+n/2+\delta/2}, \tag{2.2.21}$$

$$\|A\| \le \|V\|\|(H_0(t)-z)^{-1/2}\|^2 \le 2\|V\|k^{-2l+n+\delta} = 2\|V\|k^{-\gamma_0-\delta}. \tag{2.2.22}$$

Thus, $\|A\| \ll 1$ for sufficiently large k. Expanding $(I-\alpha A)^{-1}$ in a series in powers of αA we obtain:

$$(H_\alpha(t)-z)^{-1}-(H_0(t)-z)^{-1} = \sum_{r=1}^{\infty} \alpha^r (H_0(t)-z)^{-1/2} A^r (H_0(t)-z)^{-1/2}. \tag{2.2.23}$$

Since $2l > n$ it is not hard to show that $(H_0(t)-z)^{-1/2} \in \mathbf{S}_2$. Taking into account estimates (2.2.22), we see that series (2.2.23) converges in the class $\mathbf{S}_1$ uniformly with respect to α in the disk $|\alpha| \le 1$.

The segment $\varepsilon(k,\delta)$ contains n_α $(n_\alpha \ge 0)$ eigenvalues of operator the $H_\alpha(t)$. The spectral projection, corresponding to them is found from the relation

$$E(\alpha,t) = -\frac{1}{2\pi i}\oint_{C_0} (H_\alpha(t)-z)^{-1}dz. \tag{2.2.24}$$

Substituting (2.2.23) into (2.2.24) and using the obvious formula

$$E_j = -\frac{1}{2\pi i}\oint_{C_0} (H_0(t)-z)^{-1}dz, \tag{2.2.25}$$

we obtain the series (2.2.15), which converges in $\mathbf{S}_1$ uniformly with respect to α α in the unit disk.

We prove estimate (2.2.17) for $G_r(k,t)$. It is easy to see that

$$G_r(k,t) = -\frac{1}{2\pi i}\oint_{C_0} (H_0(t)-z)^{-1/2}A^r(H_0(t)-z)^{-1/2}dz \tag{2.2.26}$$

We introduce the operator $A_0 = (I-E_j)A(I-E_j)$. It is obvious that $\|A_0\| \le \|A\|$. In addition

$$\oint_{C_0} (H_0(t)-z)^{-1/2}A_0^r(H_0(t)-z)^{-1/2}dz = 0, \tag{2.2.27}$$

since the integrand is holomorphic inside the circle. Thus,

$$G_r(k,t) = -\frac{1}{2\pi i}\oint_{C_0} (H_0(t)-z)^{-1/2}(A^r - A_0^r)(H_0(t)-z)^{-1/2}dz. \tag{2.2.28}$$

It is not hard to show that

$$\|A^r - A_0^r\|_1 \le r\|A-A_0\|_1\|A\|^{r-1}. \tag{2.2.29}$$

Since $A - A_0 = E_jA(I-E_j) + (I-E_j)AE_j$, $E_jAE_j = 0$ and E_j is a one-dimensional projection, we get

$$\|A-A_0\|_1 \le \|A\|. \tag{2.2.30}$$

Using inequalities (2.2.29), (2.2.22) we obtain from relation (2.2.30) that

$$||A^r - A_0^r||_1 \leq r(2||V||)^r k^{-(\gamma_0+\delta)r} \leq k^{-\gamma_0 r-\delta/2}. \tag{2.2.31}$$

Noting that the length of C_0 is equal to $2\pi k^{2l-n-\delta}$, we obtain from formula (2.2.28):

$$||G_r(k,t)||_1 \leq 2\pi k^{2l-n-\delta}||(H_0(t)-z)^{-1/2}||^2||A^r - A_0^r||_1 \tag{2.2.32}$$

Using inequalities (2.2.21), (2.2.31), we get (2.2.17).

We show now that inside the contour C_0, i.e., on the segment $\varepsilon(k,\delta)$, operator H_α has a single eigenvalue ($n_\alpha = 1$). It is easy to see that n_α is determined from the formula

$$n_\alpha - n_0 = -\frac{1}{2\pi i}\oint_{C_0} \mathrm{Tr}((H_\alpha(t)-z)^{-1} - (H_0(t)-z)^{-1})dz = \tag{2.2.33}$$

$$= \mathrm{Tr}(E(t) - E_j).$$

We prove the relation $n_\alpha - n_0 =_{k\to\infty} o(1)$. Indeed, considering formula (2.2.15) for the spectral projection, corresponding to segment $\varepsilon(k,\delta)$, we obtain

$$| n_\alpha - n_0 | < \sum_{r=1}^{\infty} ||G_r(k,t)||_1 < 2k^{-\gamma_0} \tag{2.2.34}$$

Since n_α and n_0 are integers, and $n_0 = 1$ by the hypothesis of the theorem, the operator $H_\alpha(t)$ for all α, $-1 \leq \alpha \leq 1$, has a single eigenvalue in $\varepsilon(k,\delta)$.

Further, we use the well-known formula:

$$\frac{\partial\lambda(\alpha,t)}{\partial\alpha} = -\frac{1}{2\pi i}\oint_{C_0} \mathrm{Tr}(V(H_\alpha(t)-z)^{-1})dz \tag{2.2.35}$$

Using formula (2.2.23) and considering that $\mathrm{Tr}(V(H_0(t)-z)^{-1}) = 0$, we obtain:

$$\frac{\partial\lambda(\alpha,t)}{\partial\alpha} = \sum_{r=2}^{\infty} r\alpha^{r-1} g_r(k,t). \tag{2.2.36}$$

Integrating the last relation with respect to α and noting that $\lambda(0,t) = p_j^{2l}(t)$, we get formula (2.2.14). To prove estimate (2.2.16) we note that

$$| g_r(k,t) | \leq k^{2l-n-\delta}||A^r - A_0^r||_1, \tag{2.2.37}$$

Using estimate (2.2.31), we obtain inequality (2.2.16).

Finally, we prove that operator $G_r(k,t)$ is nonzero only on the finite-dimensional subspace $(\sum_{i:|i-j|<rR_0} E_i)l_2^n$. Indeed, it is easy to see that $G_r(k,t)$ can be represented in the form:

$$G_r(k,t)_{i_0 i_r} = \sum_{i_1,\ldots,i_{r-1}\in Z^n} I_{i_0 i_1 \ldots i_{r-1} i_r}, \tag{2.2.38}$$

where

$$I_{i_0 i_1 \ldots i_{r-1} i_r} = \frac{(-1)^{r+1}}{2\pi i} \oint_{C_0} \frac{v_{i_0-i_1} \ldots v_{i_{r-1}-i_r}}{(p_{i_0}^{2l}(t)-z)\ldots(p_{i_r}^{2l}(t)-z)} dz.$$

The last integral can be nonzero only when the integrand has a pole inside the contour, i.e., when at least one of the indices $i_0, \ldots, i_r$ is equal to j. Considering that V is a trigonometric polynomial, we obtain $| i_q - j | \leq rR_0$, $q = 0, \ldots r$. From this it follows immediately that $G_r(k,t)_{i_0 i_r}$ can be nonzero only when $| i_0 - j | \leq rR_0$ and $| i_r - j | \leq rR_0$, i.e., operator $G_r(k,t)$ is equal to zero on the orthogonal complement of $(\sum_{i:|i-j|<rR_0} E_i) l_2^n$. The theorem is proved.

We use the notation (1.0.30).

Theorem 2.2 . *Under the conditions of Theorem 2.1 the series (2.2.14), (2.2.15), can be differentiated termwise with respect to t any number of times, and they retain their asymptotic character. For the coefficients $g_r(k,t)$ and $G_r(k,t)$ in the $(k^{-n+1-2\delta})$-neighborhood in C^n of the nonsingular set the following estimates hold:*

$$| T(m) g_r(k,t) | < m! k^{2l-n-\delta-\gamma_0 r+|m|(n-1+\delta)} \tag{2.2.39}$$

$$\|T(m) G_r(k,t)\| < m! k^{-\gamma_0 r+|m|(n-1+\delta)} \tag{2.2.40}$$

Corollary 2.3 . *There are the estimates for the perturbed eigenvalue and its spectral projection:*

$$| T(m)(\lambda(\alpha,t) - p_j^{2l}(t)) | < 2m! \alpha^2 k^{2l-n-\delta-2\gamma_0+|m|(n-1+\delta)}, \tag{2.2.41}$$

$$\|T(m)(E(\alpha,t) - E_j)\|_1 < 2m! \alpha k^{-\gamma_0+|m|(n-1+\delta)}. \tag{2.2.42}$$

Proof of the theorem. Since (2.2.6) is valid in the complex $(2k^{-n+1-2\delta})$-neighborhood of the nonsingular set, it is easy to see that the coefficients $g_r(k,t)$, and $G_r(k,t)$ can be continued from the real $(2k^{-n+1-2\delta})$-neighborhood of t to the complex $(2k^{-n+1-2\delta})$-neighborhood as holomorphic functions of n variables, and inequalities (2.2.16), (2.2.17) are hereby preserved. Estimating by means of the Cauchy integral the value the derivative at t in terms of the value of the function itself on the boundary of the $(2k^{-n+1-2\delta})$-neighborhood of t (formulas (2.2.16), (2.2.17)), we obtain (2.2.39), (2.2.40). The theorem is proved.

2.3 Construction of the Nonsingular Set.

We introduce the following notations:

1. S_r is the sphere of radius r in R^n centered at the origin.

2. $\Pi_m(\mu)$ is the plane layer in R^n given by:

$$\Pi_m(\mu) = \{x : x \in R^n, || x |^2 - | x - \mathbf{p}_m(0) |^2| < \mu\}. \tag{2.3.1}$$

3. $S_m(k,\xi)$, $\xi > 0$, $m \neq 0$, is the spherical layer:

$$S_m(k,\xi) = S_k \cap \Pi_m(4k^{-\xi}). \tag{2.3.2}$$

It is clear that $S_m(k,\xi) = \emptyset$ for sufficiently large m i.e., for $m : p_m(0) > 2k + 16k^{-\xi-1}$.

4. $S(k,\xi)$ is the union of all the spherical layers $S_m(k,\xi)$:

$$S(k,\xi) = \cup_{m\in Z^n, m\neq 0} S_m(k,\xi) = S_k \cap (\cup_{m\in Z^n, m\neq 0} \Pi_m(4k^{-\xi})) \tag{2.3.3}$$

5. We note that any point x in R^n can be represented uniquely in the form $x = \mathbf{p}_m(t)$, where $m \in Z^n$ and $t \in K$.

Let $\mathcal{K}$ be the mapping:

$$\mathcal{K} : R^n \to K, \quad \mathcal{K}\mathbf{p}_m(t) = t. \tag{2.3.4}$$

Suppose $\Omega \subset R^n$. In order to obtain $\mathcal{K}\Omega$ it is necessary to partition Ω by the lattice with nodes at the points $x = \mathbf{p}_m(0)$, $m \in Z^n$ and to shift all parts in a parallel manner into a single cell. It is obvious that for a surface Ω

$$s(\mathcal{K}\Omega) \leq s(\Omega). \tag{2.3.5}$$

If $\Omega = S_k$, then $\mathcal{K}S_k = S_0(k)$ and

$$s(S_0(k)) = s(S_k). \tag{2.3.6}$$

We define the nonsingular set as follows:

$$\chi_0(k,\delta) = S_0(k) \setminus \mathcal{K}S(k, n-2+\delta). \tag{2.3.7}$$

Note that $S_0(k)$ is sphere S_k packed by mapping $\mathcal{K}$ into the cell K and $\mathcal{K}S_k(n-2+\delta)$ is the $(k^{-n+2-\delta})$- neighborhood on $S_0(k)$ of the selfintersections of $S_0(k)$. Thus, $\chi_0(k,\delta)$ is obtained from $S_0(k)$ by deleting the $(k^{-n+2-\delta})$-neighborhood of its selfintersections, which, in fact, are described by the von Laue diffraction conditions.

Thus, we proceed to the proof of Geometric Lemma 2.1. It contains two steps. First, using the formula for the nonsingular set $\chi_0(k,\delta)$, we check that for t in $\chi_0(k,\delta)$ conditions 1,2 of Geometrical Lemma 2.1 are satisfied and for all t in the complex $(k^{-n+1-2\delta})$-neighborhood of $\chi_0(k,\delta)$ there exists a unique $j \in Z^n$ such that $| p_j^{2l}(t) - k^{2l} | < k^{2l-n-\delta}$ and the second condition of Geometric Lemma is satisfied. Secondly, we prove that the nonsingular set $\chi_0(k,\delta)$ is nonempty; moreover it is a subset of an asymptotically full measure on $S_0(k)$.

Lemma 2.2 contains the proof of the first step. Lemmas 2.3–2.4 are of auxiliary character. The remainder of the proof of the second step is contained in Lemma 2.5.

Lemma 2.2 . *If $t \in \chi_0(k,\delta)$, then*

1. *There exists a unique $j \in Z^n$, such that $p_j(t) = k$.*

2.

$$\min_{i \neq j} \mid p_j^2(t) - p_i^2(t) \mid > 2k^{-n+2-\delta} \tag{2.3.8}$$

Moreover, for any t in the $(2k^{-n+1-2\delta})$-neighborhood of the nonsingular set in C^n there exists a unique $j \in Z^n$, such that $\mid p_j^2(t) - k^2 \mid < 5k^{-n+2-2\delta}$ and relation (2.3.8) is satisfied.

Proof. Since $t \in S_0(k)$, there exists at least one j, such that $p_j(t) = k$. We prove that it is unique; moreover, the condition

$$\min_{i \neq j} \mid p_i^2(t) - p_j^2(t) \mid geq 4k^{-n+2-\delta} \tag{2.3.9}$$

is satisfied. We suppose that this is not the case, i.e., there exists $i \neq j$ such that

$$\min_{i \neq j} \mid p_i^2(t) - p_j^2(t) \mid < 4k^{-n+2-\delta}. \tag{2.3.10}$$

Let $x = \mathbf{p}_j(t)$ and $m = j - i$. Then $x \in S_k$, $\mathcal{K}x = t$ and the last estimate implies that

$$\mid\mid x \mid^2 - \mid x - \mathbf{p}_m(t) \mid^2 \mid < 4k^{-n+2-\delta},$$

i.e.

$$x \in S_m(k, n-2+\delta) \subset S(k, n-2+\delta).$$

Hence, $t \in \mathcal{K}S(k, n-2+\delta) \subset S_0(k) \setminus \chi_0(k,\delta)$. This contradiction proves that conditions 1,2 are satisfied.

Suppose t is in the $(k^{-n+1-2\delta})$-neighborhood of $\chi_0(k,\delta)$. Then, there exists t_0 such that $t_0 \in \chi_0(k,\delta)$ and $\mid t - t_0 \mid < k^{-n+2-2\delta}$. We consider $j : p_j(t_0) = k$. It is easy to see that $\mid p_j^2(t) - p_j^2(t_0) \mid < 5k^{-n+1-2\delta}$ i.e., $\mid p_i^2(t) - k^2 \mid < 5k^{-n+1-2\delta}$. Further, it is obvious that estimate (2.3.9) is stable with respect to a perturbation of order $2k^{-n+1-2\delta}$. The lemma is proved.

In $R^n, n \geq 2$ we consider the intersection $S_m(k,\xi)$ of the sphere S_k with the plane layer $\Pi_m(4k^{-\xi})$. The next lemma estimates the area of this spherical layer in dependence on m, k and ξ. If $p_m(0)/2 > k + 8k^{-\xi-1}$, then the sphere and the plane layer do not intersect, and therefore $S_m(k,\xi) = 0$. If $p_m(0)/2 < k - 8k^{-\xi-1}$, then their intersection is a "nondegenerate" spherical layer whose area can be estimated in terms of its thickness and the radius and can be expressed by (2.3.11). In the case $k - 8k^{-\xi-1} \leq p_m(0)/2 \leq k + 8k^{-\xi-1}$, the set $S_m(k,\xi)$ is a spherical "cap". Its area can be estimated in terms of the radius of the sphere and the solid angle which is subtends from the center of the sphere. An upper bound is given by (2.3.12).

Lemma 2.3 . *Suppose $n \geq 2$ and $\xi > -1$. Then* [2]

$$s(S_m(k,\xi)) \approx p_m^{-1}(0) k^{-\xi+1} (k^2 - p_m^2(0)/4)^{(n-3)/2}, \tag{2.3.11}$$

[2] We write $a(k) \approx b(k)$ when the inequalities $c_1 b(k) < a(k) < c_2 b(k), c_1 \neq c_1(k), c_2 \neq c_2(k)$ are valid.

if $p_m(0)/2 < k - 8k^{-\xi-1}$,

$$s(S_m(k,\xi)) < ck^{-\xi(n-1)/2}, \quad c \neq c(k), \tag{2.3.12}$$

if $k - 8k^{-\xi-1} \le p_m(0)/2 \le k + 8k^{-\xi-1}$; *and*

$$s(S_m(k,\xi)) = 0, \tag{2.3.13}$$

if $p_m(0)/2 > k + 8k^{-\xi-1}$.

Proof. In R^n we introduce a coordinate system so that the axis x_1 is directed alone $\mathbf{p}_m(0)$, and we introduce spherical coordinates $(R, \vartheta, \boldsymbol{\nu})$: $x_1 = R\cos\vartheta$, $x_i = (R\sin\vartheta)\nu_i$, $i = 2, ..., n$, here $\boldsymbol{\nu}$ are orthogonal coordinates on the unit sphere S_1^{n-2} in the space of dimension $n-1$ spanned by $x_2, ..., x_n$; $0 \le \vartheta \le \pi$, and $R =| x |$. In the new coordinates the volume element is

$$dV_n = R(R\sin\vartheta)^{n-2} d\vartheta dS_1^{n-2}(\boldsymbol{\nu}) dR, \quad n > 2 \tag{2.3.14}$$

$$dV_2 = RdRd\vartheta, \quad n = 2,$$

here $dS_1^{n-2}(\boldsymbol{\nu})$ is the element of the surface area of S_1^{n-2}. The surface element of the sphere S_R is given by

$$dS_R = dV_n/dR = R(R\sin\vartheta)^{n-2} d\vartheta dS_1^{n-2}(\boldsymbol{\nu}), \quad n > 2, \tag{2.3.15}$$

$$dS_R = Rd\vartheta, \quad n = 2.$$

1) Suppose $p_m(0)/2 < k - 8k^{-\xi-1}$. We represent $p_m(0)/2$ in the form $p_m(0)/2 = k\cos\vartheta_0$, $\cos\vartheta_0 = p_m(0)/2k < 1 - 8k^{-\xi-2}$, $\sin^2(\vartheta_0/2) > 4k^{-\xi-2}$, $\pi/2 > \vartheta_0 > 4k^{-(\xi+2)/2}$. The set $S_m(k,\xi)$ in the new notation can be represented as follows:

$$S_m(k,\xi) = \{x :| x |= k, 2p_m(0) \mid x_1 - p_m(0)/2 \mid< 4k^{-\xi}\}, \quad c_m = 2p_m(0)^{-1}.$$

We consider the set $\theta_{\beta,\vartheta_0,c} : 0 < \beta \neq \beta(k),\ 0 < c \neq c(k),\ \vartheta_0 = \vartheta_0(k)$:

$$\theta_{\beta,\vartheta_0,c} = \{\vartheta :| \cos\vartheta - \cos\vartheta_0 |< ck^{-\beta}\}.$$

It turns out that

$$\{\vartheta :| \vartheta - \vartheta_0 |< (c/2)k^{-\beta} \mid \sin\vartheta_0 |^{-1}\} \subset \theta_{\beta,\vartheta_0,c} \subset$$

$$\{\vartheta :| \vartheta - \vartheta_0 |< 2ck^{-\beta} \mid \sin\vartheta_0 |^{-1},\} \tag{2.3.16}$$

if $\pi/2 > \vartheta_0 > 2c^{1/2}k^{-\beta/2}$. Indeed, in the case where $\sin\vartheta_0$ as $k \to \infty$ is bounded away from zero (2.3.16) is obvious. Suppose $\vartheta_0 \to 0$. Then

$$| \cos\vartheta - \cos\vartheta_0 |= 2 \mid \sin^2(\vartheta/2) - \sin^2(\vartheta_0/2) |< ck^{-\beta}.$$

Since $\vartheta_0 > 2c^{1/2}k^{-\beta/2}$, it follows that $\sin^2(\vartheta_0/2) < 2\sin^2(\vartheta/2) < 3\sin^2(\vartheta_0/2)$; i.e.,

$$\vartheta_0/\sqrt{2} < \vartheta < \vartheta_0\sqrt{3/2}. \tag{2.3.17}$$

Using this relation, we obtain

$$2 \mid \sin^2(\vartheta/2) - \sin^2(\vartheta_0/2) \mid = \mid \vartheta^2 - \vartheta_0^2 \mid (1 + O(\vartheta_0^2))/2 < ck^{-\beta}.$$

Hence

$$\mid \vartheta - \vartheta_0 \mid < 2ck^{-\beta}(\vartheta + \vartheta_0)^{-1}(1 + O(\vartheta_0^2)) <$$
$$< 3ck^{-\beta}(2\sin\vartheta_0)^{-1}(1 + O(\vartheta_0^2)) < 2ck^{-\beta}(\sin\vartheta_0)^{-1}$$

and, similarly,

$$\mid \vartheta - \vartheta_0 \mid > (c/2)(\sin\vartheta_0)^{-1}.$$

Formula (2.3.16) is proved. Using the definitions of $S_m(k,\xi)$ and $\theta_{\beta,\vartheta_0,c}$ we obtain

$$S_m(k,\xi) = \{x : R = k, \vartheta \in \theta_{\xi+1,\vartheta_0,c_m}\}, c_m = 2p_m^{-1}(0).$$

We note that $\vartheta_0 > 4k^{-(\xi+2)/2} = 4\sqrt{c_m p_m(0)/2}k^{-(\xi+2)/2} > 2\sqrt{c_m}k^{-(\xi+1)/2}$ (since $p_m(0) \approx k$ for small ϑ_0); it is therefore possible to apply (2.3.16). As a result we obtain

$$S_m(k,\xi) \subset \{x : R = k, \mid \vartheta - \vartheta_0 \mid < \varepsilon_1\}, \quad \varepsilon_1 = 4p_m(0)^{-1}k^{-\xi-1}(\sin\vartheta_0)^{-1}.$$

Further, using (2.3.15) for $n > 2$, we verify:

$$s(S_m(k,\xi)) \leq \int_{|\vartheta-\vartheta_0|<\varepsilon_1, \nu\in S_1^{n-2}} dS_k = k\int_{|\vartheta-\vartheta_0|<\varepsilon_1,\nu\in S_1^{n-2}} (k\sin\vartheta)^{n-2}d\vartheta dS_1^{n-2}(\nu)$$

$$= ck\int_{|\vartheta-\vartheta_0|<\varepsilon_1} (k\sin\vartheta)^{n-2}d\vartheta.$$

Similarly,

$$s(S_m(k,\xi)) \geq ck\int_{|\vartheta-\vartheta_0|<\varepsilon_1/4} (k\sin\vartheta)^{n-2}d\vartheta.$$

It is easy to verify that $\sin\vartheta < \sqrt{3/2}\sin\vartheta_0$. Indeed, in the case where ϑ_0 is bounded away from zero as $k \to \infty$ we have $\sin\vartheta = (1 + o(1))\sin\vartheta_0$. For small ϑ the required relation immediately follows from (2.3.17). Using it, we obtain:

$$s(S_m(k,\xi)) \approx k(k\sin\vartheta_0)^{n-2}\varepsilon_1 \approx p_m^{-1}(0)k^{-\xi+1}(k^2 - p_m^2(0)/4)^{(n-3)/2}. \tag{2.3.18}$$

The case $n = 2$ can be considered in a similar way.

2)Suppose $k - 8k^{-\xi-1} \leq p_m(0)/2 \leq k + 8k^{-\xi-1}$. Then

$$S_m(k,\xi) = \{x :\mid x \mid = k, 2p_m(0) \mid x_1 - p_m(0)/2 \mid < 4k^{-\xi}\} \subset$$

$$\subset \{x :\mid x \mid = k, 0 < k - x_1 < 10k^{-\xi-1}\} \subset \{x :\mid x \mid = k, 1 - \cos\vartheta < 10k^{-\xi-2}\} \subset$$
$$\{x :\mid x \mid = k, 0 < \vartheta < \varepsilon_2\}, \quad \varepsilon_2 = 5k^{-(\xi+2)/2};$$

Therefore

$$s(S_m(k,\xi)) < \int_{0<\vartheta<\varepsilon_2, \nu\in S_1^{n-2}} k(k\sin\vartheta)^{n-2} d\vartheta dS_1^{n-2}(\nu) \le ck^{n-1}\varepsilon_2^{n-1} =$$

$$ck^{-\xi(n-1)/2}.$$

The case $n = 2$ can be considered in a similar way.

We suppose that $p_m(0)/2 > k + 8k^{-\xi-1}$. Then, for all $x \in S_k$

$$\mid \mathbf{p}_m(0) - x \mid - \mid x \mid > 16k^{-\xi-1}.$$

Hence

$$\left|\mid \mathbf{p}_m(0) - x \mid^2 - \mid x \mid^2\right| > 16k^{-\xi},$$

i.e., $\Pi_m(4k^{-\xi}) \cap S_k = \emptyset$. The lemma is proved.

The next lemma estimates the area of the union $S(k,\xi)$ of all layers $S_m(k,\xi), m \in Z^n \setminus \{0\}$.

Lemma 2.4 . *The following estimates for $S(k,\xi)$ are valid:*

$$s(S(k,\xi)) < ck^{n-1-\mu}, \quad n > 2, \quad \mu = \xi - n + 2; \tag{2.3.19}$$

$$s(S(k,\xi)) < ck^{1-\xi/2}, \quad n = 2 \tag{2.3.20}$$

Proof. Suppose $n > 2$. Then

$$s(S(k,\xi)) \le \sum_{m\in Z^n, m\neq 0} s(S_m(k,\xi)) = \Sigma_1 + \Sigma_2, \tag{2.3.21}$$

where

$$\Sigma_1 = \sum_{m:|p_m(0)/2-k|>8k^{-\xi-1}, m\neq 0} s(S_m(k,\xi)),$$

$$\Sigma_2 = \sum_{m:|p_m(0)/2-k|\le 8k^{-\xi-1}} s(S_m(k,\xi)).$$

We estimate Σ_1. Using inequality (2.3.11), we obtain:

$$\Sigma_1 < ck^{-\xi+1} \sum_{m:0<p_m(0)/2<k-8k^{-\xi-1}} p_m(0)^{-1}(k^2 - p_m^2(0)/4)^{(n-3)/2} <$$

$$< ck^{-\mu} \sum_{m:0<p_m(0)<2k} p_m(0)^{-1}.$$

The series on the right can be estimated in terms of the integral over the ball $K_{2k} : \int_{x\in R^n, |x|<2k} \mid x \mid^{-1} dx$, since it is an integral sum. It can be shown that

$$\sum_{m:0<p_m(0)<2k} p_m(0)^{-1} < c\int_{x\in R^n, |x|<2k} \mid x \mid^{-1} dx < ck^{n-1}.$$

Hence

$$\Sigma_1 < ck^{n-1-\mu}. \tag{2.3.22}$$

In order to estimate Σ_2, we compute N, the number of terms in this sum. It is less than the number of points $\mathbf{p}_m(0)$ in the spherical shell $\| x \mid -k \mid < p, p = 8\pi \max\{a_1^{-1}, ..., a_n^{-1}\}$. It is not hard to see that $N < ck^{n-1}$. Lemma 2.3 enables us to estimate the terms of the series Σ_2. Using (2.3.12), we obtain

$$\Sigma_2 < Nck^{-(n-1)\mu/2} < ck^{n-1-\mu}. \tag{2.3.23}$$

Adding (2.3.22) and (2.3.23), we obtain (2.3.19). We consider the case $n = 2$. We decompose the sum $\sum_{m \in Z^n, m \neq 0} s(S_m(k,\xi))$ into three terms $\Sigma_1, \Sigma_2, \Sigma_3$; here

$$\Sigma_1 = \sum_{m \in \sigma_1} s(S_m(k,\xi)), \quad \sigma_1 = \{m : m \in Z^n, 0 < p_m(0) < p, m \neq 0\};$$

$$\Sigma_2 = \sum_{m \in \sigma_2} s(S_m(k,\xi)), \quad \sigma_2 = \{m : m \in Z^n, p \leq p_m(0) < 2(k-p)\};$$

$$\Sigma_3 = \sum_{m \in \sigma_3} s(S_m(k,\xi)), \quad \sigma_3 = \{m : m \in Z^n, \mid p_m(0)/2 - k \mid \leq p\}.$$

Using (2.3.11), it is easy to find that $\Sigma_1 < ck^{-\xi}$. We estimate Σ_2:

$$\Sigma_2 < ck^{-\xi+1} \sum_{m \in \sigma_2} p_m(0)^{-1}(k^2 - p_m^2(0)/4)^{-1/2}.$$

It is not hard to show that if $m \in \sigma_2$ and $x = \mathbf{p}_m(t), t \in K$, then we have $\mid x \mid < 2p_m(0)$ and $\sqrt{k^2 - \mid x \mid^2 /4} < 2\sqrt{k^2 - p_m^2(0)/4}$. From this we obtain an estimate in terms of integral over the ball K_{2k-p}:

$$\sum_{m \in \sigma_2} p_m(0)^{-1}(k^2 - p_m^2(0)/4)^{-1/2} < c \int_{K_{2k-p}} \mid x \mid^{-1} \mid k^2 - \mid x \mid^2 /4 \mid^{-1/2} dx < c_1,$$

i.e., $\Sigma_2 < ck^{-\xi+1}$.

We now estimate Σ_3. The number of nonzero terms in this sum is equal to the number of points $\mathbf{p}_m(0)$ in the ring $\| x \mid -k \mid < p$, which is obviously less than ck. From (2.3.11) and (2.3.12) it is not hard to find that $s(S_m(k,\xi)) < ck^{-\xi/2}$ if $m \in \sigma_3$. Hence $\Sigma_3 < ck^{1-\xi/2}$.

Adding the estimates for Σ_1, Σ_2, and Σ_3, we prove (2.3.20).

The lemma is proved.

Lemma 2.5 . *The nonsingular set $\chi_0(k,\delta)$ has an asymptotically full measure on $S_0(k)$. Moreover,*

$$\frac{s(S_0(k) \setminus \chi_0(k,\delta))}{s(S_0(k))} =_{k \to \infty} O(k^{-\delta/2}). \tag{2.3.24}$$

Proof. Considering relations (2.3.7), (2.3.5) it is easy to show that

$$s(S_0(k) \setminus \chi_0(k,\delta)) = s(\mathcal{K}S(k, n-2+\delta)) \leq s(S(k, n-2+\delta)).$$

Using estimates (2.3.19), (2.3.20), we obtain:

$$s(S_0(k) \setminus \chi(k,\delta)) = O(k^{n-1-\delta/2}).$$

Relation (2.3.24) immediately follows from the last inequality.

The lemma is proved.

The combination of Lemmas 2.2, 2.5 just proved gives the assertion of Lemma 2.1.

2.4 Perturbation Series for the Singular Set.

We have shown that perturbation formulae have the simplest form for t not in a vicinity of the von Laue diffraction plane:

$$p_i^{2l}(t) = p_j^{2l}(t), i \neq j. \tag{2.4.1}$$

More precisely, the perturbation formulae are valid, when

$$\min_{i \neq j} | p_i^2(t) - p_j^2(t) | > k^{-n+2-\delta}, \tag{2.4.2}$$

$0 < 2\delta < 2l - n$, $\delta \neq \delta(k)$, that is when t belongs to $\chi_0(k,\delta)$ (see (2.3.7)).

Now we construct the formulae for an eigenvalue and its spectral projection in the case of t not in $\chi_0(k,\delta)$. It should be noted that perturbation series converge in $\chi_0(k,\delta)$ for $0 < 2\delta < 2l - n$. When δ is smaller, the set $\chi_0(k,\delta)$ is reacher, but the convergence is slower. We define the singular set as follows

$$A_0(k,\delta) \equiv S_0(k) \setminus \chi_0(k, 2l - n - \delta). \tag{2.4.3}$$

It is clear that the perturbation series (2.2.14), (2.2.15) converge outside the set $A_0(k,\delta)$, $0 < 2\delta < 2l - n$. If t belongs to $A_0(k,\delta)$ then the inequality

$$| p_{j-q}^2(t) - p_j^2(t) | < k^{-n+2-\delta}, \quad \mathbf{p}_j(t) \in S_k, \tag{2.4.4}$$

is valid at least for one q. The last estimate means that $t \in \mathcal{K}S_q(n-2+\delta)$. Conversely, for almost all $t \in \mathcal{K}S_q(n-2+\delta)$ there exists a unique j such that $p_j(t) = k$ and inequality (2.4.4) holds. We shall construct the perturbation series on a set $\chi_q \subset \mathcal{K}S_q$, which has an asymptotically full measure on $\mathcal{K}S_q(n-2+\delta)$. We assume that $| q | < R_0, v_q \neq 0$.

Thus, we represent $H(t)$ in the form:

$$H(t) = \hat{H}_q(t) + \hat{V}, \tag{2.4.5}$$

$$\hat{V} = V - P_q V P_q,$$

where P_q is the two-dimensional diagonal projection, given by formula (2.1.8). We construct the perturbation series for operator $H(t)$ with respect to $\hat{H}_q(t)$. In fact, they are expansions of eigenvalue $\lambda(t)$ and its spectral projection $E(t)$ in powers of $\hat{V}$. Therefore, $\lambda(t)$ and $E(t)$ are analytic functions of $\hat{V}$, but not, generally speaking, of V. Hence, it would be natural now to place a multiplier α before $\hat{V}$ (just as in the preceding considerations it was helpful to place α before V). However, we omit it to make formulae less cumbersome.

The results for the singular set were proved in [K12].

First of all we describe more in detail operator $\hat{H}_q(t)$. As noted above, operator $\hat{H}_q(t)$ has only two nondiagonal terms not equal to zero:

$$\hat{H}_q(t)_{j,j-q} = \overline{\hat{H}_q(t)_{j-q,j}} = v_q. \tag{2.4.6}$$

Thus, for the spectral analysis of $\hat{H}_q(t)$ it suffices to consider the block (2.1.10). Its eigenvalues are:

$$\hat{\lambda}^+(t) = a + b \quad \hat{\lambda}^-(t) = a - b, \tag{2.4.7}$$

$$2a = p_j^{2l}(t) + p_{j-q}^{2l}(t), \quad 2b = (4 \mid v_q \mid^2 + (p_j^{2l}(t) - p_{j-q}^{2l}(t))^2)^{1/2}.$$

The corresponding eigenvectors are:

$$\hat{e}^+ = c^+(p_{j-q}^{2l}(t) - \lambda^+(t), \; -v_q), \quad \hat{e}^- = c^-(-v_{-q}, \; p_j^{2l}(t) - \lambda^-(t)), \tag{2.4.8}$$

c^+, c^- being the normalizing factors:

$$c^+ = (\mid v_q \mid^2 + \mid p_{j-q}^{2l}(t) - \hat{\lambda}^+(t) \mid^2)^{-1/2},$$

$$c^- = (\mid v_q \mid^2 + \mid p_j^{2l}(t) - \hat{\lambda}^-(t) \mid^2)^{-1/2}.$$

Thus, $\hat{H}_q(t)$ has the spectrum $\{\{p_i^{2l}(t)\}_{i \neq j, j-q}, \hat{\lambda}^+(t), \hat{\lambda}^-(t)\}$. The spectral projections are $\{\{E_i\}_{i \neq j, j-q}, \hat{E}^+, \hat{E}^-\}$, where $\hat{E}^+, \hat{E}^-$ correspond to $\hat{\lambda}^+(t)$, $\hat{\lambda}^-(t)$ and satisfy the relation:

$$(I - P_q)\hat{E}^\pm = \hat{E}^\pm(I - P_q) = 0 \tag{2.4.9}$$

the matrices $P_q\hat{E}^+P_q$, $P_q\hat{E}^-P_q$ being the spectral projections of the two-dimensional operator (2.1.10).

It is easy to see that $|\hat{\lambda}^+(t) - p_j^{2l}(t)| < |v_q|$, when $p_{j-q}^2 - p_j^2(t) < 0$, i.e., $\hat{\lambda}^+(t)$ is closer to k^{2l}, than $\hat{\lambda}^-(t)$. In the case of the opposite inequality we see that $|\hat{\lambda}^-(t) - p_j^{2l}(t)| < |v_q|$, i.e., $\hat{\lambda}^-(t)$ is closer to k^{2l}, than $\hat{\lambda}^+(t)$. We denote by $\hat{\lambda}_0^q(t)$ the nearest to k^{2l} of the eigenvalues $\hat{\lambda}^+(t)$, $\hat{\lambda}^-(t)$, i.e., $\hat{\lambda}_0^q(t) = \hat{\lambda}^+(t)$, when $p_{j-q}^2 - p_j^2(t) < 0$, and $\hat{\lambda}_0^q(t) = \hat{\lambda}^-(t)$, when $p_{j-q}^2 - p_j^2(t) > 0$. It is easy to see that

$$\hat{\lambda}_0^q(t) = p_j^2(t) - \Delta + (\operatorname{sgn}\Delta)\sqrt{|v_q|^2 + \Delta}, \tag{2.4.10}$$

where $2\Delta = p_j^{2l}(t) - p_{j-q}^{2l}(t)$. Denote by $\hat{E}_0^q$ the corresponding to $\hat{\lambda}^q(t)$ the spectral projection.

In the case of the singular set we prove the following Geometric Lemma:

Lemma 2.6 . *For an arbitrarily small positive δ, $2\delta < 2l - n$, and sufficiently large k, $k > k_0(\delta, a_1, ...a_n)$ there exists a subset $\chi_q(k,\delta)$ of $\mathcal{K}S_q(k, n-2+\delta)$, such that for any its point t the following conditions hold:*

1. *There exists a unique $j \in Z^n$, such that $p_j = k$.*

2.
$$| p_{j-q}^2(t) - p_j^2(t) | < k^{-n+2-\delta}, \tag{2.4.11}$$

3.
$$\min_{i \neq j, j-q} | p_j(t) - p_i^2 | > 2k^{-n+2-6\delta} \tag{2.4.12}$$

For any t in the $(2k^{-n+1-7\delta})$-neighborhood of $\chi_q(k,\delta)$ in C^n there exists a unique $j \in Z^n$, such that $| p_j^2(t) - k^2 | < 5k^{-n+2-6\delta}$ and relation (2.4.12) is satisfied.

Set $\chi_q(k,\delta)$ has an asymptotically full measure on $\mathcal{K}S_q(k, n-2+\delta)$. Moreover,

$$\frac{s(\mathcal{K}S_q(k,n-2+\delta) \setminus \chi_q(k,\delta))}{s(\mathcal{K}S_q(k,n-2+\delta))} =_{k\to\infty} O(k^{-\delta/2}) \tag{2.4.13}$$

Corollary 2.4 . *If t belongs to the $(2k^{-n+1-7\delta})$-neighborhood of $\chi_q(k,\delta)$, then for all z lying on the circle C_1, $C_1 = \{z :| z - \hat{\lambda}_0^q(t) |= k^{-\delta}\}$ the following inequalities are valid :*

$$2 | p_i^{2l}(t) - z | \geq k^{2l-n-6\delta}, \quad i \neq j, j-q, \tag{2.4.14}$$

$$| \hat{\lambda}^+(t) - z | \geq k^{-\delta}, \tag{2.4.15}$$

$$| \hat{\lambda}^-(t) - z | \geq k^{-\delta}. \tag{2.4.16}$$

Proof of the corollary. Suppose for the definiteness that $\hat{\lambda}_0^q = \hat{\lambda}^+$. Estimate (2.4.14) is proved just as (2.2.6). Inequality (2.4.15) immediately follows from the definition of C_1. Relation (2.4.16) we obtain, using the definition of C_1 and the obvious inequality $|\hat{\lambda}^+ - \hat{\lambda}^-| \geq 2 | v_q |$. The corollary is proved.

We define functions $\tilde{g}_r(k,t)$ and operator-functions $\tilde{G}_r(k,t)$, $r \in N$, $t \in \chi_q(k,\delta)$, as follows:

$$\tilde{g}_r(k,t) = \frac{(-1)^r}{2\pi i r} \mathrm{Tr} \oint_{C_1} ((\hat{H}_q(t) - z)^{-1}\hat{V})^r dz \tag{2.4.17}$$

$$\tilde{G}_r(k,t) = \frac{(-1)^{r+1}}{2\pi i} \oint_{C_1} ((\hat{H}_q(t) - z)^{-1}\hat{V})^r (\hat{H}_q(t) - z)^{-1} dz. \tag{2.4.18}$$

We shall obtain the explicit expressions for $\tilde{g}_0(k,t)$, $\tilde{g}_1(k,t)$, $\tilde{g}_2(k,t)$, $\tilde{G}_0(k,t)$, $\tilde{G}_1(k,t)$.

It follows from Corollary 2.4 that the operator $(\hat{H}_q(t) - z)^{-1}$ has a single pole within contour C_1 at the point $z = \hat{\lambda}_0^q(t)$, just as the integrands on the rights of (2.4.17), (2.4.18). Determining the residue at this point, we obtain that

$$\tilde{g}_0(k,t) = 0, \tag{2.4.19}$$

$$\tilde{G}_0(k,t) = \hat{E}_0^q. \tag{2.4.20}$$

Lemma 2.7 . *If t belongs to the $k^{-n+l-7\delta}$-neighbourhood of $\chi_q(k,\delta)$ in K, then*

$$\tilde{g}_1(k,t) = 0, \tag{2.4.21}$$

$$\tilde{G}_1(k,t) = -(H_0(t) - \lambda_0^q(t))^{-1}\hat{V}E_0^q - E_0^q\hat{V}(H_0(t) - \lambda_0^q(t))^{-1}, \tag{2.4.22}$$

$$\tilde{g}_2(k,t) = Tr((H_0(t) - \lambda_0^q(t))^{-1}\hat{V}E_0^q\hat{V}) \tag{2.4.23}$$

Proof.The operator $(\hat{H}(t) - z)^{-1}$, just as the integrands on the rights of (2.4.17), (2.4.18), has a single pole within C_1. We obtain the explicit expressions for $\tilde{g}_1(k,t)$, $\tilde{g}_2(k,t)$, $\tilde{G}_1(k,t)$ by determining the residue at this point. Let us represent $\tilde{g}_1(k,t)$ in the form: $\tilde{g}_1(k,t) = I_1 + I_2$, where

$$I_1 = -\frac{1}{2\pi i}\oint_{C_1} \mathrm{Tr}(I - P_q)(\hat{H}_q(t) - z)^{-1}\hat{V}dz, \tag{2.4.24}$$

$$I_2 = -\frac{1}{2\pi i}\oint_{C_1} \mathrm{Tr}P_q(\hat{H}_q(t) - z)^{-1}\hat{V}dz.$$

Noting that $\mathrm{Tr}(I - P_q)(\hat{H}_q(t) - z)^{-1}\hat{V}$ is holomorthic inside the circle yields $I_1 = 0$. Then, taking into account that

$$\mathrm{Tr}(P_q(\hat{H}_q(t) - z)^{-1}V) = \mathrm{Tr}(\hat{H}_q(t) - z)^{-1}P_q\hat{V}P_q,$$

$$P_q\hat{V}P_q = 0,$$

we obtain $I_2 = 0$. Thus, $\tilde{g}_1(k,t) = 0$.

Next, we represent $\tilde{G}_1(k,t)$ in the form: $\tilde{G}_1(k,t) = \sum_{i=1}^4 I_i$, where

$$I_1 = (I - P_q)\tilde{G}_1(k,t)(I - P_q), \quad I_2 = P_q\tilde{G}_1(k,t)P_q,$$

$$I_3 = (I - P_q)\tilde{G}_1(k,t)P_q, \quad I_4 = P_q\tilde{G}_1(k,t)(I - P_q).$$

Since $(I - P_q)(\hat{H}_q(t) - z)^{-1}\hat{V}V$ is holomorthic inside the circle, we conclude $I_1 = 0$. Item I_2 is zero too, becouse P_q commutes with $\hat{H}_q(t)$ and $P_q\hat{V}P_q = 0$. We see that the operator-valued function

$$(I - P_q)(\hat{H}_q(t) - z)^{-1}\hat{V}(\hat{H}_q(t) - z)^{-1}P_q$$

has a single pole within C_1, this pole being simple. Determining the residue, we get:

$$I_3 = -(H_0(t) - \lambda_0^q(t))^{-1}\hat{V}\hat{E}_0^q.$$

Similarly, we verify that

$$I_4 = -\hat{E}_0^q\hat{V}(H_0(t) - \hat{\lambda}_0^q(t))^{-1}.$$

Adding I_3 and I_4, we obtain formula (2.4.22). Using the obvious relation $-r\tilde{g}_r = \mathrm{Tr}(\tilde{G}_{r-1}\hat{V})$ we get the expression for $\tilde{g}_2$. The lemma is proved.

We denote by $\hat{\varepsilon}(k,\delta)$ the interval $(\lambda_0^q(t) - k^{-\delta}, \lambda_0^q(t) + k^{-\delta})$.

Theorem 2.3 . *Suppose t belongs to the $(k^{-n+1-7\delta})$-neighbourhood in K of the set $\chi_q(k,\delta)$, $0 < 7\delta < 2l - n$. Then for sufficiently large k, $k > k_0(V,\delta)$, in the interval $\hat{\varepsilon}(k,\delta)$ there exists a single eigenvalue of the operator $H(t)$. It is given by series:*

$$\hat{\lambda}^q(t) = \hat{\lambda}_0^q(t) + \sum_{r=2}^{\infty} \tilde{g}_r(k,t), \tag{2.4.25}$$

The spectral projection, corresponding to $\hat{\lambda}^q(t)$ is given by

$$\hat{E}^q(t) = \hat{E}_0^q + \sum_{r=1}^{\infty} \tilde{G}_r(k,t), \tag{2.4.26}$$

which converges in the class $\mathbf{S}_1$.

For coefficients $\tilde{g}_r(k,t)$, $\tilde{G}_r(k,t)$ the next estimates hold:

$$| \tilde{g}_r(k,t) | < k^{-\gamma_1 r - \delta}, \tag{2.4.27}$$

$$\| \tilde{G}_r(k,t) \|_1 < k^{-\gamma_1 r}, \tag{2.4.28}$$

$$\gamma_1 = (2l-n)/2 - 4\delta.$$

Corollary 2.5 . *For the perturbed eigenvalue and its spectral projection the following estimates are valid:*

$$| \hat{\lambda}^q(t) - \hat{\lambda}_0^q(t) | \le 2k^{-2\gamma_1 - \delta}, \tag{2.4.29}$$

$$\|\hat{E}^q(t) - \hat{E}_0^q\|_1 \le 2k^{-\gamma_1}. \tag{2.4.30}$$

Proof. The main point is to prove the estimate:

$$\|\tilde{A}\| < k^{\gamma_1}, \tag{2.4.31}$$

where

$$\tilde{A} = (\hat{H}_q(t) - z)^{-1/2} \hat{V} (\hat{H}_q(t) - z)^{-1/2}. \tag{2.4.32}$$

Further the proof is just similar to that of Theorem 2.1 after replacing $H_0(t)$ by $\hat{H}_q(t)$, αV by $\hat{V}$, E_j by $\hat{E}_0^q$, $p_j^{2l}(t)$ by $\hat{\lambda}_0^q(t)$, A by $\tilde{A}$ and using Lemma 2.6 instead of Lemma 2.1.

We prove estimate (2.4.31). Noting that $P_q \tilde{A} P_q = 0$ we represent $\tilde{A}$ in the form:

$$\tilde{A} = \sum_{i=1}^{3} \tilde{A}_i \tag{2.4.33}$$

$$\tilde{A}_1 = (I - P_q)\tilde{A}(I - P_q), \quad \tilde{A}_2 = P_q \tilde{A}(I - P_q), \quad \tilde{A}_3 = (I - P_q)\tilde{A}P_q.$$

Taking into account that

$$(I - P_q)(\hat{H}_q(t) - z)^{-1/2} = (I - P_q)(H_0(t) - z)^{-1/2}$$

and considering the obvious equality

$$\|(I-P_q)(H_0(t)-z)^{-1/2}\| = \left(\min_{i\neq j, j-q} | p_i^2(t) - z |\right)^{-1/2}$$

and estimate (2.4.14), we obtain:

$$\|(I-P_q)(\hat{H}_q(t)-z)^{-1/2}\|^2 < 2k^{-2\gamma_1-2\delta}. \tag{2.4.34}$$

Using the last relation, we get

$$\|\tilde{A}_1\| \leq 2\|V\| k^{-2\gamma_1-2\delta}.$$

Next, from estimates (2.4.15), (2.4.16) it follows that:

$$\|P_q(\hat{H}_q(t)-z)^{-1/2}\| < 2k^{-\delta/2}. \tag{2.4.35}$$

From inequalities (2.4.34), (2.4.35) we get:

$$\|\tilde{A}_2\| < 4\|V\| k^{-\gamma_1-\delta/2}$$

and similar inequality for $\tilde{A}_3$. Adding the estimates for $\tilde{A}_1$, $\tilde{A}_2$, $\tilde{A}_3$ we obtain relation (2.4.31). To prove estimates (2.4.27), (2.4.28) we note in addition that the radius of C_1 is equal to $k^{-\delta}$. The theorem is proved.

Remark 2.2 Using relation (2.4.34), it is not hard to show that the eigenvalue $\lambda_0^q(t)$ can be replaced by $p_j^{2l}(t)$ in formulae (2.4.22), (2.4.23) for $\tilde{G}_1(t)$, $\tilde{g}_2(t)$, the arrising errow being less than $k^{-4\gamma_1}$. This errow is, generally speaking, small as compared with $\tilde{G}_1(t)$, $\tilde{g}_2(t)$. Thus, the following relations are valid:

$$\tilde{G}_1(t) = \tilde{G}_1^0(t) + O(k^{-4\gamma_1}), \tag{2.4.36}$$

$$\tilde{g}_2(t) = \tilde{g}_2^0(t) + O(k^{-4\gamma_1})), \tag{2.4.37}$$

where

$$\tilde{G}_1^0(t) = -(H_0(t) - p_j^{2l}(t))^{-1}\hat{V}\hat{E}_0^q - \hat{E}_0^q\hat{V}(H_0(t) - p_j^{2l}(t))^{-1}, \tag{2.4.38}$$

$$\tilde{g}_2^0(t) = \mathrm{Tr}((H_0(t) - p_j^{2l}(t))^{-1}\hat{V}\hat{E}_0^q\hat{V}) \tag{2.4.39}$$

Using the estimates (2.4.36), (2.4.37) we obtain, that under the conditions of Theorem 2.3 the following estimates are valid:

$$\hat{\lambda}^q(t) = \hat{\lambda}_0^q(t) + \tilde{g}_2^0(t) + O(k^{-3\gamma_1-\delta}), \tag{2.4.40}$$

$$\hat{E}^q(t) = \hat{E}_0^q + \tilde{G}_1^0(t) + O(k^{-2\gamma_1}). \tag{2.4.41}$$

We have denoted by $\hat{\lambda}_0^q$ the eigenvalue (2.4.10). This eigenvalue is the nearest to $p_j^2(t)$ of $\hat{\lambda}^+$, $\hat{\lambda}^-$. Let $\hat{\lambda}_1^q$ be the eigenvalue nearest to $p_{j-q}^{2l}(t)$:

$$\hat{\lambda}_1^q(t) = p_{j-q}^2(t) + \Delta - (\mathrm{sgn}\Delta)\sqrt{|v_q|^2 + \Delta}, \tag{2.4.42}$$

We denote by $\hat{E}_1^q$ the spectral projection corresponding to $\hat{\lambda}_1^q(t)$.

Remark 2.3 Suppose we have instead of inequality (2.4.4) a little stronger inequality $|p_j^2(t) - p_{j-q}^2(t)| < k^{-n+2-7\delta}$. In this case for p_{j-q}^2 the estimate similar to (2.4.12) is valid:

$$\min_{i \neq j, j-q} 4 \mid p_{j-q}^2(t) - p_i^2(t) \mid > k^{-n+2-6\delta}.$$

Therefore for $\hat{\lambda}_1^q(t)$, $\hat{E}_0^q(t)$ formulae similar to (2.4.21), (2.4.22) hold.

Theorem 2.4 . *Under the conditions of Theorem 2.3 the series (2.4.25), (2.4.26) can be differentiated termwise with respect to t any number of times, and they retain their asymptotic character. For the coefficients $\tilde{g}_r(k,t)$ and $\tilde{G}_r(k,t)$ the following estimates hold in the $(k^{-n+1-7\delta})$-neighbourhood in C^n of set $\chi_q(k,\delta)$:*

$$\mid T(m)\tilde{g}_r(k,t) \mid < m! k^{-\gamma_1 r - \delta + |m|(n-1+7\delta)} \tag{2.4.43}$$

$$\|T(m)\tilde{G}_r(k,t)\| < m! k^{-\gamma_1 r + |m|(n-1+7\delta)} \tag{2.4.44}$$

Corollary 2.6 . *The following estimates for the perturbed eigenvalue and its spectral projection are valid:*

$$\mid T(m)(\hat{\lambda}^q(t) - \hat{\lambda}_0^q(t)) \mid < 2m! 2k^{-2\gamma_1 - \delta + |m|(n-1+7\delta)}, \tag{2.4.45}$$

$$\|T(m)(\hat{E}^q(t) - \hat{E}_0^q(t)\| < 2m! k^{-\gamma_1 + |m|(n-1+7\delta)}, \tag{2.4.46}$$

Proof. We prove that the coefficients $\tilde{g}_r(k,t)$, $\tilde{G}_r(k,t)$ can be continued from the real $(k^{-n+1-7\delta})$-neighbourhood of $\chi_q(k,\delta)$ to the complex one as holomorphic functions (operator-valued functions) of n variables and inequalities (2.4.27), (2.4.28) are hereby preserved. In fact, let $t_0 \in \chi_q(k,\delta)$. It is clear that

$$((\hat{H}_q(t) - z)^{-1}\hat{V})^r(\hat{H}_q(t) - z)^{-1}$$

depends analitically on t in the complex $(k^{-n+1-7\delta})$- neighbourhood of t_0 for any fixed z lying on $C_1(t_0)$. It is easy to see the stability of estimates (2.4.27), (2.4.28) with respect to a perturbation of order $k^{-n+1-2\delta}$. Therefore we arrive to the inequality:

$$\| \int_{C_1(t_0)} ((\hat{H}_q(t) - z)^{-1}V)^r(\hat{H}_q(t) - z)^{-1} dz\|_1 < k^{-\gamma_1 r}.$$

The integrand has a single pole at the point $z = \hat{\lambda}_0^q(t)$. It is inside both $C_1(t_0)$ and $C_1(t)$. Therefore, if we replace $C_1(t_0)$ for $C_1(t)$, then the result of integration preserved. Thus, $\|\tilde{G}_r(k,\delta)\| < k^{-\gamma_1 r}$ in the complex $(k^{-n+1-7\delta})$-neighbourhood of t_0 that is in the complex $(k^{-n+1-7\delta})$-neighbourhood of each simply connected component of $\chi_q(k,\delta)$. The estimate for $\tilde{g}_r(k.t)$ is proved similarly.

Remark 2.4 We have proved that, if in formulae (2.4.17), (2.4.18) the contour $C_1(t)$ is slightly changed, then the result of the integration is preserved. Hence the operation of differentation with respect to t commutes with the operation of integration over the contour.

2.5 Geometric Constructions for the Singular Set.

We have constructed the perturbation series for the singular set in the assumption that there exists subset $\chi_q(k,\delta)$ of $\mathcal{K}S_q(k,n-2+\delta)$, satisfying the conditions of Geometric Lemma 2.6. Now we prove that the set, defined by the formula

$$\chi_q(k,\delta) = \mathcal{K}S_q(k,n-2+\delta) \setminus \mathcal{K}A_q(k,\delta), \tag{2.5.1}$$

$$A_q(k,\delta) = \cup_{m\in Z^n, m\neq q} \left(S_q(k,n-2+\delta) \cap S_m(k,n-2+6\delta)\right) \tag{2.5.2}$$

satisfies all the conditions of this lemma.

Lemma 2.8 . *Suppose $t \in \chi_q(k,\delta)$. Then there exists a unique $j \in Z^n$, such that $p_j^{2l}(t) = k^{2l}$ and inequalities (2.4.11), (2.4.12) are satisfied. Moreover, for any t in the $(k^{-n+1-7\delta})$-neighbourhood of $\chi_q(k,\delta)$ in C^n there exists a unique j such that $\mid p_j^2(t) - k^2 \mid < 5k^{-n+2-7\delta}$ and inequalities (2.4.11), (2.4.12) are satisfied.*

Proof. If $t \in \chi_q(k,\delta)$, then $t \in \mathcal{K}S_q(k,n-2+\delta)$ i.e., there exists at least one j, such that $p_j^2(t) = k^2$ and (2.4.11) holds. We prove inequality (2.4.12). Suppose for some $i, i \neq j, j-q$

$$\mid p_j^2(t)(t) - p_i^2 \mid \leq k^{-n+2-6\delta}.$$

This estimate means that $\mathbf{p}_j(t) \in S_{j-i}(k,n-2+6\delta)$ i.e., $t \in \mathcal{K}S_{j-i}(k,n-2+6\delta)$, $j-i \neq q$. But this is not the case, because $t \notin \mathcal{K}A_q(k,\delta)$ by the hypothesis of the lemma. Thus, estimate (2.4.12) is proved.

Suppose t in the $(k^{-n+1-7\delta})$-neighbourhood of $\chi_q(k,\delta)$ in C^n. Then there exists t_0 such that $\mid t - t_0 \mid < k^{-n+1-7\delta}$, $t_0 \in \chi_q(k,\delta)$. Hence, we have $\mid p_j^2(t) - k^2 \mid < 5k^{-n+2-7\delta}$. It is easy to show the stability of estimate (2.4.12) with respect to a perturbation of order $k^{-n+1-7\delta}$. Suppose there are two j, satisfying the conditions of the lemma. We denote them by j_1, j_2. Since, $p_{j_1}(t) = p_{j_2}(t)$, we have $j_1 - j_2 = q$; otherwise (2.4.12) is not valid for $j = j_1$, $i = j_2$. Similarly, $j_2 - j_1 = q$. Therefore, $j_1 = j_2$. The lemma is proved.

Next, we prove that $\chi_q(k,\delta)$ has an asymptotically full measure on $\mathcal{K}S_q(k,n-2+\delta)$. It suffices to check that $A_q(k,\delta)$ has an asymptotically small measure on $\mathcal{K}S_q(k,n-2+\delta)$. Estimating above $s(S_m \cap S_q)$ for all $m \neq q$ and adding the inequalities verify that the sum

$$\sum_{m\in Z^n, m\neq q} s(S_m \cap S_q)$$

is asymptotically small as compared with $s(S_q)$. Noting that $s(S_q) = s(\mathcal{K}S_q)$ and $s(S_q \cap S_m) = s(\mathcal{K}(S_q \cap S_m))$ we obtain the result to prove. But, practically, this way is different for the cases of $n = 2, n = 3$ and $n > 3$ because in two and three dimensional situations one cann't estimate $s(S_m \cap S_q)$ for all q by the similar way. This means that estimates have different forms for the cases where S_m, S_q have a "perfect" intersection and where they "almost touch". Thus, we have to consider the cases of $n = 2$, $n = 3$, $n > 3$ separately. We begin with the case $n = 2$.

Lemma 2.9 . *If $n = 2$, $-1 < \xi_1 \leq \delta < 0$, $p_m(0) < k$, then the following estimate for the area of the intersection of $S_q(k,\xi)$ and $A_q(k,\delta)$ is valid:*

$$s(S_q(k,\xi_1) \cap A_q(k,\delta)) < ck^{-\delta} s(S_q(k,\xi_1)). \tag{2.5.3}$$

Proof. Firstly, we consider the case of $\xi_1 = \delta$. It is clear that

$$s(S_q(k,\delta) \cap A_q(k,\delta)) \leq \Sigma_1 + \Sigma_2,$$

$$\Sigma_1 = \sum_{m \in Q_1} s(S_q(k,\delta) \cap S_m(k,6\delta)),$$

$$\Sigma_2 = \sum_{m \in Q_2} s(S_q(k,\delta) \cap S_m(k,6\delta)),$$

where

$$Q_1 = \{m : m \neq 0, | p_m(0)/2 - k | \geq p, S_q(k,\xi_1) \cap S_m(k,6\delta)) \neq \emptyset\},$$

$$Q_2 = \{m : m \neq 0, | p_m(0)/2 - k | < p, S_q(k,\xi_1) \cap S_m(k,6\delta)) \neq \emptyset\},$$

$$p = 8\pi(a_1^{-1} + a_2^{-1}).$$

It is obvious that

$$\Sigma_i \leq \sum_{m \in Q_i} s(S_m(k,6\delta)), \quad i = 1,2.$$

We estimate the each sum. Using formula (2.3.11), we obtain

$$\Sigma_1 < ck^{1-6\delta} \sum_{m \in Q_1} p_m^{-1}(0)(k^2 - p_m^2(0)/4)^{-1/2} \tag{2.5.4}$$

By elementary geometrical considerations one can show that the points $\mathbf{p}_m(0)$, satisfying the condition $S_q \cap S_m \neq \emptyset$, are in an asymptotically small neighbourhood of two circles $R_\pm$ with radii k cenred at the points $\mathbf{x}_\pm = (p_q(0)/2, \pm(k^2 - p_q^2(0)/4)^{1/2})$. Hence, the sum on the right can be estimated from above by the sum of integrals I_+ and I_-:

$$I_\pm = a_1 a_2 \int_{\tilde{R}_\pm} | x |^{-1} (k^2 - | x |^2 /4)^{-1/2} dx,$$

where $\tilde{R}_\pm$ are the rings of the width $A/2$ near the circles $R_\pm$ without the points $x : | x | \leq A/4$. It is not hard to show that $I_\pm < c(a_1,a_2)k^{-1+\delta}$. Therefore,

$$\Sigma_1 < c(a_1,a_2)k^{-5\delta}.$$

Next, we estimate Σ_2. It is easy to see that the number of points in Q_2 is bounded, when k goes to infinity, because they lie in the unit neighbourhood of a bounded curve. Hence,

$$\Sigma_2 < c(V) \max_{m \in Q_2} s(S_m(k,6\delta)) \tag{2.5.5}$$

Using estimate (2.3.12) for $s(S_m(k,6\delta))$ we get

$$\Sigma_2 < ck^{-3\delta}.$$

Thus,

$$s(S_q(k,\delta) \cap A_q(k,\delta)) \leq ck^{-3\delta}.$$

Taking into account that $s(S_q(k,\delta)) \approx k^{-\delta}p_m^{-1}(0)$, we obtain inequality (2.5.3) for $\xi_1 = \delta$. If $-1 < \xi_1 < \delta$ then we break the layer $S_q(k,\xi_1)$ into a number of parallel layers. Estimate (2.5.3) holds for each of them. Adding inequalities (2.5.3) over the all layers yields estimate (2.5.3) for ξ_1, $-1 < \xi_1 \leq \delta$. The lemma is proved.

Let now $n = 3$. As we noted above the form of the estimate for the area of $S_q(k,\xi_1) \cap S_m(k,\xi_2)$ depends strongly of the type of the intersection. We consider this question in detail now.

Let T_q be the body of the torus with the radii $k' \equiv \sqrt{k^2 - p_q^2(0)/4}$ and k, the main circle O_q of the radius k' being centred at the point $p_q(0)/2$ and lying in the plane orthogonal to $\mathbf{p}_q(0)$. Thus,

$$T_q = \{x : | x - x_s | \leq k, x_s \in O_q\}, \tag{2.5.6}$$

$$O_q = \{x : | x - \mathbf{p}_q(0)/2 |= k', \ (x - \mathbf{p}_q(0)/2, \mathbf{p}_q(0)) = 0\}. \tag{2.5.7}$$

It is clear that the sphere $| x - \mathbf{p}_m(0) |= k$ intersects with the circle O_q if and only if $\mathbf{p}_m(0)$ belongs to T_q. Noting that the plane-spherical layer $S_q(k,\xi_1)$ is formed by circles close to O_q, we conclude that $S_q(k,\xi_1)$ intersects with the sphere $| x - \mathbf{p}_m(0) |= k$ and, therefore, with $S_m(k,\xi_2)$ only if $\mathbf{p}_m(0)$ is inside a neighbourhood of T_q.

It turns out that the area of the intersection depends essentially on ρ_{qm}, ρ_{qm} being the distance from point $\mathbf{p}_m(0)$ to the torus. It is easy to see that

$$(k - \rho_{mq})^2 = (k' - p_m(0)_\perp)^2 + z_m'^2,$$

where $z_m' = z_m - p_q(0)/2$, $p_m(0)_\perp$ is the absolute value of the projection of $\mathbf{p}_m(0)$ onto the plane orthogonal to $\mathbf{p}_q(0)$. When $\mathbf{p}_m(0)$ is close to the torus ($\rho_{qm} \approx 0$) the layers $S_q(k,\xi_1)$ and $S_m(k,\xi_2)$ intersect "weakly". The estimates for $s(S_q(k,\xi_1) \cap S_m(k,\xi_2))$ get worse (this change to the worse has no place in the case of $n > 3$). The following lemma estimates the area of $s(S_q(k,\xi_1) \cap S_m(k,\xi_2))$ in dependence of ρ_{mq}.

Lemma 2.10 . *If $p_q(0) < c$, $c \neq c(k)$, $\xi_1 > -1$, $\xi_2 > -1$ and $\mathbf{p}_q(0)$ and $\mathbf{p}_m(0)$ are linearly independent, then*

$$s(S_q(k,\xi_1) \cap S_m(k,\xi_2)) < ck^{-\xi_1-\xi_2}p_q^{-1}(0)\rho_{qm}^{-1/2}p_m(0)_\perp^{-1/2}, \tag{2.5.8}$$

when $\rho_{qm} > c_1 \max\{k^{-\xi_1-1}p_m(0),\ k^{-\xi_2-1}$, $c_1 \neq c_1(k)$ and c_1 is positive and large enough;

$$s(S_q(k,\xi_1) \cap S_m(k,\xi_2)) < ck^{-\xi_1/2-\xi_2+3/2}p_q^{-1/2}(0)p_m^{-3/2}(0)(k^2 - p_m^2(0)/4)^{-1/4}, \tag{2.5.9}$$

when $0 < \rho_{qm} \leq k,\ p_m(0) < 2k-1$; *and*

$$s(S_q(k,\xi_1) \cap S_m(k,\xi_2)) < ck^{-\xi_1-\xi_2/2}p_q^{-1}(0), \tag{2.5.10}$$

when $p_m(0) > 2k-1$.

Proof. In R^3 we introduce a coordinate system so that the axis x_1 is directed alone $\mathbf{p}_q(0)$, axis x_2 belongs to the plane of $\mathbf{p}_q(0)$, $\mathbf{p}_m(0)$, and we introduce spherical coordinates (R,ϑ,φ): $x_1 = R\cos\vartheta, x_2 = R\sin\vartheta\cos\varphi, x_3 = R\sin\vartheta\sin\varphi$. Considering as in the proof of Lemma 2.3 we show that

$$S_q(k,\xi_1) \cap S_m(k,\xi_2) \subset$$

$$\subset \{x : \mid \cos\vartheta - \cos\vartheta_0 \mid < \varepsilon_1, \mid \sin\vartheta\cos\varphi - \sin\vartheta\cos\varphi_0(\vartheta) \mid \leq \varepsilon_2, R = k\}, \tag{2.5.11}$$

where

$$\cos\vartheta_0 = p_q(0)(2k)^{-1}, \varepsilon_1 = k^{-\xi_1-1}p_q^{-1}(0), \varepsilon_2 = k^{-\xi_2-1}p_m(0)_\perp^{-1},$$

$$\cos\varphi_0 = (p_m^2(0) - 2p_m(0)_{||}k\cos\vartheta)(2kp_m(0)_\perp \sin\vartheta)^{-1},$$

where $p_m(0)_{||}$, $p_m(0)_\perp$ are the absolute values of the projections of $\mathbf{p}_m(0)$ on the direction $\mathbf{p}_q(0)$ and on the orthogonal plane, correspondingly.

By the hypothesis of the lemma $2p_q(0) < c$. Hence, $\pi/3 \leq \vartheta_0 \leq \pi/2$, Since ϑ lies in a small neighbourhood of ϑ_0, the following estimates hold: $\mid \vartheta - \vartheta_0 \mid < 4\varepsilon_1$, $\sin\vartheta > 1/2$. Suppose $\rho_{qm} > c_1 \max\{k^{-\xi_1-1}p_m(0),\ k^{-\xi_2-1}$. We shall show that

$$1 - \cos\varphi_0 > \rho_{mq}/2p_m(0)_\perp. \tag{2.5.12}$$

Hence, $4\varphi_0^2 > \rho_{qm}/p_m(0)_\perp$. Using inequalities $\rho > ck^{-\xi_2-1}$, $2p_m(0)_\perp > \rho$, it is not hard to show that $\varepsilon_2 \ll \varphi_0$. From the relation $\mid \sin\vartheta(\cos\varphi - \cos\varphi_0) \mid < \varepsilon_2$, we obtain $\mid \varphi - \varphi_0 \mid < 8\varepsilon_3, \varepsilon_3 = k^{-\xi_2-1}(\rho_{qm}p_m(0)_\perp)^{-1/2}$. Thus,

$$S_q(k,\xi_1) \cap S_m(k,\xi_2) \subset \{x : \mid \vartheta - \vartheta_0 \mid < 4\varepsilon_1, \mid \varphi - \varphi_0 \mid < \varepsilon_3\}.$$

Therefore,

$$s(S_q(k,\xi_1) \cap S_m(k,\xi_2)) \leq k^2 \int_{|\vartheta-\vartheta_0|<4\varepsilon_1} \sin\vartheta \int_{|\varphi-\varphi_0|<\varepsilon_3} d\varphi d\vartheta \leq$$

$$\leq 8k^{-\xi_1-\xi_2}(p_q^{-1}(0)(\rho_{qm}p_m(0)_\perp)^{-1/2}.$$

From estimate (2.5.12) we see that $\rho_{mq} < 2p_m(0)_\perp$. Using the last two inequalities, we obtain (2.5.8). To finish the proof of the estimate it suffices to verify inequality (2.5.12). Indeed, it is obvious that:

$$1 - \cos\varphi_0 = \frac{2k'p_m(0)_\perp - p_m^2(0) + p_m(0)_{||}p_q(0) + O(k\varepsilon_1)p_m(0)_{||}}{2k'p_m(0)_\perp(1 + O(\varepsilon_1))}.$$

By the definition of ρ_{qm}

$$(k-\rho_{qm})^2 = (p_m(0)_{||} - p_q(0)/2)^2 + (p_m(0)_\perp - k')^2. \tag{2.5.13}$$

Considering the last relation, we obtain for $\rho_{mq} > c_1 k^{-\xi_1 - 1} p_m(0)$:

$$1 - \cos\varphi_0 = \frac{k^2 - (k-\rho_{qm})^2 + O(k^{-\xi_1+1})}{2k' p_m(0)_\perp (1 + O(\varepsilon_1))} > \rho_{qm}(2p_m(0)_\perp)^{-1}.$$

Thus, estimate (2.5.8) is proved.

Using formula (2.5.11) we easily show that

$$s(S_q(k,\xi_1)\cap S_m(k,\xi_2)) \le k^2 \int_{|\cos\vartheta - \cos\vartheta_0|<\varepsilon_1} \int_{|\cos\varphi - \cos\varphi_0|<\varepsilon_2(\vartheta)} d\varphi d(-\cos\vartheta) \le$$
$$\le 2k^2\varepsilon_1\varepsilon_2^{1/2} = k^{-\xi_1-\xi_2/2+1/2} p_q^{-1}(0) p_m(0)_\perp^{-1/2}. \tag{2.5.14}$$

Suppose $p_m(0) > 2k-1$. It is easy to see that in this case $p_m(0)_\perp > k$. Using (2.5.14) yields inequality (2.5.10).

Finally, let $p_m(0) \le 2k-1$. We introduce a coordinate system so that the axis x_1 is directed alone $\mathbf{p}_m(0)$, axis x_2 belongs to the plane of $\mathbf{p}_q(0)$, $\mathbf{p}_m(0)$,and we introduce spherical coordinates (R, ϑ, φ): $x_1 = R\cos\vartheta, x_2 = R\sin\vartheta\cos\varphi, x_3 = R\sin\vartheta\sin\varphi$. Considering as in the proof of (2.5.11) we show that

$$S_q(k,\xi_1)\cap S_m(k,\xi_2) \subset$$
$$\{x : |\cos\vartheta - \cos\vartheta_1| < \varepsilon_4, |\sin\vartheta\cos\varphi - \sin\vartheta\cos\varphi_1(\vartheta)| \le \varepsilon_5, R = k\}, \tag{2.5.15}$$

where

$$\cos\vartheta_1 = \frac{p_m(0)}{2k},\ \varepsilon_4 = k^{-\xi_2-1}p_m^{-1}(0),\ \varepsilon_5 = k^{-\xi_1-1}p_q(0)_\perp^{-1},$$
$$\cos\varphi_1 = (p_q^2(0) - 2p_q(0)_{||}k\cos\vartheta)(2kp_q(0)_\perp\sin\vartheta)^{-1},$$

$p_q(0)_{||}, p_q(0)_{||}$ being the absolute values of the projections of $\mathbf{p}_q(0)$ on the direction $\mathbf{p}_m(0)$ and on the orthogonal plane, correspondingly.

From this we obtain:

$$S_q(k,\xi_1)\cap S_m(k,\xi_2) \subset$$
$$\{x : |\cos\vartheta - \cos\vartheta_1| < \varepsilon_4, |\varphi - \varphi_1(\vartheta)| \le c(\varepsilon_5/\sin\vartheta_1)^{1/2}, R = k\}, \tag{2.5.16}$$

From this relation it follows that

$$s(S_q(k,\xi_1)\cap S_m(k,\xi_2)) < ck^2\varepsilon_4\varepsilon_5^{1/2}(\sin\vartheta_1)^{-1/2}.$$

Substituting the expressions for $\varepsilon_4, \varepsilon_5$ and $\sin\vartheta_1$, we obtain

$$s(S_q(k,\xi_1)\cap S_m(k,\xi_2)) < ck^{-\xi_1/2-\xi_2+1}p_q(0)_\perp^{-1/2}p_m^{-1}(0)(k^2 - p_m^2(0)/4)^{-1/4}, \tag{2.5.17}$$

We prove that

$$p_q(0)_\perp \geq p_q(0)p_m(0)(2k)^{-1}. \tag{2.5.18}$$

It is obvious that

$$p_q(0)_\perp = p_m(0)_\perp p_q(0)p_m(0)^{-1}, \tag{2.5.19}$$

where $p_m(0)_\perp$ is the absolute value of the projection of $\mathbf{p}_m(0)$ onto the plane orthogonal to $\mathbf{p}_q(0)$. From (2.5.13) it follows that

$$2p_m(0)_\perp k' = 2\rho k - \rho^2 + p_m^2(0) - 2p_m(0)_{||}p_q(0).$$

Suppose $p_m(0) > 10p_q(0)$. It is not hard to show that in this case $\rho > 0$ and $4p_m(0)_\perp k > p_m^2(0)$. Using (2.5.19), we get (2.5.18). If $p_m(0) \leq 10p_q(0)$, then considering that vectors $\mathbf{p}_m(0)$, $\mathbf{p}_q(0)$ are linearly independent, we obtain

$$\frac{p_m(0)_\perp}{p_m(0)} > cp_q^{-1}(0).$$

Taking into account (2.5.19), we get $p_q(0)_\perp > c$. From the last relation it follows (2.5.18), because $p_q(0) < k^{1/2}$ by the hypothesis. Thus, we prove (2.5.18). Using it in inequality (2.5.17) we obtain (2.5.9). The lemma is proved.

Lemma 2.11 . *Suppose $n = 3$, $p_m(0) < c$, $c \neq c(k)$. Then the following estimate is valid:*

$$s(S_q(k, 1+\delta) \cap A_0(k,\delta)) < k^{-\delta}s(S_q(k, 1+\delta)). \tag{2.5.20}$$

Proof. It is clear that

$$s(S_q(k, 1+\delta) \cap A_0(k,\delta)) \leq \sum_{m \in Z^n, m \neq 0} s(S_q(k, 1+\delta) \cap S_m(k, 1+6\delta))$$

It is easy to see that $S_q(k, 1+\delta) \cap S_m(k, 1+6\delta)$, when $\mathbf{p}_m(0)$, $\mathbf{p}_q(0)$ are linearly dependent. Further, we break the summation set into three, accordingly to the fields of the validity of estimates (2.5.8)–(2.5.10). Let

$$Q_1 = \{m : \rho_{qm} > 1\},$$

$$Q_2 = \{m : \rho_{qm} \leq 1,\ p_m(0) > 2k - 1\},$$

$$Q_3 = \{m : \rho_{qm} \leq 1\ p_m(0) \leq 2k - 1\}.$$

We estimate the sum over the each set. It is easy to see from formula (2.5.13) that $p_m(0) < 2k - 1$, when $\rho > 1$. Thus,

$$Q_1 = \{m : \rho_{mq} > 1, p_m(0) < 2k - 1\}.$$

For $\rho > 1$ inequality (2.5.8) is valid ($\xi_1 = 1+\delta$, $\xi_2 = 1+6\delta$). Using it, we obtain:

$$\Sigma_1 < cp_q^{-1}(0)k^{-2-7\delta} \sum_{m:\rho_{qm}>1} \rho_{qm}^{-1/2} p_m(0)_\perp^{-1/2}.$$

It follows from (2.5.12) that $4p_m(0)_\perp > \rho_{mq}$. Therefore,

$$\Sigma_1 < cp_q^{-1}(0)k^{-2-7\delta} \sum_{m:\rho_{qm}>1} \rho_{qm}^{-1}.$$

It is easy to show that the last sum can be estimated by the integral

$$I_1 = \int_{T_1} \rho^{-1}(x)dx, \quad T_1 = \{x : x \in T_k, \rho(x) > 1\},$$

$\rho(x)$ being the distance from point x to the torus.

It is not hard to see that $I_1 \approx k^2 \ln k$. Thus, we get the following estimate for Σ_1:

$$\Sigma_1 < cp_q^{-1}(0)k^{-7\delta} \ln k.$$

Next, we estimate Σ_2. Using (2.5.9), we obtain:

$$\Sigma_2 < ck^{-6\delta}p_q^{-1/2}(0) \sum_{m:\rho_{qm}<1} p_m^{-3/2}(0)(k^2 - p_m^2(0)/4)^{1/4}.$$

Considering that $m \neq 0$, it is not hard to show that the sum on the right can be estimated by the integral I_2:

$$I_2 = c\int_{T_2} \mid x \mid^{-3/2} (k^2 - \mid x \mid^2 /4)^{-1/4}dx, \; c \neq c(k),$$

$$T_2 = \{x : \mid x \mid > p, \; x_2 < 2k - 1/2, \; \rho(x) < 2, \; x \in T_k\}$$

It is not hard to show that $I_2 < ck^\delta$. Therefore,

$$\Sigma_2 < ck^{-4\delta}p_q^{-1/2}(0).$$

Taking into account that $p_q(0) < k^\delta$ by the hypothesis of the lemma, we get:

$$\Sigma_2 < k^{-3\delta}p_q^{-1}(0).$$

We consider Σ_3. Using estimate (2.5.10), we get:

$$\Sigma_3 < k^{-3/2-4\delta}p_q^{-1} \sum_{m\in Q_3} 1.$$

The sum on the right can be estimated by the following integral:

$$I_3 = \int_{T_3} dx, \quad T_3 = \{x : \; x \in T_q, \; x_2 > 2k - 1, \}.$$

It is not hard to show that $I_3 < ck^{3/2}$. Therefore,

$$\Sigma_3 < ck^{-3\delta}p_q^{-1}(0).$$

Adding the estimates for $\Sigma_1, \Sigma_2, \Sigma_2$, we obtain:

$$s(S_q(k, 1+\delta) \cap A_0(k,\delta)) < 3k^{-3\delta}p_q^{-1}(0) \leq k^{-\delta}s(S_q(k, 1+\delta)).$$

The lemma is proved.

Finally, we consider the case of $n > 3$.

Lemma 2.12 . *If $n > 3$ and $\mathbf{p}_q(0)$ and $\mathbf{p}_m(0)$ are linearly independent, then the area of the intersection of two plane-spherical layers $S_m(k,\xi_1)$ and $S_q(k,\xi_2)$ satisfies the following estimate:*

$$s(S_q(k,\xi_1)\cap S_m(k,\xi_2)) < ck^{-\xi_1-\xi_2+n-2}p_q^{-1}(0)p_m^{-2}(0). \tag{2.5.21}$$

Proof. In R^n we introduce a coordinate system so that the axis x_1 is directed alone $\mathbf{p}_q(0)$, axis x_2 belongs to the plane of $\mathbf{p}_q(0)$, $\mathbf{p}_m(0)$,and we introduce spherical coordinates $(R,\vartheta,\varphi,\nu)$: $x_1 = R\cos\vartheta, x_2 = R\sin\vartheta\cos\varphi, x_i = R\sin\vartheta\sin\varphi\nu_i, i = 3,...,n$, here $\nu = (\nu_3,...,\nu_n)$ are orthogonal coordinates on the unit sphere S_1^{n-3} in the space of the dimension $n-2$, spanned by $x_3,...,x_n$. The surface element of the sphere S_R is given by the formula:

$$dS_R = R^{n-1}\sigma(\vartheta,\varphi)d(-\cos\vartheta)d(\sin\vartheta\cos\varphi)dS_1^{n-3},$$

$$\sigma(\vartheta,\varphi) = (\sin\vartheta\sin\varphi)^{n-4}.$$

Noting that $\sigma(\vartheta,\varphi) < 1$, we obtain:

$$s(S_q(k,\xi_1)\cap S_m(k,\xi_2)) < ck^{n-1}\int \chi(\vartheta,\varphi,\nu)d(-\cos\vartheta)d(\sin\vartheta\cos\varphi)dS_1^{n-3}, \tag{2.5.22}$$

$\chi(\vartheta,\varphi,\nu)$ being the characteristic function of the set $S_q(k,\xi_1)\cap S_m(k,\xi_2)$. Considering as in the proof of Lemma 2.10, we show that

$$S_q(k,\xi_1)\cap S_m(k,\xi_2) \subset \{x :\mid \cos\vartheta - \cos\vartheta_0 \mid < \varepsilon_1, \mid \sin\vartheta\cos\varphi - \alpha_1 \mid \le \varepsilon_2, R = k\}, \tag{2.5.23}$$

where $\varepsilon_1 = k^{-\xi_1-1}p_q^{-1}(0)$, $\varepsilon_2 = k^{-\xi_2-1}p_m(0)_\perp^{-1}$, $\vartheta_0 = \vartheta_0(k,p_q(0))$, $\alpha_1 = \alpha_1(k,\mathbf{p}_m(0),\mathbf{p}_q(0))$. Using inequality (2.5.22), we get

$$s(S_q(k,\xi_1)\cap S_m(k,\xi_2)) \le 2k^{n-1}\varepsilon_1\varepsilon_2.$$

Substituting the expressions for $\varepsilon_1,\varepsilon_2$ we obtain the estimate

$$s(S_q(k,\xi_1)\cap S_m(k,\xi_2)) \le 4k^{-\xi_1-\xi_2+n-3}p_q^{-1}(0)p_m(0)_\perp^{-1}. \tag{2.5.24}$$

Considering as in the proof of (2.5.18), we show that $p_m(0)_\perp \ge p_m^2(0)(4k)^{-1}$. Using this estimate in (2.5.24), we obtain (2.5.21). The lemma is proved.

Lemma 2.13 . *If $n > 3, \delta > 0$, then for sufficiently large k the following estimate is valid:*

$$s(S_q(k,n-2+\delta)\cap A_0(k,\delta)) < k^{-\delta}s\left(S_q(k,n-2+\delta)\right). \tag{2.5.25}$$

Proof. Using inequality (2.5.21) for $\xi_1 = n-2-\delta, \xi_2 = n-2-6\delta$, we get:

$$s(S_q(k,n-2+\delta)\cap A_0(k,\delta)) < ck^{-n+2-7\delta}\sum_{m\in Z^n, m\neq 0, p_m(0)<2k} p_m^{-2}(0).$$

Estimating the last sum by the integral yields:

$$s(S_q(k,n-2+\delta)\cap A_0(k,\delta)) < k^{-7\delta} \approx k^{-6\delta}s(S_q(k,n-2+\delta)).$$

The lemma is proved.

Proof of Geometric Lemma. According to Lemma 2.8 for any t of $\chi_q(k,\delta)$ there exists a unique j, such that $p_j(t) = k$ and estimate (1.0.12) is valid. If t in the $(k^{-n+1-7\delta})$-neighbourhood of $\chi_q(k,\delta)$, then there exists a unique j, such that $\mid p_j^2(t) - k^2 \mid < 5k^{-n+2-7\delta}$, and estimate (1.0.12) holds. It remains only to verify relation (1.0.13). It is clear that $s(S_q(k,n-2+\delta)\setminus\chi_q(k,\delta)) = s(A_0(k,\delta))$. According to Lemma 2.9 ($n=2$), Lemma 2.11 ($n=3$) and Lemma 2.13 ($n>3$)

$$s(A_0(k,\delta)) < k^{-\delta}s(S_q(k,n-2+\delta)).$$

Hence, estimate (1.0.13) is valid. The lemma is proved.

2.6 Proof of the Bethe-Sommerfeld Conjecture. Description of the Isoenergetic Surface.

According to the famous Bethe-Sommerfeld conjecture there exists only a finite number of gaps in the spectrum of H. The first rigorous proof of this conjecture was given by M. M. Skriganov. He used complicated methods from geometrical and arithmetical theory of lattices. One may suppose that Bethe and Sommerfeld were guided by the ideas of the perturbation theory for n- dimensional case. However, mathematical foundation of this conjecture is a complicated matter. Here it is represented the proof of Bethe-Sommerfeld conjecture as a simple consequence of the asymptotic formula for a eigenvalue. The first proof of this conjecture with using of an asymptotic formula in the case of a smooth potential belongs to O. A. Veliev (for l=1, n=2,3).

Another application of the perturbation formulae is connected with an isoenergetic surface of the perturbed operator. As described above, the isoenergetic surface $S_0(k)$ of the free operator is the sphere, "packed into the bag K". We consider the nonsingular part of this surface. This subset has an asymptotically full measure on $S_0(k)$. We shall show that near each simply connected component of the nonsingular set there exists a unique simply connected component of the perturbed isoenergetic surface. How can be described the rest of the perturbed isoenergetic surface? Near every piece $\mathcal{K}S_q$ of the unperturbed isoenergetic surface there exists the another one $\mathcal{K}S_{-q}$. Considering the most simple part $\chi_q \cup \chi_{-q}$ of $\mathcal{K}S_q \cup \mathcal{K}S_{-q}$, we show that there exist a part of isoenergetic surface of perturbed operator H near $\chi_q \cup \chi_{-q}$. They are close to ones of the model operators $\hat{H}_q$ and $\hat{H}_{-q}$. The corresponding normals are close too (Theorem 2.8). In fact, these pieces of the perturbed isoenergetic surface can be approximated with higher accuracy by corresponding ones of H_q (or H_{-q}).

Taking the size of the block in operator $\hat{H}$ greater than two, one can describe the isoenergetic surface near the rest of the singular set. However, now for the finitness of the considerations we restrict them only by description of the perturbed isoenergetic surface in vicinities of the nonsingular set and the simple part of the singular set.

Theorem 2.5 . *(proof of the Bethe-Sommerfeld conjecture) There only a finite number of gaps in the spectrum of operator H.*

Proof. In this proof of the Bethe-Sommerfeld conjecture the central features are the trivial validity of this conjecture on the case of $V = 0$, estimate (2.2.18) for the eigenvalue and the continuity of the function $\lambda(\alpha, t)$ in t in the $(k^{-n+1-2\delta})$- neighbourhood of the nonsingular set. Obviously, it is enough to prove that all points of the real axis to the right of some point k_0^{2l} belong to the spectrum. Suppose $0 < 2\delta < 2l - n$ and k is so large that the nonsingular set $\chi_0(k, \delta)$ is not empty, and $t_0 \in \chi_0(k, \delta)$. By Theorem 2.1 for all t in the ball $A = \{t : | t - t_0 |< k^{-n+1-2\delta}, t \in K\}$ in the disk $| \alpha |< 1$ there exists an analytic eigenvalue $\lambda(\alpha, t)$, such that

$$\lambda(0, t_0) = p_j^{2l}(t_0) = k^{2l}, \quad \lambda(0, t) = p_j^{2l}(t),$$

where j is uniquely determined. Let

$$\Lambda(\alpha) \equiv \{\lambda : \lambda = \lambda(\alpha, t), t \in A\}.$$

It is obvious that

$$\Lambda(0) \supset [k^{2l} - c_1 k^{2l-n-2\delta}, k^{2l} + c_1 k^{2l-n-2\delta}], \quad 0 < c_1 \neq c_1(k).$$

Since $\lambda(\alpha, t)$ is continuous in α, t and (2.2.18) is valid, we clearly have:

$$\Lambda(\alpha) \supset [k^{2l} - c_1 k^{2l-n-2\delta} + c_1 k^{-2\gamma_0}, k^{2l} + c_1 k^{2l-n-2\delta} - c_1 k^{-2\gamma_0}].$$

From the relations $2l > n, \gamma_0 = 2l - n - \delta$, it follows that $2l - n - 2\delta > -2\gamma_0$. Therefore,

$$\Lambda(\alpha) \supset [k^{2l} - (c_1/2) k^{2l-n-2\delta}, k^{2l} + (c_1/2) k^{2l-n-2\delta}],$$

i.e., $k^{2l} \in \Lambda(\alpha)$. The theorem is proved.

One can see from the proof of the Bethe-Sommerfeld conjecture that there exists a point of the perturbed isoenergetic surface $S_H(k)$ in the $(k^{-n+1-2\delta})$-neighbourhood of any point t_0, belonging to the nonsingular set. We prove now that in the $(k^{-n+1-2\delta})$-neighbourhood of each simply connected component $\chi_0'(k, \delta)$ of the nonsingular set $\chi_0(k, \delta)$ there exists a unique smooth piece $S_H(k)'$ of the perturbed isoenergetic surface $S_H(k)$. Points of piece $S_H(k)'$ can be uniquely parametrized by the projection on any plane $t_i = 0$ $(i = 1, ..., n)$ i.e., $S_H(k)'$ can be described for any given i in the form

$$t = t_0 + \varphi_i(k^2, t_0), \quad t_0 \in \chi_0'(k, \delta),$$

where $\varphi_i(k^2, t_0)$ is a smooth vector-valued function with nonzero single component $(\varphi_i)_i$. We prove that the normal vectors of $S_H(k)'$ are close to corresponding those of $\chi_0'(k, \delta)$. Hence the part of the isoenergetic surface situated in a vicinity of $\chi_0(k, \delta)$ has the measure which is close to the measure of $\chi_0(k, \delta)$, i.e., to $s(S_k)$. Let us prove this results rigorously.

Theorem 2.6 . *There exists a single piece $S_H(k)'$ of the perturbed isoenergetic surface in the $(k^{-n+1-2\delta})$-neighbourhood of each simply connected component of the nonsingular set $\chi_0(k,\delta)$. For any given i, $i = 1, ..., n$, points t of $S_H(k)'$ can be represented in the form:*

$$t = t_0 + \varphi_i(k, t_0), \quad t_0 \in \chi_0'(k, \delta), \tag{2.6.1}$$

where $\varphi_i(k, t_0)$ is a continiously differentiable vector-valued function with the single nonzero component $(\varphi_i)_i$. The following asymptotic estimates are fulfilled for $(\varphi_i)_i$:

$$| \varphi_i(k, t_0)_i |=_{k\to\infty} O(k^{-n+1-2\gamma_0}), \tag{2.6.2}$$

$$| \nabla\varphi_i(k, t_0)_i |=_{k\to\infty} O(k^{-2\gamma_0}). \tag{2.6.3}$$

Estimates (2.6.2), (2.6.3) are uniform in t_0, $t_0 \in \chi_0(k, \delta)$.

Remark. The piece of the isoenergetic surface $S_H(k)'$ is a unique one in the $(k^{-n+1-2\delta})$-neighbourhood of $\chi_0(k,\delta)$. But as a matter of fact $S_H(k)'$ lies in a smaller $(k^{-n+1-2\gamma_0})$-neighbourhood of $\chi_0(k,\delta)$.

Proof. Eigenvalue $\lambda(\alpha, t)$ is determined by formula (2.2.14) in the $(2k^{-n+1-2\delta})$-neighbourhood of the nonsingular set $\chi_0(k,\delta)$. Function $\lambda(\alpha, t)$ is continuously differentiable with respect to t in such neighbourhood of each simply connected component of the nonsingular set, the estimates being valid:

$$| \lambda(\alpha, t) - p_j^{2l}(t) |=_{k\to\infty} O(k^{2l-n-\delta-2\gamma_0}), \tag{2.6.4}$$

$$\left| \frac{\partial\lambda(\alpha, t)}{\partial t_i} - \frac{\partial p_j^{2l}(t)}{\partial t_i} \right| =_{k\to\infty} O(k^{2l-1+\delta-2\gamma_0}) = o(k^{2l-1-2\delta}), \tag{2.6.5}$$

Taking into acoount that t is in the $(k^{-n+1-2\delta})$-neighbourhood of the nonsingular set, it is not hard to check the inequality:

$$\left| \frac{\partial p_j^{2l}(t)}{\partial t_i} \right| > k^{2l-1-2\delta}. \tag{2.6.6}$$

Now, using (2.6.5) we obtain:

$$\left| \frac{\partial\lambda(\alpha, t)}{\partial t_i} \right| > k^{2l-1-2\delta} \tag{2.6.7}$$

We consider the solution t of the equation

$$\lambda(\alpha, t) = p_j^{2l}(t_0), \quad t_0 \in \chi_0(k, \delta), \tag{2.6.8}$$

where $t_r = t_{0r}$ for $i \neq j$. Using relation (2.6.6) and Implicit function theorem, we obtain that in the $(k^{-n+1-\delta-2\gamma_0})$-neighbourhood of each simply connected component of $\chi_0(k,\delta)$, there exists a unique solution of equation (2.6.8). It can be represented in the form (2.6.1), where a vector-valued function $\varphi_i(t_0)$ is smooth and has the single nonzero component $(\varphi_i)_i$. Asymptotic formulae (2.6.2), (2.6.3) are valid for $\varphi_i(k,t_0)$. If $t_0 \in \chi_0(k,\delta)$, then $p_j^{2l}(t_0) = k^2$, therefore formula (2.6.1) determines the piece of the perturbed isoenergetic surface. We prove that in the $(k^{-n+1-2\delta})$-neighbourhood of $S_H(k)'$ there are no other pieces of the isoenergetic surface. It suffices to verify that any point of $S_H(k)$ being in the $(k^{-n+1-2\delta})$-neighbourhood of $\chi_0'(k,\delta)$ is determined by the equation $\lambda(\alpha,t) = k^{2l}$. Indeed, suppose that this is not so, i.e., there exists another eigenvalue $\tilde{\lambda}(t)$ of operator H, such that for some $\tilde{t}$, belonging to the $(k^{-n+1-2\delta})$-neighbourhood of $\chi_0'(k,\delta)$, the relation $\tilde{\lambda}(t) = k^{2l}$ is satisfied. According to Corollary 2.1 the inequality $|\lambda(\alpha,\tilde{t}) - k^{2l}| < k^{2l-n-\delta}$ holds. Thus, we have two eigenvalues in the interval $\varepsilon(k,\delta)$ for quasimomentum $\tilde{t}$. But this contradicts to the assertion of Theorem 2.1, because $\tilde{t}$ belongs to the $(k^{-n+1-2\delta})$-neighbourhood of $\chi_0(k,\delta)$. This contradiction finishes the proof of the theorem.

Let $S_H(k)_0$ be the part of the perturbed isoenergetic surface being in the $(k^{-n+1-2\delta})$-neighbourhood of $\chi_0(k,\delta)$. According to Theorem 2.6 it is a union of smooth pieces. Let $e(t)$ be the normal to $S_H(k)_0$ at point t, and $e_0(t_0), t_0 \in \chi_0(k,\delta)$, be the normal to the isoenergetic surface of free operator. It is clear that $e_0(t_0) = \mathbf{p}_j(t_0)/p_j(t_0)$, j being determined uniquely from the relation $p_j^{2l}(t_0) = k^{2l}$. If t is in the $(k^{-n+1-2\delta})$- neighbourhood of $\chi_0(k,\delta)$, then j is uniquely determined from the relation $p_j^{2l}(t_0) \in \varepsilon(k,\delta)$. Hence, the vector $\mathbf{p}_j(t_0)/p_j(t_0)$ is correctly defined in the $(k^{-n+1-2\delta})$- neighbourhood of $\chi_0(k,\delta)$ too.

Theorem 2.7 . *If $t \in S_H(k)_0$, then*

$$e(t) =_{k\to\infty} e_0(t) + O(k^{-2\gamma_0+2\delta}). \tag{2.6.9}$$

The measure of surface $S_H(k)_0$ is asymptotically close to those of $S_0(k)$:

$$\frac{s(S_H(k)_0)}{s(S_0(k)}\to_{k\to\infty} 1 \tag{2.6.10}$$

Corollary 2.7 . *For the area of the perturbed isoenergetic surface $S_H(k)$ the following estimate is valid:*

$$\lim_{k\to\infty} \frac{s(S_H(k))}{s(S_0(k))} \geq 1. \tag{2.6.11}$$

The estimate (2.6.11) is fulfilled because $S_H(k)_0 \subset S_H(k)$ and relation (2.6.10) is satisfied.

Proof of the theorem. Since t belongs to the $(k^{-n+1-2\delta})$-neighbourhood of the nonsingular set $\chi_0(k,\delta)$, then in a vicinity of point t the perturbed isoenergetic surface is determined by the equation $\lambda(\alpha,t) = k^{2l}$ (see Theorem 2.6). Taking into account that

$$dS_H(k)_0 = p(t_2,...,t_n)dt_2...dt_n, \tag{2.6.12}$$

$$p(t_2,...,t_n) = \left|\nabla\lambda(\alpha,t)/\frac{\partial\lambda(\alpha,t)}{\partial t_1}(t)\right|,$$

$$e(t)=\frac{\nabla\lambda(\alpha,t)}{|\nabla\lambda(\alpha,t)|} \tag{2.6.13}$$

and considering asymptotic formulae (2.6.5) and (2.6.6) we obtain relations (2.6.9), (2.6.10). The theorem is proved.

Let us study the perturbed isoenergetic surface near the simple part of the singular set.

First, it is not hard to show that in the $(k^{-n+1-7\delta})$- neighbourhood of $\mathcal{K}S_q \cup S_{-q}$ there exist two surfaces S_+ and S_- determined by the formulae:

$$\lambda_+(t)=k^{2l},$$

$$\lambda_-(t)=k^{2l}.$$

The distance between these surfaces is more than ck^{-1}. Each point of $S_+ \cup S_-$ belongs to the isoenergetic surface of $\hat{H}_q$ (or $\hat{H}_{-q}$, which has the same isometric surface). We call this surface the model isoenergetic surface.

The normal to $S_+ \cup S_-$ is given by the formula

$$\hat{e}(t)=\frac{\nabla\lambda_\pm(t)}{|\nabla\lambda_\pm(t)|},\quad t\in S_\pm. \tag{2.6.14}$$

It is easy to see that $\hat{e}(t)$ can be determined by this formula also in the $(k^{-n+1-7\delta})$-neighbourhood of $S_+ \cup S_-$ and $\hat{e}(t)$ is close to the normal to $S_q \cup S_{-q}$, i.e.,

$$\hat{e}(t)=\frac{\mathbf{p}_j(t)}{p_j(t)}+O(k^{-1})=\frac{\mathbf{p}_{j\pm q}(t)}{p_{j\pm q}(t)}+O(k^{-1}).$$

It turns out that there exist parts of the isoenergetic surface of H, which are very close to $S_+ \cup S_-$, namely they are in the $(k^{-2l+1-\delta-2\gamma_1})$-neighbourhood of $S_+ \cup S_-$. Note that the size $k^{-2l+1-\delta-2\gamma_0}$ of this neighbourhood is small with respect to $k^{-n+1-7\delta}$.

Theorem 2.8 . *There exists two parts of the isoenergetic surface of H in $(k^{-n+1-7\delta})$-neighbourhood of $\chi_q(k,\delta) \cup \chi_{-q}(k,\delta)$. They are, in fact, in the $(k^{-2l+1-\delta-2\gamma_1})$-neighbourhood of $S_+ \cup S_-$ and are determined by the following equations:*

$$\hat{\lambda}^q(t)=k^{2l}, \tag{2.6.15}$$

$$\hat{\lambda}^{-q}(t)=k^{2l}. \tag{2.6.16}$$

The normal $e(t)$, to the isoenergetic surface of H at the point t satisfying one of the equations (2.6.15), (2.6.16) satisfies the asymptotic:

$$e(t)=\hat{e}(t)+O(k^{-\gamma_1-\gamma_0}).$$

Proof. The proof of the theorem is similar to that of Theorem 2.6. We use Theorem 2.3 instead of Theorem 2.1. The theorem is proved. The isoenergetic surface near singular and nonsingular sets is described in [K13].

2.7 Formulae for Eigenfunctions on the Isoenergetic Surface.

Let t belong to the nonsingular part $S_H(k)_0$ of the isoenergetic surface situated near the nonsingular set $\chi_0(k,\delta)$. Formula (2.2.15) for the perturbed spectral projection is valid at this point. However, this formula is directly applicable, only when the all components of quasimomentum t are given. To use the formula formula in the case, when only $n-1$ coordinates of t, say, $t_2,...,t_n$ are known and instead of t_1 the value of the perturbed eigenvalue $\lambda(\alpha,t) = k_0^{2l}$ is given, we have previously to determine t_1 from the equation $\lambda(\alpha,t) = k^{2l}$. According to Theorem 2.6 $t_1 = t_{01} + O(k^{-n+1-2\gamma_0})$, t_{01} being the solution of the equation $p_j^{2l}(t_0) = k^{2l}$, $t_0 = (t_{01}t_2,...,t_n)$. Substituting this approximate solution to the formulae for $E_j(t)$ and $\nabla\lambda(\alpha,t)$ we prove the following theorem:

Theorem 2.9 . *Suppose t_0 belongs to $\chi_0(k,\delta)$. Then at the point $t(t_0)$, determined by the formula (2.6.1) ($i=1$) the following formulae for $\nabla\lambda(\alpha,t)$ and $E(t)$ are valid:*

$$\nabla\lambda(\alpha,t) = \mathbf{p}_j(t_0)p_j^{2l-2}(t_0)(1+O(k^{-2\gamma_0})), \tag{2.7.1}$$

$$E(t) = E_j + G_1(k,t_0) + G_2(k,t_0) + O(k^{-3\gamma_0+2\delta}). \tag{2.7.2}$$

Proof. Using the series for $\nabla\lambda(\alpha,t)$ and the relation $\mid t-t_0 \mid < k^{-n+1-2\gamma_s}$, we obtain (2.7.1). According to Corollary 3.4

$$\|\partial E(t)\partial t_1\|_1 < ck^{n-1-\gamma_0+2\delta}. \tag{2.7.3}$$

Therefore,

$$\|E(t) - E(t_0)\| < k^{-3\gamma_0}.$$

Using the formula for $E(t_0)$, we get (2.7.2). The theorem is proved.

We can calculate $T(m)\lambda(\alpha,t)$, $T(m)E(t)$ by the similar way. However, the accuracy of the formulae is restricted by that of approximation $t_1 \approx t_{01}$. To write out the asymptotic terms of higher orders it is nessasary to resolve more precisely the equation $\lambda(\alpha,t) = k^{2l}$. But it seems to be not effective way. There exists another way of the construction of the formula for the eigenfunction. This way is not connected with the resolving the equation $\lambda(\alpha,t) = k^{2l}$. Now, we describe it.

Suppose $t \in S_H(k)_0$. Accorging Theorem 2.6 t can be represented in the form (2.6.1). We set $i=1$. Let us consider the integral:

$$I(k,t_0) = \frac{1}{2\pi i}\int_{C_1}(H(t)-k^{2l})^{-1}dt_1, \tag{2.7.4}$$

where $t = (t_1,t_{02},...,t_{0n})$, C_1 is the circle of radius $k^{-n+1-2\delta}$ centred at the point t_{01}.

Lemma 2.14 . *Suppose t_0 belongs to $\chi_0(k,\delta)$. Then operator $(H(t)-k^2)^{-1}$ has a unique pole inside C_1 at the point t, given by relation (2.6.1), i being equal to 1. For $I(k,t_0)$ the following formula is valid:*

$$I(k,t_0)=\frac{E(\alpha,t)}{(\partial\lambda(\alpha,t)/\partial t_1)}|_{t=t(t_0)}. \qquad (2.7.5)$$

Proof. The operator $(H(t)-k^{2l})^{-1}$ has a pole at the point t if and only if t belongs to the isoenergetic surface of operator H. Let $t=t_0+\varphi_1(t_0)$ (see (2.6.1)). Then $|t_1-t_{01}|<k^{-n+1-2\delta}$, i.e., t is inside C_1. Thus, there exists at least one pole inside C_1. Suppose, there exist two poles: t and $\tilde{t}$. Since $t_1,\tilde{t}_1\in C_1$, then $t_1,\tilde{t}_1$ are in the $(k^{-n+1-2\delta})$ -neighbourhood of $\chi_0(k,\delta)$. According to Theorem 2.6 the points of the isoenergetic surface in this neighbourhood can be represented in the form (2.6.1). Considering that $t_i=\tilde{t}_i, i\neq 1$ we obtain $t=\tilde{t}$, i.e., there exists only a single pole inside C_1.

Now we prove relation (2.7.5). It is clear that the operator-valued function

$$(H(t)-k^{2l})^{-1}-E(t)(\lambda(\alpha,t)-k^{2l})^{-1}$$

is holomorphic inside C_1. The corresponding integral over C_1 is equal zero. Let us consider the operator-valued function $E(t)(\lambda(\alpha,t)-k^{2l})^{-1}$. It analytically depends on t and has the single pole at the point $t=t(t_0)$ inside C_1. Therefore,

$$\frac{1}{2\pi i}\oint_{C_1}E(t)(\lambda(\alpha,t)-k^{2l})^{-1}dt=E(\alpha,t)(\partial\lambda(\alpha,t)/\partial t_1)^{-1}(t(t_0)). \qquad (2.7.6)$$

The lemma is proved.

Expanding formally $(H(t)-k^{2l})^{-1}$ in the series in powers of αV, we obtain

$$I(k,t_0)=\sum_{r=0}^{\infty}\alpha^r D_r(k,t_0), \qquad (2.7.7)$$

where

$$D_r(k,t_0)=\frac{(-1)^r}{2\pi i}\oint_{C_1}(H_0(t)-k^{2l})^{-1}(V(H_0(t)-k^{2l})^{-1})^r dt. \qquad (2.7.8)$$

It is clear that $D_0(k,t_0)$ is equal to the integral $I(k,t_0)$ for the free operator. According Lemma 2.15 we have:

$$D_0(k,t_0)=E_j k^{-2l+2}p_{j_1}^{-1}(t_1), \qquad (2.7.9)$$

$$p_{j_1}(t_1)=t_1+2\pi j_1 a_1^{-1}.$$

To justify the asymptotic formula for the operator

$$E(\partial\lambda(\alpha,t)/\partial t_1)^{-1}(t(t_0)). \qquad (2.7.10)$$

it suffices to prove power estimates similar to (2.2.17) for $D_r(k,t_0)$. In this formula the asymptotic terms explicitly depend on k^2 and $t_{02},\dots,t_{0n}$. We do not need to resolve the equation $\lambda(\alpha,t)=k^2$.

Now, we prove the estimates for $D_r(k,t_0)$.

Theorem 2.10 . *Suppose t_0 belongs to $\chi_0(k,\delta)$. Then at the point $t(t_0)$, defined by formula (2.6.1), the operator $E(\partial\lambda(\alpha,t)/\partial t_1)^{-1}(t(t_0))$ can be represented by the series:*

$$E(\partial\lambda(\alpha,t)/\partial)^{-1}(t(t_0)) = \sum_{r=0}^{\infty} \alpha^r D_r(k,t_0), \tag{2.7.11}$$

which converges in class $\mathbf{S}_1$. Moreover, $D_r(k,t_0)$ satisfy the estimates:

$$\|D_r(k,t_0)\|_1 < k^{-(\gamma_0-8\delta)r-2l+1}. \tag{2.7.12}$$

Proof. It remains only to prove estimate (2.7.12). Suppose, we have verified the following inequalities:

$$\max_{t\in C_1} \|A(k,t)\|_1 < k^{-(\gamma_0-8\delta)r}, \tag{2.7.13}$$

$$\max_{t\in C_1} \|(H_0(t)-k^{2l})^{-1}\| < k^{-2l+n+2\delta} \tag{2.7.14}$$

Estimating $D_r(k,t_0)$ above by the maximum of the multiplication of the norm of the integrand and the length $2\pi k^{-n+1-2\delta}$ of the circle C_1, we obtain:

$$\|D_r(k,t_0)\|_1 < (\max_{t\in C_1} \|(H_0(t)-k^{2l})^{-1}\| \|A(k,t)\|^r) 2\pi k^{-n+1-2\delta} < k^{-2l+1-(\gamma_0-8\delta)r}. \tag{2.7.15}$$

Thus, it suffices to verify estimates (2.7.13), (2.7.14). We begin with (2.7.13). Note that in the formulae for g_r, G_r the contour C_0 is the same for all t, being in the $(k^{-n+1-2\delta})$-neighbourhood of the nonsingular set. Let us take for every t its special contour $C(t)$ to be centred at the point $z = p_j^{2l}(t)$ and to have the radius $r(t)k^{2l-n-\delta}$, $k^{-2\delta} < r(t) < 1$. The results of the integrations preserve, when we replace C_0 by $C(t)$, because the integrands have no poles between C_0 and $C(t)$. Moreover, the estimates for the integrands are stable relative to such perturbation of the contour. Indeed, repeating after replacing δ by 3δ all consideration of Theorem 2.1, we obtain instead estimate (2.2.22) the similar one:

$$\max_{z\in C(t)} \|A(z,t)\| < k^{-\gamma_0+8\delta}. \tag{2.7.16}$$

Suppose $t_0 \in \chi_0(k,\delta)$, $t \in C_1(t_0)$. It is clear that t belongs to the complex $(k^{-n+1-2\delta})$-neighbourhood of $\chi_0(k,\delta)$. Hence t satisfies estimate (2.7.16). It is not hard to show that

$$| k^{2l} - p_j^{2l}(t) | = | p_j^{2l}(t) - p_j^{2l}(t_0) | = r(t)k^{2l-n-\delta},$$

where $r(t) \approx k^{-\delta}$. Therefore, $k^2 \in C(t)$ for this $r(t)$. Estimate (2.7.16) for $z = k^2$ implies inequality (2.7.13). Using inequality (2.2.6) we similarly verify (2.7.14). The theorem is proved.

We consider the following function:

$$\Psi(k^{2l}, t_{02}, \ldots, t_{0n}, x) = \sum_{q\in Z^n} \oint_{C_1} (H(t)-k^2)^{-1}_{qj} \exp(i(\mathbf{p}_q(t), x))dt_1, \tag{2.7.17}$$

where $t_0 \in \chi_0(k,\delta)$, j is uniquely determined from the relation $p_j^{2l}(t) = k^{2l}$. It is easy to see that function $\Psi(k^{2l}, t_{02}, ..., t_{0n}, x)$ satisfies the equation $(-\Delta + V)\Psi = k^2\Psi$ and the quasiperiodic conditions with the quasimomentum $t = (t_1, t_{02}, ..., t_{0n})$, the value of t_1 being given by formula (2.6.1). Thus, $\Psi(k^{2l}, t_0, x)$ is an eigenfunction of the operator $H(t)$, $t \in S_H(k)_0$. This function can be formally expanded in powers of αV:

$$\Psi(k^{2l}, t_0, x) = \sum_{r=0}^{\infty} B_r(k^{2l}, t_0, x), \tag{2.7.18}$$

function $B_r(k^{2l}, t_0, x)$ being defined by the formula:

$$B_r(k^{2l}, t_0, x) = \tag{2.7.19}$$

$$\frac{(-1)^r}{2\pi i} \sum_{m \in Z^n} \oint_{C_1} [(H_0(t) - k^2)^{-1}(V(H_0(t) - k^2)^{-1})^r]_{mj} \exp(i(\mathbf{p}_m(t), x)) dt_1,$$

$$B_0(k^{2l}, t_0, x) = \frac{\exp(i(\mathbf{p}_j(t), x))}{k^{2l-2} p_{j_1}(t_1)}, \tag{2.7.20}$$

$$p_{j_1}(t_1) = 2\pi j_1 a_1^{-1} + t_1.$$

$$t = (t_1, t_{02}, ..., t_{0n}).$$

It is not hard to show that functions $B_r(k^{2l}, t_0, x)$ satisfy quasiperiodic conditions in the directions orthogonal to x_1. The corresponding components of the quasimomentum are $t_{02}, ..., t_{0n}$. We introduce the notation:

$$\|B_r(k^{2l}, t_0, x)\|_{2,M} \equiv \max_{|x_1|<M} \int_{K_{\|}} |B_r|^2 \, dx_2 ... dx_n, \tag{2.7.21}$$

where $K_{\|} = [0, a_2) \times ... \times [0, a_n)$.

Theorem 2.11 . *If $t_0 \in \chi_0(k,\delta)$, then for sufficiently large k, $k > k_0(\|V\|, \delta)$, the functions $B_r(k^{2l}, t_0, x)$ can be represented as the sum:*

$$B_r(k^{2l}, t_0, x) = \sum_{q:|q-j|<rR_0, 0\le m\le r} a_{qm}^{(r)} \exp(i(\mathbf{p}_q(t), x)) x_1^m, \tag{2.7.22}$$

where the coefficients $a_{qm}^{(r)}$ satisfy the estimates:

$$| a_{qm}^{(r)} | < k^{-2l+1-(\gamma_0-8\delta)r-(n-1+2\delta)m}. \tag{2.7.23}$$

For all r the relation

$$B_r(k^{2l}, t_0, x_1)_{m_{\|}} = 0 \tag{2.7.24}$$

holds, when $|m_{\|} - j_{\|}| > rR_0$.

Proof. From formula (2.7.19) it is easily follows that

$$B_r(k^{2l}, t_0, x_1)_{m_{\|}} = \tag{2.7.25}$$

$$\sum_{m_1 \in Z} \oint_{C_1} [(H_0(t) - k^2)^{-1}(V(H_0(t) - k^2)^{-1})^r]_{mj} \exp(i(\mathbf{p}_m(t), x)) dt_1,$$

$$m = (m_1, m_{\|}).$$

Using the assumption that V is a trigonometric polynomial we easily get the integrand in (2.7.25) to be zero, when $|j - m| > rR_0$ and all the more when $|j_{\|} - m_{\|}| > rR_0$.

We represent $\exp(\mathbf{p}_q(t), x))$ in the form:

$$\exp(\mathbf{p}_q(t), x) = \exp(\mathbf{p}_q(t_0), x)) \sum_{m=0}^{\infty} (t_1 - t_{01})^m x_1^m / m!.$$

Note that the function

$$\left((H_0(t) - k^2)^{-1}(V(H_0(t) - k^2)^{-1})^r\right)_{mj} \varphi_r,$$

$$\varphi_r = \sum_{m=r+1}^{\infty} (t_1 - t_{01})^m x_1^m / m!,$$

is holomorphic incide the circle. Therefore, $B_r(k^{2l}, t_{02}, ..., t_{0n}, x)$ can be represented in the form (2.7.22), where

$$a_{qm}^{(r)} = \oint_{C_1} [(H_0(t) - k^2)^{-1}(V(H_0(t) - k^2)^{-1})^r]_{qj} (t_1 - t_{01})^m / m!. \tag{2.7.26}$$

Estimate (2.7.23) follows from inequalities (2.7.13) and (2.7.14).
The theorem is proved.

Theorem 2.12 . *Under the conditions of Theorem 2.11 there are the following estimates for the functions* $B_r(k^{2l}, t_0, x)$*:*

$$\|B_r\|_{2,M} < ck^{-2l+1-(\gamma_0 - 9\delta)r}(1+ \mid M \mid^r), \tag{2.7.27}$$

$$\left\| \frac{\partial B_r}{\partial x_1} \right\|_{2,M} < ck^{-2l+2-(\gamma_0 - 10\delta)r}(1+ \mid M \mid^r). \tag{2.7.28}$$

Proof. Using (2.7.22, it is not hard to show that

$$\left\| B_r(k^2, t_0, x) \right\|_{L_2(K_{\|})} \leq (rR_0)^{3/2}(1+ \mid M \mid^r) \max_{|q-j|<rR_0, 0 \leq m \leq r} |a_{qm}^{(r)}|. \tag{2.7.29}$$

Now, taking into account (2.7.23) and the obvious estimate $(rR_0)^{3/2} < k^{\delta r}$, we get (2.7.27).

To obtain (2.7.28) we note that

$$\left\| \frac{\partial B_r(k^2, t_0, x)}{\partial x_1} \right\|_{L_2(K_{\|})} \leq c(rR_0)^{3/2}(1+ \mid M \mid^r) \max_{|q-j|<rR_0, 0 \leq m \leq r} |q_1 a_{qm}^{(r)}|. \tag{2.7.30}$$

Noting that $|q_1| < |j_1| + rR_0 < k + k^{r\delta}$, we get (2.7.28). The theorem is proved.
The results of this section was published in [K13].

3. Perturbation Theory for the Polyharmonic Operator in the Case $4l > n+1$

3.1 Introduction. Generalized Laue Diffraction Conditions.

In this chapter we consider the operator

$$H_\alpha = (-\Delta)^l + \alpha V \tag{3.1.1}$$

in $L_2(R^n)$, $n \geq 2$, $4l > n+1$. As shown in Section 2.1, the average distance between eigenvalues $p_i^{2l}(t)$ of the free operator in the interval $[k^{2l}, (2k)^{2l}]$ is of order k^{2l-n}. In the case $2l > n$, they are situated rather far from each other when k is large and regular perturbation theory works. In the case $2l < n$,, the denseness of the eigenvalues increases infinitely with the energy. Under a perturbation, eigenvalues influence strongly on each other and the regular perturbation theory does not work. Nevertheless, we construct here the perturbation series for eigenvalues and their spectral projections, when t belongs to a nonsingular set. These series have the same form as in the simpler case $2l > n$, but the diffraction conditions, the structure of the nonsingular set and, therefore, the proof of convergence of the series are more complicated.

Now, using nonrigorous considerations, we find out the conditions of the convergence of the perturbation series for an eigenvalue and its spectral projection.

First, let us find the conditions on t, under which the perturbation of an eigenvalue $p_j^{2l}(t)$ is small as $p_j^{2l}(t)$ becomes large. Indeed, under the action of potential $V(x)$, an eigenvalue $p_j^{2l}(t)$ interacts with other eigenvalues $p_{j+q}^{2l}(t)$, $q \neq 0$. As the first approximation, we consider that the "interaction" of each pair $\{p_j^{2l}(t), p_{j+q}^{2l}(t)\}, q \neq 0$ does not depend on other eigenvalues $p_i^{2l}(t), i \neq j, j+q$. It is described by a simple matrix $\hat{H}_q(t)$ (see (2.1.7)), i.e., is defined only by two nondiagonal elements (see (1.0.36)). The eigenvalues of $\hat{H}_q(t)$ are $\hat{\lambda}^+, \hat{\lambda}^-, \{p_i^{2l}(t)\}_{i \neq j, j+q}$; here $\hat{\lambda}^+, \hat{\lambda}^-$ are the eigenvalues of matrix (1.0.37) and given by formula (2.4.7). Note that the pair $\hat{\lambda}^+, \hat{\lambda}^-$ is the result of the interaction of $p_j^{2l}(t)$ and $p_{j+q}^{2l}(t)$.

Let us denote by $\delta_q p_j^{2l}(t)$ the perturbation of $p_j^{2l}(t)$, which is caused by the "interaction" with $p_{j+q}^{2l}(t)$:

$$\delta_q p_j^{2l}(t) = \min\{|\hat{\lambda}^+ - p_j^{2l}(t)|, |\hat{\lambda}^- - p_j^{2l}(t)|\}.$$

We assume that the perturbation $\delta p_j^{2l}(t)$ of $p_j^{2l}(t)$ by the potential $V(x)$ is approximately equal to the sum of the perturbations, $\delta_q p_j^{2l}(t)$:

$$\delta p_j^{2l}(t) \approx \sum_q \delta_q p_j^{2l}(t) \tag{3.1.2}$$

(actually, it turns out to be true when the $\delta_q p_j^{2l}(t)$ are small). We suppose for simplicity that $V(x)$ is a trigonometric polynomial, more precisely, $v_0 = 0$ and $v_q = 0$, when $|q| \geq R_0, R_0 < \infty$. It is clear that $\delta_q p_j^{2l}(t) = 0$ when $|q| \geq R_0$. Therefore, the sum on the right of (3.1.2) contains only a finite number of terms. The resulting perturbation is asymptotically small, if each term of the sum is small. Using formula (1.0.37) for $\hat{H}_q(t)$ and regular perturbation considerations, it is easy to see that

$$|\delta_q p_j^{2l}(t)| \approx \frac{|v_q|^2}{\rho_{jq}}, \qquad \rho_{jq} = |p_{j+q}^{2l}(t) - p_j^{2l}(t)|, \text{ when } \rho_{jq} \geq |v_q|, \tag{3.1.3}$$

and

$$|\delta_q p_j^{2l}(t)| \approx |v_q|, \text{ when } \rho_{jq} \leq |v_q|. \tag{3.1.4}$$

From the last two relations, we obtain that

$$|\delta_q p_j^{2l}(t)| \leq |v_q|^2 k^{-\gamma_2},$$

when

$$|p_j^{2l}(t) - p_{j+q}^{2l}(t)| > k^{\gamma_2}, \qquad \gamma_2 > 0.$$

It is easy to show that $|p_j^{2l}(t) - p_{j+q}^{2l}(t)| < ck^{2l-1}$, when $|q| < R_0$. Hence, the best we can hope for is $\gamma_2 = 2l - 1 - \delta$ for some δ, $0 < \delta \ll 1$. Thus, one can expect that $\delta p_j^{2l}(t) \approx k^{-\gamma_2}$, if the following condition is satisfied:

a)

$$\min_{q \neq 0, |q| < R_0} |p_j^{2l}(t) - p_{j+q}^{2l}(t)| > k^{2l-1-\delta}. \tag{3.1.5}$$

Next, we clear up which conditions are sufficient for the perturbed eigenvalue to be simple. First of all we suppose the unperturbed eigenvalues $p_{j+q}^{2l}(t)$, $q \neq 0$, to be rather far from $p_j^{2l}(t)$. As shown in Chapter 2, the strongest condition which can possibly be satisfied is:

b)

$$\min_{q \neq 0} |p_j^{2l}(t) - p_{j+q}^{2l}(t)| > k^{2l-n-\delta}. \tag{3.1.6}$$

Under the perturbation, the eigenvalue $p_j^{2l}(t)$ stays simple, if its displacement and also the displacements of its neighbors $p_i^{2l}(t)$, $i \neq j$ are sufficiently small. Namely, let us show that it is simple if

$$|\delta p_j^{2l}(t)| < \frac{1}{4} \min_{q \neq 0} |p_j^{2l}(t) - p_{j+q}^{2l}(t)|, \tag{3.1.7}$$

$$|\delta p_i^{2l}(t)| < \frac{1}{4}|p_j^{2l}(t) - p_i^{2l}(t)|. \tag{3.1.8}$$

Indeed, on the one hand, the perturbation of the eigenvalue $p_j^{2l}(t)$ is less than a quarter of the distance between $p_j^{2l}(t)$ and the nearest eigenvalue $p_i^{2l}(t)$, $i \neq j$. On the other hand, the perturbation of any eigenvalue $p_i^{2l}(t)$, $i \neq j$ is less than a quarter of the distance between $p_j^{2l}(t)$ and $p_i^{2l}(t)$. Thus, if relations (3.1.5) - (3.1.8) hold, then the distance between $p_j^{2l}(t)+\delta p_j^{2l}(t)$ and the nearest eigenvalue $p_i^{2l}(t) + \delta p_i^{2l}(t)$, $i \neq j$ is greater than $\frac{1}{2}k^{2l-n-\delta}$.

Now we describe the set of t, where relations (3.1.7) and (3.1.8) hold. Suppose, estimate (3.1.5) holds. Then, it follows from inequality (3.1.3) that

$$\delta p_j^{2l}(t) \approx k^{-2l+1+\delta} = o(k^{2l-n-\delta}). \tag{3.1.9}$$

The last relation holds for sufficiently small positive δ, $0 < \delta < 4l - n - 1$. Using relations (3.1.9) and (3.1.6), we obtain (3.1.7). Let us consider (3.1.8). If $|p_i^{2l}(t) - p_j^{2l}(t)| \geq 4\|V\|$, then inequality (3.1.8) follows from the obvious relation $|\delta p_i^{2l}(t)| \leq \|V\|$. In the case $|p_i^{2l}(t) - p_j^{2l}(t)| < 4\|V\|$ we will use an estimate similar to (3.1.3). Indeed, let us suppose that

c)

$$|p_i^{2l}(t) - p_{i+q}^{2l}(t)| > k^{4l-n-1-8\delta}|p_i^{2l}(t) - p_j^{2l}(t)|^{-1}, \tag{3.1.10}$$

$0 < 8\delta < 4l - n - 1$. The right-hand side of this inequality is much greater than $\|V\|$. So we can apply (3.1.3) for $p_i^{2l}(t)$ and obtain:

$$\delta p_i^{2l}(t) \approx k^{-4l+n+1+8\delta}|p_i^{2l}(t) - p_j^{2l}(t)|.$$

From the last relation (3.1.8) follows.

Thus, based on general considerations, we have formulated conditions a) – c) (see (3.1.5), (3.1.6), (3.1.10)), which we expect to be sufficient for the convergence of the perturbation series. In Section 3.2, we prove rigorously that this is actually true: under conditions a) – c) the perturbation series for the eigenvalue and corresponding spectral projection converge and have an asymptotic character in a high energy region. Proving this result in Section 3.2, we set $R_0 = k^\beta, 0 < \beta < 1$. This enables us to expand our results easily from the case of a trigonometric polynomial to the case of a smooth potential (Section 3.3).

Conditions a) – c) upon t, as a matter of fact, represent the analytical definition of the nonsingular set. The geometrical construction of this set is made in Section 3.4. We show that it has an asymptotically full measure on $S_0(k)$.

Note that the singular set is a neighborhood of the planes

$$p_j^{2l}(t) = p_{j+q}^{2l}(t), \qquad q \neq 0 \tag{3.1.11}$$

and the planes

$$p_i^{2l}(t) = p_{i+q}^{2l}(t), \qquad q \neq 0, \qquad |q| < R_0, \tag{3.1.12}$$

where $i : |p_i^2(t) - p_j^{2l}(t)| \leq 4\|V\|$. Relations (3.1.11) are well-known as the von Laue Diffraction Conditions. Note that relations (3.1.12) are not the same as (3.1.11), because they are written not only for the eigenvalue $p_j^{2l}(t)$, whose perturbation is described, but for its neighbors. These new diffraction conditions we call the *Associated Laue Diffraction Conditions*.

In the case of a nonsmooth potential, additional analytical and geometrical considerations (in particular, the introduction of a function of the number of states) are necessary(Sections 3.5 - 3.7).

It should be mentioned that the perturbation series for an eigenvalue and its spectral projection, constructed in this chapter, are infinitely differentiable with respect to t, their asymptotic character preserving.

The perturbation formula for the eigenvalue enable us to prove the Bethe-Sommerfeld conjecture for a general class of potentials and to describe an essential part of the perturbed isoenergetic surface (Section 3.8). Moreover, in Section 3.8, a converging asymptotic series for an eigenfunction at a point of the perturbed isoenergetic surface, corresponding to a given energy k^{2l}, is constructed. The convenience of this formula is that it does not require the solution of the dispersion equation $\lambda(\alpha, t) = k^{2l}$, as it would be necessary for using the series with a fixed quasimomentum.

In Section 3.9, the procedure of the determination of the potential from the asymptotic of an eigenfunction in a high energy region is described.

In this chapter we don't construct the perturbation theory series on the singular set. Note that these constructions are similar to those in the case $l = 1$, $n = 3$, which are described in Chapter 4. The results proved in this chapter were published in [K4] - [K8] and [K13].

3.2 Analytic Perturbation Theory for the Nonsingular Set.

Now we formulate and prove the main results of the perturbation theory for the case of $4l > n + 1$. We construct the perturbation series for an eigenvalue and its spectral projection for t in the $(k^{-n+1-2\delta})$-neighborhood of the nonsingular set. These series have the same form as in the case of $2l > n$, but the structure of the nonsingular set and, therefore, the proof of the convergence of the series are more complicated.

Now we formulate the main Geometric Lemma. It will be proved in section 3.4.

Lemma 3.1 . *For any β, $0 < \beta < 1$, and an arbitrarily small positive δ, $2\delta < (n-1)(1-\beta)$, and sufficiently large k, $k > k_0(\beta, \delta, a_1, ... a_n)$ there exists a nonsingular set $\chi_1(k, \beta, \delta)$, belonging to the isoenergetic surface $S_0(k)$ of the free operator H_0 such that for any point t of it the following conditions hold:*

1) There exists a unique $j \in Z^n$ such that $p_j(t) = k$.

2)

$$\min_{i \neq j} | p_j^2(t) - p_i^2(t) | > 2k^{-n+2-\delta} \tag{3.2.1}$$

3)

$$\min_{|q|<k^{\beta},q\neq 0} |p_j^2(t) - p_{j+q}^2(t)| > k^{1-\beta(n-1)-\delta}, \tag{3.2.2}$$

4) For all $i \neq j$ *such that* $|p_i^2(t) - k^2|^2 < k^{-\zeta}$, $\zeta = n - 3 + \beta(n-1) + 2\delta$, *the inequality*

$$\min_{|q|<k^{\beta},q\neq 0} |p_i^2(t) - p_{i+q}^2(t)| > k^{-\zeta}|p_i^2(t) - k^2|^{-1}, \tag{3.2.3}$$

holds.

Moreover, for any t *in the* $(2k^{-n+1-2\delta})$*-neighborhood of the nonsingular set in* C^n *there exists a unique* $j \in Z^n$ *such that* $\mid p_j^2 - k^2 \mid < 5k^{-n+2-2\delta}$ *and conditions 2–4 are satisfied.*

The nonsingular set has an asymptotically full measure on $S_0(k)$. *Moreover,*

$$\frac{s(S_0(k) \setminus \chi_1(k,\beta,\delta))}{s(S_0(k))} =_{k\to\infty} O(k^{-\delta/8}) \tag{3.2.4}$$

Corollary 3.1 . *If* t *belongs to the* $(2k^{-n+1-2\delta})$*-neighborhood of the nonsingular set in* C^n, *then, for all* z *lying on the circle* $C_0 = \{z :\mid z - k^{2l} \mid = k^{2l-n-\delta}\}$, *the inequality*

$$200 \mid p_i^{2l}(t) - z \mid\mid p_{i+q}^{2l}(t) - z \mid > k^{2\gamma_2}, \tag{3.2.5}$$

$$2\gamma_2 = 4l - n - 1 - \beta(n-1) - 2\delta,$$

holds simultaneously for all $i \in Z^n$ *and* $|q| < k^{\beta}$, $q \neq 0$.

Proof of the corollary. Using the obvious relation

$$2|(p_i^{2l}(t) - z)(p_i^2(t) - z^{1/l})| > lk^{2l-2}$$

, we easily see that (3.2.5) can be obtained from the inequality

$$36l^2 \mid p_i^2 - z^{1/l} \mid\mid p_{i+q}^2 - z^{1/l} \mid > k^{2\gamma_2 - 4l + 4} \tag{3.2.6}$$

We prove the last inequality. Let $t \in \chi_1(k,\beta,\delta)$. If

$$6l \min\{|p_i^2(t) - z^{1/l}|, |p_{i+q}^2(t) - z^{1/l}|\} > k^{\gamma_2 - 2l + 2},$$

then (3.2.6) is obvious. Suppose that

$$6l|p_i^2(t) - z^{1/l}| \le k^{\gamma_2 - 2l + 2}. \tag{3.2.7}$$

From the definition of C_0 and estimate (3.2.1) it is easy to see that

$$6l|p_i^2(t) - z^{1/l}| > |p_i^2(t) - k^2| \qquad (l > 1/2) \tag{3.2.8}$$

Using relations (3.2.3) and (3.2.8) when $i \neq j$ (3.2.2) and the relation $l|p_j^2(t) - z^{1/l}| > k^{-n+2-\delta}$ when $i = j$ and , we show that

$$6l|p_{i+q}^2(t) - p_i^2(t)| > k^{2\gamma_2-4l+4}|p_i^2(t) - z^{1/l}|^{-1}.$$

From (3.2.7) it follows that

$$6l|p_i^2(t) - z^{1/l}| \le k^{2\gamma_2-4l+4}(6l|p_i^2(t) - z^{1/l}|)^{-1}.$$

From the last two inequalities we obtain

$$|p_{i+q}^2(t) - z^{1/l}| \ge k^{2\gamma_2-4l-2}(10l|p_i^2(t) - z^{1/l}|)^{-1},$$

i.e., (3.2.6) is satisfied. In the case, where $6l|p_{i+q}^2(t) - z^{1/l}| \le k^{\gamma_2-2l+2}$, by making the transformation $i' = i+q$, $i'+q' = i$, we arrive at the case (3.2.7). It is easy to see, that all the estimates are stable under a perturbation of t of order $2k^{-n+1-2\delta}$; therefore (3.2.6) holds not only on a nonsingular set, but also in the $(2k^{-n+1-2\delta})$-neighborhood of it. The corollary is proved.

Note, that conditions 2-4 are somewhat stronger, than conditions a-c represented in the introduction (inequalities (3.2.1) and (3.1.6) are equivalent up to a constant factor; estimate (3.2.2) is stronger than (3.1.5) because $k^\beta > R_0$ for sufficiently large k. The fourth condition in the lemma is a little stronger, than c because it is imposed on a richer set of indices i). The intensification of the conditions enable us to generalize the theorems to the case of a smooth potential and to obtain the optimal estimates for asymptotic terms.

Note, that the nonsingular set does not depend on l and V. However, this independence, as it will be shown below, can be guaranteed only for a smooth potential.

Theorem 3.1 . *Suppose t belongs to the $(k^{-n+1-2\delta})$-neighborhood in K of the nonsingular set $\chi_1(k,\beta,\delta)$, $0<\beta<1$, $0<2\delta<(1-\beta)(n-1)$, $2\delta+\beta(n-1) < 4l-n-1$. Then for sufficiently large k, $k > k_0(V,\beta,\delta)$, for all α, $-1 \le \alpha \le 1$ in the interval $\varepsilon(k,\delta) \equiv [k^{2l}-k^{2l-n-\delta}, k^{2l}+k^{2l-n-\delta}]$ there exists a unique eigenvalue of the operator $H_\alpha(t)$. It is given by the series:*

$$\lambda(\alpha,t) = p_j^{2l}(t) + \sum_{r=2}^{\infty} \alpha^r g_r(k,t), \tag{3.2.9}$$

converging absolutely in the disk $|\alpha| \le 1$, where the index j is uniquely determined from the relation $p_j^{2l}(t) \in \varepsilon(k,\delta)$. The spectral projection, corresponding to $\lambda(\alpha,t)$ is given by

$$E(\alpha,t) = E_j + \sum_{r=1}^{\infty} \alpha^r G_r(k,t), \tag{3.2.10}$$

which converges in the class S_1 uniformly with respect to α in the disk $|\alpha| \le 1$.

For coefficients $g_r(k,t)$, $G_r(k,t)$ the following estimates hold:

$$|g_r(k,t)| < k^{2l-n-\delta}(\hat{v}k^{-\gamma_2})^r, \tag{3.2.11}$$

$$\| G_r(k,t) \|_1 < (\hat{v}k^{-\gamma_2})^r, \tag{3.2.12}$$

where $\hat{v} = cvR_0^n$, $v = \max|v_m|$, $c \ne c(k,V)$.

Remark 3.1. Here we discuss only the selfadjoint case, and hence all arguments are carried out only for real α, $-1 \leq \alpha \leq 1$. Actually, the majority of the formulas are preserved also for complex α, although completely new spectral phenomena arise here which we shall discuss in detail elsewhere.

Proof of the theorem. The proof of the theorem just as that of Theorem 1.1 is based on expanding the resolvent $(H(t)-z)^{-1}$ in a perturbation series for z, lying on the contour C_0 about the unperturbed eigenvalue $p_j^{2l}(t)$. Then, integrating the resolvent yields the formulae for the perturbed eigenvalue and its spectral projection. Now we prove that

$$\|A(z,t)\| \leq \hat{v}k^{-\gamma_2}. \tag{3.2.13}$$

Taking into account that $A_{i,i+q}=0$, when $q=0$ or $|q|>R_0$, we obtain:

$$\|A(z,t)\| < cR_0^n \max_{i,q\in Z^n,|q|<R_0,q\neq 0} |A_{i,i+q}|. \tag{3.2.14}$$

It is clear that

$$|A_{i,i+q}(z,t)| < v|(p_i^2(t)-z)(p_{i+q}^2(t)-z)|^{-1/2}.$$

Using inequality (3.2.5) we obtain that

$$|A_{i,i+q}| < 200vk^{-\gamma_2}.$$

From the last relation and estimate (3.2.14), we get (3.2.13). Thus, we have obtained the main estimate needed for expanding the resolvent in a series. Furthermore, arguing as in the proof of Theorem 1.1 we obtain (3.2.9) - (3.2.12). The theorem is proved.

It turns out that stronger estimates are valid for $g_r(k,t)$ and $G_r(k,t)$ when $r<k^{\beta}R_0^{-1}$.

Theorem 3.2 . *Under the conditions of Theorem 3.1 for $g_r(k,t)$ and $G_r(k,t)$ the following estimates hold when $r<M_0$, $M_0=k^{\beta}R_0^{-1}$:*

$$|g_r(k,t)| < vR_0^n r^{n-1}(\hat{v}k^{-\gamma_3})^{r-1}, \tag{3.2.15}$$

$$\|G_r(k,t)\| < (\hat{v}k^{-\gamma_3})^r, \tag{3.2.16}$$

$$\|G_r(k,t)\|_1 < (rR_0)^n(\hat{v}k^{-\gamma_3})^r, \tag{3.2.17}$$

$$|g_2(k,t)| < v^2R_0^{n+2}k^{-2l+2\epsilon}. \tag{3.2.18}$$

$$\gamma_3 = 2l-1-\epsilon,\ \epsilon=\beta(n-1)+\delta.$$

The operator $G_r(k,t)$ can differ from zero only on the finite-dimensional subspace $(\sum_{i\in Z^n,|i-j|<rR_0} E_i)l_2^n$ for any $r\in Z^n$.

Corollary 3.2 . *For the perturbed eigenvalue and its spectral projection the following estimates are valid:*

$$|\lambda(\alpha,t)-p_j^{2l}(t)| \leq c\alpha^2v^2R_0^{n+2}k^{-2l+2\epsilon}, \quad \text{if } l>1 \tag{3.2.19}$$

$$|\lambda(\alpha,t)-p_j^{2l}(t)| \leq c\alpha^2v^2 2(R_0^2+v)R_0^4k^{-2\gamma_3}, \quad \text{if } l\leq 1\ (n=2), \tag{3.2.20}$$

$$\|E(\alpha,t)-E_j\|_1 \leq c\alpha vR_0^{2n}k^{-\gamma_3}. \tag{3.2.21}$$

Proof of the theorem. We represent $G_r(k,t)$ in the form:

$$G_r(k,t)_{i_0 i_r} = \sum_{i_1,\dots,i_{r-1}\in Z^n} I_{i_0 i_1 \dots i_{r-1} i_r}, \tag{3.2.22}$$

where

$$I_{i_0 i_1 \dots i_{r-1} i_r} = \frac{(-1)^{r+1}}{2\pi i} \oint_{C_0} \frac{v_{i_0-i_1}\dots v_{i_{r-1}-i_r}}{(p_{i_0}^{2l}(t)-z)\dots(p_{i_r}^{2l}(t)-z)} dz.$$

The last integral can be nonzero only when the integrand has a pole inside the contour, i.e., when at least one of the indices $i_0, \dots, i_r$ is equal to j. Considering that V is a trigonometric polynomial, we obtain $| i_q - j | \leq rR_0$, $q = 0, \dots r$. From this it follows immediately that $G_r(k,t)_{i_0 i_r}$ can differ from zero only when $| i_0 - j | \leq rR_0$ and $| i_r - j | \leq rR_0$, i.e., the operator $G_r(k,t)$ is equal to zero on the orthogonal complement of $(\sum_{i\in Z^n, |i-j|<rR_0} E_i) l_2^n$. Let $r < k^{\beta} R_0^{-1}$. We write the denominator of the integrand of the integral $I_{i_0 i_1 \dots i_{r-1} i_r}$ in the form of a product of the function $(p_j^{2l}(t)-z)^{m+1}, m \geq 0$ and a function not vanishing inside C_0, i.e.,

$$(p_{i_0}^{2l}(t)-z)\dots(p_{i_r}^{2l}(t)-z) = (p_j^{2l}(t)-z)^{m+1}(p_{q_1}^{2l}(t)-z)\dots(p_{q_{r-m}}^{2l}(t)-z),$$

here $q_1, \dots, q_{r-m} \neq j$. Since $|q_s - j| < rR_0 < k^{\beta}$, $s = 1, \dots, r-m$, from condition 3 of Geometric Lemma we get

$$2|p_{q_s}^{2l}(t) - z| > 2|p_{q_s}^{2l}(t) - p_j^{2l}(t)| - 2k^{2l-n-\delta} > k^{\gamma_3}. \tag{3.2.23}$$

The last inequality implies that in the disk C_0' with center at the point $z = p_j^{2l}$ and radius $k^{\gamma_3}/4$ the integrand has a single pole at the center of the disk, and hence in the integral $I_{i_0 i_1 \dots i_{r-1} i_r}$ the contour of integration C_0 may be replaced by C_0'. It is obvious that $4|p_{q_s}^{2l}(t) - z| > k^{\gamma_3}$, if $z \in C_0'$, $s = 0, \dots, r-m$. Hence,

$$|I_{i_0 i_1 \dots i_{r-1} i_r}| < \tilde{V}_{i_0 i_1} \dots \tilde{V}_{i_{r-1} i_r} (ck^{-\gamma_3})^r,$$

$$|(G_r)_{i_0 i_r}| < (\tilde{V})_{i_0 i_r}^r (ck^{-\gamma_3})^r,$$

where $\tilde{V}$ is the operator in l_2^n given by the matrix $\tilde{V}_{qq'}$:

$$\tilde{V}_{qq'} = \begin{cases} |v_{q-q'}|, & \text{if} |q-j| < rR_0,\ |q'-j| < rR_0 \\ 0 & \text{otherwise.} \end{cases}$$

Using obvious properties of $\tilde{V}$, we obtain (3.2.16) and (3.2.17). Relation (3.2.15) is proved similarly. Inequality (3.2.18) follows from formula (1.0.24) and the obvious estimate $|2p_j^{2l}(t) - p_{j+q}^{2l}(t) - p_{j-q}^{2l}(t)| < c|q|^2 p_j^{2l-2}(t)$. The theorem is proved.

Theorem 3.3 . *Under the conditions of Theorem 3.1 the functions $g_r(k,t)$ and the operator-valued functions $G_r(k,t)$ depend analytically on t in the complex $(k^{-n+1-2\delta})$-neighborhood of each simply connected component of the nonsingular set $\chi_1(k,\beta,\delta)$. They satisfy the estimates:*

$$| T(m)g_r(k,t) | < m! k^{2l-n-\delta}(\hat{v}k^{-\gamma_2})^r k^{(n-1+2\delta)|m|} \tag{3.2.24}$$

$$\| T(m)G_r(k,t) \|_1 < m!(\hat{v}k^{-\gamma_2})^r k^{(n-1+2\delta)|m|} \tag{3.2.25}$$

For $r < M_0 = k^\beta R_0^{-1}$ *there are the stronger estimates:*

$$|T(m)g_r(k,t)| < m!(ck^\epsilon)^{|m|} v R_0^n r^{n-1} (\hat{v}k^{-\gamma_3})^{r-1}, \tag{3.2.26}$$

$$\|T(m)G_r(k,t)\| < m!(ck^\epsilon)^{|m|}(\hat{v}k^{-\gamma_3})^r, \tag{3.2.27}$$

$$\|T(m)G_r(k,t)\|_1 < m!(ck^\epsilon)^{|m|}(rR_0)^n(\hat{v}k^{-\gamma_3})^r, \tag{3.2.28}$$

$$|T(m)g_2(k,t)| < m!(ck^\epsilon)^{|m|} v^2 R_0^{n+2} k^{-2l+2\epsilon}. \tag{3.2.29}$$

The operator $T(m)G_r(k,t)$ *can be nonzero only on the finite-dimensional subspace* $(\sum_{i\in Z^n, |i-j|<rR_0} E_i) l_2^n$ *for any* $r \in Z^n$.

Corollary 3.3 . *For the perturbed eigenvalue and its spectral projection the following estimates are valid:*

$$| T(m)(\lambda(\alpha,t) - p_j^{2l}(t)) | \le cm!\alpha^2 v^2 R_0^{n+2} k^{-2l+2\epsilon+(n-1+2\delta)|m|}, \quad if\, l > 1 \tag{3.2.30}$$

$$| T(m)(\lambda(\alpha,t) - p_j^{2l}(t)) | \le cm!\alpha^2 v^2 (R_0^2+v) R_0^4 k^{-2\gamma_3+(n-1+2\delta)|m|}, \quad if\, l < 1\ (n=2) \tag{3.2.31}$$

$$\|T(m)(E(\alpha,t) - E_j)\|_1 \le cm!\alpha v R_0^{2n} k^{-\gamma_3+(n-1+2\delta)|m|}. \tag{3.2.32}$$

If $m < \gamma_3 R_0^{-1} n^{-1} k^\beta$, *then there are the stronger estimates:*

$$| T(m)(\lambda(\alpha,t) - p_j^{2l}(t)) | \le cm!(ck^\epsilon)^{|m|}\alpha^2 v^2 R_0^{n+2} k^{-2l+2\epsilon}, \quad if\, l > 1 \tag{3.2.33}$$

$$| T(m)(\lambda(\alpha,t) - p_j^{2l}(t)) | \le cm!(ck^\epsilon)^{|m|}\alpha^2 v^2 (R_0^2 + v) R_0^2 k^{-2\gamma_3}, \quad if\, l < 1, (n=2) \tag{3.2.34}$$

$$\|T(m)(E(\alpha,t) - E_j)\|_1 \le cm!(ck^\epsilon)^{|m|}\alpha v R_0^{2n} k^{-\gamma_3}. \tag{3.2.35}$$

The proof of the corollary. Estimates (3.2.30)–(3.2.32) we easily obtain from inequalities (3.2.24), (3.2.25). We verify relations (3.2.33) – (3.2.35) using for the first terms of the series estimates (3.2.26), (3.2.28), and (3.2.29). The first term has the weakest above estimate as compared with all other ones when $m < \gamma_3 R_0^{-1} n^{-1} k^\beta$. from this (3.2.33) – (3.2.35) follow. The corollary is proved.

Proof of Theorem 3.3. Since estimate (3.2.5) is valid in the complex $(2k^{-n+1-2\delta})$-neighborhood of the nonsingular set, it is easy to see that the functions $g_r(k,t)$ and the operator-valued functions $G_r(k,t)$ can be continued from the real $(k^{-n+1-2\delta})$-neighborhood of t to the complex $(k^{-n+1-2\delta})$-neighborhood as analytic functions of n variables and inequalities (3.2.11), (3.2.12) are hereby preserved. Estimating by means of the Cauchy integral the value of the derivative at t in terms of the value of the function itself on the boundary of the $(k^{-n+1-2\delta})$-neighborhood of t (formulae (3.2.11), (3.2.12)) we obtain (3.2.24) and (3.2.25). Considering that the functions $g_r(k,t)$ and the operator-valued functions $G_r(k,t)$ depend analytically on t in the complex $(2k^{-\epsilon})$-neighborhood of each simply connected component of the nonsingular set when $r < M_0(k)$, we obtain (3.2.26)–(3.2.29). The theorem is proved.

3.3 The Case of a Smooth Potential.

We now show that with the help of simple considerations it is possible to extend the proof of Theorems 3.1 and 3.2 to the case of a smooth potential

$$|v_q|_{|q|\to\infty} = O(|q|^{-\zeta_0}), \qquad \zeta_0 > \max\{2n-2l, n\}. \tag{3.3.1}$$

We consider the set

$$\tilde{\chi}_1(k,\delta) = \cap_{p=1}^{p_0}\chi_1(k,p\delta,\delta), \tag{3.3.2}$$

here p_0 and positive δ are chosen so that $(p_0+2)\delta < 1$ and $(p_0+3)\delta > 1$. It is easy to verify that $\tilde{\chi}_1(k,\delta)$ is a set of asymptotically full measure on $S_0(k)$. We consider the norm of operator A:

$$\|A\| \le \sup_{i\in Z^n} \sum_{q\in Z^n} |v_q| \left(|p_i^{2l}(t)-z||p_{i+q}^{2l}(t)-z|\right)^{-1/2} \le \sum_{p=1}^{p_0} \Sigma_p + \Sigma_0,$$

$$\Sigma_p = \sup_{i\in Z^n} \sum_{k^{\delta(p-1)}\le|q|<k^{\delta p}} |v_q| \left(|p_i^{2l}(t)-z||p_{i+q}^{2l}(t)-z|\right)^{-1/2},$$

$$\Sigma_0 = \sup_{i\in Z^n} \sum_{|q|>k^{1-3\delta}} |v_q| \left(|p_i^{2l}(t)-z||p_{i+q}^{2l}(t)-z|\right)^{-1/2}.$$

The sum Σ_p contains in order $ck^{\delta pn}$ terms. From (3.3.1) we obtain

$$\tilde{v}_p \equiv \sup_{|q|\ge k^{\delta(p-1)}} |v_q| < ck^{-\zeta_0(p-1)\delta}.$$

Moreover, since $t \in \chi_1(k,p\delta,\delta)$, it follows that

$$\varepsilon_p \equiv \inf_{i\in Z^n|q|<k^{\delta p},q\ne0} \left(|p_i^{2l}(t)-z||p_{i+q}^{2l}(t)-z|\right) > k^{4l-n-1-\delta p(n-1)-2\delta}.$$

Considering what has been said above, we obtain

$$\Sigma_p < ck^{\delta pn}\tilde{v}_p\varepsilon_p^{-1/2} < ck^{-\gamma_p'},$$

$$2\gamma_p' = (2\zeta_0-3n+1)p\delta + (4l-n-1-2\zeta_0\delta-2\delta), \qquad \gamma_p' \ge \gamma_4,$$
$$2\gamma_4 = \min\{2\zeta_0-4n+4l-2\delta(\zeta_0+1), 4l-n-1-2\delta(\zeta_0+1)\}.$$

It is obvious, that by choosing a sufficiently small δ we can always make γ_4 positive. We estimate Σ_0. Noting, that $\min\{|p_i^{2l}(t)-z|, |p_{i+q}^{2l}(t)-z|\} \ge k^{2l-n-\delta}$, we obtain

$$\Sigma_0 < k^{n-2l+\delta} \sum_{|q|>k^{1-\delta}} |v_q| < ck^{n-2l+\delta}\int_{r>k^{1-\delta}} r^{n-1-\zeta_0}dr < ck^{2n-2l-\zeta_0+\delta} < k^{-\gamma_4}.$$

Thus, $\|A\| < (p_0+1)k^{-\gamma_4}$. This enables us to expand the resolvent in a series converging in $\mathbf{S}_1$ in a manner similar to the way this was done in the proof of

Theorem 2.1. We prove that for sufficiently large k, $k > k_0(V,\beta,\delta)$, for all α, $-1 \le \alpha \le 1$ in the interval $\varepsilon(k,\delta)$ there exists a unique eigenvalue of the operator $H_\alpha(t)$. It is given by the series (3.2.9), converging absolutely in the disk $|\alpha| \le 1$, where the index j is uniquely determined from the relation $p_j^{2l}(t) \in \varepsilon(k)$. The spectral projection corresponding to $\lambda(\alpha,t)$ is determined by the series (3.2.10) which converges in the class $\mathbf{S}_1$ uniformly with respect to α in the disk $|\alpha| \le 1$. The following estimates hold

$$| T(m)g_r(k,t) | < m! k^{2l-n-\delta-\gamma_4 r)+|m|(n-1+\delta)}, \tag{3.3.3}$$

$$\|T(m)G_r(k,t)\| < m! k^{-\gamma_4 r+|m|(n-1+\delta)}. \tag{3.3.4}$$

From the last inequalities the estimates for the perturbations of the eigenvalue and the spectral projection follow:

$$| T(m)(\lambda(\alpha,t) - p_j^{2l}(t)) | < 2m!\alpha^2 k^{2l-n-\delta-2\gamma_4+|m|(n-1+\delta)}, \tag{3.3.5}$$

$$\|T(m)(E(\alpha,t) - E_j)\| < 2m!\alpha k^{-\gamma_4+|m|(n-1+\delta)}, \tag{3.3.6}$$

The more general case of a nonsmooth potential is considered in Sections 3.5-3.7.

3.4 Construction of the Nonsingular Set.

We proceed to the proof of the Geometric Lemma. It contains three steps:

1) By formally describing the nonsingular set $\chi_1(k,\beta,\delta)$, we prove that for t in $\chi_1(k,\beta,\delta)$ conditions 1-4 of Lemma 3.1 are satisfied.

2) We prove that the nonsingular set $\chi_1(k,\beta,\delta)$ is nonempty; moreover, it is a subset of an asymptotically full measure on $S_0(k)$.

3) We prove that for all t in the $(k^{-n+1-2\delta})$-neighborhood of $\chi_1(k,\beta,\delta)$ in C^n there exists a unique $j \in Z^n$ such that $|p_j^2(t) - k^2| < 5k^{-n+2-2\delta}$; moreover, conditions 2-4 of Lemma 3.1 are satisfied.

Lemma 3.2 contains the proof of step 1. The remainder of the proof of step 2 is contained in Lemma 3.6. Lemmas 3.3–3.6 are of auxiliary character. Step 3 essentially asserts the stability of all the estimates encountered in the definition of the nonsingular set $\chi_1(k,\beta,\delta)$ relative to a perturbation of order $2k^{-n+1-2\delta}$. Lemma 3.7 corresponds to this.

We introduce the following notations:

1. K_r is the ball of radius r in R^n.

2. K_{rR} is the spherical shell between S_r and S_R.

3. $\Pi_q(k,\xi,\varepsilon)$ is the intersection of the spherical shell $K_{k-2k^{-\varepsilon},k+2k^{-\varepsilon}}$ with the plane layer $\Pi_q(4k^{-\xi})$:

$$\Pi_q(k,\xi,\varepsilon) = K_{k-2k^{-\varepsilon},k+2k^{-\varepsilon}} \cap \Pi_q(4k^{-\xi}).$$

4. $\Pi(k,\xi,\varepsilon,\beta)$ is the union of $\Pi_q(k,\xi,\varepsilon)$ over $q: |q| < k^\beta$:

$$\Pi(k,\xi,\varepsilon,\beta) = \cup_{|q|<k^\beta}\Pi_q(k,\xi,\varepsilon) = K_{k-2k^{-\varepsilon},k+2k^{-\varepsilon}} \cap (\cup_{|q|<k^\beta}\Pi_q(4k^{-\xi})).$$

We consider the set: $\chi_1(k,\beta,\delta) = S_0(k) \setminus \mu_1(k,\beta,\delta)$, where

$$\mu_1(k,\beta,\delta) = \chi_0(k,\delta) \cup (\cup_{m=0}^{m=M} T_m), \tag{3.4.1}$$

$$T_m = \mathcal{K}\Pi(k,\zeta_m,\varepsilon_m,\beta), \tag{3.4.2}$$

$$\zeta_m = -1 + (m+1)\delta + \beta(n-1), \quad \varepsilon_m = n - 1 - m\delta, \quad M = [(n-1)(1-\beta)/(2\delta)].$$

Lemma 3.2 . *If $t \in \chi_1(k,\beta,\delta)$, then*

1. *There exists a unique $j \in Z^n$ such that $p_j(t) = k$.*

2.
$$\min_{i\neq j} | p_j^2(t) - p_i^2(t) |> 4k^{-n+2-\delta} \tag{3.4.3}$$

3.
$$\min_{|q|<k^\beta, q\neq 0} |p_j^2(t) - p_{j+q}^2(t)| > 4k^{1-\beta(n-1)-\delta}, \tag{3.4.4}$$

4. *For all $i \neq j$ such that $|p_i^2(t) - k^2|^2 > k^{-\zeta}$, $\zeta = n - 3 + \beta(n-1) + 2\delta$, the inequality*
$$\min_{|q|<k^\beta, q\neq 0} |p_i^2(t) - p_{i+q}^2(t)| > k^{-\zeta}|p_i^2(t) - k^2|^{-1}, \tag{3.4.5}$$
holds.

Remark 3.2. Conditions 2-4 are somewhat stronger than conditions 2-4 of Lemma 3.1. We need this so that the inequalities of Lemma 3.1 will hold not only on the nonsingular set but also in a small neighborhood of it.

Proof. Since $t \in S_0(k)$, there exists at least one j such that $p_j(t) = k$. We shall prove that it is unique; moreover, the condition 2 is satisfied. We suppose that this is not case, i.e., there exists $i \neq j$ such that

$$\min_{i\neq j} | p_i^2(t) - p_j^2(t) |< 4k^{-n+2-\delta} \tag{3.4.6}$$

Let $x \equiv \mathbf{p}_j(t)$ and $m = j - i$. Then $x \in S_k$, $\mathcal{K}x = t$ and the last estimate implies, that

$$|| x |^2 - | x - \mathbf{p}_m(t) |^2|< 4k^{-n+2-\delta},$$

i.e.,

$$x \in S_m(k, n-2+\delta) \subset S(k, n-2+\delta).$$

Hence $t \in \mathcal{K}S(k,n-2+\delta) \subset S_0(k) \setminus \chi_1(k,\delta)$. This contradiction proves that conditions 1,2 are satisfied.

We prove now that conditions 3 and 4 are satisfied. Thus, suppose i is such that $|p_i^2(t) - k^2|^2 < 2k^{-\zeta}$ (in particular, i can be equal to j). This is equivalent to

$$\mathbf{p}_i(t) \in K_{k-r,k+r} \subset \Upsilon_0 \cap \ldots \cap \Upsilon_M; \tag{3.4.7}$$

here $r = 2k^{-(n-1)(1+\beta)/2-\delta}$, and

$$\Upsilon_0 = K_{k-k^{-n+1},k+k^{-n+1}},$$

$$\Upsilon_m = K_{k-R_m,k-r_m} \cup K_{k+r_m,k+R_m}, \tag{3.4.8}$$

$$2r_m = k^{-n+1+(m-1)\delta}, \qquad 2R_m = k^{-n+1+m\delta}, \qquad m = 1, \ldots, M.$$

Suppose $\mathbf{p}_i(t) \in \Upsilon_m$, $m = m(i)$. We shall prove that in this case

$$\min_{|q|<k^\beta, q\neq 0} |p_i^2(t) - p_{i+q}^2(t)| > 4k^{1-\delta(m+1)-\beta(n-1)}. \tag{3.4.9}$$

Indeed, suppose this is not the case. We introduce the notation $x = \mathbf{p}_i(t)$. Then $\mathcal{K}x = t$, $x \in K_{k-k^{-\varepsilon},k+k^{-\varepsilon}}$, $\varepsilon = n - 1 - m\delta$, and

$$\min_{|q|<k^\beta, q\neq 0} \left||x|^2 - |x + \mathbf{p}_q(0)|\right| < 4k^{1-\delta(m+1)-\beta(n-1)}.$$

This implies that $x \in \Pi(k, -1 + (m+1)\delta + \beta(n-1), n - 1 - m\delta, \beta)$, i.e., $t \in T_m \subset \chi_1(k, \beta, \delta)$. This contradiction proves that inequality (3.4.9) is satisfied in the case where $\mathbf{p}_i(t) \in \Upsilon_m$.

Suppose $i = j$. Then m=0, i.e., condition 3 is satisfied. If $i \neq j$, then from the definition of Υ_m for $m \neq 0$ or from condition 2 we find that

$$2|p_i^2(t) - k^2| > k^{-n+2+(m-1)\delta}, \tag{3.4.10}$$

if $\mathbf{p}_i(t) \in \Upsilon_m$, $m = 0, \ldots, M$. From (3.4.9) and (3.4.10) it follows that

$$|p_i^2(t) - p_{i+q}^2(t)| > 2k^{-\zeta}|p_i^2(t) - k^2|^{-1}$$

for all $q \neq 0, |q| < k^\beta$, i.e., condition 4 is satisfied. The lemma is proved.

After we have explicitly described the nonsingular set $\chi_1(k, \beta, \delta)$ we shall prove that it has an asymptotically full measure on $S_0(k)$. The main part of the proof is contained in Lemma 3.6; Lemmas 3.3–3.5 are of auxiliary character.

Lemma 3.3 . *Suppose $\xi > -1, \varepsilon \geq -1$. Then*

$$s(S(k, \xi, \beta)) < ck^{(n-1)\beta-\mu}, \tag{3.4.11}$$

$$V(\Pi(k, \xi, \varepsilon, \beta)) < ck^{\beta(n-1)-\mu-\varepsilon}, \tag{3.4.12}$$

$$\mu = \xi - n + 2.$$

Proof. The proof of estimate (3.4.11) is a simplified proof of Lemma 2.4. Indeed, it is clear that

$$s(S(k,\xi,\beta)) < \sum_{|m|<k^{\beta}} s(S_m(k,\xi)).$$

Now considering in the same way as in Lemma 2.4 for Σ_1, we get inequality (3.4.11). The relation (3.4.12) is obtained by integrating of the estimate (3.4.11) with respect to radius of the sphere. Indeed, let us introduce polar coordinates R, ϑ, where ϑ is the vector defining a point on the unit sphere in R^n. It is obvious that

$$V(\Pi(k,\xi,\varepsilon,\beta)) = \int_0^{\infty} \int_{\vartheta \in S_1^{n-1}} \chi(\Pi(k,\xi,\varepsilon,\beta), R\vartheta) dS_R(R,\vartheta). \tag{3.4.13}$$

It follows from the definition of $\Pi(k,\xi,\varepsilon,\beta)$ that

$$\chi(\Pi(k,\xi,\varepsilon,\beta), R\vartheta) =$$

$$= \chi\left(K_{k-2k^{-\varepsilon},k+2k^{-\varepsilon}}, R\vartheta\right) \chi\left(\cup_{|m|<k^{\beta}} \Pi_m(4k^{-\xi}), R\vartheta\right).$$

Substituting this into (3.4.13), we obtain

$$V(\Pi(k,\xi,\varepsilon,\beta)) = \int_{k-2k^{-\varepsilon}}^{k+2k^{-\varepsilon}} \int_{\vartheta \in S_1^{n-1}} \chi\left(\cup_{|m|<k^{\beta}} \Pi_m(4k^{-\xi}), R\vartheta\right) dS_R(R,\vartheta). \tag{3.4.14}$$

We note that

$$\int_{\vartheta \in S_1^{n-1}} \chi\left(\cup_{|m|<k^{\beta}} \Pi_m(4k^{-\xi}), R\vartheta\right) dS_R(R,\vartheta) = s(S(k,\xi,\beta)) < ck^{\beta(n-1)-\mu}.$$

Using the last estimate in relation (3.4.14) we obtain (3.4.12).

The lemma is proved.

Let Ω be a set in the cube K with the opposite faces identified. We denote an ε-neighborhood of it in K by $\Gamma(\Omega,\varepsilon)$. If $\Omega_0 \subset R^n$, then we denote an ε-neighborhood of it in R^n by $\Gamma'(\Omega_0,\varepsilon)$.

Lemma 3.4 . *Suppose $\varepsilon \geq -1$, $\xi \geq -1$, $r > \max\{\varepsilon, \xi+1\}$. Then for sufficiently large k the volume of the (k^{-r})- neighborhood of $\mathcal{K}\Pi(k,\xi,\varepsilon,\beta)$ can be bounded above by the quantity $ck^{\beta(n-1)-\mu-\varepsilon}$:*

$$V(\Gamma(\mathcal{K}\Pi(k,\xi,\varepsilon,\beta), k^{-r})) < ck^{\beta(n-1)-\mu-\varepsilon}. \tag{3.4.15}$$

Proof. The proof is based on the fact that the value k^{-r} is small with respect to sizes of the set $\mathcal{K}\Pi(k,\xi,\varepsilon,\delta)$. Therefore, the volume of the neighborhood can be bounded above by the volume of the set $\mathcal{K}\Pi(k,\xi,\varepsilon,\delta)$ itself. Indeed,

$$\Gamma'(K_{k-k^{-\varepsilon},k+k^{-\varepsilon}}, k^{-r}) \subset K_{k-2k^{-\varepsilon},k+2k^{-\varepsilon}}, \tag{3.4.16}$$

$$\Gamma'(\Pi_m(4k^{-\xi}), k^{-r}) \subset \Pi_m(4k^{-\xi} + 2k^{-r+1}) \subset \Pi_m(8k^{-\xi}), \qquad |m| < k^{\beta}. \tag{3.4.17}$$

Further, considering the identification of the faces of cube K, it is not hard to see that

$$\Gamma(\mathcal{K}\Pi(k,\xi,\varepsilon,\beta), k^{-r}) \subset \mathcal{K}\Gamma'(\Pi(k,\xi,\varepsilon,\beta), k^{-r}). \tag{3.4.18}$$

The definition of $\Pi(k,\xi,\varepsilon,\beta)$ and relations (3.4.16)–(3.4.18) together give

$$\Gamma(\mathcal{K}\Pi(k,\xi,\varepsilon,\beta), k^{-r}) \subset \mathcal{K}\left(K_{k-2k^{-\epsilon},k+2k^{-\epsilon}} \cap \left(\cup_{|m|<k^{\beta}} \Pi_m(8k^{-\xi})\right)\right).$$

Since on the right we have a set similar to $\mathcal{K}\Pi(k,\xi,\varepsilon,\beta)$, using Lemma 3.3, we obtain estimate (3.4.15). The lemma is proved.

Lemma 3.5 . *Suppose S is a subset of $S_0(k)$ such that $s(S) > ck^{-n-1-\nu}$, $0 < \nu < n-1$, $c \neq c(k)$. Then for all $r > n-1+2\nu$ and sufficiently large k, $k > k_0(S)$, the volume of the (k^{-r})-neighborhood of S can be bounded below by the quantity $ck^{n-1-\nu-r}$:*

$$V(\Gamma'(S, k^{-r})) > c_1 k^{n-1-\nu-r}, \quad c_1 \neq c_1(k). \tag{3.4.19}$$

Proof. Suppose $n - 2 + 2\nu < \xi < r - 1$. It is obvious that

$$\Gamma(S, k^{-r}) \supset \Gamma(S \setminus S(k,\xi), k^{-r}).$$

The surface $S_0(k)$ consists of "tiles". We denote the intersection of the i-th "tile" with S by S_i. Thus,

$$\Gamma(S, k^{-r}) \supset \Gamma(S \setminus S(k,\xi), k^{-r}) = \cup_i \Gamma(S_i \setminus S(k,\xi), k^{-r}). \tag{3.4.20}$$

It is obvious that

$$V\left(\cup_i \Gamma(S_i \setminus S(k,\xi), k^{-r})\right) \leq \sum_i V\left(\Gamma(S_i \setminus S(k,\xi), k^{-r})\right). \tag{3.4.21}$$

We shall prove that equality actually holds here. For this we must show that

$$\Gamma(S_i \setminus S(k,\xi), k^{-r}) \cap \Gamma(S_j \setminus S(k,\xi), k^{-r}) = \emptyset, \tag{3.4.22}$$

if $i \neq j$. We suppose that this is not so. Then there exist t', t'' such that

$$t' \in S_i \setminus S(k,\xi), \qquad t'' \in S_j \setminus S(k,\xi), \qquad |t' - t''| < 2k^{-r}.$$

We note that in this case

$$|p_i^2(t') - p_j^2(t')| = |p_j^2(t'') - p_j^2(t')| < ck^{-r+1} < 4k^{-\xi}.$$

We introduce the notation $x = \mathbf{p}_j(t')$. Then $x \in S_k$, $\mathcal{K}x = t'$, and

$$||x|^2 - |x - p_q(0)|^2| < 4k^{-\xi}, \qquad q = j - i.$$

This implies that $x \in S_q(k,\xi) \subset S(k,\xi)$. Hence, $t' \in \mathcal{K}S(k,\xi)$. This contradiction proves (3.4.22), and (3.4.21) actually contains an equality sign:

$$V\left(\cup_i \Gamma(S_i \setminus S(k,\xi), k^{-r})\right) = \sum_i V\left(\Gamma(S_i \setminus S(k,\xi), k^{-r})\right). \tag{3.4.23}$$

From (3.4.20) and (3.4.23) it is easy to see that

$$V\left(\Gamma(S_i \setminus S(k,\xi), k^{-r})\right) \geq ck^{-r} s(S_i \setminus S(k,\xi))$$

Therefore,

$$V(\Gamma(S,k^{-r})) > ck^{-r} \sum_i s(S_i \setminus S(k,\xi)) = ck^{-r} s(S(k) \setminus KS(k,\xi)) \geq$$

$$ck^{-r}\left(s(S(k) - s(S(k,\xi))\right).$$

Since $\xi > n - 2 + 2\nu$, by Lemma 2.4 $s(S(k,\xi)) = O(k^{n-1-2\nu}) = o(S(k))$. Formula (3.4.19) follows directly from this. The lemma is proved.

Lemma 3.6 . *The nonsingular set $\chi_1(k,\beta,\delta)$ is a set of an asymptotically full measure on $S_0(k)$. Moreover,*

$$\frac{s(S_0(k) \setminus \chi_1(k,\beta,\delta))}{s(S_0(k))} =_{k\to\infty} O(k^{-\delta/8}). \tag{3.4.24}$$

Proof. Since $\chi_1(k,\beta,\delta) = S_0(k) \setminus \mu_1(k,\beta,\delta)$ (see (3.4.1)), it suffices to prove that

$$s(S_0(k) \cap \mu_1(k,\beta,\delta)) = o(k^{n-1-\delta/8}). \tag{3.4.25}$$

From the definition of $\mu_1(k,\beta,\delta)$ we obtain

$$s(S_0(k) \cap \mu_1(k,\beta,\delta)) \leq s(S(k, n-2+\delta)) + \sum_{m=0}^{M} s(S_0(k) \cap T_m). \tag{3.4.26}$$

By Lemma 2.4

$$s(S(k, n-2+\delta)) < ck^{n-1-\delta/2}. \tag{3.4.27}$$

It remains to show that $s(S_0(k) \cap T_m) = O(k^{n-1-\delta/8})$. We suppose that this is not so, i.e., the estimate $s(S_0(k) \cap T_m) > c_0 k^{n-1-\delta/8}$ holds with some positive c_0 not depending on k. We consider the (k^{-r})-neighborhood of this set: $\Gamma(S_0(k) \cap T_m, k^{-r})$ for $r = n - 1 + \delta/2$. The previous lemma makes it possible to obtain a lower bound for its volume:

$$V(\Gamma(S_0(k) \cap T_m, k^{-r})) > ck^{-5\delta/8}. \tag{3.4.28}$$

On the other hand, by the definition of T_m (see (3.4.2))

$$T_m = \mathcal{K}\Pi(k, \zeta_m, \varepsilon_m, \beta),$$

$$\zeta_m = -1 + (m+1)\delta + \beta(n-1), \quad \varepsilon_m = n - 1 - m\delta, \quad M = [(n-1)(1-\beta)/(2\delta)].$$

We shall prove that we are in the conditions of Lemma 3.4, i.e., $r > \varepsilon_m$ and $r > \zeta_m + 1$. The first relation is obvious. Noting that $m \le M = [(n-1)(1-\beta)/2\delta]$, we obtain

$$\zeta_m + 1 = (m+1)\delta + \beta(n-1) \le [(n-1)(1-\beta)/2\delta]\delta + \delta + \beta(n-1) \le$$

$$\le n - 1 + \delta - (1-\beta)(n-1)/2 < n - 1.$$

Lemma 3.4 gives an upper bound for the volume of the (k^{-r})-neighborhood of T_m:

$$V(\Gamma(S_0(k) \cap T_m, k^{-r})) \le V(\Gamma(T_m, k^{-r})) \le ck^{(n-1)\beta - (\zeta_m - n + 2 + \varepsilon_m)} = ck^{-\delta}. \tag{3.4.29}$$

Inequalities (3.4.28), (3.4.29) are contradictory; therefore,

$$s(S_0(k) \cap T_m) = O(k^{n-1-\delta/8}). \tag{3.4.30}$$

Relations (3.4.26), (3.4.27), (3.4.30) together give (3.4.25). The lemma is proved.

Lemma 3.7 . *If t in the $(2k^{-n+1-2\delta})$-neighborhood of the nonsingular set $\chi_1(k,\beta,\delta)$ in C^n there exists a unique $j \in Z^n$, such that $|p_j^2 - k^2| < 5k^{-n+2-2\delta}$ and conditions 2-4 of Geometric Lemma 3.1 hold.*

Proof. We must essentially prove the stability of all estimates encountered in the determination of the nonsingular set $\chi_1(k,\beta,\delta)$ with respect a perturbation of order $2k^{-n+1-2\delta}$. Thus, suppose $t_0 \in \chi_1(k,\beta,\delta)$ and $|t - t_0| < 2k^{-n+1-2\delta}$. By Lemma 3.2 there exists a unique j such that $p_j^2(t_0) = k^2$ and $\min_{i \ne j} |p_i^2(t_0) - p_j^2(t_0)| > 4k^{-n+1-\delta}$. We consider $p_j^2(t)$. It is easy to show that $|p_j^2(t) - p_j^2(t_0)| < k^{-n+1-\delta}$, i.e., $|p_j^2(t) - k^2| < k^{-n+1-\delta}$. Further, it is obvious that

$$\min_{i \ne j} |p_i^2(t) - p_j^2(t)| >$$

$$\min_{i \ne j} \left(|p_i^2(t_0) - p_j^2(t_0)| - |p_i^2(t_0) - p_i^2(t_0)| - |p_j^2(t_0) - p_j^2(t)|\right) > 2k^{-n+1-\delta}.$$

This proves that conditions 1 and 2 are satisfied. Moreover,

$$|p_j^2(t) - p_{j+q}^2(t)| > |p_j^2(t_0) - p_{j+q}^2(t_0)| - |p_j^2(t_0) - p_j^2(t)| -$$

$$|p_{j+q}^2(t_0) - p_{j+q}^2(t)| > k^{1-\beta(n-1)-\delta}.$$

This shows that condition 3 is satisfied. We consider the last condition. Since $t_0 \in \chi_1(k,\beta,\delta)$, it follows that

$$|p_j^2(t_0) - p_{j+q}^2(t_0)| > 2k^{-\zeta} |p_i^2(t_0) - k^2|^{-1}, \tag{3.4.31}$$

if $|q| < k^\beta, q \ne 0$ and $|p_i^2(t_0) - k^2|^2 < 2k^{-\zeta}$. Just as in the verification of conditions 2 and 3, it is possible to show that

$$|p_i^2(t_0) - p_{i+q}^2(t_0)| < \sqrt{2}|p_i^2(t_0) - p_{i+q}^2(t_0)|, \tag{3.4.32}$$

$$2|p_i^2(t_0) - k^2| > \sqrt{2}|p_i^2(t) - k^2| > |p_i^2(t_0) - k^2|. \tag{3.4.33}$$

Relations (3.4.31)- (3.4.33) together prove that condition 4 of Lemma 3.1 is satisfied. The lemma is proved.

The combination of Lemmas 3.2, 3.6 and 3.7 just proved gives the assertion of the geometric lemma.

3.5 The Main Result of the Analytic Perturbation Theory in the Case of a Nonsmooth Potential.

Now we consider the case when all smoothness requirements for $V(x)$ are contained in the condition:

$$\sum_{m \in Z^n} | v_m |^2 | m |^{-4l+n+\beta} < \infty, \tag{3.5.1}$$

for some β satisfying the inequality:

$$\begin{array}{ll} \beta > 0 & \text{if } 2l \leq n, \ n \neq 2; \\ \beta \geq 2l - n & \text{if } 2l > n, \ n \neq 2 \text{ or } 2l > 3, \ n = 2; \\ \beta > 1 & \text{if } 2l < 3, \ n = 2. \end{array} \tag{3.5.2}$$

The potential does not need to be smooth to satisfy this condition. For example, in the case $n = 3$, a function which behaves in a neighborhood of some point x_0 as $| x - x_0 |^{-b}$, $b < 2l$, in particular, a Coulomb potential satisfies (3.5.1).

We shall prove that the formulae of the analytic perturbation theory are valid in the case of a nonsmooth potential on a rich set of quasimomenta. We construct a nonsingular set $\chi_2(k, V, \delta)$, and prove the convergence of the perturbation series. We show that the theorems similar to Theorems 3.1, 3.2 hold. However, in this case there is a new value of γ_0 (we denote it by γ_5) in the power estimates for $g_r(k,t), G_r(k,t)$. The value of γ_5 is given by the following formulae:

$$\gamma_5 = \tilde{\gamma}_5 - 8l\delta,$$

$$2\tilde{\gamma}_5 = \min\{4l - n - 1, \beta\}, \quad \text{if } n \geq 2,$$

$$2\tilde{\gamma}_5 = \min\{4l - n - 1, \beta - 1\}, \quad \text{if } n = 2.$$

Now, we represent the main result of the perturbation theory for a nonsmooth potential. The following theorem is the union of the theorems similar to Theorems 3.1, 3.2.

Theorem 3.4 . *Suppose t belongs to the $(k^{-n+1-2\delta})$-neighborhood in K of the nonsingular set $\chi_2(k, V, \delta)$, $0 < 8l\delta < \tilde{\gamma}_5$. Then for sufficiently large k, $k > k_0(V, \delta)$ and for all α, $-1 \leq \alpha \leq 1$ there exists a unique eigenvalue of the operator $H_\alpha(t)$ in the interval $\varepsilon(k, \delta) \equiv [k^{2l} - k^{2l-n-\delta}, k^{2l} + k^{2l-n-\delta}]$. It is given by the series:*

$$\lambda(\alpha,t) = p_j^{2l}(t) + \sum_{r=2}^{\infty} \alpha^r g_r(k,t), \tag{3.5.3}$$

converging absolutely in the disk $|\alpha| \leq 1$, *where the index* j *is uniquely determined from the relation* $p_j^{2l}(t) \in \varepsilon(k,\delta)$. *The spectral projection corresponding to* $\lambda(\alpha,t)$ *is given by the series:*

$$E(\alpha,t) = E_j + \sum_{r=1}^{\infty} \alpha^r G_r(k,t), \tag{3.5.4}$$

which converges in the class $\mathbf{S}_1$ *uniformly with respect to* α *in the disk* $|\alpha| \leq 1$. *The series (3.5.3), (3.5.4) can be differentiated termwise any number of times with respect to* t, *and retain their asymptotic characters.*

For the functions $g_r(k,t)$ *and the operator-valued functions* $G_r(k,t)$, *the following estimates hold:*

$$|T(m)g_r(k,t)| < m! k^{2l-n-\delta-\gamma_s r} k^{(n-1+2\delta)|m|}, \tag{3.5.5}$$

$$\| T(m)G_r(k,t) \|_1 < m! k^{-\gamma_s r} k^{(n-1+2\delta)|m|}. \tag{3.5.6}$$

Corollary 3.4 . *For the perturbed eigenvalue and its spectral projection the following estimates are valid:*

$$|T(m)(\lambda(\alpha,t) - p_j^{2l}(t))| \leq cm!\alpha^2 k^{2l-n-\delta-2\gamma_s+(n-1+2\delta)|m|}, \tag{3.5.7}$$

$$\|T(m)(E(\alpha,t) - E_j)\|_1 \leq cm!\alpha k^{-\gamma_s+(n-1+2\delta)|m|}. \tag{3.5.8}$$

Remark 3.3. Here we discuss only the selfadjoint case, and hence all arguments are carried out only for real $\alpha, -1 \leq \alpha \leq 1$. Actually, the majority of the formulas are preserved also for complex α, although completely new spectral phenomena arise here, which we shall discuss in detail elsewhere.

3.6 Construction of the Nonsingular Set for a Nonsmooth Potential.

This section describes the nonsingular set $\chi_2(k,V,\delta)$. The main result is contained in Lemma 3.10. However, first we prove a several auxiliary assertions.

Let Ω be a bounded, closed domain in R^n. We introduce an integer-valued function $N(\Omega,t)$ which is the number of points $\mathbf{p}_j(t)$, $j \in Z^n$, belonging to Ω. We call $N(\Omega,t)$ the number of states.

Proposition 3.1 . *The following equality holds for the function* $N(\Omega,t)$:

$$V(\Omega) = \int_K N(\Omega,t)dt. \tag{3.6.1}$$

Proof. It is clear that

$$V(\Omega) = \int_{R^n} \chi(\Omega, x)dx.$$

Any point x can be uniquely represented in the form $x = \mathbf{p}_j(t)$, $j \in Z^n$, $t \in K$. Therefore,

$$V(\Omega) = \sum_{j \in Z^n} \int_K \chi(\Omega, \mathbf{p}_j(t))dt.$$

Note that

$$N(\Omega, t) = \sum_{j \in Z^n} \chi(\Omega, \mathbf{p}_j(t))dt,$$

where the series contains only a finite number of nonzero terms. Exchanging summation and integration, we obtain (3.6.1). The proposition is proved.

We denote by $P(\Omega, \mu)$, the set:

$$P(\Omega, \mu) = \{t : t \in K,\ N(\Omega, t) \geq \mu V(\Omega)\}, \quad \mu > 1.$$

Proposition 3.2 . *The set $P(\Omega, \mu)$ is measurable. Its volume does not exceed μ^{-1}.*

Proof. The measurability of $P(\Omega, \mu)$ directly follows from those of function $N(\Omega, t)$. We obtain the estimate for the volume from relation (3.6.1). Indeed,

$$V(\Omega) \geq \int_{P(\Omega,\mu)} N(\Omega, t)dt \geq \mu V(\Omega) V(P(\Omega, \mu)).$$

Now it is clear that $V(P(\Omega, \mu)) < \mu^{-1}$. The proposition is proved.

Let $\Omega = \{\Omega_q\}$ be a finite number of closed bounded domains ($q \in Z^n, q \in \Gamma_1 \subset Z^n$). We define function $M(\Omega, t)$ by the formula:

$$M(\Omega, t) = \sum_{q \in \Gamma_1} |v_q|^2 N(\Omega_q, t). \tag{3.6.2}$$

We denote by $R(\Omega, \mu)$ the set:

$$R(\Omega, \mu) = \{t : t \in K,\ M(\Omega, t) \geq \mu \sum_q |v_q|^2 V(\Omega_q)\}. \tag{3.6.3}$$

Proposition 3.3 . *The following equality holds for the function $M(\Omega, t)$:*

$$\int_K M(\Omega, t)dt = \sum_q |v_q|^2 V(\Omega_q). \tag{3.6.4}$$

The set $R(\Omega, \mu)$ is measurable and its volume does not exceed μ^{-1}.

Proof. Formula (3.6.4) follows from relations (3.6.2), (3.6.1). The measurability of $R(\Omega, \mu)$ follows from those of $N(\Omega, t)$, $M(\Omega, t)$. Its volume can be easily estimated by using the relation (3.6.4). The proposition is proved.

Proposition 3.4 . *If* $|t - t_0| < \xi$, $\xi > 0$, *then*

$$N(\Omega, t_0) \leq N(\Omega_\xi, t), \tag{3.6.5}$$

$$M(\Omega, t_0) \leq M(\Omega_\xi, t), \tag{3.6.6}$$

where Ω_ξ *is the* ξ*-neighborhood of* Ω.

Proof. If $\mathbf{p}_j(t_0) \in \Omega$, then $\mathbf{p}_j(t) \in \Omega_\xi$. From this we immediately obtain (3.6.5), and from (3.6.5), relation (3.6.6) follows. The proposition is proved.

We denote by $\Omega(k, R)$ the spherical shell in R^n:

$$\Omega(k, R) = \{x : ||x| - k| < k^{-n+1} R\}.$$

Let us consider the set:

$$\Upsilon_1(k, k^\epsilon, k^\delta) = S_0(k) \setminus P(\Omega(k, k^\epsilon + 1), k^\delta),$$

i.e, $\Upsilon_1(k, k^\epsilon, k^\delta)$ is such a subset of $S_0(k)$ that

$$N(\Omega(k, k^\epsilon + 1), t) < k^\delta < V(\Omega(k, k^\epsilon + 1)).$$

Evaluating the volume of the spherical shell we obtain that, when $t \in \Upsilon_1(k, k^\epsilon, k^\delta)$,

$$N(\Omega(k, k^\epsilon + 1), t) < 3n\omega_n k^{n-1+\epsilon+\delta}, \tag{3.6.7}$$

with ω_n being the volume of the unit ball.

Lemma 3.8 . *If* t *is in the* (k^{-n+1})*-neighborhood of the set* $\Upsilon_1(k, k^\epsilon, k^\delta)$, *then*

$$N(\Omega(k, k^\epsilon), t) < 3\omega_n k^{\epsilon+\delta}. \tag{3.6.8}$$

The set $\Upsilon_1(k, k^\epsilon, k^\delta)$ *has an asymptotically full measure on* $S_0(k)$. *Moreover,*

$$s(S_0(k) \setminus \Upsilon_1(k, k^\epsilon, k^\delta))/s(S_0(k)) < ck^{-\delta/4}. \tag{3.6.9}$$

Proof. We denote the (k^{-n+1})-neighborhood of $\Upsilon_1(k, k^\epsilon, k^\delta)$ by Q. Suppose $t \in Q$. Then, there exists $t_0 \in \Upsilon_1(k, k^\epsilon, k^\delta)$, such that $|t - t_0| \leq k^{-n+1}$. According to Proposition 3.1,

$$N(\Omega(k, k^\epsilon), t) \leq N(\Omega(k, k^\epsilon + 1), t_0).$$

Using (3.6.7), we obtain (3.6.8).

It remains to verify inequality (3.6.9). Suppose (3.6.9) does not hold. Then,

$$s(S_0(k) \cap P(\Omega(k, k^\epsilon + 1), k^\delta)/s(S_0(k)) > g(k)k^{-\delta/4}, \tag{3.6.10}$$

$$g(k) \to_{k\to\infty} \infty.$$

We denote by $\tilde{Q}$ the $(k^{-n+1-\delta/2})$-neighborhood of $S_0(k) \cap P(\Omega(k, k^\varepsilon + 1), k^\delta)$, and by $\tilde{Q}_0$ the $(k^{-n+1-\delta/2})$-neighborhood of $P(\Omega(k, k^\varepsilon + 1), k^\delta)$. It is clear that $\tilde{Q} \subset \tilde{Q}_0$. Taking into account relation (3.6.10) and considering in the same way as in prooving Lemma 3.5, we obtain that

$$V(\tilde{R}_0) \geq V(\tilde{R}) > c_1 k^{-3\delta/4}, \qquad c_1 \neq c_1(k). \tag{3.6.11}$$

Let us prove the opposite inequality. Using Proposition 3.4 we easily show that $\tilde{Q}_0 \subset P(\Omega(k, k^\varepsilon + 2), k^\delta)$. Considering the obvious estimate $V(\Omega(k, k^\varepsilon + 1)) \approx 2n\omega_n k^\varepsilon$, and Proposition 3.2, we obtain $V(\tilde{R}_0) < k^{-\delta}$, $c \neq c(k)$. This inequality contradicts estimate (3.6.11). The lemma is proved.

Let ε, ε'; $\varepsilon < \varepsilon'$, be positive values, and $\Omega_q(k, \varepsilon, \varepsilon')$ be the intersection of the spherical shells K_1 and K_2:

$$K_1 = \{x : ||x| - k| < k^{-n+1+\varepsilon}\},$$

$$K_2 = \{x : ||x + \mathbf{p}_q(0)| - k| < k^{-n+1+\varepsilon'}\}.$$

Let us consider

$$\Omega(k, \varepsilon, \varepsilon') = \{\Omega_q(k, \varepsilon, \varepsilon')\}_{q \in \Gamma_1},$$

where $\Gamma_1 = \{q : |q| < k^{1+\delta}\}$, $0 < \delta < 1$. Let $M(\Omega_q(k, \varepsilon, \varepsilon'), t)$ be given by (3.6.2).

We consider the set

$$\Upsilon_2(k, k^\varepsilon, k^{\varepsilon'} k^\delta) = S_0(k) \setminus R(\Omega(k, k^\varepsilon + 1, k^{\varepsilon'} + 1), k^\delta),$$

where R is given by (3.6.3). Taking into account that $V(\Omega_q(k, k^\varepsilon + 1, k^{\varepsilon'} + 1), k^\delta) \approx V(\Omega_q(k, k^\varepsilon, k^{\varepsilon'}), k^\delta)$, we obtain that

$$M(\Omega(k, k^\varepsilon + 1), k^{\varepsilon'} + 1, t_0) < 2k^\delta \sum_{q \in \Gamma_1(k)} |v_q|^2 V(\Omega_q(k, k^\varepsilon, k^{\varepsilon'})), \tag{3.6.12}$$

when $t_0 \in \Upsilon_2(k, k^\varepsilon, k^{\varepsilon'}, k^\delta)$.

Lemma 3.9 . *If ε and ε' are in the interval $(0, n)$, and $\delta > 0$, then for t in the (k^{-n+1})-neighbourhood of $\Upsilon_2(k, k^\varepsilon, k^{\varepsilon'}, k^\delta)$ the following estimate is valid:*

$$M(\Omega(k, \varepsilon, \varepsilon'), t) < k^{-n+1+\varepsilon+\varepsilon'+\delta} \sum_{q \in \Gamma_1} |v_q|^2 |q|^{-\tau}, \tag{3.6.13}$$

where $\tau = 1$ when $n \geq 3$ and $\tau = 0$ when $n = 2$.

The set $\Upsilon_2(k, k^\varepsilon, k^{\varepsilon'}, k^\delta)$ has an asymptotically full measure on $S_0(k)$. Moreover,

$$s(S_0(k) \setminus \Upsilon_2(k, k^\varepsilon, k^{\varepsilon'}, k^\delta))/s(S_0(k)) < ck^{-\delta/4}, \qquad c \neq c(k). \tag{3.6.14}$$

Proof. Considering as in Lemma 3.8, we obtain

$$M(\Omega(k,k^{\epsilon}),k^{\epsilon'},t)<2k^{\delta}\sum_{q\in\Gamma_0(k)}|v_q|^2V(\Omega_q(k,k^{\epsilon},k^{\epsilon'})). \tag{3.6.15}$$

It is clear that $\Omega_q(k,k^{\epsilon},k^{\epsilon'})$ belongs to the intersection of the spherical shell K_1 and the plane layer $\Pi_{-q}(2k^{-n+2+\epsilon'})$. Integrating estimates (2.3.11) - (2.3.13) for a spherical shell with respect to radius, we obtain:

$$V(\Omega_q(k,\varepsilon,\varepsilon'))<ck^{-n+1+\epsilon+\epsilon'}|q|^{-\tau}, \qquad c\neq c(k).$$

Therefore, inequality (3.6.15) can be written in the form (3.6.13). Arguing in the same way as in the proof of Lemma 3.8, we obtain (3.6.14). The lemma is proved.

We define the nonsingular set $\chi_2(k,V,\delta)$ by the formula:

$$\chi_2(k,V,\delta)=\chi_0(k,\delta)\cap\hat{\Upsilon}_1(k,\delta)\cap\Upsilon_2(k,\delta), \tag{3.6.16}$$

where

$$\hat{\Upsilon}_1(k,\delta)=\cap_{r=1}^{m_0}\Upsilon_2(k,r\delta), \tag{3.6.17}$$

$$\hat{\Upsilon}_2(k,\delta)=\cap_{r,m=1}^{m_0}\Upsilon_2(k,r\delta,m\delta), \tag{3.6.18}$$

$$m_0=[n\delta^{-1}]+1.$$

Lemma 3.10 . *If t is in the $(k^{-n+1-2\delta})$-neighbourhood of $\chi_2(k,V,\delta)$, then there exists a unique j such that $|p_j^{2l}(t)-k^{2l}|<8lk^{2l-n-2\delta}$ $(p_j^{2l}(t)=k^{2l})$ if $t\in\chi_2(k,V,\delta)$ and the following estimates hold:*

1.

$$2\min_{q\neq0,q\in Z^n}|p_j^{2l}(t)-p_{j+q}^{2l}(t)|>k^{2l-n-\delta}, \tag{3.6.19}$$

2.

$$N(\Omega(k,r\delta),t)<ck^{(1+r)\delta}, \qquad r=1,...,m_0, \tag{3.6.20}$$

$$M(\Omega(k,r\delta,m\delta),t)<k^{-n+1+(r+m+1)\delta}\sum_{q\in\Gamma_1(k)}|v_q|^2|q|^{-\tau}, \tag{3.6.21}$$

$$m,r=1,...,m_0,$$

$\tau=1$, *if* $n\geq3$, $\tau=0$, *if* $n=2$.

The set $\chi_2(k,V,\delta)$ has an asymptotically full measure on $S_0(k)$. Moreover,

$$\frac{s(S_0(k)\setminus\chi_2(k,V,\delta))}{s(S_0(k))}=_{k\to\infty}O(k^{-\delta/4}). \tag{3.6.22}$$

Proof. Taking into account that $\chi_2(k,V,\delta)\subset\chi_0(k,\delta)$ and Corollary 2.1 (see (2.2.6)), we obtain inequality (3.6.19). Considering that $\hat{\Upsilon}_1(k,\delta)\subset\Upsilon_1(k,r\delta)$, $\hat{\Upsilon}_2(k,\delta)\subset\Upsilon_2(k,r\delta,m\delta)$, $m,r=1,...m_0$, and using relations (3.6.8), (3.6.13) in the case of $\varepsilon=r\delta$, $\varepsilon'=m\delta$, we get estimates (3.6.19), (3.6.21).

It remains to prove inequality (3.6.22). It is clear that

$$s(S_0(k) \setminus \chi_2(k,\delta)) \leq$$

$$s(S_0(k) \setminus \chi_0(k,\delta)) + \sum_{r=1}^{m=m_0} s(\Upsilon_1(k, m\delta)) + \sum_{r,m=1}^{m_0} s(\Upsilon_2(k, r\delta, m\delta)).$$

Using relations (2.2.5), (3.6.9), (3.6.14) we immediately obtain estimate (3.6.22). The lemma is proved.

3.7 Proof of the Main Result in the Case of a Nonsmooth Potential.

Proof of the theorems is again based on the formula expanding the resolvent $(H(t)-z)^{-1}$ in a perturbation series for z lying on the contour C_0 around unperturbed eigenvalue $p_j^{2l}(t)$. Then, integrating the resolvent yields the formulae for the perturbed eigenvalue and the spectral projection. The main point is to check the estimate:

$$\sup_{z \in C_0} \|A(z,t)\|_2 < k^{-\gamma_5}, \tag{3.7.1}$$

where $\|\cdot\|_2$ is the norm in $\mathbf{S_2}$. The proof of this estimate is based on properties of the nonsingular set $\chi_2(k, V, \delta)$. It is clear that

$$\|A(z,t)\|_2^2 = \mathrm{Tr} AA^* = \sum_{i,q \in Z^n} a_{iq}, \tag{3.7.2}$$

$$a_{iq} \equiv |v_q|^2 |(p_i^{2l}(t) - z)(p_{i+q}^{2l}(t) - z)|^{-1}.$$

Note that in estimating the terms a_{iq}, the values of $|p_i^{2l}(t) - z|$, $|p_{i+q}^{2l}(t) - z|$ play the main role. We will show that $|p_i^{2l}(t) - z|$ is of order $k^{2l-1}|p_i(t) - k|$, and use the fact that in geometric sense $|p_i(t) - k|$ is the distance from the point $\mathbf{p}_i(t)$ to the sphere $|x| = k$.

Let us break Z^n into subsets P_r: $Z^n = \cup_{r=-1}^{m_0} P_r \cup P_\infty$, in accordance with distances from points $\mathbf{p}_i(t)$, $i \in Z^n$ to the sphere $|x| = k$:

$$P_1(k,t) = \{i : |p_i(t) - k| < 2k^{-n+1-\delta}\},$$

$$P_r(k,t) = \{i : 2k^{-n+1+(r-1)\delta} \leq |p_i(t) - k| < 2k^{-n+1+r\delta}\}, \tag{3.7.3}$$

$$r = 0, 1, ..., m_0;$$

$$P_\infty(k,t) = \{i : |p_i(t) - k| \geq 2k^{-n+1+m_0\delta}\}.$$

It is clear that

$$(\cup_{r=-1}^{m_0} P_r) \cup P_\infty = Z^n,$$

$$\sum_{i,q \in Z^n} a_{iq} = \sum_{r,m=-1}^{N} \Sigma_{rm} + 2 \sum_{r=-1}^{N} \Sigma_r + \Sigma_\infty,$$

$$\Sigma_{rm} = \sum_{i\in P_r, i+q\in P_m} a_{iq},$$

$$\Sigma_r = \sum_{i\in P_r, i+q\in P_\infty} a_{iq},$$

$$\Sigma_\infty = \sum_{i,q\in P_\infty} a_{iq}.$$

We estimate Σ_{rm}. From the relations

$$i \in P_r, \qquad i+q \in P_m, \qquad r \le m_0, \quad m \le m_0,$$

the inequality $|q| < 4k^{1+\delta}$ follows. Indeed, it is obvious that

$$|q| < p_{i+q}(0) + p_i(0) < |p_{i+q}(0) - k| + |p_i(0) - k| + 2k.$$

Taking into account the definition of P_r, we obtain $|q| < 4k^{1+\delta}$. Thus,

$$\Sigma_{rm} = \sum_{i\in P_r, i+q\in P_m, |q|<4k^{1+\delta}} a_{iq}.$$

It is not hard to show that

$$4|p_i^{2l}(t) - z| > k^{2l-n+(r-1)\delta}, \tag{3.7.4}$$

when $z \in C_0, r = -1, 0, ..., m_0$. Indeed, in the case of $r \neq -1$, the inequality (3.7.4) immediately follows from the definitions of P_r and C_0. Let $r = -1$. Since inequality (3.6.19) holds, then the set P_{-1} contains a unique point j : $|p_j^{2l}(t) - k^{2l}| < 8lk^{2l-n-2\delta}$. Therefore,

$$4|p_i^{2l}(t) - z| = 4|p_j^{2l}(t) - z| > k^{2l-n-\delta}, \quad i \in P_{-1},$$

i.e., inequality (3.7.4) is valid for $r = -1$, too. Using (3.7.4), we obtain

$$|a_{iq}| < c|v_q|^2 k^{-4l+2n-(r+m-2)\delta}.$$

Taking into account the last estimate, and the definitions of $N(\Omega_q(k,\varepsilon,\varepsilon'),t)$ and $M(\Omega(k,\varepsilon,\varepsilon'),t)$, we get:

$$\Sigma_{rm} < k^{-4l+2n-(r+m-2)\delta} \sum_{|q|<k^{1+\delta}} |v_q|^2 N(\Omega_q(k,k^{r\delta},k^{m\delta}),t) =$$

$$k^{-4l+2n-(r+m-2)\delta} M(\Omega(k,k^{r\delta},k^{m\delta}),t). \tag{3.7.5}$$

Applying Lemma 3.8, we show that

$$\Sigma_{rm} < k^{-4l+n+1+4\delta} \sum_{|q|<k^{1+\delta}} |v_q|^2 |q|^{-\tau}.$$

It is clear that

$$\Sigma_{rm} < k^{-4l+n+1+4\delta}$$

if $4l - n - \beta \leq \tau$. Suppose $4l - n - \beta > \tau$. Then

$$\Sigma_{rm} < k^{1-\beta-\tau+6\delta l} \sum_{|q|<k^{1+\delta}} |v_q|^2 |q|^{-4l+n+\beta}. \tag{3.7.6}$$

Thus,

$$\Sigma_{rm} < k^{-\mu_0}, \tag{3.7.7}$$

where

$$\mu_0 = -1 + \beta + \tau - 6\delta l, \quad \text{if } 4l - n - \beta > \tau;$$

$$\mu_0 = 4l - n - 1 - 5\delta, \quad \text{if } 4l - n - \beta \leq \tau.$$

Note that we used condition (3.5.1) for the first time in proving estimate (3.7.6). It is important that μ_0 is positive for sufficiently small positive δ. It is not hard to show that $\mu_0 \geq \gamma_5 + \delta$. Therefore,

$$\Sigma_{rm} < k^{-2\gamma_5-\delta}. \tag{3.7.8}$$

Next, we consider Σ_r. We represent it in the form $\Sigma_r = \Sigma_{r1} + \Sigma_{r2}$, where

$$\Sigma_{r1} = \sum_{i \in P_r, i+q \in P_\infty, |q|<k^{1-2\delta}} a_{iq},$$

$$\Sigma_{r2} = \sum_{i \in P_r, i+q \in P_\infty, |q| \geq k^{1-2\delta}} a_{iq}.$$

First, we consider Σ_{r1}. The relations $i + q \in P_\infty$, $|q| < k^{1-2\delta}$ together give that $i \in P_N \cup P_\infty$. Therefore,

$$\Sigma_{r1} < ck^{-4l+6\delta l} \sum_{|q|<k^{1-2\delta}} |v_q|^2. \tag{3.7.9}$$

Using (3.5.1) we obtain:

$$\Sigma_{r1} < ck^{-\mu_1}, \tag{3.7.10}$$

where $\mu_1 = n + \beta - 6l\delta$ if $4l - n - \beta > 0$, and $\mu_1 = 4l - 6l\delta$ if $4l - n - \beta \leq 0$. It is not hard to show that $\mu_1 \geq 2\tilde{\gamma}_5 + \delta$. Therefore,

$$\Sigma_{r1} < ck^{-2\tilde{\gamma}_5+\delta}. \tag{3.7.11}$$

To estimate Σ_{r2}, we represent it in the form:

$$\Sigma_{r2} = \Sigma'_{r2} + \Sigma''_{r2},$$

where

$$\Sigma'_{r2} = \sum_{T_1} a_{iq}, \qquad \Sigma''_{r2} = \sum_{T_2} a_{iq};$$

$$T_1 = \{i, q : \ i \in P_r, \ i + q \in P_\infty, \ k^{1-2\delta} < |q| < k^{1+\delta}\},$$

$$T_2 = \{i, q : \ i \in P_r, \ i + q \in P_\infty, \ |q| \geq k^{1+\delta}\}.$$

We consider Σ'_{r2}. If $i + q \in P_\infty, k^{1-2\delta} < |q| < k^{1+\delta}$, then, obviously,

$$4|p_{i+q}^{2l}(t) - z| > c|q|^{2l} k^{-8l\delta}, \quad 0 < c \neq c(k).$$

Using this estimate and the definition of $N(\Omega(k,\varepsilon), t)$, we obtain:

$$\Sigma'_{r2} < k^{-2l+n-r\delta+10l\delta} N(\Omega(k, k^{r\delta}), t) \sum_{k^{1-2\delta} < |q| < k^{1+\delta}} |v_q|^2 |q|^{-2l}. \tag{3.7.12}$$

To estimate Σ''_{r2}, we note that $|p_{i+q}^{2l}(t) - z| > c|q|^{2l}$, $0 < c \neq c(k)$, when $i \in P_r$, $|q| \geq k^{1+\delta}$. Therefore,

$$\Sigma''_{r2} < k^{-2l+n-r\delta} N(\Omega(k, k^{r\delta}), t) \sum_{|q| \geq k^{1+\delta}} |v_q|^2 |q|^{-2l}. \tag{3.7.13}$$

Adding the estimates for Σ'_{r2}, Σ''_{r2} and using Lemma 3.10 (see (3.6.20)), we obtain

$$\Sigma_{r2} < k^{-2l+n+10l\delta} \sum_{|q| > k^{1-2\delta}} |v_q|^2 |q|^{-2l}. \tag{3.7.14}$$

Now, considering (3.5.1) and the estimate $2l > 4l - n - \beta$, which easily follows from (3.5.2), we get:

$$\Sigma_{r2} < k^{-\beta+20l\delta} \sum_{|q| > k^{1-2\delta}} |v_q|^2 |q|^{-4l+n+\beta} < ck^{-\beta+10l\delta}. \tag{3.7.15}$$

Now it is easy to see that

$$\Sigma_{r2} < ck^{-\mu_2}, \tag{3.7.16}$$

where $\mu_2 = 2l - n - 10l\delta$, if $2l > n, n \geq 3$ or $2l > 3$, $n = 2$ and $\mu_2 = \beta - 10l\delta$, if $2l \leq 3$, $n = 2$ or $2l \leq n$, $n \geq 3$.

Adding estimates (3.7.11), (3.7.16), we obtain

$$\Sigma_r < ck^{-2\gamma_5 - \delta}. \tag{3.7.17}$$

Let us consider Σ_∞. If $i \in P_\infty$, then obviously $2|p_i^{2l}(t) - z| > p_i^{2l}(t)$. Hence,

$$\Sigma_\infty < 4 \sum_{i, i+q \in P_\infty} p_i^{-2l} p_{i+q}^{-2l} |v_q|^2 = \sum_{q \in Z^n} |v_q|^2 \Sigma_{\infty q},$$

$$\Sigma_{\infty q} = \sum_{i, i+q \in P_\infty} p_i^{-2l} p_{i+q}^{-2l}.$$

It is clear that

$$\Sigma_{\infty q} < cI,$$

$$I = \int_\Omega |x|^{-2l} |x + \mathbf{p}_q(0)|^{-2l} dx,$$

$$\Omega = \{x : x \in R^n, |x| > k, |x + \mathbf{p}_q(0)| > k\}.$$

We consider the integral. Suppose $2|q| < k$; then

$$I < 2^{2l} \int_{|x|>k} |x|^{-4l} dx < ck^{-4l+n}.$$

In the case $2p_q(0) > k$ we produce the change of the variables: $x' = xp_q^{-1}(0)$. Thus,

$$I = p_q^{n-4l}(0) \int_P |x'|^{-2l} |x' + q/|q||^{-2l} dx',$$

$$P = \{x' : x' \in R^n, |x'| > kp_q^{-1}(0), |x' + e| < kp_q^{-1}(0)\}, \quad e = \frac{\mathbf{p}_q(0)}{p_q(0)}.$$

It is easy to see that $I < c|q|^{-4l+n+\delta}$, if $2l \le n$, and $I < ck^{n-2l}|q|^{-2l}$, if $2l > n$. Using the estimates for I and condition (3.5.1) yields:

$$\Sigma_\infty < ck^{-4l+n} \left(\sum_{|q|<k/2} |v_q|^2 + \sum_{|q|\ge k/2} |v_q|^2 |q|^{-4l+n+\delta} \right) \tag{3.7.18}$$

when $2l \le n$; and

$$\Sigma_\infty < ck^{-4l+n} \big(\sum_{|q|<k/2} |v_q|^2 + k^{n-2l} \sum_{|q|\ge k/2} |v_q|^2 |q|^{-2l} \big) \tag{3.7.19}$$

when $2l > n$. Using condition (3.5.1) we get

$$\Sigma_\infty < ck^{-\mu_2}, \tag{3.7.20}$$

where $\mu_2 = \min\{4l - n, \beta\}$. It is easy to see that $\mu_3 > 2\gamma_5 + \delta$. Therefore,

$$\Sigma_\infty < ck^{-2\gamma_5 - \delta}. \tag{3.7.21}$$

Adding estimates (3.7.7), (3.7.17) and (3.7.21) we obtain (3.7.1).
The theorem is proved.

3.8 Proof of the Bethe-Sommerfeld Conjecture. The Description of the Isoenergetic Surface.

Theorem 3.5 . *(Proof of the Bethe-Sommerfeld conjecture.) Suppose potential $V(x)$ satisfies condition (3.5.1). Then, there exists only a finite number of gaps in the spectrum of operator H.*

The proof is similar to that in the case $2l > n$ (Theorem 2.5), it being based on Theorem 3.4 instead of Theorem 2.1.

The Bethe-Sommerfeld conjecture for a smooth potential was proved in [Sk1]-[Sk7] in the following cases: $2l \ge n$; $n = 3, l = 1$; and $4l > n + 1$ only for a rational lattice of periods. Another proof in the two-dimensional situation was represented in [DahTr]. In [Ve1]- [Ve6] for the first time an asymptotic formula for an eigenvalue was used, which contained the first few terms of the asymptotic. The Bethe-Sommerfeld conjecture was proved for the cases where $l = 1$, $n = 2, 3$ and V is a smooth potential.

Here we represent the first proof of a stronger result – we describe the behavior of the perturbed isoenergetic surface near the nonsingular set (see the following theorems). We prove the Bethe-Sommerfeld conjecture for a general class of potential (see (3.5.1)). This is the first of existing proofs which is valid for nonsmooth potentials.

Theorem 3.6 . *Suppose potential $V(x)$ satisfies condition (3.5.1). Then, for $0 < \delta < 10^{-2}$ and sufficiently large k, $k > k_0(V, \delta)$, there exists a unique piece $S_H(k)'_0$ of the perturbed isoenergetic surface in the $(k^{-n+1-2\delta})$-neighbourhood of each simply connected component of the nonsingular set $\chi_2(k, V, \delta)$. For any given i, $i = 1, ..., n$, points t belonging to $S'(k)$ can be represented in the form:*

$$t = t_0 + \varphi_i(k, t_0), \quad t_0 \in \chi_2(k, V, \delta), \tag{3.8.1}$$

where $\varphi_i(k, t_0)$ is a continuously differentiable vector-valued function with the single nonzero component $(\varphi_i)_i$. The following asymptotic estimates are fulfilled for $(\varphi_i)_i$:

$$| \varphi_i(k, t_0)_i |=_{k\to\infty} O(k^{-n+1-2\gamma_5}), \tag{3.8.2}$$

$$| \nabla \varphi_i(k^2, t_0)_i |=_{k\to\infty} O(k^{-2\gamma_5}). \tag{3.8.3}$$

Estimates (3.8.2), (3.8.3) are uniform in t_0, $t_0 \in \chi_2(k, V, \delta)$.

Remark 3.4. The piece of the isoenergetic surface $S_H(k)'$ is a unique one in the $(k^{-n+1-2\delta})$-neighbourhood of $\chi_2(k, V, \delta)$. However, as a matter of fact, $S_H(k)'_0$ lies in a smaller $(k^{-n+1-2\gamma_5})$-neighbourhood of $\chi_2(k, V, \delta)$.

Proof. The proof is similar to that of Theorem 2.6. Indeed, the eigenvalue $\lambda(\alpha, t)$ is determined by formula (3.5.3) in the $(2k^{-n+1-2\delta})$-neighbourhood of the nonsingular set $\chi_2(k, V, \delta)$. The function $\lambda(\alpha, t)$ is continuously differentiable with respect to t in the same neighbourhood of each simply connected component of the nonsingular set. The following estimates are valid:

$$\left|\lambda(\alpha, t) - p_j^{2l}(t)\right|_{k\to\infty} = O(k^{2l-n-2\gamma_5}), \tag{3.8.4}$$

$$\left|\frac{\partial\lambda(\alpha, t)}{\partial t_i} - \frac{\partial p_j^{2l}(t)}{\partial t_i}\right|_{k\to\infty} = O(k^{2l-1-2\gamma_5+2\delta}). \tag{3.8.5}$$

Taking into account that t is in the $(k^{-n+1-2\delta})$-neighbourhood of the nonsingular set, it is not hard to check the inequality:

$$\left|\frac{\partial p_j^{2l}(t)}{\partial t_i}\right| > k^{2l-1-2\delta}.$$

Further, considering in the same way as in proving Theorem 2.6, we verify estimates (3.8.2), (3.8.3), and prove the uniqueness of $S'(k)$. The theorem is proved.

As in Chapter 2, we denote by $S_H(k)_0$ the part of the perturbed isoenergetic surface being in the $(k^{-n+1-2\delta})$-neighbourhood of $\chi_2(k,V,\delta)$. According to Theorem 3.6, it is the union of smooth pieces. Let $e(t)$ be a normal to $S_H(k)_0$ at the point t, and $e_0(t_0)$, $t_0 \in \chi_2(k,V,\delta)$, be a normal to the isoenergetic surface of free operator. It is clear that $e_0(t_0) = \mathbf{p}_j(t_0)/p_j(t_0)$, j being determined uniquely from the relation $p_j^{2l}(t_0) = k^{2l}$. If t is in the $(k^{-n+1-2\delta})$- neighbourhood of $\chi_2(k,V,\delta)$, then j is uniquely determined from the relation $p_j^{2l}(t_0) \in \varepsilon(k,\delta)$. Hence, the vector $\mathbf{p}_j(t_0)/p_j(t_0)$ is correctly defined in the $(k^{-n+1-2\delta})$- neighbourhood of $\chi_2(k,V,\delta)$, too.

Theorem 3.7 . *Suppose $t \in S_H(k)$. Then,*

$$e(t) =_{k\to\infty} e_0(t) + O(k^{-2\gamma_5+2\delta}). \tag{3.8.6}$$

The measure of surface $S_H(k)_0$ is asymptotically close to that of $S_0(k)$:

$$\frac{s(S_H(k)_0)}{s(S_0(k)} \to_{k\to\infty} 1. \tag{3.8.7}$$

Corollary 3.5 . *The following estimate is valid for the measure of the perturbed isoenergetic surface $S_H(k)$:*

$$\lim_{k\to\infty} \frac{s(S_H(k))}{s(S_0(k))} \geq 1. \tag{3.8.8}$$

The estimate (3.8.8) is fulfilled because $S_H(k)_0 \subset S_H(k)$ and relation (3.8.7) is valid.

The proof of the theorem is based on formula (3.8.1) and is completely analogous to that of Theorem 2.7 .

3.9 Formulae for Eigenvalues on the Perturbed Isoenergetic Surface.

Suppose t belongs to $S_H(k)_0$, i.e., to the part of the perturbed isoenergetic surface being in a vicinity of the nonsingular set. Suppose we know the energy k^2 corresponding to this surface and $n-1$ components, say, $t_2, ..., t_n$, of t. These dates define the point t uniquely. The question is to find the eigenfunction corresponding to given k^2, $t_2, ..., t_n$. This problem is important for physical calculations. Here we will apply it for a semicrystal problem (see Chapter 5). The obvious way of solving this problem is to define t_1 from the equation $\lambda(\alpha, t) = k^{2l}$, and then to use formula (3.5.3) for the spectral projection. We do this in Theorem 3.8. However, this way is not particularly effective, because we can solve the equation $\lambda(\alpha, t) = k^{2l}$ only approximately. We use another approach to this problem already represented in Section 2.7. It uses the integration of the resolvent with respect to quasimomentum (Theorem 3.9).

Theorem 3.8 . *Suppose t_0 belongs to $\chi_2(k, V, \delta)$. Then at the point $t(t_0)$, determined by the formula*

$$\lambda(\alpha, t) = k^{2l}, \qquad t = (t_1, t_{02}, ..., t_{0n}), \tag{3.9.1}$$

the following formulae for $\nabla\lambda(\alpha, t)$ and $E(\alpha, t)$ are valid:

$$\nabla\lambda(\alpha, t) = \mathbf{p}_j(t_0)p_j^{2l-2}(t_0)(1 + O(k^{-2\gamma_5+2\delta})), \tag{3.9.2}$$

$$E(t) = E_j + G_1(k, t_0) + G_2(k, t_0) + O(k^{-3\gamma_5+2\delta}). \tag{3.9.3}$$

The proof is similar to that of Theorem 2.10. Indeed, using $|t - t_0| < k^{-n+1-2\gamma_5}$, and formulae for $\nabla\lambda(\alpha, t)$ (Theorem 3.4), we get (3.9.2).

According to Corollary 3.4,

$$||\partial E(t)\partial t_1||_1 < ck^{n-1-\gamma_5+2\delta}. \tag{3.9.4}$$

Therefore:

$$||E(t) - E(t_0)||ck^{-3\gamma_5+2\delta}.$$

Using the formula for $E(t_0)$ (Theorem 3.4), we obtain (3.9.3).
The theorem is proved.

We can calculate $T(m)\lambda(\alpha, t)$ and $T(m)E(t)$ in a similar way. However, the accuracy of the formulae is restricted by that of the approximation $t_1 \approx t_{01}$. To write out the next asymptotic terms, it is necessary to solve the equation $\lambda(\alpha, t) = k^{2l}$ more precisely. However, it seems to be impossible in the explicit form. In Chapter 2 (Section 7), we described another way of constructing a formula for the spectral projection. It is not connected with solving the equation $\lambda(\alpha, t) = k^{2l}$. The approach developed in Section 2.7 is valid in the case here. Let us describe its main points.

Suppose $t \in S_H(k)_0$. According to Theorem 3.6, t can be represented in the form (3.8.1). We set $i = 1$. Considering the integral $I(k, t_0)$ and the functions $D_r(k, t_0)$ (see (2.7.4), (2.7.8), (2.7.9)), and arguing in the same way as in Chapter 2, we prove the following theorem:

Theorem 3.9 . *Suppose t_0 belongs to $\chi_2(k, V, \delta)$. Then at the point $t(t_0)$, defined by formula (3.8.1), the following formula holds:*

$$\left.\frac{E(\alpha, t)}{(\partial\lambda(\alpha, t)/\partial t_1)}\right|_{t=t(t_0)} = \sum_{r=0}^{\infty} \alpha^r D_r(k, t_0), \tag{3.9.5}$$

where the series converges in class $\mathbf{S}_1$. Moreover, the operator-valued coefficients $D_r(k, t_0)$ satisfy the estimates:

$$||D_r(k, t_0)||_1 < k^{-(\gamma_4-8\delta)r-2l+1}. \tag{3.9.6}$$

Now we consider the eigenfunction of the operator H_α, defined by formula (2.7.17). It is easy to show that the corresponding quasimomentum is $t = (t_1, t_{02}, ..., t_{0n})$, t_1 being given by formula (3.8.1). We formally expand the function $\Psi(k^{2l}, t_{02}, ..., t_{0n}, x)$ in powers of αV (see (2.7.18), (2.7.19)). It is not hard to show that functions $B_r(k^{2l}, t_{02}, ..., t_{0n}, x)$ satisfy the quasiperiodic conditions in the directions orthogonal to x_1. The corresponding components of the quasimomentum are $t_{02}, ..., t_{0n}$.

Theorem 3.10 . *If $t_0 \in \chi_2(k, V, \delta)$, then for sufficiently large k, $k > k_0(V, \delta)$, the function $\Psi(k^{2l}, t_{02}, ..., t_{0n}, x)$ admits the expansion in series (2.7.18). There are estimates for the functions $B_r(k^{2l}, t_{02}, ..., t_{0n}, x)$:*

$$\|B_r\|_{2,M} < ck^{-2l+1-(\gamma_5-8\delta)r}(1+ \mid M \mid^r), \tag{3.9.7}$$

$$\left\| \frac{\partial B_r}{\partial x_1} \right\|_{2,M} < ck^{-2l+2-(\gamma_5-8\delta)r}(1+ \mid M \mid^r). \tag{3.9.8}$$

The proof is similar to that of Theorem 2.12 up to using Theorem 3.4 instead of Theorem 2.1.

The latter theorem is valid for a potential all of whose smoothness requirements are contained in the conditions (3.5.1). The next theorem is valid for $V(x)$ being a trigonometric polynomial.

Theorem 3.11 . *Suppose $0 < \delta < \delta_0$, $\delta_0 = (4l - n - 1)/2(n + 1)$, and $t_0 \in \chi_1(k, \delta, \delta)$. Then, for sufficiently large $k > k_0(V, \delta)$, the function $\Psi(k^{2l}, t_{02}, ..., t_{0n}, x)$ admits the expansion in series (2.7.18). Each function $B_r(k^{2l}, t_{02}, ..., t_{0n}, x)$ can be represented as the sum:*

$$B_r(k^{2l}, t_{02}, ..., t_{0n}, x) = \sum_{q: q\in Z^n, |q-j|<rR_0, 0\le m\le r} a_{qm}^{(r)} \exp i(\mathbf{p}_q(t), x) x_1^r, \tag{3.9.9}$$

where coefficients $a_{qm}^{(r)}$ satisfy the estimates:

$$\mid a_{qm}^{(r)} \mid < k^{-2l+1-(\gamma_5-8\delta)r-(n-1+2\delta)m}. \tag{3.9.10}$$

Corollary 3.6 . *There are estimates for the functions $B_r(k^{2l}, t_{02}, ..., t_{0n}, x)$:*

$$\|B_r\|_{2,M} < ck^{-2l+1-(\gamma_2-8\delta)r}(1+ \mid M \mid^r), \tag{3.9.11}$$

$$\left\| \frac{\partial B_r}{\partial x_1} \right\|_{2,M} < ck^{-2l+2-(\gamma_2-8\delta)r}(1+ \mid M \mid^r). \tag{3.9.12}$$

There are stronger estimates when $r < k^\delta R_0$:

$$\|B_r\|_{2,M} < ck^{-(2l-1+3\delta)(r+1)}(1+ \mid M \mid^r), \tag{3.9.13}$$

$$\left\| \frac{\partial B_r}{\partial x_1} \right\|_{2,M} < ck^{-(2l-2-3\delta)(r+1)}(1+ \mid M \mid^r). \tag{3.9.14}$$

Proof of the theorem is similar to that of Theorem 2.11. with the radius $k^{-2\delta}$ and the same centre and argueing as in

3.10 Determination of the Potential from the Asymptotic of the Eigenfunction.

The first term $G_1(k,t)$ in the asymptotic formula (3.2.10) for the spectral projection depends on V in a linear way. It is easy to show, using the expression (1.0.25) for $G_1(k,t)$, that

$$v_q = \lim_{k\to\infty}(E_{j,j-q} - E_j)(k^{2l} - |\mathbf{k} - \mathbf{p}_q^{2l}(0)|^{2l}), \tag{3.10.1}$$

where $\mathbf{k}$ is such that $t \equiv \mathcal{K}\mathbf{k}$ is in $\chi_2(k,V,\delta)$, and j is uniquely determined from the relation $p_j(t) = k$. In this section we assume $V(x)$ to be a trigonometric polynomial. Formula (3.10.1) enables us to determine the potential from the second term of the asymptotic expansion of the eigenfunction in a high energy region . Now we develop this scheme rigorously. Let $P_\alpha(k_1,k_2)$ be the set of the eigenfunctions of the operator H_α, whose eigenvalues belong to the interval $[k_1^{2l}, k_2^{2l}]$. If k_1 is sufficiently large, then the set $P_\alpha(k_1,k_2)$ contains a rich subset $Q_\alpha(k_1,k_2)$ of the eigenfunctions, which can be expanded together with their eigenvalues in the series (3.2.9), (3.2.10). Let us prove that $Q_\alpha(k_1,k_2)$ is rich. We define measure $\beta(Q)$ of some subset Q of the eigenfunctions as follows:

$$\beta(Q) = \int_K R(Q,t)dt,$$

where $R(Q,t)$ is the number of linearly independent eigenfunctions with the quasimomenta t belonging to Q. Note that the eigenfunction $\exp i(\mathbf{p}_j(t),x)$ of the free operator belongs to $P_0(k_1,k_2)$, if and only if $\mathbf{p}_j(t)$ is in the spherical shell $K_{k_1k_2}$. It is not hard to show that

$$R(P_0(k_1,k_2),t) = N(K_{k_1k_2},t),$$

N being the number of states (see Section 3.6). From this it follows that

$$\beta(P_0(k_1,k_2)) = V(K_{k_1,k_2}) = \omega_n(k_2^n - k_1^n), \tag{3.10.2}$$

ω_n being the volume of a unit ball in R^n.

<u>DEFINITION</u> We say set $Q_\alpha(k_1,k_2)$ is rich if its measure is close to the measure of the set $P_0(k_1,k_2)$, namely

$$\beta(Q_\alpha(k_1,k_2))/V(K_{k_1,k_2}) > 1 - (\ln k_1)^{-1}. \tag{3.10.3}$$

Let $Q_1(k_1,k_2)$ be a subset of $P_0(k_1,k_2)$, such that the quasimomenta of any function from this subset belongs to the nonsingular set $\chi_1(k,\delta,\delta)$, i.e., the perturbations of this function and corresponding eigenvalue are described by formulae (3.2.9), (3.2.10).

Lemma 3.11 . *The set $Q_0(k_1,k_2)$ is rich for a sufficiently large k_1.*

Proof. Suppose $k_2 - k_1 > k_2^{-n+1-2\delta}$. Let us represent $[k_1, k_2)$ as the union of the nonintersecting intervals $[\hat{k}_m, \hat{k}_{m+1})$ of identical length of the order $k_2^{-n+1-2\delta}$, $0 < \delta \le 1$, $\delta \neq \delta(k)$. Suppose the assertion of the lemma is proved for each of these intervals, i.e.,

$$\beta(Q_1(\hat{k}_m, \hat{k}_{m+1}))/V(K_{\hat{k}_m \hat{k}_{m+1}}) > 1 - (\ln \hat{k}_m)^{-1}. \tag{3.10.4}$$

Then, considering the obvious relations

$$1 - (\ln \hat{k}_m)^{-1} \ge 1 - (\ln k_1)^{-1};$$

$$\int_K R(Q_1(k_1, k_2), t)dt = \sum_{m=1}^{M} \int_K R(Q_1(\hat{k}_m, \hat{k}_{m+1}), t)dt, \quad M = (k_2 - k_1)k_2^{n-1+2\delta};$$

$$\sum_{m=1}^{M} V(K_{\hat{k}_m \hat{k}_{m+1}}) = V(K_{k_1 k_2})$$

and estimate (3.10.4), we obtain (3.10.3) for $\alpha = 1$. It remains to prove the assertion of the lemma for the small interval $[\hat{k}_m, \hat{k}_{m+1})$. Firstly we show that

$$R(Q_1(\hat{k}_m, \hat{k}_{m+1})) \le 1. \tag{3.10.5}$$

Indeed, suppose there exist two linearly independent functions, corresponding to quasimomentum t in $Q_1(\hat{k}_m, \hat{k}_{m+1})$. Then, there exist i, j $(i \neq j)$, such that $|p_i(t) - p_j(t)| < k^{-n+1-2\delta}$. But this is not the case, because t belongs to the nonsingular set. This contradiction proves inequality (3.10.5). Thus, $R(Q_1(\hat{k}_m, \hat{k}_{m+1}))$ is the indicator of Ω_0; here Ω_0 is the set of quasimomenta belonging to the nonsingular set $\chi_1(k, \delta, \delta)$ for some k, such that corresponding $\lambda(\alpha, t)$ is in the interval $[\hat{k}_m^{2l}, \hat{k}_{m+1}^{2l})$. Therefore,

$$\beta(Q_1(\hat{k}_m, \hat{k}_{m+1})) = \int_K R(Q_1(\hat{k}_m, \hat{k}_{m+1}), t)dt = V(\Omega_0). \tag{3.10.6}$$

Taking into account that $\chi_1(k, \delta, \delta)$ has an asymptotically full measure on $S_0(k)$, and is stable with respect to the perturbation of order $k^{-n+1-2\delta}$, we show

$$V(\Omega_0) = \int_{[\hat{k}_m, \hat{k}_{m+1})} s(\chi_1(k, \delta, \delta))dk = V(K_{\hat{k}_m \hat{k}_{m+1}})(1 + O(k^{-\delta})). \tag{3.10.7}$$

From relations (3.10.6), (3.10.7) we obtain (3.10.4). When $k_2 - k_1 < k_2^{-n+1-2\delta}$, we don't need to break the interval $[k_2, k_1]$ into small pieces. The proof is the same as for $[\hat{k}_m, \hat{k}_{m+1})$. The lemma is proved.

Lemma 3.12 . *The set $Q_\alpha(k_1, k_2)$ is rich for sufficiently large k_1, if $1 > |k_2 - k_1| \ge k_1^{-n+1}$.*

Proof. To begin with, we consider $Q_1(k_1^+, k_2^-)$, $k_1^+ = k_1 + k_1^{-n+1-\delta}$, $k_2^- = k_2 - k_1^{-n+1-\delta}$, $0 < \delta \ll 1$. If t is in $\Omega_0(k_1^+, k_2^-)$, then formulae (3.2.9), (3.2.10) hold, and $\lambda(\alpha, t) \in (k_1^{2l}, k_2^{2l})$ when $|\alpha| \leq 1$. This means that the corresponding eigenfunctions belong to $Q_\alpha(k_1, k_2)$. Therefore, $Q_1(k_1^+, k_2^-) \subset Q_\alpha(k_1, k_2)$, and

$$R(Q_\alpha(k_1, k_2), t) \geq R(Q_1(k_1, k_2), t).$$

Applying Lemma 3.11, we obtain inequality (3.10.3). The lemma is proved. We consider the sequence of the sets $P_1(k_1^{(m)}, k_2^{(m)})$, $m \in N$, $k_1^{(m)} \to_{m\to\infty} \infty$, $k_2^{(m)} \to_{m\to\infty} \infty$, $1 > k_2^{(m)} - k_1^{(m)} > (k_1^{(m)})^{-n+1}$. It follows from Lemma 3.12 that there exists a rich set of functions $\Psi_m(x, t)$ belonging to $P_1(k_1^{(m)}, k_2^{(m)})$, such that they admit the asymptotic expansion:

$$\Psi_m(x, t) = \exp(i(\mathbf{k}_m(t), x))(1 + f_m(x, t) + O(|\mathbf{k}_m(t)|^{-4l+2+6\delta n})), \qquad (3.10.8)$$

where $\mathbf{k}_m(t) \in R^n$, $|\mathbf{k}_m(t)| \to_{m\to\infty} \infty$, $f_m(x, t) \to_{m\to\infty} 0$, $0 < \delta \ll 1$, $\delta \neq \delta(k)$. Note that the coefficients $f_m(x, t)$ are determined up to the value of order $|\mathbf{k}_m(t)|^{-4l+2+6\delta}$.

Theorem 3.12 . *Coefficients $f_m(x, t)$ can be chosen in such a way that they would have the same periods. The common periods are the periods of potential $V(x)$. Fourier coefficients of the potential are determined by the formula:*

$$v_r = \lim_{m\to\infty} a_r^{(m)}, \qquad (3.10.9)$$

$$a_r^{(m)} = b_r^{(m)}(|\mathbf{k}_m(t)|^{2l} - |\mathbf{k}_m(t) + \mathbf{p}_r(0)|^{2l}),$$

where $b_r^{(m)}$ are the Fourier coefficients of functions $f_m(x, t)$ with respect to x.

Remark 3.5. To determine v_r, it suffices not to know all the Fourier coefficients of f_m, but only $b_r^{(m)}$ with the same m.

Proof. From the assertion of Theorem 3.1, we immediately obtain the estimates:

$$|b_0^{(m)}|^2 = 1 + O(|\mathbf{k}_m(t)|^{-4l+2+6\delta n}), \qquad (3.10.10)$$

$$b_r^{(m)} b_0^{(m)} = v_r(|\mathbf{k}_m(t)|^{2l} - |\mathbf{k}_m(t) + \mathbf{p}_r(0)|^{2l})^{-1} + O(|\mathbf{k}_m(t)|^{-4l+2+6\delta n}). \quad (3.10.11)$$

Suppose r satisfies the estimate $|r| < |\mathbf{k}_m(t)|^\delta$. Taking into account that

$$||\mathbf{k}_m(t)|^{2l} - |\mathbf{k}_m(t) + \mathbf{p}_r(0)|^{2l}| < |\mathbf{k}_m(t)|^{2l-1+\delta},$$

we obtain

$$v_r = b_r^{(m)}(|\mathbf{k}_m(t)|^{2l} - |\mathbf{k}_m(t) + \mathbf{p}_r(0)|^{2l}) + O(|\mathbf{k}_m(t)|^{-2l+1+7\delta n}).$$

Considering that any fixed r satisfies the inequality $|r| < |\mathbf{k}_m(t)|^\delta$, we prove formula (3.10.9). The theorem is proved.

4. Perturbation Theory for Schrödinger Operator with a Periodic Potential.

4.1 Introduction. Modified Laue Diffraction Conditions.

From the physical point of view, the Schrödinger operator

$$H = -\Delta + V \tag{4.1.1}$$

in $L_2(R^3)$ is most interesting, because it describes a motion of a particle in a bulk matter. However, the mathematical study of this operator is more complicated than that of the operators considered above. The perturbation series converge in the case of $4l > n+1$, because the parameter γ_2 in the power estimates (2.2.12), (2.2.13) for the asymptotic terms is positive ($2\gamma_2 = 4l-n-1-8l\delta$) . In the case of the Schrödinger operator, we have $l=1, n=3$ and, hence, γ_2 is negative ($\gamma_2 = -8\delta$). The power estimates (2.2.12), (2.2.13) hold for this negative γ_2 too, but they can not insure the convergence of the series. As a matter of fact, this difficulty has a physical nature. As can be seen from the previous chapters, the convergence of the series is directly connected with diffraction processes inside the crystal. When a quasimomentum satisfies a diffraction condition, a refracted wave arises inside the crystal. It interferes with the initial wave and distorts it strongly. Thus, the perturbed eigenfunction is not close to the initial one. This means that the corresponding perturbation series diverges. When a quasimomentum is sufficiently far from the diffraction planes, then unperturbed and perturbed eigenfunctions are close – the perturbation series converge. This indicates that the expressions corresponding to the diffraction conditions do control the convergence of the series. The series diverge when at least one of them takes a value close to zero. So, they have to be in the denominators of terms of the properly constructed perturbation series. These conditions do not depend on the potential. They can be described in terms of $p_m^{2l}(t)$ – matrix elements of the free operator. So, it is natural that this operator is in the denominators of the terms of the series. It is possible to control the convergence of the series only in terms of this operator.

New diffraction conditions depending on the potential arise in the case of the Schrödinger operator (see Section 4.2). This means that H_0, not depending on the potential, is no longer able to control the convergence of the series. To construct properly the perturbation series, one has to take for the initial operator some operator $\hat{H}(t)$ instead of $H_0(t)$. The operator $\hat{H}(t)$ is already described in the Introduction (see (1.0.4)). It corresponds to the new diffraction conditions. Of course, $\hat{H}$ depends on V; therefore, the asymptotic terms depend on the potential, not by a power but in a more complicated way. These series are not convenient for being use, because the calculation of their terms is rather complicated. Fortunately, it turns out that the modified series can be replaced with high accuracy by sufficiently long segments of the series for H_0. Furthermore, the power estimates of the type (1.0.27), (1.0.29) for these terms hold. Roughly speaking, the difference between series constructed with respect to H_0 and $\hat{H}$ is only in their far tails. Thus, the simplified formulae of the perturbation theory arise. These formulae are, in fact, the aim of our considerations. However, the justification of the simplified formulae requires the construction of the modified series (with respect to $\hat{H}$), which, moreover, give the precise result.

The main theorems for the case where $V(x)$ is a trigonometric polynomial are formulated in Section 4.2. The general considerations about the diffraction conditions can be founded in Section 4.3. The nonsingular set for the Schrödinger operator is constructed in Section 4.4, by deleting a neighborhood of the curves satisfying the modified Laue diffraction conditions from the nonsingular set for the case of $4l > n + 1$, $n = 3$. This new set also has an asymptotically full measure on the isoenergetic surface $S_0(k)$ of the free operator. The proof of the convergence of the perturbation series can be founded in Section 4.5. All considerations are based on expanding the resolvent in the series::

$$(H(t) - z)^{-1} = (\hat{H} - z)^{-1/2}\Big(\sum_{r=0}^{\infty} \hat{A}^r\Big)(\hat{H} - z)^{-1/2}, \tag{4.1.2}$$

$$\hat{A} \equiv (\hat{H} - z)^{-1/2} W (\hat{H} - z)^{-1/2}, \tag{4.1.3}$$

W being given by formula (1.0.45). Proving the convergence of the series for z on the contour about unperturbed eigenvalue $p_j^2(t)$, and integrating the resolvent over this contour, we obtain the series for the eigenvalue and the spectral projection. The main difficulties in this scheme are in proving the convergence of the series in powers of $\hat{A}$. These problems have two aspects. First, the operator $\hat{H}(t)$ is not diagonal. To estimate $\hat{A}$ one has to reduce $\hat{H}(t)$ to the diagonal form $U\hat{H}(t)U^*$. However, after this the structure of the perturbing operator UWU^* becomes rather complicated. We describe this structure (Lemmas 4.5 – 4.10) in order to estimate $\|\hat{A}\|$. Second, estimating $\|\hat{A}\|$, we find that, generally speaking, the estimate $\|\hat{A}\| < 1$ does not hold. To verify the convergence of the series we prove the weaker estimates: $\|\hat{A}\| < k^{3\delta}$, $\|\hat{A}^3\| \ll 1$.

The next interesting problem is to construct the perturbation formulae on the singular set. As it was noted before, the perturbed and unperturbed eigenvalues and eigenfunctions are not close for quasimomenta of this set. We prove that, on the essential part of the singular set, the perturbed eigenvalues and eigenfunctions are close to those of the model operator $\hat{H}(t)$. This operator has a block structure. Its diagonal part coincides with the corresponding part of free operator $(I - \sum P_q)\hat{H}(t) = (I - \sum P_q)H_0$, while the nondiagonal blocks of $\hat{H}(t)$ are determined by the orthogonal projections P_q: $P_q\hat{H}(t) = P_q\hat{H}(t)P_q = \hat{H}(t)P_q$. In the case of the nonsingular set, the eigenvalue and the corresponding spectral projection of $H(t)$ are asymptotically close to those of the free operator, and so are the spectral projections. In the case of the singular set (of its essential part), the eigenvalues and the spectral projections of $H(t)$ are close to those of the nondiagonal part of $\hat{H}(t)$. Thus, the diagonal part of $\hat{H}(t)$ gives the first approximation for the perturbed eigenvalues and eigenfunctions on the nonsingular set, while its block part gives the first approximation on the singular set. Section 4.6 contains the summary of the results for the singular set. The perturbed series for eigenvalues and spectral projections are constructed there. The conditions $1'$ – $5'$ of the convergence of the series on the singular part (see Section 4.6) are similar to the conditions $1^0 - 5^0$ on the nonsingular part (Section 4.3). They are, roughly speaking, obtained by the replacement of eigenvalue $p_j^2(t)$ of the free operator, in the conditions $1^0 - 5^0$ for the nonsingular set, by an eigenvalue $\hat{\lambda}(t)$, corresponding to the block part of $\hat{H}$. In fact, the conditions $1'$ – $5'$ describe the situation when there is a real diffraction inside a crystal, but the picture of the diffraction is relatively simple. Thus, conditions $1'$ – $5'$ determine the part of the singular set where the perturbation series converge with respect to $\hat{H}$, which roughly takes into account the processes of diffraction inside the crystal. The geometrical description of this set is in Section 4.7. It is proved that the part, where conditions $1'$ – $5'$ hold, is an essential part of the singular set. The proof of the convergence of the perturbation series is contained in Section 4.8. It is similar to the proof for the nonsingular set.

The case of a nonsmooth potential is considered in Sections 4.9 – 4.11. The restriction on the potential is contained in the condition of convergence of the series

$$\sum_{j \in Z^3 \setminus \{0\}} | v_j |^2 | j |^{-1+\beta} \tag{4.1.4}$$

for some positive β. This class is, obviously, wider than L_2. A general class of singularities are allowed. For example, a function, with a singularity of the form x_0 as $| x - x_0 |^{-\zeta}$, $\zeta < 2$, in particular, the Coulomb potential, is contained in this class. The perturbation series on the nonsingular set are constructed. Section 4.9 contains the summary of the results. The series 4.1.2) converges on the nonsingular set. The main technical work here is in the analytical part, that is in the proof of the convergence of the series (Section 4.11). For proving we represent the potential $V(x)$ in the form $V = V_1 + V_2$, where V_1 is a trigonometric polynomial:

$$V_1 = \sum_{|j| \le k^\rho} v_j \exp(i(\mathbf{p}_j(0), x)), \quad 0 < \rho < 1.$$

Naturally,

$$V_2 = \sum_{|j| \ge k^\rho} v_j \exp(i(\mathbf{p}_j(0), x)).$$

We consider V_2 as a relatively small perturbation of V_1, because

$$\sum_{|j|\geq k^{\rho}} |v_j|^2 |j|^{-1+\beta} \to_{k\to\infty} 0, \tag{4.1.5}$$

since the series (4.1.4) converge. We construct $\hat{H}(t)$ for V_1, just as we did it in the case of a trigonometric polynomial (Section 4.4). Then, we represent the resolvent $(H(t)-z)^{-1}$ in the form (4.1.2), (4.1.3), where

$$W = W_1 + V_2, \quad W_1 = V_1 - \hat{V}_1 \tag{4.1.6}$$

(for the definition of $\hat{V}_1$ see in Introduction). To prove the convergence of the series (4.1.3), we justify the estimates $\|\hat{A}\| < k^{100\delta}$, $\|\hat{A}^3\| \ll 1$ on the nonsingular set. To verify the latter inequality we estimate (Lemma 4.35) the operator $(H_0 - z + \Delta\Lambda)^{-1/2}U(H_0(t)-z)^{1/2}$, U being the matrix reducing $\hat{H}$ to the diagonal form $H_0(t) + \Delta\Lambda(t) - z$. Integrating the resolvent over the contour yields the series for the eigenvalue and the spectral projection (Theorem 4.8, formulae (4.10.5) and (4.10.6)). Then, we prove (Theorem 4.9) that the initial terms $\hat{g}_r(k,t)$ and $\hat{G}_r(k,t)$ of these series are close to $g_r(k,t)$ and $G_r(k,t)$ with the same r. Thus, in the case of a nonsmooth potential, we also can simplify the modified series and obtain the approximate formulae (4.10.12), (4.10.13) (Theorem 4.9). Note that the accuracy of the approximation increases together with increasing smoothness of the potential, that is when β increases. Roughly speaking, the nonsingular set for a nonsmooth potential is the intersection of the nonsingular set $\chi_2(k,V,\delta)$ for the polyharmonic operator in the case $n = 3$ (see Section 3.3) and the nonsingular set for Schrödinger operator with a smooth potential (Section 4.4). We show that it has an asymptotically full measure on $S_0(k)$ (Section 4.10).

The proof of the Bethe-Sommerfeld conjecture (Theorem 4.10) and the investigation of the isoenergetic surface of operator H are contained in Section 4.12. They are consequences of the perturbation formulae on the nonsingular set. We prove that there exists a unique piece of the perturbed isoenergetic surface in the $(k^{-3-\delta})$-neighborhood of each simply connected component of the nonsingular set (Theorem 4.11). Their normals are close to each other (Theorem 4.12). Thus, the isoenergetic surface of H is close to the isoenergetic surface of the free operator near the nonsingular set. How does the perturbed isoenergetic surface behave near the singular set? Note that an essential part of the singular set is the union of vicinities of selfintersections of the unperturbed isoenergetic surface. By the perturbation, these intersections are tranformed to quasiintersections. The operator $\hat{H}$ describes these quasiintersections approximately. We prove that there exists a unique piece of the perturbed isoenergetic surface near each simply connected component of the isoenergetic surface of $\hat{H}$, satisfying conditions $1' - -5'$. Their normals are close to each other.

Perturbation formulae (4.10.6), (4.10.13) can be used for the calculation of the spectral projection of H for a given quasimomentum t. However, these formulae are not directly suitable in the situation when the value of perturbed eigenvalue $\lambda(t) = k^{2l}$ and two coordinates, say, t_2, t_3 of the point t, are given. To use formulae (4.10.6), (4.10.13) in this case, one has previously to determine t_1 from the equation

$$\lambda(t) = k^{2l}, \quad t = (t_1, t_2, t_3). \tag{4.1.7}$$

However, we can do this only approximately, taking $t_1 \approx t_{01}$, where t_{01} is the solution of the unperturbed equation $p_j^2(t) = k^2$. Substituting $t_1 \approx t_{01}$ in formula (4.10.13), we obtain a rather rough asymptotic formula (4.14.2) for the spectral projection. Similarly, the approximate formula (4.14.1) for the normal vector to the isoenergetic surface of H is obtained.

It turns out that there exists another way to construct the eigenfunction for t belonging to the isoenergetic surface. In this way we do not solve the equation (4.1.7). We construct the perturbation series for the eigenfunction with fixed $\lambda(t) = k^{2l}, t_2, t_3,$, in an explicit form by integrating the perturbation series for the resolvent over some contour on the complex plane of t. Thus obtained, formula (4.14.39) gives the eigenfunction for given k^2, t_2, t_3, to any accuracy. It will be used in Chapter 5 for solving a semicrystal problem. The results proved in this chapter are published in [K5], [K9] – [K11], [K13] – [K15].

4.2 The Main Results for the Case of a Trigonometric Polynomial.

To describe the main results we recall the definition of the model operator $\hat{H}(t)$ given in the Introduction (Chapter I). First, we define the set $\Gamma(R_0)$. Let us consider $j : j \in Z^3, |j| < R_0$. In this set some of the j are scalar multipliers of others. Let us keep from every family of scalar multipliers only the minimal representative, i.e., the representative having the minimal length. We denote by $\Gamma(R_0)$ the union of these minimal representatives (see Fig. 2, page 14). In other words, each $j : j \in Z^3, |j| < R_0$ can be uniquely represented in the form $j = mj_0$, where $m \in Z$, $j_0 \in \Gamma(R_0)$. It is easy to see that potential $V(x)$ can be represented in the form

$$V = \sum_{q \in \Gamma(R_0)} V_q, \tag{4.2.1}$$

where V_q depends only on the single variable $(x, \mathbf{p}_q(0))$:

$$V_q(x) = \sum_{|nq| < R_0, n \in Z} v_{nq} \exp(in(x, \mathbf{p}_q(0))), \tag{4.2.2}$$

We shall use this representation of the potential.

We consider the following sets in Z^3:

$$\Pi_q(k^{1/5}) = \left\{ j :| (\mathbf{p}_j(0), \mathbf{p}_q(0)) |< k^{1/5}, \right\} \tag{4.2.3}$$

[1]

$$T(k, R_0) = \{j : \exists q, q' \in \Gamma(R_0), q \neq q' :$$
$$| (\mathbf{p}_j(0), \mathbf{p}_q(0)) |< k^{1/5}, | (\mathbf{p}_j(0), \mathbf{p}_q(0)) |< k^{3/5}\} \tag{4.2.4}$$

[1] In fact, there is an auxiliary coefficient in front of $k^{1/5}$, which is equal to 1/5 or 5. This coefficient arises for technical reasons; we drop it here to describe the principal scheme.

Let us define a diagonal projection P_q as follows:

$$(P_q)_{jj} = \begin{cases} 1, & \text{if } j \in \Pi_q(k^{1/5}) \setminus T(k, R_0); \\ 0, & \text{otherwise.} \end{cases} \tag{4.2.5}$$

We define the model operator $\hat{H}(t)$ by the formula

$$\hat{H}(t) = H_0(t) + \sum_{q \in \Gamma(R_0)} P_q V_q P_q. \tag{4.2.6}$$

Let

$$W = V - \sum_{q \in \Gamma(R_0)} P_q V_q P_q, \tag{4.2.7}$$

i.e., $H(t) = \hat{H}(t) + W$. Further,

$$\hat{A}(z,t,W) = (\hat{H}(t) - z)^{-1/2} W (\hat{H}(t) - z)^{-1/2},$$

$$\hat{g}_r(k,t) = \frac{(-1)^r}{2\pi i r} \mathrm{Tr} \oint_{C_0} ((\hat{H}(t) - z)^{-1} W)^r dz, \tag{4.2.8}$$

$$\hat{G}_r(k,t) = \frac{(-1)^{r+1}}{2\pi i} \oint_{C_0} ((\hat{H}(t) - z)^{-1} W)^r (\hat{H}(t) - z)^{-1} dz,$$

C_0 being the circle of the radius $k^{-1-\delta}$ about the point $z = k^2$. In Section 4.4 we describe a nonsingular set $\chi_3(k, V, \delta)$ of an asymptotically full measure on $S_0(k)$ such that for any t of this set the operator $\hat{H}(t)$ has a unique eigenvalue $p_j^2(t)$ inside C_0, j being uniquely determined from the relation $p_j^2(t) = k^2$. This assertion is stable with respect to t: If t is of the $(k^{-2-2\delta}$-neighborhood of $\chi_3(k, V, \delta)$, then the operator $\hat{H}(t)$ has a unique eigenvalue $p_j^2(t)$ inside C_0, j being uniquely determined from the relation $p_j^2(t) \in \varepsilon(k, \delta)$[2] The spectral projection E_j (the same as for the free operator) corresponds to this eigenvalue. For the operator $\hat{A}$ we have:

$$\|\hat{A}\| < k^{2\delta}, \quad \|\hat{A}^3\| < k^{-1/5+16\delta}. \tag{4.2.9}$$

and

$$\hat{g}_r = g_r, \quad \hat{G}_r = G_r \tag{4.2.10}$$

when $|r| < k^{k^{1-5\delta} R_0^{-1}}$. The following theorem holds:

Theorem 4.1 . *Suppose t is in the $(k^{-2-2\delta})$-neighborhood of the nonsingular set $\chi_3(k, V, \delta)$, $0 < 2\delta < 1/100$. Then for sufficiently large k, $k > k_0(V, \delta)$, in the interval $\varepsilon(k, \delta) \equiv [k^2 - k^{-1-\delta}, k^2 + k^{-1-\delta}]$ there exists a single eigenvalue of the operator H. It is given by series:*

$$\lambda(t) = p_j^2(t) + \sum_{r=2}^{M_1} g_r(k,t) + \varphi(k,t), \tag{4.2.11}$$

[2] $\varepsilon(k, \delta) \equiv [k^2 - k^{-1-\delta}, k^2 + k^{-1-\delta}]$.

$$\varphi(k,t) = \sum_{m=M_1+1}^{\infty} \hat{g}_r(k,t),$$

$$M_1 \equiv [k^{1-5\delta} R_0^{-1}],$$

where j is uniquely determined from the relation $p_j^2(t) \in \varepsilon(k,\delta)$. The spectral projection corresponding to $\lambda(t)$ is determined by the series:

$$E(t) = E_j + \sum_{r=2}^{M_1} G_r(k,t) + \psi(k,\delta), \tag{4.2.12}$$

$$\psi(k,t) = \sum_{m=M_1+1}^{\infty} \hat{G}_r(k,t),$$

which converges in the class $\mathbf{S}_1$.

For the functions $g_r(k,t)$ and the operator-valued functions $G_r(k,t)$ the estimates

$$| g_r(k,t) | < k^{-1-\delta-r/20}, \tag{4.2.13}$$

$$\|G_r(k,t)\|_1 < k^{-r/20} \tag{4.2.14}$$

hold. The similar estimates are fullfiled for $\hat{g}_r(k,t)$, $\hat{G}_r(k,t)$:

$$| \hat{g}_r(k,t) | < k^{-1-\delta-r/20} \tag{4.2.15}$$

$$\|\hat{G}_r(k,t)\|_1 < k^{-r/20} \tag{4.2.16}$$

Corollary 4.1 . *For the function $\varphi(k,t)$ and the operator-valued function $\psi(k,t)$ there are the estimates:*

$$|\varphi(k,t)| < k^{-k^{1-6\delta}}, \tag{4.2.17}$$

$$\|\psi(k,\delta)\|_1 < k^{-k^{1-6\delta}}. \tag{4.2.18}$$

We obtain estimates (4.2.17), (4.2.18) adding estimates (4.2.15), (4.2.16) for $\hat{g}_r(k,t)$, $\hat{G}_r(k,t)$ for $r > M_1$.

It turns out that estimates (4.2.13) and (4.2.14) can be improved when $r < k^\delta R_0$. This is what the following theorem about:

Theorem 4.2 . *Under the conditions of Theorem 3.1 there are the estimates:*

$$|g_r(k,t)| < \hat{v} r^2 (\hat{v} k^{-1+3\delta})^{r-1}, \tag{4.2.19}$$

$$\|G_r(k,t)\| < (\hat{v} k^{-1+3\delta})^r, \tag{4.2.20}$$

$$\|G_r(k,t)\|_1 < (rR_0)^3 (\hat{v} k^{-1+3\delta})^r, \tag{4.2.21}$$

$$|g_2(k,t)| < \hat{v}^2 R_0^{-1} k^{-2+6\delta}. \tag{4.2.22}$$

$$\hat{v} \equiv c_0 (\max_{|q|<R_0} |v_q|) R_0^3, c_0 \neq c_0(k,V)$$

for $r < M_2$, $M_2 = k^\delta R_0^{-1}$.

The operator $G_r(k,t)$, $r \in N$ is nonzero only on the finite-dimensional subspace $(\sum_{i \in Z^n, |i-j|<rR_0} E_i) l_2^n$.

Corollary 4.2 . *The perturbed eigenvalue and its spectral projection satisfy the following estimates:*

$$| \lambda(t) - p_j^2(t) | \leq c\hat{v}^2(\hat{v} + R_0^{-1})k^{-2+6\delta}, \tag{4.2.23}$$

$$\|E(t) - E_j\|_1 \leq c\hat{v}R_0^3k^{-1+3\delta}. \tag{4.2.24}$$

Theorem 4.3 . *Under the conditions of Theorem 3.1 the functions $g_r(k,t)$, $\hat{g}_r(k,t)$ and the operator-valued functions $G_r(k,t)$, $\hat{G}_r(k,t)$ depend analytically on t in the complex $(k^{-2-2\delta})$-neighborhood of each simply connected component of the nonsingular set $\chi_3(k,V,\delta)$. They satisfy the estimates:*

$$| T(m)\hat{g}_r(k,t) | < m!k^{-1-\delta-r/20+2(1+\delta)|m|}, \tag{4.2.25}$$

$$\|T(m)\hat{G}_r(k,t)\| < m!k^{-r/20+2(1+\delta)|m|}, \tag{4.2.26}$$

$$| T(m)g_r(k,t) | < m!k^{-1-\delta-r/20+2(1+\delta)|m|}, \tag{4.2.27}$$

$$\|T(m)G_r(k,t)\| < m!k^{-r/20+2(1+\delta)|m|}. \tag{4.2.28}$$

If $r < M_2$, then the estimates can be improved:

$$|T(m)g_r(k,t)| < m!(c_0k^{3\delta})^{|m|}\hat{v}r^2(\hat{v}k^{-1+3\delta})^{r-1}, \tag{4.2.29}$$

$$\|T(m)G_r(k,t)\| < m!(c_0k^{3\delta})^{|m|}(\hat{v}k^{-1+3\delta})^r, \tag{4.2.30}$$

$$\|T(m)G_r(k,t)\|_1 < m!(c_0k^{3\delta})^{|m|}(rR_0)^3(\hat{v}k^{-1+3\delta})^r, \tag{4.2.31}$$

$$|g_2(k,t)| < m!(c_0k^{3\delta})^{|m|}\hat{v}^2R_0^{-1}k^{-2+6\delta}. \tag{4.2.32}$$

$$c_0 \neq c_0(k,V).$$

Corollary 4.3 . *The function $\lambda(t)$ and the operator-valued function $E(t)$ depend analytically on t in the complex $(k^{-2-2\delta})$-neighborhood of each simply connected component of the nonsingular set $\chi_3(k,V,\delta)$. They admit the estimates:*

$$| T(m)(\lambda(t) - p_j^2(t)) | \leq cm!\hat{v}^2(\hat{v} + R_0^{-1})k^{-2+6\delta+2(1+\delta)|m|}, \tag{4.2.33}$$

$$\|T(m)(E(t) - E_j)\|_1 \leq cm!\hat{v}R_0^3k^{-1+3\delta+2(1+\delta)|m|}. \tag{4.2.34}$$

If $|m| < k^\delta(60R_0)^{-1}$, then the estimates can be improved:

$$| T(m)(\lambda(t) - p_j^2(t)) | \leq cm!(c_0k^{3\delta})^{|m|}\hat{v}^2(\hat{v} + R_0^{-1})k^{-2+6\delta}, \tag{4.2.35}$$

$$\|T(m)(E(t) - E_j)\|_1 \leq cm!(c_0k^{3\delta})^{|m|}\hat{v}R_0^3k^{-1+3\delta}. \tag{4.2.36}$$

Using estimates (4.2.25) – (4.2.28) yields inequalities (4.2.33) and (4.2.34). To obtain (4.2.35) and (4.2.36) we consider estimates (4.2.29), (4.2.31) and (4.2.32) for the initial terms of the asymptotics. If $|m| < k^\delta(60R_0)^{-1}$, then the first term of asymptotic is estimated from above by the largest value. Thus, we obtain (4.2.35) and (4.2.36).

Corollary 4.4 . *The function $\varphi(k,t)$ and the operator-valued function $\psi(k,t)$ depend analytically on t in the complex $(k^{-2-2\delta})$-neighborhood of each simply connected component of the nonsingular set $\chi_3(k,V,\delta)$. They satisfy the following estimates:*

$$|\varphi(k,t)| < m!k^{-k^{1-6\delta}+2(1+\delta)|m|}, \tag{4.2.37}$$

$$\|\psi(k,\delta)\|_1 < m!k^{-k^{1-6\delta}+2(1+\delta)|m|}. \tag{4.2.38}$$

4.3 Preliminary Consideration.

Let $t \in S_0(k)$. Then the operator $H_0(t)$ has an eigenvalue equal to k^2, i.e., there exists a $j \in Z^3$ such that $p_j^2(t) = k^2$. In Lemma 4.1 we formulate the condition 1^0, under which the perturbed operator $H(t)$ has an eigenvalue $\lambda(t)$ in the interval $\varepsilon(k, \delta)$ for sufficiently large k. Further, arguing on a qualitative level, we get the conditions $2^0 - 5^0$ below, which ensure the uniqueness of the eigenvalue $\lambda(t)$ in the interval $\varepsilon(k, \delta)$. We suppose $V(x)$ to be a trigonometric polynomial.

Lemma 4.1 . *Suppose that $t \in S_0(k)$ and let j be such that $p_j(t) = k$. Let $0 < \delta < 1/200$, $\delta \neq \delta(k)$ and t satisfy the condition*

$$1^0 \qquad |p_{j+q}^2(t) - p_j^2(t)| > k^{1-3\delta} \tag{4.3.1}$$

for all $q \neq 0$, $|q| < k^\delta$. Then for sufficiently large k, $k > k_0(V, \delta)$, there exists at least one eigenvalue of $H(t)$ given by the asymptotic formula

$$\lambda(t) = p_j^2(t) + \sum_{r=2}^{r_0(k)} g_r(k, t) + O(k^{-(1-4\delta)r_0(k)}), \tag{4.3.2}$$

where $r_0(k) = [k^\delta R_0^{-1}]$. For $g_r(k, t)$ the estimates hold:

$$|g_r(k, t)| < r^2 \hat{v}(c\hat{v}R_0^3 k^{-(1-3\delta)})^{(r-1)}, \tag{4.3.3}$$

$$|g_2(k, t)| < c\hat{v}^2(R_0^{-1}k^{-2+6\delta}), \quad c \neq c(k, V). \tag{4.3.4}$$

Corollary 4.5 . *For the perturbed eigenvalue $\lambda(t)$ the estimate holds:*

$$| \lambda(t) - p_j^2(t) | \leq c\hat{v}^2(\hat{v} + R_0^{-1})k^{-2+6\delta}. \tag{4.3.5}$$

Proof. The operator $H(t)$ is given by the infinite matrix (1.0.9) (l=1). We "excise" from it the finite matrix with the center at the point (j, j) and row of order k^δ, i.e., we consider the matrix of the operator $PH(t)P$, where P is a diagonal projection, $P_{ii} = 1$ if $|i - j| < k^\delta$, and $P_{ii} = 0$ otherwise. It is clear from condition 1^0 that the eigenvalues $p_{j+q}^2(t)$, $q \neq 0$, of the operator $PH_0(t)P$ are at a distance greater than $k^{1-3\delta}$ from $p_j^2(t)$. Therefore, when operator $PH_0(t)P$ is perturbed by the operator PVP, the interval $[p_j^2(t) - \|V\|, p_j^2(t) + \|V\|]$ turns out to contain a unique eigenvalue $\tilde{\lambda}(t)$ of $PH(t)P$, which arises from the eigenvalue $p_j^2(t)$ of the unperturbed problem. Following the method described in Chapter 2 (Theorem 2.1), we show that $\tilde{\lambda}(t)$ and the spectral projection $\tilde{E}(t)$ corresponding to $\tilde{\lambda}(t)$ can be expressed by the formulas

$$\tilde{\lambda}(t) = p_j^2(t) + \sum_{r=2}^{\infty} \tilde{g}_r(k, t),$$

$$\tilde{E}(t) = E_j + \sum_{r=1}^{\infty} \tilde{G}_r(k, t),$$

$$\tilde{g}_r(k,t) = \frac{(-1)^r}{2\pi i r}\mathrm{Tr}\oint_{C_2}(P(H_0(t)-z)^{-1}VP)^r dz$$

$$\tilde{G}_r(k,t) = \frac{(-1)^{r+1}}{2\pi i}\oint_{C_2}(P(H_0(t)-z)^{-1}V)^r(H_0(t)-z)^{-1}Pdz,$$

where C_2 is a contour of radius $(1/2)k^{1-3\delta}$ about the point $z = k^2$ on the complex plane. Furthermore,

$$|\tilde{g}_r(k,t)| < r^2\hat{v}(c\hat{v}R_0^3k^{-1+3\delta})^{(r-1)}, \tag{4.3.6}$$

$$\|\tilde{G}_r(k,t)\| < (c\hat{v}k^{-1+3\delta})^r. \tag{4.3.7}$$

From this,

$$\tilde{\lambda}(t) = p_j^2(t) + \sum_{r=2}^{r_0(k)}\tilde{g}_r(k,t) + O(k^{-(1-3\delta)r_0(k)}),$$

$$\tilde{E}(t) = E_j + \sum_{r=1}^{r_0(k)}\tilde{G}_r(k,t) + O(k^{-(1-3\delta)r_0(k)}).$$

We prove that $\tilde{g}_r(k,t) = g_r(k,t)$ if $r \le r_0(k)$. Indeed, the contour C_2 contains only one pole of the integrand, at the point $z = k^2$; therefore C_2 can be replaced by the contour C_0 of radius $k^{-1-\delta}$. Further,

$$g_r(k,t) = \sum_{i_0,\dots,i_{r-1}\in Z^3} I_{i_0 i_1\dots i_{r-1}},$$

where

$$I_{i_0 i_1\dots i_{r-1}} = \frac{(-1)^r}{2\pi i r}\oint_{C_0}\frac{v_{i_0-i_1}\dots v_{i_{r-1}-i_0}}{(p_{i_0}^2(t)-z)\dots(p_{i_{r-1}}^2(t)-z)}dz.$$

Obviously, the contour integral can be nonzero only when the integrand has a pole inside the contour, i.e., when at least one of the indices $i_0,\dots,i_r$ is equal to j. By considering that V is a trigonometric polynomial we obtain $|i_q - j| \le rR_0 \le r_0(k)R_0 \le k^\delta$, $q = 0,\dots r-1$. But in this case the projection P acts as I, and therefore, $\tilde{g}_r(k,t) = g_r(k,t)$ if $r \le r_0(k)$. Similarly, $\tilde{G}_r(k,t) = G_r(k,t)$ if $r \le r_0(k)$. Thus, (4.3.3) follows from (4.3.6). For $l = 1$, $n = 3$ formula (2.2.12) has the form:

$$g_2(k,t) = \sum_{q\in Z^3, q\neq 0} |v_q|^2 (p_j^2(t) - p_{j+q}^2(t))^{-1} =$$

$$\sum_{q\in Z^3, q\neq 0}\frac{|v_q|^2(2p_j^2(t) - p_{j+q}^2(t) - p_{j-q}^2(t))}{2(p_j^2(t) - p_{j+q}^2(t))(p_j^2(t) - p_{j-q}^2(t))}. \tag{4.3.8}$$

Using inequality (4.3.1) yields (4.3.4). It is now obvious that $\tilde{\lambda}(t)$ and $\tilde{E}(t)$ can be represented in the form

$$\tilde{\lambda}(t) = p_j^2(t) + \sum_{r=2}^{r_0} g_r(k,t) + O(k^{-(1-3\delta)r_0(k)}),$$

$$\tilde{E}(t) = E_j + \sum_{r=1}^{r_0} G_r(k,t) + O(k^{-(1-3\delta)r_0(k)}),$$

and

$$\|\tilde{E}(t) - E_j\| < c\hat{v}k^{-1+3\delta}. \tag{4.3.9}$$

We introduced in Pl_2^3 a new basis composed of the eigenfunctions of the operator $PH(t)P$, and we see how the matrix of $H(t)$ then change. Let $U_0 : Pl_2^3 \to Pl_2^3$, $U_0^{-1} = U_0^*$, be an isometry reducing $PH(t)P$ to diagonal form. We consider in l_2^3 the unitary operator $U(t)$ acting in Pl_2^3 as U_0, and in the orthogonal complement of Pl_2^3 as I. Let $H'(t) = UH(t)U^*$. We represent $H'(t)$ in the form:

$$H'(t) = H''(t) + H'''(t),$$

$$H'''(t)_{nr} = \begin{cases} H'(t)_{jr} & \text{if } n = j,\ r \neq j; \\ H'(t)nj & \text{if } n \neq j,\ r = j; \\ 0 & \text{if } n \neq j,\ r \neq j, \quad \text{or } n = r = j. \end{cases}$$

We prove that $\|H'''(t)\| = O(k^{-(1-4\delta)r_0(k)})\|$. It is follows from the definitions of U and $H'(t)$ that $H'(t)_{jn} = H'(t)_{nj} = 0$ if $|n-j| < k^\delta$, $n \neq j$. We consider the case $|n-j| \geq k^\delta$. Let $\tilde{\psi}_m$ be the mth coordinate of the eigenvector $\tilde{\psi}$ of $PH(t)P$ corresponding to $\tilde{\lambda}(t)$. Then, by the definition of $H'(t)$,

$$H'(t)_{jn} = \sum_{m \in Z^3} H(t)_{mn}\bar{\tilde{\psi}}_m = \tilde{\psi}_j^{-1} \sum_{m \in Z^3} \tilde{E}(t)_{jm} H(t)_{mn} =$$

$$= \tilde{\psi}_j^{-1}\left(\sum_{r=1}^{\infty} (\tilde{G}_r H)_{jn} + H_{jn}\right).$$

Since $|j-n| \geq k^\delta > R_0$, we have $H_{jn} = 0$. Estimate (4.3.9) means that $|\tilde{\psi}_j|^2 = 1 + O(k^{-1+3\delta})$. From the formula for $\tilde{G}_r$,

$$\tilde{G}_r(k,t)_{jm} = \sum_{i_1,\ldots,i_{r-1} \in Z^3} \frac{(-1)^{r+1}}{2\pi i} \oint_{C_0} \frac{v_{j-i_1} \ldots v_{i_{r-1}-m}}{(p_j^2(t) - z)(p_{i_1}^2(t) - z)\ldots(p_m^2(t) - z)} dz, \tag{4.3.10}$$

it follows that $\tilde{G}_r(k,t)_{jm} = 0$ if $|j-m| > rR_0$, and $(\tilde{G}_r V)_{jm} = 0$ if $|j-m| > (r+1)R_0$. Thus, for $|j-n| > (r+1)R_0$

$$(\tilde{G}_r H)_{jn} = (\tilde{G}_r)_{jn} p_n^2(t) + (\tilde{G}_r V)_{jn} = 0.$$

Hence,

$$H'(t)_{jn} =$$

$$= \tilde{\psi}_j^{-1} \left(\sum_{r > |n-j| R_0^{-1} - 1} (\tilde{G}_r V)_{jn} + p_n^2(t) \sum_{r > |n-j| R_0^{-1}} (\tilde{G}_r)_{jn} \right).$$

Considering (4.3.7) and the obvious estimate $p_n^2(t) < k^2 + c|n-j|^2$, $c \neq c(k)$, we obtain that

$$|H'(t)_{jn}| < (k^2 + c|n-j|^2)(ck^{-1+3\delta})^{|n-j|R_0^{-1}} < k^{-(1-4\delta)|n-j|R_0^{-1}}.$$

Since $|n-j| \geq k^\delta$, it is easy to see that $\|H'''(t)\| = O(k^{-(1-4\delta)r_0(k)})$. Since $H'(t)_{jj} = \tilde{\lambda}(t)$, the latter is an eigenvalue of the operator $H''(t)$, and the operator $H(t)$ has at least one eigenvalue $\lambda(t)$ lying in the $(\|H'''\|)$-neighborhood $\tilde{\lambda}(t)$, i.e., such that (4.3.2) holds. The lemma is proved.

The lemma similar to Lemma 4.1 was proved in [Ve2]. in another way (with asymptotic terms defined by a recurrent procedure.)

Assume that $\lambda(t)$ is the only eigenvalue in the interval $\varepsilon(k,\delta)$. Then it follows from the general theory of perturbations (see f.e. [Kato]) that in some neighborhood of t the function $\lambda(t)$ and the corresponding spectral projection $E(t)$ depend analytically on t , i.e., can be represented in the form $\lambda(t) = \tilde{\lambda}(t) + \varphi_1(t)$ and $E(t) = \tilde{E}(t) + \psi_1(t)$, with an analytic function φ_1 and an analytic operator-valued function ψ_1 such that $|\varphi_1(t)| + \|\psi_1(t)\|k^{-1-\delta} = O(k^{-(1-4\delta)r_0(k)})$. Therefore, the hope arises of obtaining expansions of $\lambda(t)$ and $E(t)$ in asymptotic series with respect to k that depend analytically on t.

We present qualitative arguments helping us to clear up when the interval $\varepsilon(k,\delta)$ contains a single eigenvalue. We consider "shifts" Δ_m in the perturbation only of the $p_m^2(t)$ such that $|p_m^2(t) - k^2| < 2\|V\|$. Obviously, the perturbed eigenvalues do not fall in $\varepsilon(k,\delta)$ otherwise. Consider first the $p_m^2(t)$ satisfying the inequality $|p_m^2(t) - k^2| < k^{-12\delta}$. Let t be such that

$$2^0 \quad |p_{m+q}^2(t) - p_m^2(t)| > k^{-4\delta}|p_m^2(t) - k^2|^{-1} \tag{4.3.11}$$

for all $m : |p_m^2(t) - k^2| < k^{-2\delta}$, $|q| < k^\delta$, $q \neq 0$.

If $|p_m^2(t) - k^2| < k^{-12\delta}$, then we have that

$$|p_{m+q}^2(t) - p_m^2(t)| > k^{-4\delta}|p_m^2(t) - k^2|^{-1} > k^{8\delta} > \|V\|.$$

Repeating the arguments in Lemma 4.1 with respect to $p_m^2(t)$, we get that the $(k^{10\delta}(p_m^2(t) - k^2)^2)$-neighborhood of $p_m^2(t)$ contains a perturbed eigenvalue that arose from $p_m^2(t)$.

Assume the condition

$$3^0 \quad \min_{m\in Z^3, m\neq j} |p_m^2(t) - p_j^2(t)| > 2k^{-1-\delta} \tag{4.3.12}$$

($p_j^2(t) = k^2$) is satisfied. Since

$$k^{10\delta}(p_m^2(t) - k^2)^2 =_{k\to\infty} o(|p_m^2(t) - k^2|),$$

the eigenvalue $(p_m^2 + \Delta_m)(t)$ of the operator $H(t)$ lies outside $\varepsilon(k,\delta)$. It remains to consider the $p_m^2(t)$ such that

$$k^{-12\delta} \leq |p_m^2(t) - k^2| < 2\|V\|. \tag{4.3.13}$$

We first make a slight degression. Consider in l_2^1 the family of operators $H_1(\tau)$, $0 \le \tau < \pi$, corresponding to the one-dimensional Schrödinger operator with a periodic potential

$$H_1(\tau)_{n_1 n_2} = (2\pi n_1 + \tau)^2 a^{-2}\delta_{n_1 n_2} + \hat{v}_{n_1 - n_2}, \quad n_1, n_2 \in Z, \tag{4.3.14}$$

where $a > 0$, $\hat{v}_n = \overline{\hat{v}_{-n}}$, $\hat{v}_n = 0$ if $n = 0$ or $|n| > N$, $0 < N < \infty$. The operator $H_1(\tau)$ has a discrete spectrum. It is well known that for each τ the eigenvalues $\lambda_l(\tau)$ of it can be enumerated by integers in such a way that

$$|\lambda_l(\tau) - (2\pi l + \tau)^2 a^{-2}| < c\|V\|^2 l^{-2}, \quad l \in Z. \tag{4.3.15}$$

Further, the function $\lambda_l(\tau)$ is piecewise continuous, and on the smooth parts

$$|d\lambda_l(\tau)/d\tau|_{l\to\infty} = O(l). \tag{4.3.16}$$

Let us now consider the three-dimensional case of a "simple" potential V_q depending only on one variable $(x, \mathbf{p}_q(0))$, $q \in R^3$. We denote by H_q the corresponding Schrödinger operator. Let us consider the matrix of $H_q(t)$ in l_2^3. We associate with each i in Z^3 the diagonal projection P_i^q in l_2^3:

$$(P_i^q)_{mm} = \begin{cases} 1, & \text{if } \mathbf{p}_i(0) - \mathbf{p}_m(0) = l\mathbf{p}_q(0),\ l \in Z, \\ 0 & \text{otherwise.} \end{cases} \tag{4.3.17}$$

Obviously, $P_i^q = P_m^q$ if $\mathbf{p}_i(0) - \mathbf{p}_m(0) = l\mathbf{p}_q(0)$, $l \in Z$. It is clear that there exists a minimal subset J_q^0 of Z^3 such that

$$\sum_{i \in J_q^0} P_i^q = I. \tag{4.3.18}$$

It is obvious that

$$H_q(t) = \sum_{i \in J_q^0} H_q(t) P_i^q. \tag{4.3.19}$$

Note that $(V_q)_{rn} = 0$ if $\mathbf{p}_r(0) - \mathbf{p}_n(0) \ne l\mathbf{p}_q(0)$, $l \in Z$, because V_q depends only on $(x, vecp_q(0))$. This at once gives us that

$$P_i^q V_q P_i^q = P_i^q V_q = V_q P_i^q, \tag{4.3.20}$$

$$P_i^q H_q(t) P_i^q = P_i^q H_q(t) = H_q(t) P_i^q. \tag{4.3.21}$$

Considering relations (4.3.18) and (4.3.21) we get

$$H_q(t) = \sum_{i \in J_q^0} P_i^q H_q(t) P_i^q. \tag{4.3.22}$$

We establish an isometric isomorphism between $P_i^q l_2^3$ and l_2^1. Denote by δ_m, $m \in Z^3$, the element of l_2^1 given by the formula $\{\delta_m\}_n = \delta_{mn}$, and by δ_l^1, $l \in Z$, the analogous element of l_2^1 given by the formula $\{\delta_l^1\}_r = \delta_{lr}$. To construct an isometric isomorphism we represent $\mathbf{p}_i(t)$, $i \in J_q^0$, in the form of a linear combination of $\mathbf{p}_q(0)$ and a vector in the orthogonal complement of $\mathbf{p}_q(0)$: $\mathbf{p}_i(t) = \tau_i(2\pi)^{-1}\mathbf{p}_q(0) + \mathbf{d}_i^q$, $\tau_i \in R$, $\tau_i = \tau_i(t,q)$, $(\mathbf{d}_i^q(t), \mathbf{p}_q(0)) = 0$. Thus, if $\delta_m \in P_i^q l_2^3$, then $\mathbf{p}_m(t)$ is uniquely representable in the form

$$\mathbf{p}_m(t) = \mathbf{d}_i^q(t) + (\tau_i + 2\pi l)(2\pi)^{-1}\mathbf{p}_q(0), \quad l \in Z, \tag{4.3.23}$$

where it can be assumed without loss of generality that $0 \le \tau_i < 2\pi$. From this,

$$(P_i^q)_{mm} = \begin{cases} 1 & \text{if } \mathbf{p}_m(t) = \mathbf{d}_i^q(t) + (\tau_i + 2\pi l)(2\pi)^{-1}\mathbf{p}_q(0), \quad l \in Z, \\ 0 & \text{otherwise.} \end{cases} \tag{4.3.24}$$

An isomorphism between $P_i^q l_2^3$ and l_2^1 is now established in the natural way with the help of the formula (4.3.23): $\delta_m \leftrightarrow \delta_l^1$, $\delta_m \in P_i^q l_2^3$, $\delta_l^1 \in l_2^1$, and it follows from (4.3.23) that

$$l = [(\mathbf{p}_m(t), \mathbf{p}_q(0))p_q(0)^{-2}], \tag{4.3.25}$$

$$\tau_i = 2\pi(\mathbf{p}_m(t), \mathbf{p}_q(0))p_q(0)^{-2} - 2\pi l. \tag{4.3.26}$$

It is easy to verify that the operator $P_i^q H_q P_i^q$ is equivalent to the operator $H_1(\tau_i) + |(\mathbf{d}_i^q(t)|^2 I$; here the operator $H_1(\tau_i)$ is of the type (4.3.14). Its matrix elements are

$$H_1(\tau_i)_{lp} = (\tau_i + 2\pi l)^2(2\pi)^{-2}p_q(0)^{-2}\delta_{lp} + v_{(l-p)q}, \quad l,p \in Z. \tag{4.3.27}$$

We denote the eigenvalues of this operator by $\lambda_l^q(\tau_i)$, $l \in Z$. Let $\Delta\lambda_l^q(\tau_i)$ be the shift of the eigenvalue in perturbation: $\Delta\lambda_l^q(\tau_i) = \lambda_l^q(\tau_i) - (\tau_i + 2\pi l)^2(2\pi)^{-2}p_q(0)^{-2}$. As already mentioned, we can always choose an enumeration such that

$$|\Delta\lambda_l^q(\tau_i)| < c_1 l^{-2}, \quad 0 < c_1 \ne c_1(l) \tag{4.3.28}$$

Thus, the spectrum of the operator $P_i^q H_q P_i^q$ can be represented in the form:

$$\begin{aligned} \{\lambda_l^q(\tau_i) + |(\mathbf{d}_i^q(t)|^2\}_{l\in Z} &= \\ = \{\Delta\lambda_l^q(\tau_i) + (\tau_i + 2\pi l)^2(2\pi)^{-2}p_q(0)^2 + |(\mathbf{d}_i^q(t)|^2\}_{l\in Z} &= \\ = \{\Delta\lambda_{l(m)}^q(\tau_i) + p_m^2(t)\}_{m:\delta_m \in P_i^q l_2^3} &= \\ = \{\Delta\lambda_{l(m)}^q(\tau_m) + p_m^2(t)\}_{m:\delta_m \in P_i^q l_2^3}&; \end{aligned} \tag{4.3.29}$$

here τ_m can be computed from the formula

$$\tau_m = 2\pi(\mathbf{p}_m(t), \mathbf{p}_q(0))p_q(0)^{-2} - 2\pi[(\mathbf{p}_m(t), \mathbf{p}_q(0))p_q(0)^{-2}] \tag{4.3.30}$$

and, as it is easily seen, τ_m coincides with τ_i for all $m : \delta_m \in P_i^q l_2^3$ and $l(m)$ is given by the formula (4.3.25). The spectrum $\Lambda_q(t)$ of the operator H_q is the union of the spectra of the operators $P_i^q H_q(t) P_i^q$:

$$\Lambda_q(t) = \cup_{i\in J_q^0}\{\Delta\lambda_{l(m)}^q(\tau_m) + p_m^2(t)\}_{m:\delta_m \in P_i^q l_2^3} = \{\Delta\lambda_{l(m)}^q(\tau_m) + p_m^2(t)\}_{m\in Z^3}.$$

Introducing the diagonal matrix $\Delta\Lambda_q(t)$ of shifts of the eigenvalues,

$$\Delta\Lambda_q(t)_{mm} = \Delta\lambda^q_{l(m)}(\tau_m), \tag{4.3.31}$$

we get that the diagonal matrix $\Lambda_q(t)$ of eigenvalues of the operator $H_q(t)$ is $H_0(t) + \Delta\Lambda_q(t)$, and

$$|\Delta\Lambda_q(t)_{mm}| < c_1|(\mathbf{p}_m(0), \mathbf{p}_q(0))|^{-2}, \quad 0 < c_1 \neq c_1(m), \tag{4.3.32}$$

$$|\nabla_t \Delta\Lambda_q(t)_{mm}| < c_1|(\mathbf{p}_m(0), \mathbf{p}_q(0))|p_q(0)^{-1}. \tag{4.3.33}$$

The last relation holds, where $\Delta\Lambda_q(t)_{mm}$ is smooth.

We return to the consideration of perturbed eigenvalues satisfying the estimate (4.3.13). First let $V = V_q$, i.e., the case of a simple potential. If the projection of $\mathbf{p}_m(0)$ on $\mathbf{p}_q(0)$ is sufficiently large, namely, $|(\mathbf{p}_m(0), \mathbf{p}_q(0))| > k^{1/5}$, [3] then $\Delta_m = \Delta\Lambda_q(t)_{mm} = O(k^{-2/5}) = o(k^{-12\delta}) = o(|p^2_m(t) - k^2|)$. Consequently, the perturbed eigenvalues $(p^2_m(t) + \Delta_m)(t)$ do not fall in $\varepsilon(k,\delta)$ for $m : |(\mathbf{p}_m(0), \mathbf{p}_q(0))| > k^{1/5}$. In the case of the general potential $V = \sum_{q\in\Gamma(R_0)} V_q$ the eigenvalues $p^2_m(t)$ of the unperturbed operator such that $|(\mathbf{p}_m(0), \mathbf{p}_q(0))| \le k^{1/5}$ for at least one q in $\Gamma(R_0)$ are most sensitive to the perturbation. The shift of the remaining ones is much less than $k^{-12\delta}$, and they do not fall in $\varepsilon(k,\delta)$. Suppose that for some m there exists a unique q such that $|(\mathbf{p}_m(0), \mathbf{p}_q(0))| \le k^{1/5}$, while for all q' in $\Gamma(R_0)$ not equal to q the estimate $|(\mathbf{p}_m(0), \mathbf{p}_{q'}(0))| > k^{3/5}$ holds. Then the shift of the eigenvalue $p^2_m(t)$ takes place mainly because of the potential V_q, which, as already mentioned, moves the eigenvalue $p^2_m(t)$ to the eigenvalue $p^2_m(t) + \Delta\Lambda_q(t)_{mm}$ of $H_q(t)$. The remaining potentials $V_{q'}$, $q' \neq q$ have a weak influence on the eigenvalue $p^2_m(t)$. They give a shift only of order $k^{-3/5}$. Thus, under perturbation of $p^2_m(t)$ by the general potential $V = \sum_{q\in\Gamma(R_0)} V_q$ it passes into $(p^2_m + \Delta_m)(t)$, where

$$\Delta_m(t) = \Delta\Lambda_q(t)_{mm} + O(k^{-3/5}). \tag{4.3.34}$$

Denote by $\Pi_q(k^\eta)$ the set $\{m\}$ of indices such that $\mathbf{p}_m(0)$ has sufficiently small ($< k^\eta$) projection on $\mathbf{p}_q(0)$:

$$\Pi_q(k^\eta) = \{m, m \in Z^3 : |(\mathbf{p}_m(0), \mathbf{p}_q(0))| \le k^\eta\}. \tag{4.3.35}$$

We assume that the following condition holds for t.

4^0. For any q in $\Gamma(R_0)$ and $m \in \Pi_q(k^{1/5})$ the eigenvalue $p^2_m(t) + \Delta\Lambda_q(t)_{mm}$ of the operator $H_q(t)$ lies sufficiently far from the point $\lambda = k^2$, namely,

$$|p^2_m(t) + \Delta\Lambda_q(t)_{mm} - k^2| > k^{-1/5-\delta}. \tag{4.3.36}$$

Under condition 4^0 the estimate (4.3.34) implies the following qualitative assertion: if $m \in \Pi_q(k^{1/5})$ for some q in $\Gamma(R_0)$, and $m \notin \Pi_{q'}(k^{3/5})$ for all q' in $\Gamma(R_0)$

[3] In these arguments the exponent 1/5 can be replaced by any exponent exceeding 6δ. However, for optimality of the estimates we take the exponent 1/5 here and below.

not equal to q, then $2|p_m^2(t) + \Delta_m(t) - k^2| > k^{-1/5-\delta}$, i.e., $p_m^2(t) + \Delta_m(t)$ does not fall in $\varepsilon(k, \delta)$.

It remains for us only to consider those m for which there exists at least one pair q, q' ($q \neq q'$) in $\Gamma(R_0)$ such that $m \in \Pi_q(k^{1/5}) \cap \Pi_{q'}(k^{3/5})$, i.e., in the set $\Pi_0(k)$,

$$\Pi_0(k) \equiv \cup_{q \in \Gamma(R_0)} \Pi_q(k^{1/5}) \left(\cap_{q' \in \Gamma(R_0), q \neq q'} \Pi_{q'}(k^{3/5}) \right). \qquad (4.3.37)$$

It turns out that t can be chosen so that the following condition holds:

$$5^0 \quad |p_m^2(t) - k^2| > k^{1/5-9\delta} \quad \text{if } m \in \Pi_0(k). \qquad (4.3.38)$$

Obviously, in this case the perturbed eigenvalue does not fall in $\varepsilon(k, \delta)$.

In the next section we shall construct the nonsingular set $\chi_3(k, V, \delta) \subset S_0(k)$ of an asymptotically full measure on $S_0(k)$, such that for any t in its $(k^{-2-2\delta})$-neighborhood there exists a unique $j : |p_j^2(t) - k^2| < 5k^{-1-2\delta}$ ($p_j^2(t) = k^2$, when $t \in \chi_3(k, V, \delta)$) and the conditions $1^0 - 5^0$ hold (we repeat them here):

1^0
$$|p_{j+q}^2(t) - p_j^2(t)| > k^{1-3\delta} \qquad (4.3.39)$$

for all $q \neq 0$, $|q| < k^\delta$.

2^0
$$|p_{m+q}^2(t) - p_m^2(t)| > k^{-4\delta}|p_m^2(t) - k^2|^{-1} \qquad (4.3.40)$$

for all $m : |p_m^2(t) - k^2| < k^{-2\delta}$, $|q| < k^\delta$, $q \neq 0$.

3^0
$$\min_{m \in Z^3, m \neq j} |p_m^2(t) - p_j^2(t)| > 2k^{-1-\delta} \qquad (4.3.41)$$

4^0. For any q in $\Gamma(R_0)$ and $m \in \Pi_q(k^{1/5})$ the eigenvalue $p_m^2(t) + \Delta \Lambda_q(t)_{mm}$ of the operator $H_q(t)$ lies sufficiently far from the point $\lambda = k^2$, namely,

$$|p_m^2(t) + \Delta \Lambda_q(t)_{mm} - k^2| > k^{-1/5-\delta}. \qquad (4.3.42)$$

5^0

$$|p_m^2(t) - k^2| > k^{1/5-10\delta}, \quad \text{if } m \in \Pi_0(k). \qquad (4.3.43)$$

Note that inequalities 1^0 and 3^0 are connected with the von Laue diffraction conditions. They are satisfied when t is far enough from the von Laue diffraction planes

$$p_{j+q}^2(t) = p_j^2(t). \qquad (4.3.44)$$

Note that for $|q| < k^\delta$ we need much stronger inequality (1^0) than for others (3^0). Such q (more precisely, q corresponding to nonzero v_q) give planes (4.3.44) producing a stronger diffraction than others. Inequality 2^0 is connected with the Associate diffraction conditions. All these conditions were already described for the case of $4l > n + 1$. They do not depend on the potential. Inequality 4^0 is connected with a new diffraction condition

$$p_m^2(t) + \Delta \Lambda_q(t)_{mm} = p_j^2(t), \qquad (4.3.45)$$

which is, obviously, depends on the potential. The inequality 4^0 is satisfied when t is far enough from the surface (4.3.45).

·Condition 5^0 do not depend on the potential. However, it is also needed in the present case (n=3) to ensure that t is far enough from the planes $p_j^2(t) = p_m^2(t)$, $m \in \Pi_0(k)$. Thus, we see that the condition 5^0 is also connected with the von Laue diffraction conditions. The point is that in the case $m \in \Pi_0(k)$ we have to keep distance which is large enough ($k^{1/5-\delta}$) from these planes, because they can provide rather strong diffraction.

4.4 Geometric Constructions.

In this section we show that for sufficiently large k there exists a subset $\chi_3(k, V, \delta)$ of $S_0(k)$ on which conditions $1^0 - 5^0$ hold and they hold even in the $(k^{-2-2\delta})$-neighborhood of this set. We construct $\chi_3(k, V, \delta)$ as the intersection of the set $\chi_1(k, \delta, \delta)$ (see Chapter 3, section 4) on which conditions $1^0 - 3^0$ hold and the set $\Upsilon_4(k, V, \delta)$ constructed below, on which condition 4^0 holds, and the set $\Upsilon_5(k, 1/5, 3/5)$, on which condition 5^0 is satisfied. We prove in Lemmas 4.2 and 4.3 that sets $\Upsilon_4(k, V, \delta)$ and $\Upsilon_5(k, 1/5, 3/5)$ have asymptotically full measures on $S_0(k)$. The main geometric Lemma 4.4 asserts that conditions $1^0 - 5^0$ hold on the set $\chi_3(k, V, \delta)$ and this set has an asymptotically full measure on $S_0(k)$.

Now, we describe $\Upsilon_4(k, V, \delta)$. Suppose that $q \in \Gamma(R_0)$, $0 < \eta < \xi < 1$, and $\Upsilon_4'(k, \xi, \eta, V_q)$ is a subset of $S_0(k)$ defined by the inequality

$$\min_{m \in \Pi_q(k^\eta)} |p_m^2(t) + \Delta A_q(t)_{mm} - k^2| > k^{-\xi}, \tag{4.4.1}$$

i.e., $t \in \Upsilon_4'(k, \xi, \eta, V_q)$ if and only if (4.4.1) holds. Let

$$\Upsilon_4(k, V, \delta) \equiv \cap_{q \in \Gamma(R_0)} \Upsilon_4'(k, 1/5 + \delta, 1/5, V_q), \tag{4.4.2}$$

$0 < \delta < 1/200$. Obviously, condition 4^0 holds for all t in $\Upsilon_4(k, V, \delta)$. We verify now that the set $\Upsilon_4'(k, 1/5 + \delta, 1/5, V_q)$ has an asymptotically full measure on $S_0(k)$. We prove this result in a somewhat more general form, which we use in Section 4.7 for the case of a nonsmooth potential.

Lemma 4.2 . *If* $0 < \eta < \xi < 1$, *then for any* $q : |q| < k^{1-(\xi-\eta)/2}$, *the set* $\Upsilon_4'(k, \xi, \eta, V_q)$ *satisfies the estimate:*

$$\frac{s(S_0(k) \setminus \Upsilon_4'(k, \xi, \eta, V_q))}{s(S_0(k))} < ck^{-(\xi-\eta)/2}, \quad c = c(\eta, \xi). \tag{4.4.3}$$

<u>Proof.</u> Let $\pi_m^q(k, \xi)$, $m \in \Pi_q(k^\eta)$ be the subset of $S_0(k)$ containing precisely those t such that $|p_m^2(t) + \Delta A_q(t)_{mm} - k^2| < 2k^{-\xi}$ It is clear that

$$\Upsilon_4'(k, \xi, \eta, V_q)) = S_0(k) \setminus \hat{\pi},$$

$$\hat{\pi}(k, \xi, \eta, V_q) = \cup_{m \in \Pi_q(k^\eta)} \pi_m^q(k, \xi),$$

The lemma will be proved if we show that

$$\frac{s(S_0(k)\cap\hat{\pi}(k,\xi,\eta,V_q))}{s(S_0(k))} < c(\xi,\eta)k^{-(\xi-\eta)/2}, \tag{4.4.4}$$

where $c(\xi,\eta)$ does not depend on k. As already mentioned, $S_0(k)$ consists of the "tiles" $S_0(k)_j$: $S_0(k)_j=\{t: t\in K,\ p_j(t)=k\}$. Let us consider the total area of the "tiles" such that $j\in\Pi_q(k^\gamma)$, here γ is in $(0,1)$ and independent of k. It is easy to see that this area is infinitesimally small in comparison with $s(S_0(k))$, namely,

$$s(\cup_{j\in\Pi_q(k^\gamma)}S_0(k)_j)=ck^{\gamma-1}s(S_0(k)),\quad c\neq c(k).$$

Therefore,

$$s(S_0(k)\cap\hat{\pi}(k,\xi,\eta,V_q))\le\sum_{j\notin\Pi_q(k^\gamma)}s(S_0(k)_j\cap\hat{\pi}(k,\xi,\eta,V_q))+ck^{\gamma-1}s(S_0(k))\le$$

$$\le cs(S_0(k))\left(\sup_{j\notin\Pi_q(k^\gamma)}s(S_0(k)_j\cap\hat{\pi}(k,\xi,\eta,V_q))+k^{\gamma-1}\right).$$

We prove that for $\gamma>\eta$

$$\sup_{j\notin\Pi_q(k^\gamma)}s(S_0(k)_j\cap\hat{\pi}(k,\xi,\eta,V_q))\le ck^{1-\gamma+\eta-\xi},\quad c\neq c(k,q). \tag{4.4.5}$$

Taking $\gamma=1-(\xi-\eta)/2$ we get (4.4.4). Thus, we prove (4.4.5). We represent $S_0(k)_j\cap\hat{\pi}(k,\xi,\eta,V_q)$ in the form

$$S_0(k)_j\cap\hat{\pi}(k,\xi,\eta,V_q)=\cup_{m\in\Pi_q(k^\eta)}\pi^q_{mj},$$

$$\pi^q_{mj}\equiv S_0(k)_j\cap\pi^q_m(k,\xi)=$$

$$=\{t,t\in K: p_j^2(t)=k^2, |p_j^2(t)-p_m^2(t)-\Delta\Lambda_q(t)_{mm}|<2k^{-\xi}\}.$$

Let $m: |p_m^2(t)-k^2|>2\|V\|$. Obviously, $\pi^q_{mj}=\emptyset$. Hence, it is sufficient to consider m such that $|p_m^2(t)-k^2|\le 2\|V\|<k$. It is easy to see that the subset $\{m\}$ of $\Pi_q(k^\eta)$ satisfying the last inequality contains at most $ck^{1+\eta}p_q(0)^{-1}$ points. This implies that

$$s(S_0(k)_j\cap\hat{\pi}(k,\xi,\eta,V_q))\le\sum_{m\in\Pi_q(k^\eta)}s(\pi^q_{mj})\le$$

$$\le ck^{1+\eta}p_q(0)^{-1}\max_{m\in\Pi_q(k^\eta)}s(\pi^q_{mj}).$$

We verify the relation

$$s(\pi^q_{mj})<cp_q(0)k^{-\gamma-\xi},\quad c\neq c(k,q). \tag{4.4.6}$$

This implies at once the estimate (4.4.5). To prove the formula (4.4.6) we consider the system of equations with respect to t:

$$p_j^2(t) = k^2,$$
$$p_j^2(t) - p_m^2(t) - \lambda(\tau) = \varepsilon, \qquad (4.4.7)$$
$$p_q(0)^2 \nabla_t \tau(t) = 2\pi \mathbf{p}_q(0);$$

here $m \in \Pi_q(k^\eta)$, $j \notin \Pi_q(k^\gamma)$, $\tau = 2\pi(\mathbf{p}_m(t), \mathbf{p}_q(0))p_q(0)^{-2}$, $\lambda(\tau) \equiv \Delta\Lambda_q(t)_{mm} = \lambda_{l(m)}(\tau)$ (see formulas (4.3.31), (4.3.25) and (4.3.30)).

Obviously, $\pi_{mj}^q = \bigcap_{\varepsilon:|\varepsilon|<2k^{-\xi}} B(\varepsilon)$, where $B(\varepsilon)$ is the set of the solutions t the system (4.4.7) in K. The function $\lambda(\tau)$ is piecewise continuous, and $|\lambda'(\tau)| < ck^\eta$ on the smooth parts, because $m \in \Pi_q(k^\eta)$. It can be assumed without loss of generality that $\lambda(\tau)$ is continuous (in the case when this does not hold it is necessary to consider each piece of continuity separately). Let $\mathbf{e}_1$ be a unit vector orthogonal to $\mathbf{p}_j(0)$ and $\mathbf{p}_m(0)$, let $\mathbf{e}_2$ coincide in direction with $\mathbf{p}_j(0) - \mathbf{p}_m(0) = \mathbf{p}_{j-m}(0)$, $|\mathbf{e}_2| = 1$, and $\mathbf{e}_3 = [\mathbf{e}_1, \mathbf{e}_2]$. We associate coordinates x_1, x_2, x_3 with these vectors. We consider solutions of the system not in K but in the cube $\tilde{K} \supset K$, $\tilde{K} = \{x : |x_1|, |x_2|, |x_3| < 4\pi a^{-1}\}$ (the function $\lambda(\tau)$ extends smoothly to K in such a way that $|\lambda'(\tau)| < ck^\eta$). In the new coordinates the system can be written in the form

$$x_1^2 + (x_2 - x_{20})^2 + (x_3 - x_{30})^2 = k^2,$$

$$2x_2 + \hat{x}_{20} - (\varepsilon + \lambda(\tau))p_{j-m}^{-1}(0) = 0$$

$$p_q(0)^2 \nabla_t \tau(t) = 2\pi \mathbf{p}_q(0); \qquad (4.4.8)$$

here $x_{20} = (\mathbf{p}_j(0), \mathbf{e}_2)$, $x_{30} = (\mathbf{p}_j(0), \mathbf{e}_3)$, $\hat{x}_{20} = (\mathbf{p}_{j-m}(0), \mathbf{p}_{j+m}(0))p_{j-m}(0)^{-1}$. Assume first that the system is uniquely solvable with respect to x_2, and x_3, and that $x_2(x_1, \varepsilon)$, $x_3(x_1, \varepsilon)$ are smooth functions admitting the following estimates:

$$|\partial x_2/\partial x_1| + |\partial x_3/\partial x_1| < c; \qquad (4.4.9)$$

$$|\partial x_2/\partial \varepsilon| + |\partial x_3/\partial \varepsilon| < cp_q(0)k^{-\gamma} \quad c \neq c(q,k). \qquad (4.4.10)$$

Then

$$\pi_{mj}^q \subset \{x : x = (x_1, x_2(x_1, \varepsilon), x_3(x_1, \varepsilon)),\ |x_1| < 4\pi a^{-1},\ |\varepsilon| < 2k^{-\xi}\}$$

$$\subset \{x :\ |x - \hat{x}| < cp_q(0)k^{-\xi-\gamma},\ \hat{x} = (x_1, x_2(x_1, 0), x_3(x_1, 0)),\ |x_1| < 4\pi^{-1}\}.$$

It is not hard to show that $s(\pi_{mj}^q) < cp_q(0)lk^{-\gamma-\xi}$, where l is the length of the curve $\{x : x = (x_1, x_2(x_1, 0), x_3(x_1, 0)), |x_1| < 4\pi a^{-1}\}$. Considering (4.4.9), we get that $l =_{k\to\infty} O(1)$. The estimate (4.4.6) we need follows from the last relations. Thus, it remains to prove that the solution of the system (4.4.8) (if it exists) can be represented in the form of a unique pair of smooth functions $x_2(x_1, \varepsilon)$, $x_3(x_1, \varepsilon)$ satisfying the estimates (4.4.9) and (4.4.10). Differentiating (4.4.8) with respect to x_1, we get a linear system with respect to $\partial x_2/\partial x_1$, $\partial x_3/\partial x_1$:

$$(x_2 - x_{20})\frac{\partial x_2}{\partial x_1} + (x_3 - x_{30})\frac{\partial x_3}{\partial x_1} = -x_1,$$

$$(1-\rho\frac{\partial\tau}{\partial x_2})\frac{\partial x_2}{\partial x_1}-\rho\frac{\partial\tau}{\partial x_3}\frac{\partial x_3}{\partial x_1}=0; \qquad (4.4.11)$$

$$2\rho\equiv\lambda'(\tau)p_{j-m}(0)^{-1}.$$

Consider $|x_{30}|:|x_{30}|=|(\mathbf{p}_j(0),\mathbf{e}_3)|=p_{j-m}(0)^{-1}|(\mathbf{p}_j(0),[\mathbf{p}_j(0)-\mathbf{p}_m(0),\mathbf{e}_1])|=p_{j-m}(0)^{-1}|[\mathbf{p}_j(0),\mathbf{p}_m(0)]|$. By using the relations

$$|(\mathbf{p}_m(0),\mathbf{p}_q(0))|<k^{\eta},\quad |(\mathbf{p}_j(0),\mathbf{p}_q(0))|\geq k^{\gamma},$$

$$\eta<\gamma,\quad p_j(t)=k,\quad |p_m(0)-k|<c_1,$$

it is not hard to show that the angle between the vectors $\mathbf{p}_m(0)$ and $\mathbf{p}_j(0)$ is greater than $cp_q(0)^{-1}k^{\gamma-1}$. From this,

$$|x_{30}|>cp_q(0)^{-1}k^{1+\gamma}p_{j-m}(0)^{-1}. \qquad (4.4.12)$$

and

$$4k>2p_{j-m}(0)>2\,|(\mathbf{p}_j(0)-\mathbf{p}_m(0),\mathbf{p}_q(0))|\,p_q(0)^{-1}>k^{\gamma}p_q(0)^{-1}, \qquad (4.4.13)$$

By taking into account (4.4.13) and the inequalities $|\lambda'(\tau)|<ck^{\eta}$ and $|\partial\tau/\partial x_2|<p_q(0)^{-1}$, we get:

$$\rho\partial\tau/\partial x_2<k^{\eta-\gamma}<1/2, \qquad (4.4.14)$$

$(\gamma=1-(\xi-\eta)/2)$, i.e.,

$$3/2>1-\rho\partial\tau/\partial x_2>1/2. \qquad (4.4.15)$$

and similarly

$$\left|\rho\frac{\partial\tau}{\partial x_3}\right|<k^{\eta}p_q(0)^{-1}p_{j-m}(0)^{-1}<ck^{\eta-\gamma}. \qquad (4.4.16)$$

Using the relations (4.4.12) and (4.4.15) it is not hard to get an estimate for the determinant D of the system (4.4.11):

$$|D|>cp_{j-m}^{-1}(0)p_q(0)^{-1}(k^{1+\gamma}+O(k^{1+\eta}))>(c/2)p_{j-m}(0)^{-1}p_q(0)^{-1}k^{1+\gamma}. \qquad (4.4.17)$$

The solution of the system (4.4.11) can be written in the form:

$$\frac{\partial x_2}{\partial x_1}=x_1D^{-1}\frac{\partial\tau}{\partial x_3}\rho,$$

$$\frac{\partial x_2}{\partial x_1}=x_1D^{-1}(1-\frac{\partial\tau}{\partial x_2}\rho).$$

Considering the estimates (4.4.14) – (4.4.17) we see easily that (4.4.9) holds; (4.4.10) is proved similarly.

It remains to see that system (4.4.8) has at most one solution $\{x:\ x=(x_1,x_2(x_1,\varepsilon),\ x_3(x_1,\varepsilon)),\ |x_1|<4\pi a^{-1}\}$. To do this first consider its second equation. By taking into account that $|\lambda'(\tau)|<ck^{\eta}$ and $p_{j-m}(0)>cp_q(0)^{-1}k^{\gamma}$ it is not hard to verify that the left hand side is a monotone function of the variable x_2. Therefore, the solution of the equation, if it exists, can be uniquely represented in the form $x_2 = x_2(x_1, x_3, \varepsilon), |x_1| < 4\pi a^{-1}, |x_3| < 4\pi a^{-1}$, and

$$\left|\frac{\partial x_2}{\partial x_3}\right| < ck^{\eta} p_{j-m}(0)^{-1} p_q(0)^{-1}. \tag{4.4.18}$$

Substituting x_2 in the first equation of the system (4.4.8), we get that

$$f(x_1, x_3, \varepsilon) \equiv x_1^2 + (x_2(x_1, x_3, \varepsilon) - x_{20})^2 + (x_3 - x_{30})^2 = k^2. \tag{4.4.19}$$

By considering (4.4.12) and (4.4.18) it is not hard to show that

$$\left|\frac{\partial f}{\partial x_3}\right| = 2\left|(x_2(x_1, x_3, \varepsilon) - x_{20})\frac{\partial x_2}{\partial x_3} + (x_3 - x_{30})\right| > cp_q(0)^{-1} k^{1+\gamma} p_{j-m}(0)^{-1} > 0.$$

Thus, the function $f(x_1, x_3, \varepsilon)$ depends monotonically on x_3. Consequently, if there is a solution of (4.4.17), then it is unique. The lemma is proved.

Next, we construct the set $\Upsilon_5(k, \delta)$, on which condition 5^0 is satisfied. Let us introduce the set $\Pi(k, \eta, \nu, \delta) \subset Z^3$ of m such that there exists a pair $q, q' \in \Gamma(k^{\delta}), q \neq q'$ for which the vector $\mathbf{p}_m(0)$ has sufficiently small projections on $\mathbf{p}_q(0)$ and $\mathbf{p}_{q'}(0)$:

$$|\ (\mathbf{p}_m(0), \mathbf{p}_q(0))\ | < k^{\eta}, \quad |\ (\mathbf{p}_m(0), \mathbf{p}_{q'}(0))\ | < k^{\nu}$$

i.e.,

$$\Pi(k, \eta, \nu, \delta) = \cup_{q \in \Gamma(k^{\delta})} \left(\Pi_q(k^{\eta}) \cap \left(\cup_{q' \in \Gamma(k^{\delta}), q \neq q'} \Pi_{q'}(k^{\nu})\right)\right).$$

Obviously, for sufficiently large k, $k^{\delta} > R_0$, the set $\Pi(k, 1/5, 3/5, \delta)$ contains the set $\Pi_0(k)$ defined in Section 4.3 (see (4.3.37)).

Let $\tilde{\Upsilon}_5(k, \eta, \nu, \varepsilon, \delta)$ be the subset of $S_0(k)$ such that for all t in $\tilde{\Upsilon}_5(k, \eta, \nu, \varepsilon, \delta)$

$$\min_{m \in \Pi(k, \eta, \nu, \delta)} |p_m^2(t) - k^2| > k^{\varepsilon}. \tag{4.4.20}$$

Lemma 4.3 . *Suppose η, ν, ε, δ are positive and $\eta + \nu + \varepsilon + 10\delta < 1$. Then the set $\tilde{\Upsilon}_5(k, \eta, \nu, \varepsilon, \delta)$ can be represented by the formula:*

$$\tilde{\Upsilon}_5(k, \eta, \nu, \varepsilon, \delta) = S_0(k) \setminus \mathcal{K}\hat{\pi}, \tag{4.4.21}$$

$$\hat{\pi} = \cup_{q, q' \in \Gamma(k^{\delta}), q \neq q'} \hat{\pi}_{qq'}, \tag{4.4.22}$$

$$\hat{\pi}_{qq'} = \{x, x \in R^3 : ||x|^2 - k^2| < k^{\varepsilon}, |(x, \mathbf{p}_q(0))| < 2k^{\eta}, |(x, \mathbf{p}_{q'}(0))| < 2k^{\nu}\}.$$

The set $\tilde{\Upsilon}_5(k, \eta, \nu, \varepsilon, \delta)$ has an asymptotically full measure on $S_0(k)$. Moreover,

$$\frac{s\left(S_0(k) \setminus \tilde{\Upsilon}_5(k, \eta, \nu, \varepsilon, \delta)\right)}{s(S_0(k))} =_{k \to \infty} O(k^{-\delta}). \tag{4.4.23}$$

Proof. It is easy to obtain that for t in $\tilde{\Upsilon}_5(k,\eta,\nu,\varepsilon,\delta)$ the inequality (4.4.20) holds for all m in $\Pi(k,\eta,\nu,\delta)$. In fact, suppose it is not so. Let $x = \mathbf{p}_m(t)$. It is easy to see that $x \in \hat{\pi}$. Hence, $t \in \mathcal{K}\hat{\pi}$. But this is not the case, since $t \in \tilde{\Upsilon}_5(k,\eta,\nu,\varepsilon,\delta)$.

It remains to prove that the set $\tilde{\Upsilon}_5(k,\eta,\nu,\varepsilon,\delta)$ has an asymptotically full measure on $S_0(k)$. Assume not. Then $s(S_0(k)\cap\hat{\pi})/s(S_0(k)) > \varepsilon_0(k)k^{-\delta}$, $\varepsilon_0(k) \to \infty$ as $k \to \infty$. Let Ω be the $(k^{-2-2\delta})$-neighborhood of $S_0(k) \cap \hat{\pi}$. As proved in Lemma 3.5, Ω occupies a volume tending to zero more slowly than $c\varepsilon k^{-2\delta}$:

$$V(\Omega) > c\varepsilon k^{-3\delta}. \qquad (4.4.24)$$

On the other hand, consider $\hat{\pi}_{qq'}$. This is the intersection of the spherical shell of inner radius $k - k^{\varepsilon-1}$ and outer radius $k + k^{\varepsilon-1}$ with the intersection of two planar layers of width $p_q(0)^{-1}k^{\eta}$ and $p_{q'}(0)^{-1}k^{\nu}$. It is easy to see that

$$V(\mathcal{K}\hat{\pi}_{qq'}) \le V(\hat{\pi}_{qq'}) < c\varphi(q,q')^{-1}k^{\eta+\nu+\varepsilon-1};$$

here $\varphi(q,q')$ is the angle between the vectors $\mathbf{p}_q(0)$ and $\mathbf{p}_{q'}(0)$. Since $q,q' \in \Gamma(k^{\delta})$ and are linearly independent, it can be shown that $\varphi(q,q') > ck^{-\delta}$. Thus, $V(\mathcal{K}\hat{\pi}_{qq'}) < ck^{\eta+\nu+\varepsilon+\delta-1}$. Obviously, the volume of the $(k^{-2-2\delta})$-neighborhood of the set $\mathcal{K}\hat{\pi}_{qq'}$ also does not exceed $ck^{\eta+\nu+\varepsilon+\delta-1}$. The number of pairs q,q' in $\Gamma(k^{\delta})$ does not exceed $ck^{6\delta}$. This implies that $V(\Omega) < ck^{\eta+\nu+\varepsilon+7\delta-1}$. Using that $\nu+\eta+\varepsilon+10\delta < 1$, we get

$$V(\Omega) < ck^{-3\delta}, \quad c \ne c(k). \qquad (4.4.25)$$

The inequalities (4.4.24) and (4.4.25) are contradictory. This means that (4.4.23) holds. The lemma is proved.

Let

$$\Upsilon_5(k,\delta) \equiv \tilde{\Upsilon}_5(k,1/5,3/5,1/5-11\delta,\delta). \qquad (4.4.26)$$

Since $\Pi_0(k) \subset \Pi(k,1/5,3/5,\delta)$, condition 5^0 holds for t in the (k^{-2})-neighborhood of $\gamma_5(k,\delta)$.

The following lemma describes the nonsingular set $\chi_3(k,V,\delta)$:

Lemma 4.4 . *Let $0 < \delta < 1/200$. Then for sufficiently large k, $k > k_0(V,\delta)$, there exists a set $\chi_3(k,V,\delta) \subset S_0(k)$, such that for any t in its $(k^{-2-2\delta})$-neighborhood there exists a unique $j \in Z^3$ such that $|p_j^2(t) - k^2| < k^{-1-\delta}$ ($p_j^2(t) = k^2$ for t of $\chi_3(k,V,\delta)$) and the conditions $1^0 - -5^0$ [4] hold. The set $\chi_3(k,V,\delta)$ is given by the formula*

$$\chi_3(k,V,\delta) = \chi_1(k,\delta,\delta) \cap \Upsilon_4(k,V,\delta) \cap \Upsilon_5(k,\delta). \qquad (4.4.27)$$

It has an asymptotically full measure on $S_0(k)$. Moreover,

$$\frac{s\left(S_0(k) \setminus \chi_3(k,V,\delta)\right)}{s(S_0(k))} =_{k\to\infty} O(k^{-\delta}). \qquad (4.4.28)$$

---

[4]see (4.3.39) – (4.3.43).

Proof. Conditions $1^0 - 5^0$ follow from the definition of the sets $\chi_1(k,\delta,\delta)$, $\Upsilon_4(k,V,\delta)$, and $\Upsilon_5(k,\delta)$ (see formulae (3.4.1), (4.4.2), (4.4.26)). Since for each of them the estimates similar to (4.4.28) holds, it is not hard to show that $\chi_3(k,V,\delta)$ satisfies (4.4.28) .

4.5 Proof of the Main Results.

To construct the convergent series corresponding to $p_j^2(t)$ we take the unperturbed operator to be not $H_0(t)$ but another operator $\hat{H}(t)$ defined by formulae (4.2.5) and (4.2.6). It is convenient to reduce $\hat{H}(t)$ to diagonal form, $\hat{H}(t) \to U\hat{H}(t)U^*$, $U = U(t)$, and then to consider the operator $H(t)$ in this representation:

$$UH(t)U^* = U\hat{H}(t)U^* + B, \tag{4.5.1}$$

$$B = UWU^*,$$

where the operator W is defined by formula (4.2.7). If t is in the nonsingular set $\chi_3(k,V\delta)$, then the operator $\hat{H}(t)$ has the eigenvalue $p_j^2(t)$ (Lemma 4.13). This eigenvalue belongs to the diagonal part of $\hat{H}(t)$ (see Corollary 4.6). Hence, the corresponding projection coincides with E_j – the spectral projection of the free operator associated with the same eigenvalue. Since $p_j^2(t)$ is on the diagonal part of $\hat{H}(t)$, the operator $U\hat{H}(t)U^*$, clearly, also has the same eigenvalue $p_j^2(t)$ with the same spectral projection E_j, i.e., $UE_jU^* = E_j$.

We prove that the perturbation series for $UH(t)U^*$ converge with respect to $U\hat{H}(t)U^*$ for t in $\chi_3(k,V\delta)$ and its $(k^{-2-2\delta})$-neighborhood (Lemma 4.14). Performing the inverse unitary transformation, we get the formulae for the perturbed eigenvalue and its spectral projection of $H(t)$. Then, we prove that the first terms of these series are equal to those constructed with respect to $H_0(t)$ (Lemma 4.15). Thus, we simplify the formulae for the eigenvalue and its spectral projection. The main result is described by Theorem 4.1 in Section 4.2. Proposition 4.1 and Lemmas 4.5 – 4.9 describe properties of the operator B. Lemmas 4.10 – 4.13 are about the main properties of operator $\hat{H}(t)$. The estimates proving the convergence of the series are obtained in Lemma 4.14. Using this lemma we prove Theorem 4.1. After this the proofs of Theorems 4.2 and 4.3 is similar to those of Theorems 3.2 and 3.3.

For convenience of the exposition we first construct U for the one-dimensional case, then for the case of a "simple" potential, and then for the general case.

4.5.1 The operator $H_1(\ )$ acting in l_2^1.

We introduce the notation: $H_{10}(\tau)$ is the operator of the free problem ($V = 0$), $U_1(\tau)$ is a unitary matrix reducing $H_1(\tau)$ to diagonal form, $\Delta\Lambda(\tau)$ is the diagonal matrix of the shifts of the eigenvalues, $\Delta\Lambda(\tau) = (U_1H_1U_1^* - H_{10})(\tau)$, and $P(M)$ is the diagonal projection,

$$P(M)_{ii} = \begin{cases} 1 & \text{if } |i| < M; \\ 0 & \text{otherwise.} \end{cases} \tag{4.5.2}$$

Next, we consider $H_1^{(M)}(\tau) = P(M)H_1(\tau)P(M)$. It is possible to enumerate the eigenvalues $\lambda_j^{(M)}(\tau)$ and the eigenfunctions $\psi_j^{(M)}(\tau)$, $(j \in Z)$ of the operator $H_1^{(M)}(\tau)$ in a way similar to what was done in the case of $H_1(\tau)$. Let $U_1^{(M)}(\tau)$ be a unitary operator reducing $H_1^{(M)}(\tau)$ to diagonal form, $\Delta\Lambda^{(M)}(\tau)$ is the diagonal matrix of the shifts of the eigenvalues, $\Delta\Lambda^{(M)}(\tau) = (U_1^{(M)}H_1^{(M)}U_1^{(M)*} - P(M)H_{10})(\tau)$.

Proposition 4.1 . *The following relations hold for sufficiently large M, $M > 4(\|V\|a^2 + 10)$, for any τ in the interval $[0, 2\pi)$:*

$$\|P(M/2)U_1(\tau)(I - P(3M/4))\| < (\varepsilon M)^{-\varepsilon M}, \tag{4.5.3}$$

$$\|P(M/2)U_1^{(M)}(\tau)(I - P(3M/4))\| < (\varepsilon M)^{-\varepsilon M}, \tag{4.5.4}$$

$$\|P(M/2)(\Delta\Lambda(\tau) - \Delta\Lambda^{(M)}(\tau))\| < (\varepsilon M)^{-\varepsilon M}, \tag{4.5.5}$$

$$\|(P(M) - P(M/2))(\Delta\Lambda(\tau) - \Delta\Lambda^{(M)}(\tau))\| < cM^{-2}, \tag{4.5.6}$$

$$\varepsilon > \varepsilon_0 R_0^{-1}, \quad 0 < \varepsilon_0 \neq \varepsilon_0(M, \tau, R_0).$$

Proof. The proof of this assertion is in large respect similar to the proof of Lemma 4.1; therefore, we give a brief sketch of it. To verify the formula (4.5.3) we consider a circle C_M about zero with radius of order $2\pi^2a^{-2}M^2$ such that its distance from the spectrum of the free operator is of order Ma^{-2}. Let Ω be the set of indices $i \in Z$ such that the points $\lambda = (\tau + 2\pi i)^2a^{-2}$ lie inside C_M. Note, that $\{i : |i| < M/2\} \subset \Omega$. The spectral projection E_Ω of the operator $H_1(\tau)$ corresponds to the set Ω, and $(E_\Omega)_{jj} = \sum_{i\in\Omega} |(\psi_i)_j|^2$. Since all the points z of the circle C_M are sufficiently far from the spectrum of the operator $H_{10}(\tau)$, the resolvent $(H_1(\tau) - z)^{-1}, z \in C_M$, can be expanded in a perturbation series. Arguing as in Lemma 4.1, we can show that

$$|E_\Omega(\tau)_{jj}| < (\varepsilon M)^{-2\varepsilon|j|},$$

if $|j| > 3M/4$. From this, $|U_{ij}| = |(\psi_i)_j| < (\varepsilon M)^{-\varepsilon M}$ for all i in Ω, and hence for all $|i| < M/2$. The relation proved easily yield (4.5.3). It is an analogous matter to get the relations (4.5.4) and (4.5.5). Formula (4.5.6) is well-known and can be proved in the same way as (4.3.2). The proposition is proved.

We introduce the unitary operator $U_1(M, \tau)$, acting as $U_1^{(M)}(\tau)$ on $P(M)l_2^1$ and as I on the orthogonal complement to $P(M)l_2^1$, i.e.,

$$U_1(M, \tau) = U_1(\tau)^{(M)} + I - P(M), \tag{4.5.7}$$

here and below we consider $U_1^{(M)}(\tau)$ to be zero on $(I - P(M))l_2^1$.

Lemma 4.5 . *If* $M > 4\|V\|a^2 + 8R_0$, *then the operator* $U_1(M,\tau)H_1(\tau)U_1(M,\tau)^*$, $\tau \in [0, 2\pi)$ *can be represented in the form:*

$$U_1(M,\tau)H_1(\tau)U_1(M,\tau)^* =$$

$$H_0(t) + P(M)\Delta\Lambda^{(M)}(\tau) + (I - P(M))V(I - P(M)) + L(\tau) + D(\tau), \quad (4.5.8)$$

where

$$L = L^*, \quad D = D^*, \quad \|L\| < \|V\|, \quad (4.5.9)$$

$$L(P(3M/2) - P(M/2)) = (P(3M/2) - P(M/2))L = L, \quad (4.5.10)$$

$$DP(3M/2) = P(3M/2)D = D, \quad (4.5.11)$$

$$\|D\| < (\varepsilon M)^{-\varepsilon M}, \quad (4.5.12)$$

$$\varepsilon > \varepsilon_0 R_0^{-1}, \quad 0 < \varepsilon_0 \neq \varepsilon_0(M, \tau, R_0).$$

Proof. By using the formula (4.5.7) for $U_1(M,\tau)$ it is not hard to show that the representation (4.5.8) is valid, with

$$L = (I - P(M))VU_1^{(M)*}(P(M) - P(M/2)) + (P(M) - P(M/2))U_1^{(M)}V(I - P(M)),$$

$$D = (I - P(M))VU_1^{(M)*}P(M/2) + P(M/2)U_1^{(M)}V(I - P(M)).$$

The relations (4.5.9) are obvious. Since V is a trigonometric polynomial and $8R_0 < M$, it follows that $P(M)V(I - P(M)) = P(M)V(P(3M/2) - P(M))$. From this, considering that $P(M)U_1^{(M)}P(M) = U_1^{(M)}P(M) = P(M)U_1^{(M)}$, we get that

$$L = (P(3M/2) - P(M))VU_1^{(M)*}(P(M) - P(M/2)) +$$
$$(P(M) - P(M/2))U_1^{(M)}V(P(3M/2) - P(M)).$$

It can be shown similarly that

$$D = (P(3M/2) - P(M))V(P(M) - P(3M/4))U_1^{(M)*}P(M/2) +$$

$$P(M/2)U_1^{(M)}(P(M) - P(3M/4))V(P(3M/2) - P(M)).$$

The relations (4.5.10) and (4.5.11) are now obvious. The estimates (4.5.4) and (4.5.5) implies (4.5.12). The lemma is proved.

4.5.2 The case of a "simple" potential

We carry over the results of the one-dimensional case to the case of a "simple" potential. As remarked in Section 4.3, $H_q(t)$ can be represented as the orthogonal sum (4.3.19) of operators, each acting in $P_i^q l_2^3$ and equivalent to the operator $H_1(\tau_i) + |\mathbf{d}_i^q(t)|^2$ (see (4.3.27)) under an isometrically isomorphic mapping of $P_i^q l_2^3$ onto l_2^1. Under this mapping, to an arbitrary operator A acting in l_2^1 there corresponds an operator A_i acting in $P_i^q l_2^3$, such that $(A_i)_{m_1 m_2} = A_{l_1 l_2}$; here

$\delta_{m_1}, \delta_{m_2} \in P_i^q l_2^3$, $l_1, l_2 \in Z$; m_1 and l_1, m_2 and l_2 are connected by the formula (4.3.25). Thus, corresponding to the projection $P(M)$ acting in l_2^1 is the diagonal projection $P_i^q(M)$ acting in $P_i^q l_2^3$:

$$P_i^q(M)_{jj} = \begin{cases} 1 & \text{if } \|[(\mathbf{p}_j(t), \mathbf{p}_q(0))p_q(0)^{-2}]\| < M, \ \delta_j \in P_i^q l_2^3; \\ 0 & \text{otherwise.} \end{cases} \tag{4.5.13}$$

The unitary operator $U_{qi}(M,t)$ acting in $P_i^q l_2^3$ and reducing $P_i^q(M)H_q(t)P_i^q(M)$ to diagonal form is given by the relation:

$$U_{qi}(M,t)_{m_1 m_2} = U_1(M,\tau_i)_{l_1 l_2}, \quad m_1 \leftrightarrow l_1, \ m_2 \leftrightarrow l_2,$$

here τ_i is defined by the formula (4.3.26), $U(M,\tau_i)$ by formula (4.5.7). It is not hard to show that $U_{qi}(M,t)$ acts as identity on $P_i^q - P_i^q(M)$, i.e.,

$$U_{qi}(M,t) = P_i^q - P_i^q(M) + P_i^q(M)U_{qi}(M,t)P_i^q(M).$$

Corresponding to the operator $\Delta\Lambda^{(M)}(\tau_i)$ acting in l_2^1 is the diagonal operator $\Delta\Lambda_{qi}^{(M)}(t)$ acting in $P_i^q l_2^3$:

$$\Delta\Lambda_{qi}^{(M)}(t)_{mm} = \Delta\Lambda^{(M)}(\tau_i)_{ll}, \quad m \leftrightarrow l. \tag{4.5.14}$$

We denote by $\Delta\Lambda_q^{(M)}(t)$ the matrix:

$$\Delta\Lambda_q^{(M)}(t) = \sum_{i \in J_q^0} \Delta\Lambda_{qi}^{(M)}(t)P_i^q, \tag{4.5.15}$$

It is clear that [5]

$$\Delta\Lambda_q^{(M)}(t)P_i^q = P_i^q\Delta\Lambda_q^{(M)}(t) = \Delta\Lambda_{qi}^{(M)}(t). \tag{4.5.16}$$

It is obvious that Lemma 4.5 takes the following form in $P_i^q l_2^3$.

Lemma 4.6 . *If $M_i > 16\pi^2\|V_q\|p_q(0)^{-2} + 8R_0$, then the operator $U_{qi}(M_i,t)H_q(t)U_{qi}(M_i,t)^*$, can be represented in the form:*

$$U_{qi}(M_i,t)H_q(t)U_{qi}(M_i,t)^* = P_i^q H_0(t) + P_i^q(M_i)\Delta\Lambda_q^{(M)}(t) +$$
$$(P_i^q - P_i^q(M_i))V_q(P_i^q - P_i^q(M_i)) + L_i(t) + D_i(t), \tag{4.5.17}$$

where

$$L_i = L_i^*, \quad D_i = D_i^*, \quad \|L_i\| < \|V_q\|, \tag{4.5.18}$$

$$L_i(P_i^q(3M_i/2) - P_i^q(M_i/2)) = (P_i^q(3M_i/2) - P_i^q(M_i/2))L_i = L_i, \tag{4.5.19}$$

$$D_iP_i^q(3M_i/2) = P_i^q(3M_i/2)D_i = D_i, \tag{4.5.20}$$

$$\|D_i\| < (\varepsilon M_i)^{-\varepsilon M_i}, \tag{4.5.21}$$

$$\varepsilon > \varepsilon_0 R_0^{-1}, \quad 0 < \varepsilon_0 \neq \varepsilon_0(i, M_i, t, R_0).$$

[5] It is assumed all the operators acting in $P_i^q l_2^3$ are extended by zero to l_2^3.

Further, let J_q be a subset of J_q^0,

$$P_{J_q} \equiv \sum_{i \in J_q} P_i^q,$$

$$\mathbf{M}_q \equiv \{M_i\}_{i \in J_q},$$

and

$$P_{J_q}(\mathbf{M}_q) \equiv \sum_{i \in J_q} P_i^q(M_i), \tag{4.5.22}$$

$$\Delta\Lambda_q^{(\mathbf{M}_q)} = \sum_{i \in J_q} \Delta\Lambda_{qi}^{(M_i)}(t) P_i^q.$$

We introduce in l_2^3 the unitary operator $U_q(\mathbf{M}_q, t)$, which acts as $U_{qi}(M_i, t)$ on the corresponding subspace $P_i^q l_2^3$, $i \in J_q$, and as I on $(I - P_{J_q}) l_2^3$, i.e.,

$$U_q(\mathbf{M}_q, t) P_i^q = P_i^q U_q(\mathbf{M}_q, t) = U_{qi}(M_i, t), \quad i \in J_q,$$

$$U_q(\mathbf{M}_q, t)(I - P_{J_q}) = I - P_{J_q},$$

$$U_q(\mathbf{M}_q, t) = \sum_{i \in J_q} U_{qi}(M_i, t) + I - P_{J_q}; \tag{4.5.23}$$

$$P_{J_q}(\mathbf{M}_q) U_q(\mathbf{M}_q, t) P_{J_q}(\mathbf{M}_q) =$$

$$P_{J_q}(\mathbf{M}_q) U_q(\mathbf{M}_q, t) = U_q(\mathbf{M}_q, t) P_{J_q}(\mathbf{M}_q), \tag{4.5.24}$$

$$(I - P_{J_q}(\mathbf{M}_q)) U_q(\mathbf{M}_q, t)(I - P_{J_q}(\mathbf{M}_q)) = (I - P_{J_q}(\mathbf{M}_q)) U_q(\mathbf{M}_q, t) =$$

$$U_q(\mathbf{M}_q, t)(I - P_{J_q}(\mathbf{M}_q)) = (I - P_{J_q}(\mathbf{M}_q)). \tag{4.5.25}$$

Let $M_q = \min_{i \in J_q} M_i$.

Lemma 4.7 . *If $J_q \subset J_q^0$, $M_q > 16\|V_q\| p_q(0)^{-2} + 8R_0$, then*

$$U_q(\mathbf{M}_q, t) H_q(t) U_q(\mathbf{M}_q, t)^* = H_0(t) + P_{J_q}(\mathbf{M}_q) \Delta\Lambda_q^{(\mathbf{M}_q)}(t) +$$

$$(I - P_{J_q}(\mathbf{M}_q)) V_q (I - P_{J_q}(\mathbf{M}_q)) + L(t) + D(t), \tag{4.5.26}$$

where

$$L = L^*, \quad D = D^*, \quad \|L\| < \|V_q\|, \tag{4.5.27}$$

$$L(P_{J_q}(3\mathbf{M}_q/2) - P(\mathbf{M}_q/2)) = (P_{J_q}(3\mathbf{M}_q/2) - P(\mathbf{M}_q/2))L = L, \tag{4.5.28}$$

$$D P_{J_q}(3\mathbf{M}_q/2) = P_{J_q}(3\mathbf{M}_q/2) D = D, \tag{4.5.29}$$

$$\|D\| < (cM_0)^{-\varepsilon M_0}, \tag{4.5.30}$$

$$\varepsilon > \varepsilon_0 R_0^{-1}, \quad 0 < \varepsilon_0 \neq \varepsilon_0(J_q, M_q, t, R_0).$$

Proof. Using the formulae (4.3.20) – (4.3.22) and (4.5.23) and considering that $P_{J_q}H_q(t)(I - P_{J_q}) = 0$, we get that

$$U_q(\mathbf{M}_q, t)H_q(t)U_q(\mathbf{M}_q, t)^* = \tag{4.5.31}$$

$$\sum_{i\in J_q} U_{qi}(M_i, t)H_q(t)U_{qi}(M_i, t)^* + (I - P_{J_q})H_0(t) + (I - P_{J_q})V_q(I - P_{J_q}).$$

Using Lemma 4.6 and noting that

$$\sum_{i\in J_q}(P_i^q - P_i^q(M_i))V_q(P_i^q - P_i^q(M_i)) + (I - P_{J_q})V_q(I - P_{J_q}) =$$

$$(I - P_{J_q}(\mathbf{M}_q))V_q(I - P_{J_q}(\mathbf{M}_q)),$$

we get formula (4.5.24), with

$$L = \sum_{i\in J_q} L_i, \quad D = \sum_{i\in J_q} D_i. \tag{4.5.32}$$

Since the operators after the summation sign act in the orthogonal subspaces $P_i^q l_2^3$, the assertion of Lemma 4.6 leads easily to the relations (4.5.27) – (4.5.29). The lemma is proved.

4.5.3 The general case

$H = H_0 + V, V = \sum_{q\in\Gamma(R_0)} V_q$. We construct the scheme described in the preceding part for each $q \in \Gamma(R_0)$. Associated with each q is the collection of the diagonal projections $P_i^q, i \in J_q^0, \sum_{i\in J_q^0} P_i^q = I$. Recall that $|(\mathbf{p}_i(0), \mathbf{p}_q(0))| < 2p_q(0)^2$ for all i in J_q^0. We define a subset J_q of J_q^0 as follows. The set J_q corresponding to a given q contains precisely those i such that the corresponding vector $\mathbf{p}_i(0)$ has a sufficiently large (greater that a given number N) projection on the remaining vectors q' in $\Gamma(R_0)$:

$$J_q(N) = \{i, i \in J_q^0 : |(\mathbf{p}_i(0), \mathbf{p}_{q'})| > N, \forall q' \in \Gamma(R_0), q \neq q'\}. \tag{4.5.33}$$

For each set $J_q(N)$ we construct the scheme of the preceding part, i.e., with each $J_q(N)$, $q \in \Gamma(R_0)$, we associate the projections $P_{J_q(N)}$ and $P_{J_q(N)}(\mathbf{M}_q)$, $\mathbf{M}_q = \{M_i^q\}_{i\in J_q(N)}$. (see (4.5.22)), $M_q = \max_{i\in J_q(N)} M_i^q$.

Lemma 4.8 . *Suppose $q, q' \in \Gamma(R_0)$, $q \neq q'$ and $N > \max\{M_q, M_{q'}\}16\pi^2 R_0^2 a^{-2}$.* [6] *Then, the projections $P_{J_q(N)}(\mathbf{M}_q)$ and $P_{J_{q'}(N)}(\mathbf{M}_{q'})$ are mutually orthogonal.*

Proof. Let $\delta_j \in P_{J_q(N)}(\mathbf{M}_q)$. This means that $\mathbf{p}_j(0)$ can be represented in the form $\mathbf{p}_j(0) = \mathbf{p}_i(0) + l\mathbf{p}_q(0)$, $i \in J_q(N)$, $l < M_q$. Since $|(\mathbf{p}_i(0), \mathbf{p}_{q'}(0))| > N$, it follows that

[6] here as below $a \equiv \min\{a_1, a_2, a_3\}$.

$$|(\mathbf{p}_j(0), \mathbf{p}_{q'}(0))| > N - M_q p_q(0) p_{q'}(0) > M_{q'} p_{q'}(0)^2. \tag{4.5.34}$$

To get the last estimate we use the hypothesis of the lemma and the obvious inequality

$$\max_{q \in \Gamma(R_0)} p_q(0) < 2\pi R_0 a^{-1}. \tag{4.5.35}$$

Inequality (4.5.34) implies that $\delta_j \notin P_{J_{q'}(N)}(\mathbf{M}_{q'}) l_2^3$. The lemma is proved.

We consider the operator

$$P(N, \mathbf{M}) = \sum_{q \in \Gamma(R_0)} P_{J_q(N)}(\mathbf{M}_q), \tag{4.5.36}$$

$\mathbf{M} = \{\mathbf{M}_q\}_{q \in \Gamma(R_0)}$. Suppose $N > M 16\pi^2 R_0^2 a^{-2}$, $M = \max_{q \in \Gamma(R_0), i \in J_q(N)} M_i^q$. It is clear that $P(N, \mathbf{M})$ is a projection, since the projections after the summation sign are mutually orthogonal. We consider the operator $U(N, \mathbf{M}, t)$, acting as $U_q(\mathbf{M}_q, t)$ on $P_{J_q(N)}(\mathbf{M}_q) l_2^3$ for all q in $\Gamma(R_0)$ and as I on $(I - P(N, \mathbf{M}) l_2^3$,

$$U(N, \mathbf{M}, t) = \sum_{q \in \Gamma(R_0)} U_q(\mathbf{M}_q, t) P_{J_q(N)}(\mathbf{M}_q) + I - P(N, \mathbf{M}). \tag{4.5.37}$$

This operator is unitary, because the projections $P_{J_q(N)}(\mathbf{M}_q)$, $q \in \Gamma(R_0)$ are mutually orthogonal, and

$$P_{J_q(N)}(\mathbf{M}_q) U_q(\mathbf{M}_q, t) = U_q(\mathbf{M}_q, t) P_{J_q(N)}(\mathbf{M}_q) = P_{J_q(N)}(\mathbf{M}_q) U_q(\mathbf{M}_q, t) P_{J_q(N)}(\mathbf{M}_q).$$

We note that $U(N, \mathbf{M})$ has the representation

$$U(N, \mathbf{M}, t) = \sum_{q \in \Gamma(R_0)} U_q(\mathbf{M}_q, t) P_{J_q(N)}(3\mathbf{M}_q/2) + (I - P(N, 3\mathbf{M}/2)), \tag{4.5.38}$$

because

$$(U_q(\mathbf{M}_q, t) - I)(P_{J_q(N)}(3\mathbf{M}_q/2) - P_{J_q(N)}(\mathbf{M}_q)) = 0.$$

We introduce the diagonal projection

$$P_z(R)_{ii} = \begin{cases} 1, & \text{if } |p_i^2(t) - z| \le R; \\ 0, & \text{otherwise.} \end{cases}$$

Lemma 4.9 . *Suppose that*

$$N^{1/2} a / 10\pi R_0 > M_i^q > 16\pi^2 |q|^{-2} \|V_q\| + 8R_0 \tag{4.5.39}$$

for all $q \in \Gamma(R_0)$ and $i \in J_q(N)$. Then

$$U(N, \mathbf{M}, t) H(t) U(N, \mathbf{M}, t)^* = H_0(t) + \sum_{q \in \Gamma(R_0)} P_{J_q(N)}(\mathbf{M}_q) \Delta \Lambda_q^{(\mathbf{M}_q)}(t) +$$

$$(I - P(N, \mathbf{M}) V (I - P(N, \mathbf{M}) + \sum_{q \in \Gamma(R_0)} \mathcal{E}_q + L + D; \tag{4.5.40}$$

here
1)

$$\mathcal{E}_q = \mathcal{E}_q^*; \quad \|\mathcal{E}_q\| < 2\|V\|; \tag{4.5.41}$$
$$\mathcal{E}_q P(N, 3\mathbf{M}/2) = P(N, 3\mathbf{M}/2)\mathcal{E}_q = \mathcal{E}_q; \tag{4.5.42}$$
$$P_z(N/2)\mathcal{E}_q P_z(N/2) = 0; \tag{4.5.43}$$
$$(\mathcal{E}_q)_{ij} = 0 \quad \textit{if} \;\; i - j \neq lq + l'q', \;\textit{for some}\; q' \neq q; \;\textit{and}\; l, l' \in Z,$$
$$|l| < 3M, \;\; 1 \leq |l'| < R_0, \tag{4.5.44}$$

here $M = \max_{i \in J_q(N), q \in \Gamma(R_0)} M_i^q$*:*
2)

$$L = L^*, \quad \|L\| < 2\|V\|; \tag{4.5.45}$$
$$L = L\,(P(N, 3\mathbf{M}/2) - P(N, \mathbf{M})) = (P(N, 3\mathbf{M}/2) - P(N, \mathbf{M})\,L; \tag{4.5.46}$$

3)

$$D = D^*; \tag{4.5.47}$$
$$DP(N, 3\mathbf{M}/2) = P(N, 3\mathbf{M}/2)D = D; \tag{4.5.48}$$
$$\|D\| < (\varepsilon M_0)^{-\varepsilon M_0}, \tag{4.5.49}$$
$$M_0 \equiv \min_{q \in \Gamma(R_0), i \in J_q(N)} M_i^q,$$
$$\varepsilon > \varepsilon_0 R_0^{-1}, \quad 0 < \varepsilon_0 \neq \varepsilon_0(N, \mathbf{M}, R_0, t).$$

Proof. By using formula (4.5.38) it is not hard to show that

$$U(N, \mathbf{M}, t)H(t)U^*(N, \mathbf{M}, t) = \sum_{q \in \Gamma(R_0)} \left(A_q + R_q + C_q + C_q^* + \sum_{q' \in \Gamma(R_0), q' \neq q} B_{qq'} \right)$$
$$+(I - P(N, 3\mathbf{M}/2))H_0(t) + (I - P(N, 3\mathbf{M}/2))V(I - P(N, 3\mathbf{M}/2)); \tag{4.5.50}$$

here

$$A_q = P_{J_q(N)}(3\mathbf{M_q}/2))U_q(\mathbf{M}_q)(H_0(t) + V_q)U_q(\mathbf{M}_q)^* P_{J_q(N)}(3\mathbf{M_q}/2)),$$
$$R_q = \sum_{q' \in \Gamma(R_0), q' \neq q} R_{qq'}, \tag{4.5.51}$$
$$R_{qq'} = P_{J_q(N)}(3\mathbf{M_q}/2))U_q(\mathbf{M}_q)V_{q'}U_q(\mathbf{M}_q)^* P_{J_q(N)}(3\mathbf{M_q}/2)), \tag{4.5.52}$$
$$B_{qq'} = P_{J_q(N)}(3\mathbf{M_q}/2))U_q(\mathbf{M}_q)(H_0(t) + V)U_{q'}(\mathbf{M}_{q'})^* P_{J_{q'}(N)}(3\mathbf{M}_q'/2),$$
$$C_q = (I - P(N, 3\mathbf{M}/2))(H_0(t) + V)U_q(\mathbf{M}_q)P_{J_q(N)}(3\mathbf{M_q}/2)).$$

The operator A_q corresponds to the case of a"simple" potential. Using Lemma 4.7, we get that

$$A_q = P_{J_q(N)}(3\mathbf{M_q}/2)H_0(t) + P_{J_q(N)}(\mathbf{M}_q)\Delta\Lambda_q^{(\mathbf{M}_q)}(t)+$$
$$+(P_{J_q(N)}(3\mathbf{M_q}/2) - P_{J_q(N)}(\mathbf{M}_q))V_q(P_{J_q(N)}(3\mathbf{M_q}/2) - P_{J_q(N)}(\mathbf{M}_q))+$$
$$+P_{J_q(N)}(3\mathbf{M_q}/2))(L_q + D_q)P_{J_q(N)}(3\mathbf{M_q}/2)), \tag{4.5.53}$$

here L_q, D_q have the properties (4.5.27) – (4.5.30).

Let us turn to the operator R_q. We prove that this operator satisfies relations similar to (4.5.41) – (4.5.43). It is clear that

$$R_q = R_q^*, \quad \|R_q\| < \|V\|. \tag{4.5.54}$$

Note that

$$(R_{qq'})_{ij} = \sum_{n,m\in Z^3} U_q(\mathbf{M}_q)_{in}(V_{q'})_{nm}U_q(\mathbf{M}_q)^*_{mj}, \tag{4.5.55}$$

Suppose that $(R_{qq'})_{ij} \neq 0$. Obviously, $\mathbf{p}_i(0) - \mathbf{p}_n(0) = l_1\mathbf{p}_q(0)\ l_1 \in Z$, $|l_1| < M_q$; otherwise we get from the definition of $U_q(\mathbf{M}_q)$ that $(U_q(\mathbf{M}_q))_{in} = 0$. Similarly, $\mathbf{p}_m(0) - \mathbf{p}_j(0) = l_2\mathbf{p}_q(0)\ l_2 \in Z$, $|l_2| < M_q$. It is clear that $\mathbf{p}_n(0) - \mathbf{p}_m(0) = l'\mathbf{p}_{q'}(0)$, $1 \leq |l'| < R_0$, otherwise $(V_{q'})_{nm} = 0$. Thus, if $(R_{qq'})_{ij} \neq 0$, then $i - j = lq + l'q'$, $|l| < 2M_q$, $|l'| < R_0$, $l' \neq 0$. This means that $(R_q)_{ij}$ can be nonzero only if for some $q' \in R_0$, $q' \neq q$ indeces i and j satisfy the relations: $i - j = lq + l'q'$, $|l| < 2M_q$, $1 \leq |l'| < R_0$, i.e.,

$$(\mathcal{R}_q)_{ij} = 0 \ \text{ if } \ i - j \neq lq + l'q', \text{ for some } q' \neq q; \ mboxand\ l, l' \in Z,$$
$$|l| < 3M, \ 1 \leq |l'| < R_0. \tag{4.5.56}$$

Let us prove that

$$P_z(N/2)R_{qq'}P_z(N/2) = 0 \tag{4.5.57}$$

for all $q \neq q'$. Obviously,

$$p_j^2(t) - p_i^2(t) = |\mathbf{p}_j(t) - \mathbf{p}_i(t)|^2 + 2\left(\mathbf{p}_i(t), \mathbf{p}_j(t) - \mathbf{p}_i(t)\right). \tag{4.5.58}$$

As we have proved $(R_{qq'})_{ij}$ can be nonzero if and only if $i - j = lq + l'q'$, $l, l' \in Z$, $|l| < 2M_q$, $1 \leq l' < R_0$. Considering inequalities (4.5.35) and (4.5.39), we get from this that

$$|\mathbf{p}_j(t) - \mathbf{p}_i(t)|^2 < 8M_q^2 p_q(0)^2 + R_0^2 p_{q'}(0)^2 < 40\pi^2 a^{-2} M_q^2 R_0^2, \tag{4.5.59}$$
$$|(\mathbf{p}_i(t), \mathbf{p}_j(t) - \mathbf{p}_i(t))| > |(\mathbf{p}_i(t), \mathbf{p}_{q'}(0))| - M_q|(\mathbf{p}_i(t), \mathbf{p}_q(0))|.$$

Since $\delta_i \in P_{J_q(N)}(3\mathbf{M_q}/2)l_2^3$, it follows that $|(\mathbf{p}_i(t), \mathbf{p}_q(0))| < 2M_q p_q(0)^2 < 8\pi^2 M_q R_0^2 a^{-2}$, while $|(\mathbf{p}_i(0), \mathbf{p}_{q'}(0))| > N$. Thus,

$$|(\mathbf{p}_i(t), \mathbf{p}_j(t) - \mathbf{p}_i(t))| > N - 16\pi^2 M_q^2 R_0^2 a^{-2}. \tag{4.5.60}$$

Estimating the right-hand side of (4.5.58) with the help of (4.5.59) and (4.5.60) and using the condition (4.5.39), we have that $|p_j^2(t) - p_i^2(t)| > N$. This implies at once that δ_i, δ_j cannot belong to $P_z(N/2)l_2^3$ at the same time, i.e., (4.5.57) holds. From this $P_z(N/2)R_qP_z(N/2) = 0$ follows.

Let us consider the operator

$$R_q^0 = \sum_{q\neq q'} (P_{J_q(N)}(3\mathbf{M_q}/2) - P_{J_q(N)}(\mathbf{M_q}))V_{q'}(P_{J_q(N)}(3\mathbf{M_q}/2) - P_{J_q(N)}(\mathbf{M_q})). \tag{4.5.61}$$

It is easy to see that R_q^0 has the same properties as R_q. Let

$$\mathcal{E}_q = R_q - R_q^0. \tag{4.5.62}$$

From the properties of R_q, R_q^0 (4.5.44) – (4.5.43) follow.

We prove that operators $B_{qq'}$ vanish. To do this it suffices to show that

$$P_{J_q(N)}(3\mathbf{M_q}/2)VP_{J_{q'}(N)}(3\mathbf{M_q}0/2) = 0. \tag{4.5.63}$$

Let

$$\delta_i \in P_{J_q(N)}(3\mathbf{M_q}/2)l_2^3, \quad \delta_j \in P_{J_{q'}(N)}(3\mathbf{M_q}0/2)l_2^3.$$

Then

$$|i-j| \geq (2\pi)^{-1}a|\mathbf{p}_j(0) - \mathbf{p}_i(0)| \geq (2\pi)^{-1}a\,|(\mathbf{p}_j(0) - \mathbf{p}_i(0), \mathbf{p}_q(0))|\,p_q(0)^{-1}$$
$$\geq (2\pi)^{-1}a\mathbf{p}_q(0)^{-1}(N - M_q p_q(0)^2) > R_0,$$

i.e., $v_{i-j} = 0$. The relation (4.5.63) is proved, and $B_{qq'} = 0$.

It is easy to verify that

$$C_q = (I - P(N, 3\mathbf{M}/2))V\left(P_{J_q(N)}(3\mathbf{M_q}/2) - P_{J_q(N)}(\mathbf{M_q})\right). \tag{4.5.64}$$

Indeed, since V is a trigonometric polynomial and $M_i^q > 8R_0$, it follows that $(I - P(N, 3\mathbf{M}/2))(H_0(t) + V) = (I - P(N, 3\mathbf{M}/2))(H_0(t) + V)(I - P(N, \mathbf{M}))$. Taking into account that

$$(I - P(N, \mathbf{M}))U_q(\mathbf{M}_q, t)^* P_{J_q(N)}(3\mathbf{M}_q/2) = P_{J_q(N)}(3\mathbf{M}_q/2) - P_{J_q(N)}(\mathbf{M}_q)),$$

and using the definition of C_q, we get (4.5.64).

Substituting formulae (4.5.53), (4.5.64), (4.5.61) and (4.5.62) in the formula (4.5.50), using the relation $B_{qq'} = 0$, and denoting by L the sum $\sum_{q\in\Gamma(R_0)} L_q$ and by D the sum $\sum_{q\in\Gamma(R_0)} D_q$, we get formula (4.5.40). The operators L and D satisfy the relations (4.5.45) – (4.5.49), because each of the terms in the sum satisfies such relations (Lemma 4.7) and they all act in the orthogonal subspaces. The lemma is proved.

Thus, we verify that the operator B (see (4.5.1)) can be represented in the form:

$$B(t) = B_0(t) + \sum_{q\in\Gamma(R_0)} \mathcal{E}_q(t) + L(t) + D(t), \tag{4.5.65}$$

where

$$B_0(t) = (I - P(N, \mathbf{M}))V(I - P(N, \mathbf{M})). \tag{4.5.66}$$

The operators $\mathcal{E}_q$, L, D satisfy relations (4.5.41) – (4.5.49).

We write the operator $U\hat{H}U^*(t)$ in the form:

$$U\hat{H}U^*(t) = H_0(t) + \Delta\hat{A}(t), \quad U = U(N, \mathbf{M}, t), \tag{4.5.67}$$

$$\Delta\hat{A}(t) = \sum_{q \in \Gamma(R_0)} P_{J_q(N)}(\mathbf{M}_q))\Delta A_q^{(M_q)}(t). \tag{4.5.68}$$

Lemma 4.10 . *For any $z \in C$, $M_0 > 10$ and $N > 8M_0$ there is a vector $\mathbf{M} = \{M_i^q\}_{i \in J_q(N), q \in \Gamma(R_0)}$ with*

$$M_0/5 < M_i^q < 5M_0, \tag{4.5.69}$$

such that

$$(P(N, 3\mathbf{M}/2) - P(N, \mathbf{M}/2))P_z(M_0^2 A^{-2}) = 0, \tag{4.5.70}$$

$$A = \max\{a_1, a_2, a_3\}.$$

Proof. It suffices to see that

$$(P_i^q(3M_i^q/2) - P_i^q(M_i^q/2))P_z(M_0^2 A^{-2}) = 0. \tag{4.5.71}$$

for all i in $J_q(N)$ and q in $\Gamma(R_0)$. Suppose that z satisfies the inequality $|z - |\mathbf{d}_i^q(t)|^2| > 2M_0^2 p_q(0)^2$ We take $M_i^q = M_0/5$ and prove that

$$P_i^q(3M_i^q/2)P_z(M_0^2 A^{-2}) = 0. \tag{4.5.72}$$

Then the relation (4.5.71) holds all the more so. Indeed, let $\delta_j \in P_i^q(3M_i^q/2)l_2^3$. Then, $|(\mathbf{p}_j(0), \mathbf{p}_q(0))| < (3M_i^q/2+1)p_q(0)^2 < (2/5)M_0 p_q(0)^2$. Taking the last relation into account, we get that $|p_j^2(t) - z| > ||z - |\mathbf{d}_i^q(t)|^2| - (\mathbf{p}_j(t), \mathbf{p}_q(0))^2 p_q(0)^{-2} > M_0^2 p_q(0)^2$, i.e., $\delta_j \notin P_z(M_0^2 A^{-2})l_2^3$. The formula (4.5.72) is proved.

It remains to consider the z such that $|z - |\mathbf{d}_i^q(t)|^2| \leq 2M_0^2 p_q(0)^2$ We take $M_i^q = 5M_0$ and prove that

$$(P_i^q - P_i^q(M_i^q/2))P_z(M_0^2 A^{-2}) = 0.$$

Then the relation (4.5.71) holds all the more so. Indeed, let $\delta_j \in (P_i^q - P_i^q(M_i^q/2))l_2^3$. Then, $|(\mathbf{p}_j(0), \mathbf{p}_q(0))| > (M_i^q/2 - 1)p_q(0)^2 > 2M_0 p_q(0)^2$. By the last relation $|p_j^2(t) - z| > (\mathbf{p}_j(t), \mathbf{p}_q(0))^2 p_q(0)^{-2} - |z - |\mathbf{d}_i^q(t)|^2| > M_0^2 A^{-2}$, i.e., $\delta_j \notin P_z(M_0^2 A^{-2})l_2^3$. The lemma is proved.

We consider the projection $P(N, \mathbf{M})$, where $N = k^{3/5}$ and $(1/25)k^{1/5} < M_i^q < k^{1/5}$, and choose $\mathbf{M}$ so that

$$(P(N, 3\mathbf{M}/2) - P(N, \mathbf{M}/2))P_{k^2}(25A^{-2}k^{2/5}) = 0 \tag{4.5.73}$$

(by Lemma 4.10, this can be done). We construct the operator $U(t)$ corresponding to given t and $P(N, \mathbf{M})$. It is easy to see that conditions of Lemma 4.8 are satisfied for sufficiently large k.

Remark 4.1. Note that the diagonal projection $P(N, \mathbf{M})$ is described by the formula:

$$(P_q)_{jj} = \begin{cases} 1 & \text{if } j \in \Pi_q(k^{1/5}) \setminus T(k, R_0); \\ 0 & \text{otherwise;} \end{cases} \tag{4.5.74}$$

where

$$\Pi_q(k^{1/5}) = \left\{ j, j \in Z^3 : \mid (\mathbf{p}_j(0), \mathbf{p}_q(0)) \mid < c(k, j, q) k^{1/5} \right\}, \tag{4.5.75}$$

$$T(k, R_0) = \{ j, j \in Z^3 : \exists q, q' \in \Gamma(R_0), q \neq q' : \mid (\mathbf{p}_j(0), \mathbf{p}_q(0)) \mid < c(k, j, q) k^{1/5},$$
$$\mid (\mathbf{p}_j(0), \mathbf{p}_q(0)) \mid < k^{3/5} \} \}, \tag{4.5.76}$$

where

$$c(k, j, q) = \begin{cases} 1/5 & \text{if } \ |k^2 - |\mathbf{d}_j^q(t)|^2| > 2M_0^2 p_q(0)^{-2}; \\ 5 & \text{if } \ |k^2 - |\mathbf{d}_j^q(t)|^2| \le 2M_0^2 p_q(0)^{-2}. \end{cases} \tag{4.5.77}$$

It is clear that $P(N, \mathbf{M})$ coincides with the projection P_q defined in Introduction, up to the auxiliary coefficient $c(k, j, q)$.

Lemma 4.11 . *If t is in the $(k^{-2-2\delta})$-neighborhood of the nonsingular set $\chi_3(k, V, \delta)$, then for $\mathbf{p}_j(t) : |p_j^2(t) - k^2| < k^{-1-\delta}$ and for all q of $\Gamma(k^\delta)$*

$$3|(\mathbf{p}_j(t), \mathbf{p}_q(0))| > k^{1-3\delta}. \tag{4.5.78}$$

Corollary 4.6 .

$$E_j P(N, \mathbf{M}) = 0, \tag{4.5.79}$$

$$U(N, \mathbf{M}, t) E_j U^*(N, \mathbf{M}, t) = E_j. \tag{4.5.80}$$

Proof of the corollary Relation (4.5.79) holds, because $P(N, \mathbf{M})_{jj} = 0$, otherwise $|(\mathbf{p}_j(t), \mathbf{p}_q(0))| < 5 p_q(0)^2 k^{1/5}$ for some q of $\Gamma(k^\delta)$. Considering that

$$U(N, \mathbf{M}, t)(I - P(N, \mathbf{M})) = (I - P(N, \mathbf{M})) U^*(N, \mathbf{M}, t) = I - P(N, \mathbf{M}) \tag{4.5.81}$$

(see (4.5.37)), we get (4.5.80). The corollary is proved.

Proof of the lemma. Let $q \in \Gamma(k^\delta)$. By the condition 1^0 inequality (4.3.39) holds. Using the obvious relation $p_{j+q}^2(t) - p_j^2(t) = 2(\mathbf{p}_j(t), \mathbf{p}_q(0)) + p_q(0)^2$, we obtain (4.5.78). The lemma is proved.

Lemma 4.12 . *If t is in the $(k^{-2-2\delta})$-neighborhood of the nonsingular set $\chi_3(k, V, \delta)$ and $z \in C_0$, then*

$$\|(I - P(N, \mathbf{M}))(\hat{H}(t) - z)^{-1/2}\| < k^{1/2+\delta/2}, \tag{4.5.82}$$

$$\|(P(N, \mathbf{M})(\hat{H}(t) - z)^{-1/2}\| < k^{1/10+\delta}. \tag{4.5.83}$$

Corollary 4.7 .

$$\|(\hat{H}(t) - z)^{-1/2}\| < k^{1/2+\delta/2}. \tag{4.5.84}$$

Proof. Since $\hat{H}(t)$ acts on $(I - P(N,\mathbf{M}))l_2^3$ as $H_0(t)$, it is clear that

$$\|(I - P(N,\mathbf{M}))(\hat{H}(t) - z)^{-1/2}\| \leq (\min_{m \in Z^3} |p_m^2(t) - z|^{1/2})^{-1}. \quad (4.5.85)$$

Considering inequality (4.3.41) and the definition of the circle C_0 and arguing as in the proof of Corollary 2.1, we obtain:

$$\min_{m \in Z^3} |p_m^2(t) - z|^{1/2} \geq k^{-1/2-\delta/2}. \quad (4.5.86)$$

Inequality (4.5.82) follows from the last relation.

Reducing $\hat{H}(t)$ to diagonal form in the subspace $P(N,\mathbf{M})l_2^3$ and using (4.5.36), we get:

$$\|P(N,\mathbf{M})(\hat{H}(t) - z)^{-1/2}\| =$$

$$\left(\min_{m:\delta_m \in P_{J_q(N)}(\mathbf{M}_q))l_2^3, q \in \Gamma(k^\delta)} |p_m^2(t) + (\Delta A_q^{(\mathbf{M}_q)})_{mm} - z|^{1/2} \right)^{-1}. \quad (4.5.87)$$

Suppose $m : \delta_m \in P(N, \mathbf{M}/2)l_2^3$. This means that $\delta_m \in P_{J_q(N)}(\mathbf{M}_q)/2)l_2^3$ for some q. It is obvious that

$$|p_m^2(t) + (\Delta A_q^{(\mathbf{M}_q)})_{mm} - z| >$$

$$|p_m^2(t) + (\Delta A_q)_{mm} - k^2| - |(\Delta A_q)_{mm} - (\Delta A_q^{(\mathbf{M}_q)})_{mm}| - |k^2 - z| \quad (4.5.88)$$

(the operator ΔA_q is defined by formula (4.3.31)). Using inequality (4.5.5) yields:

$$|(\Delta A_q)_{mm} - (\Delta A_q^{(M)})_{mm}| < k^{-\varepsilon k^{1/5}}, \quad (4.5.89)$$

$$\varepsilon > \varepsilon_0 R_0^{-1}, \quad 0 < \varepsilon_0 \neq \varepsilon_0(k, R_0).$$

By the definition of C_0 we have:

$$|k^2 - z| = k^{-1-\delta}. \quad (4.5.90)$$

Since $\delta_m \in P_{J_q(N)}(\mathbf{M}_q)/2)l_2^3$, we have $m \in \Pi_q(k^{1/5})$. From this it follows inequality (4.3.42). Using relations (4.3.42), (4.5.89) and (4.5.90), we get:

$$\min_{m:\delta_m \in P_{J_q(N)}(\mathbf{M}_q/2)l_2^3, q \in \Gamma(k^\delta))l_2^3} |p_m^2(t) + \Delta A_{mm}^{(\mathbf{M}_q)}(t) - z| > k^{-1/5-4\delta}. \quad (4.5.91)$$

Now suppose that $m : \delta_m \in (P(N,\mathbf{M}) - P(N,\mathbf{M}/2))\, l_2^3$. This means that $\delta_m \in \left(P_{J_q(N)}(\mathbf{M}_q)) - P_{J_q(N)}(\mathbf{M}_q)/2)\right) l_2^3$ for some q. From (4.5.73) it follows that $|p_m^2(t) - k^2| > 25A^{-2}k^{2/5}$. Using that $|\Delta A_{mm}^{(\mathbf{M}_q)}(t)| \leq \|V\|$, we get

$$\min_{m:\delta_m \in \left(P_{J_q(N)}(\mathbf{M}_q) - P_{J_q(N)}(\mathbf{M}_q/2)\right)l_2^3, q \in \Gamma(k^\delta))l_2^3} |p_m^2(t) + \Delta A_{mm}^{(\mathbf{M}_q)}(t) - z| > k^{2/5-\delta}. \quad (4.5.92)$$

Considering inequality (4.5.87), (4.5.91) and (4.5.92) we obtain relation (4.5.83). The lemma is proved.

Lemma 4.13 . *Suppose t is in the $(k^{-2-2\delta})$-neighborhood of the nonsingular set $\chi_3(k, V, \delta)$. Then the operator $\hat{H}(t)$ has a unique eigenvalue inside the contour C_0. It is equal to $p_j^2(t)$.*

Proof. The matrix of the eigenvalues of the operator $\hat{H}(t)$ is given by the formula (4.5.67). Let $t \in \chi_3(k, V, \delta)$. We consider m such that:

$$p_m^2(t) + (\Delta \Lambda_q^{(\mathbf{M}_q)})_{mm}(t) = k^2. \tag{4.5.93}$$

If $m : \delta_m \notin P(N, \mathbf{M})l_2^3$, then $(\Delta \Lambda_q^{(\mathbf{M}_q)})_{mm}(t) = 0$, therefore equality (4.5.93) has the form:

$$p_m^2(t) = k^2.$$

According to Lemma 4.4 there exists a unique m, which we denote by j, such that $p_j^2(t) = k^2$. Estimate (4.3.41) yields:

$$\min_{m \in Z^3, m \neq j} |p_m^2(t) - k^2| > 2k^{-1-\delta}. \tag{4.5.94}$$

Suppose $m : \delta_m \in P(N, \mathbf{M})l_2^3$. Then relation (4.5.93) is not valid, because inequalities (4.3.42) and (4.5.89) hold.

If t is in the $(k^{-2-2\delta})$-neighborhood of nonsingular set $\chi_3(k, V\delta)$, then there exists a unique j such that $|p_j^2(t) - k^2| < k^{-1-2\delta}$, i.e., being inside C_0. For $m \neq j$ the inequalities (4.5.94), (4.3.42) and (4.5.89) are also satisfied. Therefore, the points $p_m^2(t) + (\Delta \Lambda_q)_{mm}(t)$, $m \neq j$ lie outside C_0. The lemma is proved.

We introduce the notations:

$$\tilde{A}(z, t) = (H_0(t) + \Delta\hat{\Lambda}(t) - z)^{-1/2} B(t)(H_0(t) + \Delta\hat{\Lambda}(t) - z)^{-1/2}, \tag{4.5.95}$$

$$A_0(z, t) = (H_0(t) - z)^{-1/2} B_0(H_0(t) - z)^{-1/2}, \tag{4.5.96}$$

here B, B_0 and $\Delta\hat{\Lambda}(t)$ are given by the formulae (4.5.1), (4.5.65), (4.5.66) and (4.5.68).

Lemma 4.14 . *If t is in the $(k^{-2-2\delta})$-neighborhood of the nonsingular set $\chi_3(k, V, \delta)$ and $z \in C_0$, then the following estimates hold for the operator $\tilde{A}(z, t)$:*

$$\|\tilde{A}(z, t)\| < k^{4\delta}, \tag{4.5.97}$$

$$\|\tilde{A}^3(z, t)\| < k^{-1/5 + 21\delta}. \tag{4.5.98}$$

Remark 4.2. The estimate (4.5.98) is weaker than $\|\tilde{A}(z, t)\|^3 < k^{-1/5 + 16\delta}$, since $\tilde{A}$ is not selfadjoint.

Proof. For the proof we verify the following three relations:
1)
$$\|A_0\| < k^{3\delta}, \tag{4.5.99}$$

2)
$$\|\tilde{A} - A_0\| < 5k^{-1/5 + 3\delta}, \tag{4.5.100}$$

3)

$$\|A_0^3\| < 8k^{-1/5+20\delta}. \tag{4.5.101}$$

It is easy to see that these estimates give the estimates (4.5.97) and (4.5.98).

1) The first relation is easy to get from conditions $1^0 - 3^0$. They imply that

$$\varepsilon = \inf_{i,j\in Z^3, i\neq j, |i-j|<R_0} |(p_j^2(t) - z)(p_i^2(t) - z)| > k^{-5\delta} \tag{4.5.102}$$

for all $z \in C_0$ (see Corollary 3.1). It is not hard to show that $\|A_0\| < c\bar{v}R_0^3\varepsilon^{-1/2}$, $\bar{v} = \max_i |v_i|$. It is clear that for sufficiently large k there are the estimates: $\bar{v} < k^{\delta/8}$, $R_0 < k^{\delta/8}$ [7] Using (4.5.102), we get inequality (4.5.99).

2) Note that

$$A_0(t) = (H_0(t) + \Delta\hat{A}(t) - z)^{-1/2} B_0(t)(H_0(t) + \Delta\hat{A}(t) - z)^{-1/2},$$

because $(I - P(N,\mathbf{M}))\Delta\hat{A}(t) = 0$. Using (4.5.65) we represent $\tilde{A} - A_0$ in the form:

$$\tilde{A} - A_0 = \varepsilon_1 + \varepsilon_2 + \varepsilon_3, \tag{4.5.103}$$

$$\varepsilon_1 = \sum_{q\in\Gamma(R_0)} \varepsilon_{1q}, \tag{4.5.104}$$

$$\varepsilon_{1q} = (H_0(t) + \Delta\hat{A}(t) - z)^{-1/2}\mathcal{E}_q(t)(H_0(t) + \Delta\hat{A}(t) - z)^{-1/2}, \tag{4.5.105}$$

$$\varepsilon_2 = (H_0(t) + \Delta\hat{A}(t) - z)^{-1/2}L(t)(H_0(t) + \Delta\hat{A}(t) - z)^{-1/2}, \tag{4.5.106}$$

$$\varepsilon_3 = (H_0(t) + \Delta\hat{A}(t) - z)^{-1/2}D(t)(H_0(t) + \Delta\hat{A}(t) - z)^{-1/2}. \tag{4.5.107}$$

Taking (4.5.43) into account, we represent ε_{1q} in the form

$$\varepsilon_{1q} = \varepsilon_{1q}^{11} + \varepsilon_{1q}^{10} + \varepsilon_{1q}^{01}, \tag{4.5.108}$$

$$\varepsilon_{1q}^{11} = (I - P_{k^2}(k^{3/5}/2))\varepsilon_{1q}(I - P_{k^2}(k^{3/5}/2)),$$

$$\varepsilon_{1q}^{10} = (I - P_{k^2}(k^{3/5}/2))\varepsilon_{1q}P_{k^2}(k^{3/5}/2),$$

$$\varepsilon_{1q}^{01} = P_{k^2}(k^{3/5}/2)\varepsilon_{1q}(I - P_{k^2}(k^{3/5}/2)).$$

The relation (4.5.42) enables us to write $\varepsilon_{1q}^{01} = \varepsilon_{1q}^{01*} + \varepsilon_{1q}^{01**}$; here

$$\varepsilon_{1q}^{01*} = P(N,\mathbf{M}/2)\varepsilon_{1q}^{01},$$

$$\varepsilon_{1q}^{01**} = (P(N,3\mathbf{M}/2) - P(N,\mathbf{M}/2))\varepsilon_{1q}^{01}. \tag{4.5.109}$$

Consider ε_{1q}^{01*}. It easily follows from the definition of $P_{k^2}(k^{3/5}/2)$:

$$\|(I - P_{k^2}(k^{3/5}/2))(H_0(t) + \Delta\hat{A}(t) - z)^{-1/2}\| < 4k^{-3/10}. \tag{4.5.110}$$

Considering (4.5.83) and the obvious relation $\|\mathcal{E}_q\| < 2\|V\| < k^\delta$, we see that

[7] The possibility of such a rough estimate in the proof of the lemma will enable us below to lower the smoothness requirement on the potential.

$$\|\varepsilon_{1q}^{01*}\| <$$

$$k^{\delta}\|P(N,\mathbf{M}/2)((H_0(t)+\Delta\hat{A}(t)-z)^{-1/2}\|\|(I-P_{k^2}(k^{3/5}/2))(H_0(t)+\Delta\hat{A}(t)-z)^{-1/2}\|$$
$$< k^{-1/5+2\delta}. \tag{4.5.111}$$

Since (4.5.41) and (4.5.73) hold, it follows that

$$\|\varepsilon_{1q}^{01**}\| = \|(I - P_{k^2}(k^{2/5}A^{-2}))\varepsilon_{1q}^{01**}\|$$

$$\leq \|(I - P_{k^2}(k^{2/5}A^{-2}))(H_0(t) + \Delta\hat{A}(t) - z)^{-1/2}\|\times$$
$$\|\mathcal{E}_q\|\|(I - P_{k^2}(k^{3/5}/2))(H_0(t) + \Delta\hat{A}(t) - z)^{-1/2}\|$$

Using again (4.5.110), we get

$$\|\varepsilon_{1q}^{01**}\| < k^{-1/2+\delta}. \tag{4.5.112}$$

Adding (4.5.111) and (4.5.112) we obtain $\|\varepsilon_{1q}^{01}\| < 2k^{-1/5+2\delta}$. Similarly, $\|\varepsilon_{1q}^{10}\| < 2k^{-1/5+2\delta}$. It is easy to see that $\|\varepsilon_{1q}^{11}\| < 2k^{-3/5+\delta}$. The last three estimates combine to yield $\|\varepsilon_{1q}\| < 5k^{-1/5+2\delta}$. Since the number of vectors q making up the set $\Gamma(R_0)$ is less than $k^{\delta/2}$ (because $R_0 < k^{\delta/8}$), we find that $\|\varepsilon_1\| < k^{-1/5+3\delta}$.

Using (4.5.46) and (4.5.73), we prove that $\varepsilon_2(I - P_{k^2}(25k^{2/5}A^{-2})) = (I - P_{k^2}(25k^{2/5}A^{-2}))\varepsilon_2 = \varepsilon_2$. From this, $\|\varepsilon_2\| < k^{-2/5+\delta}$. Taking into account inequalities (4.5.49) ($M_0 \approx k^{1/5}$) and (4.5.84) it is easy to show that $\|\varepsilon_3\| < (\varepsilon k)^{-\varepsilon k^{1/5}}$, $\varepsilon > \varepsilon_0 R_0^{-1}$, $0 < \varepsilon_0 \neq \varepsilon_0(k)$. Summing the estimates for $\|\varepsilon_1\|$, $\|\varepsilon_2\|$, $\|\varepsilon_3\|$, we get estimate (4.5.100).

3) For the proof we first verify that

$$P_{k^2}(k^{1/5-10\delta})A_0 P_{k^2}(k^{1/5-10\delta}) = 0. \tag{4.5.113}$$

The relation (4.5.113) holds if and only if $v_{i-l} = 0$ for all δ_i and $\delta_l \in P_{k^2}(k^{1/5-10\delta})(I-P(N,\mathbf{M}))l_2^3$. Assume that the indicated relation does not hold. Obviously, $\mathbf{p}_i(t) - \mathbf{p}_l(t) = n\mathbf{p}_q(0)$, $n \in Z$, $q \in \Gamma(R_0)$, $1 \leq |n| < R_0$, otherwise $v_{i-l} = 0$. Let i be such that $25|(\mathbf{p}_i(t), \mathbf{p}_q(0))| > k^{1/5}$ for all q in $\Gamma(R_0)$. Then for sufficiently large k

$$|p_i^2(t) - p_l^2(t)| = |2n(\mathbf{p}_i(t), \mathbf{p}_q(0)) - |\mathbf{p}_i(t) - \mathbf{p}_l(t)|^2| > 2k^{1/5} - R_0^2 p_q(0)^2 > k^{1/5}. \tag{4.5.114}$$

Thus, if $|p_i^2(t) - k^2| < k^{1/5-10\delta}$, then the opposite inequality holds for $p_l^2(t)$, i.e., $\delta_l \notin P_{k^2}(k^{1/5-10\delta})l_2^3$, and hence it must be assumed that the relation $25|(\mathbf{p}_i(t), \mathbf{p}_q(0))| \leq k^{1/5}$ for some q in $\Gamma(R_0)$ holds. Let $|(\mathbf{p}_i(t), \mathbf{p}_{q'}(0))| \geq k^{3/5}$ for all $q' \neq q$, $q' \in \Gamma(R_0)$. Then, $\delta_i \in P(N,\mathbf{M})l_2^3$, but this cannot be, by the original assumption. Hence, there exists a q', $q' \neq q$, such that $|(\mathbf{p}_i(t), \mathbf{p}_{q'}(0))| < k^{3/5}$, and, therefore, $|p_i^2(t) - k^2| > k^{1/5-10\delta}$ by condition 5^0; thus $\delta_i \notin P_{k^2}(k^{1/5-10\delta})l_2^3$. This contradiction proves the formula (4.5.113). Using this formula, we represent A_0 in the form

$$A_0 = M_0 + M_+ + M_-, \tag{4.5.115}$$

$$M_0 = (I - P_{k^2}(k^{1/5-10\delta}))A_0(I - P_{k^2}(k^{1/5-10\delta})),$$
$$M_+ = P_{k^2}(k^{1/5-10\delta})A_0(I - P_{k^2}(k^{1/5-10\delta})) = P_{k^2}(k^{1/5-10\delta})A_0,$$
$$M_- = (I - P_{k^2}(k^{1/5-10\delta}))A_0P_{k^2}(k^{1/5-10\delta}) = A_0P_{k^2}(k^{1/5-10\delta}).$$

Obviously, $M_+^2 = M_-^2 = M_0M_+ = M_-M_0 = 0$. From this, $A_0^3 = M_0^3 + M_0^2M_- + M_+M_0^2 + M_0M_-M_+ + M_+M_0M_- + M_-M_+M_0 + M_+M_-M_+ + M_-M_+M_-$. By the obvious relation $\|M_0\| < ck^{-1/5+10\delta}$, $\|M_-\| < k^{3\delta}$, $\|M_+\| < k^{3\delta}$, we obtain

$$\|M_0^3\| + \|M_0^2M_-\| + \|M_+M_0^2\| + \|M_0M_-M_+\| + \|M_+M_0M_-\| + \|M_-M_+M_0\|$$
$$< ck^{-1/5+16\delta}. \tag{4.5.116}$$

Assume that we have verified the inequality

$$\|M_+M_-\| < k^{-1/5+17\delta} \tag{4.5.117}$$

Then, it turns out that

$$\|M_+M_-M_+\|, \|M_-M_+M_-\| < k^{-1/5+17\delta}, \tag{4.5.118}$$

and summing the estimates (4.5.116) and (4.5.118), we get (4.5.101). Thus, to prove the necessary relation it remains to check the estimate (4.5.117). First we prove that the right-hand projection $(I - P(N, \mathbf{M}))(I - P_{k^2}(k^{1/5-8\delta}))$ in the definition of M_+ can be deleted without changing the result, i.e.,

$$M_+ = (I - P(N, \mathbf{M}))P_{k^2}(k^{1/5-10\delta})(H_0(t) - z)^{-1/2}V(H_0(t) - z)^{-1/2}, \tag{4.5.119}$$

This formula follows from equality (4.5.113) and the relation

$$(I - P(N, \mathbf{M}))P_{k^2}(k^{1/5-10\delta})VP(N, \mathbf{M}) = 0 \tag{4.5.120}$$

to verify. Suppose, it does not hold. Then, there exist i, m: $\delta_m \in P(N, \mathbf{M})l_2^3$, $\delta_i \in (I - P(N, \mathbf{M}))P_{k^2}(k^{1/5-10\delta})l_2^3$, $|i - m| < R_0$. This means that there exists $q \in \Gamma(R_0)$ such that $|(\mathbf{p}_m(0), \mathbf{p}_q(0))| < M$, and $|(\mathbf{p}_m(0), \mathbf{p}_{q'}(0))| > N$ for all $q' \neq q$, $q' \in \Gamma(R_0)$. Since, $|i - m| < R_0$, it follows that $|(\mathbf{p}_i(0), \mathbf{p}_q(0))| < 3M/2$, $2|(\mathbf{p}_i(0), \mathbf{p}_{q'}(0))| > N$. Suppose that $|(\mathbf{p}_i(0), \mathbf{p}_q(0))| < M$, $|(\mathbf{p}_i(0), \mathbf{p}_{q'}(0))| > N$. This means, that $\delta_i \in P(N, \mathbf{M})l_2^3$. But, the last relation contradicts to the initial hypothesis. Next, suppose $M/2 < |(\mathbf{p}_i(0), \mathbf{p}_q(0))| < 3M/2$, $|(\mathbf{p}_i(0), \mathbf{p}_{q'}(0))| > N$. In this case $\delta_i \in (P(N, 3\mathbf{M}/2) - P(N, \mathbf{M}/2))l_2^3$. It follows from relation (4.5.73) that $\delta_i \notin P_{k^2}(k^{1/5-10\delta})l_2^3$. This contradicts to the initial hypothesis. If, at last, $|(\mathbf{p}_i(0), \mathbf{p}_{q'}(0))| < N$, then by condition 5^0 (see (4.3.43)) we have $\delta_i \notin P_{k^2}(k^{1/5-10\delta})l_2^3$. The obtained contradictions proves that relation (4.5.120) holds.

Now, we consider the diagonal part of the operator M_+M_-:

$$(M_+M_-)_{ii} = \sum_{|m|<R_0} \frac{|v_m|^2}{(p_i^2(t) - z)(p_{i+m}^2(t) - z)} =$$

$$\sum_{|m|<R_0} \frac{|v_m|^2(p_m^2(0)+p_i^2(t)-z)}{(p_i^2(t)-z)(p_{i+m}^2(t)-z)(p_{i-m}^2(t)-z)}. \tag{4.5.121}$$

We prove that

$$2|(M_+M_-)_{ii}| < k^{-1/5+17\delta}. \tag{4.5.122}$$

Let $i : |p_i^2(t)-k^2| > k^{-1/5+4\delta}$. Since

$$\delta_{i\pm m} \in (I - P_{k^2}(k^{1/5-10\delta}))l_2^3, \quad p_m(0) < k^{\delta/8},$$

we get (4.5.122). If $i : |p_i^2(t)-k^2| \le k^{-1/5+4\delta}$, then conditions $1^0 - 3^0$ give us that $36|(p_i^2(t)-z)(p_{i\pm m}^2(t)-z)| > k^{-4\delta}$. This easily yields (4.5.122).

Consider $(M_+M_-)_{il}$, $i \ne l$,

$$(M_+M_-)_{il} = \sum_{j:|i-j|<R_0,|l-j|<R_0} \frac{v_{i-j}v_{j-l}}{(p_i^2(t)-z)^{1/2}(p_l^2(t)-z)^{1/2}(p_j^2(t)-z)}. \tag{4.5.123}$$

By conditions $1^0 - 3^0$ $36|(p_i^2(t)-z)(p_l^2(t)-z)| > k^{-4\delta}$. Using the inequality $|p_j^2(t)-k^2| > k^{1/5-10\delta}$, we arrive at the estimate

$$|(M_+M_-)_{il}| < k^{-1/5+12\delta}, \quad i \ne l. \tag{4.5.124}$$

Since $(M_+M_-)_{il} = 0$ if $|i-l| > 2R_0$, it follows that

$$\|M_+M_-\| < k^{\delta} \max_{i,j} |(M_+M_-)_{ij}|.$$

Taking into account (4.5.122) and (4.5.124), we get (4.5.117).

The lemma is proved.

Lemma 4.15 . *If t is in the $(k^{-2-2\delta})$-neighborhood of $\chi_3(k,V,\delta)$ and $0 \le r \le M_1$, $M_1 = k^{1-4\delta}R_0^{-1}$, then*

$$G_r(k,t) = \hat{G}_r(k,t). \tag{4.5.125}$$

$$g_r(k,t) = \hat{g}_r(k,t), \tag{4.5.126}$$

Proof. We prove (4.5.125). First we show that

$$G_r(k,t) = U\hat{G}_rU^*(k,t). \tag{4.5.127}$$

We represent $G_r(k,t)$ in the form:

$$G_r(k,t)_{i_0i_r} = \sum_{i_1,\dots,i_{r-1}\in Z^3} I_{i_0i_1\dots i_{r-1}i_r}, \tag{4.5.128}$$

where

$$I_{i_0i_1\dots i_{r-1}i_r} = \frac{(-1)^{r+1}}{2\pi i}\oint_{C_0} \frac{v_{i_0-i_1}\dots v_{i_{r-1}-i_r}}{(p_{i_0}^2(t)-z)\dots(p_{i_r}^2(t)-z)}dz.$$

Using (4.2.8) and (4.5.1), we easily get

$$U\hat{G}_r U^*(k,t)_{i_0 i_r} = \sum_{i_1,\dots,i_{r-1}\in Z^n} \hat{I}_{i_0 i_1 \dots i_{r-1} i_r}, \tag{4.5.129}$$

where

$$\hat{I}_{i_0 i_1 \dots i_{r-1} i_r} = \frac{(-1)^{r+1}}{2\pi i} \oint_{C_0} \frac{B_{i_0,i_1} \dots B_{i_{r-1},i_r}}{(p_{i_0}^2(t) + \Delta\hat{A}(t)_{i_0 i_0} - z)\dots(p_{i_r}^2(t) + \Delta\hat{A}(t)_{i_r i_r} - z)} dz.$$

Suppose $i = (i_0, \dots, i_r)$ is such that $P(N, 3\mathbf{M}/2)\delta_{i_l} = 0$ for all $l = 0, \dots, r$. In this case $B_{i_l i_{l+1}} = v_{i_l - i_{l+1}}$, $l = 0, \dots, r-1$ (see formulae (4.5.65), (4.5.66) and Lemma 4.9). Therefore,

$$I_{i_0 i_1 \dots i_{r-1} i_r} = \hat{I}_{i_0 i_1 \dots i_{r-1} i_r}, \tag{4.5.130}$$

Suppose, there exists at least one l such that

$$P(N, 3\mathbf{M}/2)\delta_l = \delta_l. \tag{4.5.131}$$

Let us prove that (4.5.130) holds, because the both sides are zero. The integrals $I_{i_0 i_1 \dots i_{r-1} i_r}$, $\hat{I}_{i_0 i_1 \dots i_{r-1} i_r}$ can be nonzero only if there exist at least one index i_l, $l = 0, \dots, r$ equal to j, otherwise the integrands turns out to be holomorphic inside the disk. Let $i_0 = j$. We denote by l_0 the smallest of l, satisfying relation (4.5.131). Then, $P(N, 3\mathbf{M}/2)\delta_l = 0$ for all $l < l_0$ and $B_{i_l i_{l+1}} = v_{i_l - i_{l+1}}$ for $l < l_0$. Therefore, the integrals $I_{i_0 i_1 \dots i_{r-1} i_r}$ and $\hat{I}_{i_0 i_1 \dots i_{r-1} i_r}$ can be nonzero only if $|i_{l-1} - i_l| \le R_0$ for all $l : 1 < l \le l_0$, i.e., only if $|j - l_0| < rR_0 < k^{1-4\delta}$. However, then, for any q in $\Gamma(R_0)$ we have $|(\mathbf{p}_{l_0}(t), \mathbf{p}_q(0))| > |(\mathbf{p}_j(t), \mathbf{p}_q(0))| - k^{1-4\delta} p_q(0) a^{-1}$, $q \in \Gamma(R_0)$. Using inequality (4.5.78) we get

$$4|(\mathbf{p}_{l_0}(t), \mathbf{p}_q(0))| > k^{1-3\delta}, \quad q \in \Gamma(R_0). \tag{4.5.132}$$

But it is not the case, because $\delta_{l_0} \in P(N, 3\mathbf{M}/2) l_2^3$. Similar arguments are valid for the case $j = i_l$, $l \ne 0$. The contradiction proves that in (4.5.130) the both sides are zero when (4.5.131) holds. Thus, we have proved (4.5.130) for any $i = (i_0, \dots, i_r)$. This means that (4.5.127) holds.

Taking into account that $\hat{I}_{i_0 i_1 \dots i_{r-1} i_r} = 0$, in particular, when (4.5.131) is satisfied for $l = 0$ or $l = r$, we get

$$U\hat{G}_r U^* = (I - P(N, 3\mathbf{M}/2)) U\hat{G}_r U^* = U\hat{G}_r U^* (I - P(N, 3\mathbf{M}/2)). \tag{4.5.133}$$

From this, considering formula (4.5.38), we obtain:

$$U\hat{G}_r U^* = \hat{G}_r. \tag{4.5.134}$$

Now, formula (4.5.125) follows from (4.5.127) and (4.5.134).

Formula (4.5.126) is proved similarly. The lemma is proved.

The proof of Theorem 4.1 Estimates (4.5.97) and (4.5.98) enable us to expand $(I - \hat{A})^{-1}$ in the series in powers of $\hat{A}$:

$$(H(t) - z)^{-1} = \sum_{r=0}^{\infty} (\hat{H}(t) - z)^{-1/2} \hat{A}^r (\hat{H}(t) - z)^{-1/2}. \tag{4.5.135}$$

Integrating the both sides of equality (4.5.135) over the contour C_0 yields:

$$-\frac{1}{2\pi i}\oint_{C_0}(H(t)-z)^{-1}dz = \sum_{r=0}^{\infty}\hat{G}_r(k,t), \qquad (4.5.136)$$

$$\|\hat{G}_r(k,t)\| < k^{-(1/5-21\delta)[r/3]+8\delta}. \qquad (4.5.137)$$

We prove that the similar estimate holds for $\|\hat{G}_r(k,t)\|_1$:

$$\|\hat{G}_r(k,t)\|_1 < k^{-(1/5-21\delta)([r/3]-2)+16\delta}. \qquad (4.5.138)$$

To prove the last estimate we consider $U\hat{G}_rU^*$. It is clear that

$$U\hat{G}_rU^*(k,t) = \oint_{C_0}(H_0(t)+\Delta\hat{A}(t)-z)^{-1/2}\tilde{A}^r(H_0(t)+\Delta\hat{A}(t)-z)^{-1/2}dz, \qquad (4.5.139)$$

$$\|U\hat{G}_rU^*(k,t)\| < k^{-(1/5-21\delta)[r/3]+8\delta}. \qquad (4.5.140)$$

It is not hard to see that

$$\tilde{A}^r = T + \tilde{A}_0^r, \qquad (4.5.141)$$

where

$$T = \sum_{i=0}^{r}((I-E_j)\tilde{A})^iE_j\tilde{A}^{r-i}, \qquad (4.5.142)$$

$$\tilde{A}_0 = (I-E_j)A(I-E_j). \qquad (4.5.143)$$

We note that

$$\oint_{C_0}(H_0(t)+\Delta\hat{A}(t)-z)^{-1/2}\tilde{A}_0^r(H_0(t)+\Delta\hat{A}(t)-z)^{-1/2}dz = 0, \qquad (4.5.144)$$

because the integrand is holomorphic inside the circle. Considering (4.5.139) and (4.5.141), we obtain

$$U\hat{G}_rU^*(k,t) = \oint_{C_0}(H_0(t)+\Delta\hat{A}(t)-z)^{-1/2}T(H_0(t)+\Delta\hat{A}(t)-z)^{-1/2}dz. \qquad (4.5.145)$$

Repeating the considerations of Lemma 4.9 it is not hard to show that the estimates similar to (4.5.97) and (4.5.98) hold for $(I-E_j)\tilde{A}$:

$$\|(I-E_j)\tilde{A}\| < k^{4\delta}, \qquad (4.5.146)$$

$$\|((I-E_j)\tilde{A})^3\| < 2k^{-1/5+21\delta}. \qquad (4.5.147)$$

Taking into account the last two estimates, we easily obtain:

$$\|T\|_1 < k^{-(1/5+21\delta)([r/3]-2)+16\delta}.$$

From this

$$\|U\hat{G}_rU^*(k,t)\|_1 < k^{-(1/5+21\delta)([r/3]-2)+16\delta}. \qquad (4.5.148)$$

Considering that operators $\hat{G}_r(k,t)$ and $U\hat{G}_rU^*(k,t)$ are unitary equivalent, we get estimate (4.5.138). It is easy to see that $(1/5+21\delta)([r/3]-2)+16\delta > r/20$ for $r > 50$. Therefore, when $r > 50$:

$$\|U\hat{G}_rU^*(k,t)\|_1 < k^{-r/20}. \tag{4.5.149}$$

Since $M_1 \gg 50$, we can rewrite estimate (4.5.138) in a somewhat weaker form (4.2.16) too. From (4.5.125) and (4.5.138) it follows:

$$\|G_r(k,t)\|_1 < k^{-(1/5+21\delta)([r/3]-2)+8\delta}, \quad r < M_1. \tag{4.5.150}$$

When $r > M_0$, $M_0 = [k^\delta/R_0]$ we can replace this estimate by a somewhat weaker one:

$$\|G_r(k,t)\|_1 < k^{-r/20}. \tag{4.5.151}$$

This estimate coincides with (4.2.14).

It remains to estimate G_r for $r < M_0$. Considering as in the proof of Theorem 3.1 and using estimate (4.3.39), we obtain that for $r < M_0$, $M_0 = [k^\delta R_0^{-1}]$ the following estimates hold:

$$\|G_r(k,t)\|_1 < (rR_0)^3(\hat{v}k^{-1+3\delta})^r. \tag{4.5.152}$$

$$\|G_r(k,t)\| < (\hat{v}k^{-1+3\delta})^r. \tag{4.5.153}$$

$$|g_r(k,t)| < r^2\bar{v}(\hat{v}k^{-(1-3\delta)})^{(r-1)} \tag{4.5.154}$$

$$|g_2(k,t)| < c\bar{v}^2R_0^{-1}k^{-2+6\delta}, c \neq c(k,V), \tag{4.5.155}$$

where $\bar{v} = \max_{|i|<R_0}|v_i|$, $\hat{v} = c_0\bar{v}R_0^3$, $c_0 \neq c_0(k,V)$. [8]

If $r < 50$, then estimates (4.5.152) – (4.5.155) are stronger, than inequalities (4.5.149) – (4.5.151). Therefore, inequalities (4.5.149) and (4.5.151) hold for $r \geq 1$, and can be made stronger up to (4.5.152) – (4.5.155) for $r < M_0$. Thus, estimates (4.2.14), (4.2.16), (4.2.20) and (4.2.21) are proved.

From relations (4.5.136) and (4.5.125) it immediately follows:

$$-\frac{1}{2\pi i}\oint_{C_0}(H(t)-z)^{-1}dz = E_j + \sum_{r=1}^{M_1} G_r(k,t) + \psi(k,t), \tag{4.5.156}$$

$$\psi(k,t) = \sum_{r=M_1+1}^{\infty} \hat{G}_r(k,t). \tag{4.5.157}$$

Considering as in the proof of Theorem 2.1 we show that there exists only a unique eigenvalue of operator $H(t)$ inside the circle C_0. Indeed, let n and $\hat{n}$ be the numbers of eigenvalues of operator $H(t)$ and $\hat{H}(t)$, correspondingly, inside the circle. Using the well-known relation

$$n = -\frac{1}{2\pi i}\mathrm{Tr}\oint_{C_0}(H(t)-z)^{-1}dz, \tag{4.5.158}$$

[8] Estimates (3.2.26) – (3.2.29) coincide with (4.5.152) – (4.5.155) for $l = 1$, $n = 3$, $\beta = \delta$.

expanding the resolvent in powers of $\hat{A}$ and taking into account that

$$\hat{n} = -\frac{1}{2\pi i}\mathrm{Tr}\oint_{C_0}(\hat{H}(t)-z)^{-1}dz, \tag{4.5.159}$$

we obtain

$$n - \hat{n} = \sum_{r=1}^{\infty}\mathrm{Tr}\hat{G}_r(k,t). \tag{4.5.160}$$

Thus,

$$n - \hat{n} = \sum_{n=1}^{M_1}\mathrm{Tr}\ G_r(k,t) + \sum_{r=M_1}^{\infty}\mathrm{Tr}\ \hat{G}_r(k,t). \tag{4.5.161}$$

According Lemma 4.13, we have $\hat{n} = 1$. Taking into account estimates (4.5.149) and (4.5.151), we get

$$|n-1| =_{k\to\infty} o(1). \tag{4.5.162}$$

Hence, $n = 1$, i.e., there exists only a unique eigenvalue of the operator $H(t)$ inside the disk C_0. The corresponding spectral projection is given by the formula:

$$E(t) = -\frac{1}{2\pi i}\oint_{C_0}(H(t)-z)^{-1}dz. \tag{4.5.163}$$

Using relation (4.5.156), we get formula (4.2.12). Arguing as in the proof of estimates (4.5.149) and (4.5.151), we obtain:

$$\|\hat{g}_r(k,t)\| < k^{-1-\delta-r/20}, \tag{4.5.164}$$

$$\|g_r(k,t)\| < k^{-1-\delta-r/20}. \tag{4.5.165}$$

When $r < M_0$ inequalities (4.5.154) and (4.5.155) hold. Thus, estimates (4.2.13), (4.2.19) and (4.2.22) are proved. Considering as in the proof of Theorem 2.1 we obtain

$$\lambda(t) = p_j^{2l}(t) + \sum_{r=1}^{\infty}\hat{g}_r(k,t). \tag{4.5.166}$$

Taking into account (4.5.126) we get (4.2.11). The theorem is proved.

The proof of Theorem 4.2 coincides with that of Theorem 3.2 up to using inequality (4.3.39) instead of estimate (2.2.3) .

Proof of Theorem 4.3 Since estimates (4.2.13) – (4.2.16) are valid in the complex $(2k^{-2-2\delta})$- neighborhood of the nonsingular set, it is easy to see that the functions $g_r(k,t)$, $\hat{g}_r(k,t)$ and the operator-valued functions $G_r(k,t)$, $\hat{G}_r(k,t)$ can be continued from the real $(2k^{-n+1-2\delta})$-neighborhood of t to the complex $(2k^{-n+1-2\delta})$-neighborhood as analytic functions of n variables, and inequalities (4.2.13) – (4.2.16) are hereby preserved. Estimating by means of the Cauchy integral the value of the derivative at t in terms of the value of the function itself on the boundary of the $(2k^{-2-2\delta})$-neighborhood of t, we obtain (4.2.25) – (4.2.28). Note that for $m < M_0(k)$ the functions $g_r(k,t)$ and operator-valued functions $G_r(k,t)$ depend analytically on t in the complex $(2k^{-2\delta})$-neighborhood of each simply connected component of the nonsingular set, and estimates (4.2.19) – (4.2.22) hold there. Therefore, inequalities(4.2.29) – (4.2.32) are valid. The theorem is proved.

4.6 The Perturbation Formulae Near the Planes of Diffraction.

In this section we consider the unstable case, i.e., the case when the perturbations of an eigenvalue and its eigenfunction are significant. Namely, we construct perturbation formulae when a quasimomentum is in a vicinity of the von Laue diffraction planes:

$$|p_j^2(t) - p_{j+q}^2(t)| \leq k^{\delta}. \tag{4.6.1}$$

In this case there is essential diffraction inside the crystal. The perturbed eigenvalues and eigenfunctions are not close to unperturbed ones. The simple operator (1.0.34), which was used to describe diffraction in the case $2l > n$ is not valid any more. This means that the picture of diffraction is, in fact, more complicated.

The unstable case was studied by J. Feldman, H. Knorrer, E. Trubowitz [FeKnTr2] in the two and three dimensional situations. In the three-dimensional case they study the eigenvalues of H, which are not close to the unperturbed ones, but can be approximated by eigenvalues of the operator $-\Delta + V_\gamma$, where γ some vector of the dual lattice and V_γ is independent of x in the direction γ i.e.,

$$V_\gamma(x) = V_\gamma(x - \gamma(x,\gamma)|\gamma|^{-2}) = \sum_{j:(\mathbf{p}_j(0),\gamma)=0} v_j \exp i(x, \mathbf{p}_j(0)).$$

It was proved that for arbitrary γ of the dual lattice and any eigenvalue of $H_\gamma(t)$ corresponding to a sufficiently large momentum in the direction γ, there exists a close eigenvalue of the operator $H(t)$ with the same quasimomentum, multiplicity being taken into account. The same result was proved for $n = 2$. Moreover, in the two-dimensional case it was shown that on the rich set of t the corresponding eigenfunctions are close too. O.A. Veliev discussed this problem [Ve6].

The approach developing here, has its own peculiarities. It provides formulae not only for unstable eigenvalues, but also for their spectral projections in three dimensional situation. The converging perturbation series with respect to the model operator $\hat{H}$ (see (1.0.44)), roughly describing also the refraction, are constructed. The series have an asympotic character in a high energy region. They can be differentiated any number of times with respect to the quasimomentum.

In the case of the polyharmonic operator with a periodic potential we constructed the perturbation series with respect to the free operator on the nonsingular set (Section 2.2) and with respect to a model operator on the singular set (Section 2.4). This model operator roughly took into account the diffraction inside the crystal. In the case of the Schrödinger operator the modification of the perturbation series ($H_0 \to \hat{H}$) is needed even for the nonsingular set (Section 4.2). The series diverge with respect to H_0, because there are not only the von Laue diffraction conditions, but also another ones, depending on the potential. Fortunately, it turns out that such modified series converge even on the essential part of the singular set, i.e., it is not necessary to reconstruct additionally these series. The perturbation series for the eigenvalue and its spectral projection converge on the essential part of the singular set, when we take the operator

$\hat{H}(t)$ as the initial one. Note that the operator $H(t)$ has a block structure. The diagonal part of $\hat{H}$ coincides with the corresponding part of the free operator: $(I - \sum_{q \in \Gamma(R_0)} P_q)\hat{H}(t) = (I - \sum_{q \in \Gamma(R_0)} P_q)H_0(t)$. The blocks of $\hat{H}(t)$ are determined by the orthogonal projections P_q, $P_q\hat{H}(t) = \hat{H}(t)P_q = P_q\hat{H}(t)P_q$. Each block is a "piece" of the matrix of the Schrödinger operator with the potential V_q, i.e., $P_q\hat{H}(t)P_q = P_q(H_0(t) + V_q)P_q$. Thus, each block is simply connected with the matrix of a periodic Schrödinger operator in the one-dimensional space, because V_q depends only on $(x, \mathbf{p}_q(0))$. In the case of the nonsingular set the perturbed eigenvalue $\lambda(t)$ is asymptotically close ($k \to \infty$) to an eigenvalue $p_j^{2l}(t)$ of the diagonal part of $\hat{H}(t)$; i.e, $\lambda(t)$ is close to the eigenvalue $p_j^{2l}(t)$ of the free operator $H_0(t)$. Accordingly, the spectral projector of $H(t)$, corresponding to $\lambda(t)$, is close to that of $H_0(t)$, corresponding to $p_j^{2l}(t)$. We prove that in the case of the singular set (more precisely of its essential part) eigenvalues and spectral projection of $H(t)$ are close to those of the block part of $\hat{H}(t)$. We construct the perturbation series for an eigenvalue and its spectral projection on the essential part of the singular set, taking $\hat{H}(t)$ as the initial operator. We obtain this result by constructing the converging perturbation series for the resolvent and integrating it over a small contour. Thus, the blocks $P_q\hat{H}P_q = P_q(H_0(t) + V_q)P_q$ describe roughly the refraction inside the crystal for t of the essential part of the singular set. We call this essential part of the singular set the simple part of the singular set, because on the relatively small rest of the singular set the picture of diffraction is even more complicated.

The proof of the formulae contains the analytical and geometrical parts. In the analytical part we formulate the conditions of the convergence of the series in the form of the inequalities for quasimomentum t (see (4.6.9)-(4.6.14)). Note that these conditions are similar to those for the nonsingular set up to the replacement of an eigenvalue $p_j^2(t)$ of the diagonal part of $\hat{H}(t)$ by an eigenvalue of the block part. In the geometrical part we prove that the conditions are satisfied on the essential part of the singular set. Note that the analytical part is similar to that for the nonsingular set up to the mentioned above replacement. The geometrical part is much more technical. The singular set is only a small part of the isoenergetic surface of the free operator. Therefore, to prove that the conditions of convergence can be satisfied on it, one has to make more subtle considerations than in the case of the nonsingular set.

Thus, we consider the spherical layer on the sphere $|x| = k$:

$$S_q(k, -\delta) = \{x : |x| = k, ||x|^2 - |x + \mathbf{p}_q(0)|^2| < k^\delta\}, \tag{4.6.2}$$

Let us recall that $\mathcal{K}$ is the mapping

$$\mathcal{K} : R^3 \to K, \quad \mathcal{K}\mathbf{p}_j(t) = t. \tag{4.6.3}$$

It is easy to see that the isoenergetic surface of the free operator $S_0(k)$ is given by the formula $S_0(k) = \mathcal{K}S_k$, where S_k is the sphere $|x| = k$. It is clear that the translation $\mu_q(k, \delta) \equiv \mathcal{K}S_q(k, -\delta)$ of the set $S_q(k, -\delta)$ to the elementary cell of the dual lattice K belongs to the isoenergetic surface $S_0(k)$ of the free operator

H_0. However, it does not belong to the nonsingular set described in Section 4.4, because condition 1^0 does not hold on it. Indeed, let $j : \mathbf{p}_j(t) = x$, $x \in S_q(k, -\delta)$ (hence, $t \in \mu_q(k, \delta)$). By the definition of $S_q(k, -\delta)$ we obtain:

$$p_j(t) = k,$$

$$|p_j^2(t) - p_{j+q}^2(t)| < k^\delta \tag{4.6.4}$$

Inequality (4.6.4) is contradictory to (4.3.39) in condition 1^0.

Let us consider the part $\hat{\mu}_q(k, \delta)$ of the isoenergetic surface of $\hat{H}$ situated in the (ck^{-1})-neighborhood of $\mu_q(k, \delta)$. It is described by the formula (see Section 4.12):

$$\hat{\mu}_q(k, \delta) = \{t : \exists j, j \in Z^3, p_j^2(t) + \Delta \hat{A}_{jj}(t) = k^2, |p_j^2(t) - p_{j+q}^2(t)| < k^\delta\}. \tag{4.6.5}$$

[9]

Note that $\hat{\mu}_q(k, \delta) = \mathcal{K}\hat{S}_q(k, -\delta)$, where

$$\hat{S}_q(k, -\delta) = \{x : |x|^2 + \varphi_0(x) = k^2, ||x|^2 - |x + \mathbf{p}_q(0)|^2| < k^\delta\}, \tag{4.6.6}$$

$\varphi_0(x)$ being uniquely determined from formulae

$$\varphi_0(x) = \Delta \hat{A}_{jj}(t), \quad x = \mathbf{p}_j(t). \tag{4.6.7}$$

It is not hard to show that $\hat{S}_q(k, -\delta)$ is the union of curve cylinders, because $\Delta \hat{A}_{jj}$, in fact, depends only on the projection of $\mathbf{p}_j(t)$ on $\mathbf{p}_q(0)$. Note that $\Delta \hat{A}_{jj} \neq 0$ in (4.6.5), because $j \in \Pi_q(k^{1/5})$. We construct perturbation series with respect to $\hat{H}(t)$. In the case of the nonsingular set the corresponding pieces of the isoenergetic surfaces of $H_0(t)$ and $\hat{H}(t)$ coincide. Clearly, describing the nonsingular set on the isoenergetic surface of one of them we, of course, describe for the other too. In the case of the nonsingular set these surfaces are not the same, i.e., $\mu_q(k, \delta)$ does not coincides with $\hat{\mu}_q(k, \delta)$. Therefore, the question is : on which isoenergetic surface the simple part of the singular set should be described? Since we construct perturbation series with respect to $\hat{H}(t)$, it is natural to describe this set on $\hat{\mu}_q(k, \delta)$. We formulate the corresponding geometric lemma here. It will be proved in Section 4.8. We also describe the simple part of the nonsingular set on $\mu_q(k, \delta)$ there. However, the first description, as it will be shown in Chapter 5, is more convenient for applications. Thus, we shall show that there exists a subset $\hat{\chi}_q^0(k, V, \delta)$ of an asymptotically full measure on $\hat{\mu}_q(k, \delta)$, such that for $t \in \hat{\chi}_q^0(k, V, \delta)$ the perturbation series with respect to $\hat{H}(t)$ converge. For such a t the operator $\hat{H}(t)$ has an eigenvalue which can be represented in the form $p_j^2(t) + \Delta \hat{A}(t)_{jj}$, j being uniquely determined from the relation $p_j^2(t) + \Delta \hat{A}(t)_{jj} = k^2$. The series for $H(t)$ converge with respect to $\hat{H}(t)$. Now we give the formulation of the main geometric lemma.

---

[9] In fact, $\hat{\mu}_q(k, \delta)$ is very close to the corresponding part of the isoenergetic surface of the operator $H_q = -\Delta + V_q$, because $p_j^2(t) + \Delta \hat{A}_{jj}(t)$ coincides with an eigenvalue $p_j^2(t) + \Delta A_{jj}(t)$, of $H_q(t)$ up to a value of order $k^{-ck^{1/5}}$ (see Proposition 4.1). All following assertions are also valid for $\tilde{\mu}_q(k, \delta)$ being the corresponding part of the isoenergetic surface of H_q.

Lemma 4.16 . *Let $0 < \delta < 1/300$. Then for sufficiently large k, $k > k_0(V,\delta)$, there exists a set $\hat{\chi}_q^0(k,V,\delta) \subset \hat{\mu}_q(k,\delta)$, such that for any t of this set there is a unique $j \in Z^3$ such that $p_j^2(t) + \Delta\hat{A}(t)_{jj} = k^2$,*

$$|p_j^2(t) - p_{j+q}^2(t)| < k^\delta, \tag{4.6.8}$$

and the following conditions hold:
$1'$

$$\min_{r \in \Omega_1} |p_j^2(t) + \Delta\hat{A}(t)_{jj} - p_{j+m}^2(t)| > k^{1-10\delta} \tag{4.6.9}$$

$$\Omega_1 = \{m : m \in Z^3, |m| < k^\delta;\ m \neq 0;\ m \neq n_0 q,\ n_0 \in Z\}. \tag{4.6.10}$$

$2'$

$$|p_{m+j+i}^2(t) - p_j^2(t) - \Delta\hat{A}(t)_{jj}| > k^{-12\delta}|p_{m+j}^2(t) - p_j^2(t) - \Delta\hat{A}(t)_{jj}|^{-1}. \tag{4.6.11}$$

for all m : $|p_{m+j}^2(t) - p_j^2(t) - \Delta\hat{A}(t)_{jj}| < k^{-20\delta}$, $m \neq n_0 q$, $n_0 \in Z$, and $|i| < k^\delta$, $i \neq 0$.
$3'$

$$\min_{m \in Z^3, m \neq n_0 q, n_0 \in Z} |p_{m+j}^2(t) - p_j^2(t) - \Delta\hat{A}(t)_{jj}| > 2k^{-1-\delta} \tag{4.6.12}$$

$4'$. *For any q' in $\Gamma(R_0)$ and $m \in \Pi_{q'}(k^{1/5})$ the eigenvalue $p_m^2(t) + \Delta\hat{A}(t)_{mm}$ of the operator $\hat{H}_q(t)$ lies sufficiently far from the point $p_j^2(t) + \Delta\hat{A}(t)_{jj}$, namely,*

$$|p_m^2(t) + \Delta\hat{A}(t)_{mm} - p_j^2(t) - \Delta\hat{A}(t)_{jj}| > |(\mathbf{p}_m(0), \mathbf{p}_{q'}(0))|^{-1+\delta}. \tag{4.6.13}$$

$5'$ *If $m \in T(k,\delta)$, then*

$$|p_m^2(t) - p_j^2(t) - \Delta\hat{A}(t)_{jj}| > k^{1/5-9\delta}. \tag{4.6.14}$$

These properties hold in a small neighbourhood of $\hat{\chi}_q^0(k,V,\delta)$: if t is in the $(k^{-2-2\delta})$-neighbourhood of $\hat{\chi}_q^0(k,V,\delta)$), then there is a unique $j \in Z^3$ such that $\left|p_j^2(t) + \Delta\hat{A}(t)_{jj} - k^2\right| < k^{-1-2\delta}$, inequality (4.6.4) and conditions $1'$ - $5'$ are satisfied.

The set $\hat{\chi}_q^0(k,V,\delta)$ has an asymptotically full measure on $\hat{\mu}_q(k,\delta)$; moreover the inequality

$$\frac{s(\hat{\mu}_q(k,\delta) \setminus \hat{\chi}_q^0(k,V,\delta))}{s(\hat{\mu}_q(k,\delta))} < k^{-3\delta} \tag{4.6.15}$$

is valid.

Conditions $1' - 5'$ coincide with conditions $1^0 - 5^0$ (see (4.3.39) – (4.3.43) up to replacement of $p_j^2(t)$ by $p_j^2(t) + \Delta\hat{A}(t)_{jj}$. Note that inequalities (4.6.12) and (4.6.13) yield that the eigenvalue $p_j^2(t) + \Delta\hat{A}(t)_{jj}$ of operator $\hat{H}(t)$ is simple. We denote its spectral projection by $\hat{E}_j$. Let

$$\hat{g}'_r(k,t) = \frac{(-1)^r}{2\pi i r}\operatorname{Tr}\oint_{C_1} \hat{A}(t)^r dz, \tag{4.6.16}$$

$$\hat{G}'_r(k,t) = \frac{(-1)^{r+1}}{2\pi i}\oint_{C_1} (\hat{H}(t)-z)^{-1/2}\hat{A}^r(t)(\hat{H}(t)-z)^{-1/2}dz, \tag{4.6.17}$$

where C_1 is the circle of radius $k^{-1-\delta}$ centered at the point $\lambda = p_j^2(t) + \Delta\hat{A}(t)_{jj}$,

$$\hat{A}(z,t,W) = (\hat{H}(t)-z)^{-1/2}W(\hat{H}(t)-z)^{-1/2},$$

W being given by formula (1.0.45).

Theorem 4.4 . *Suppose $0 < \delta < 1/300$, t is in the $(k^{-2-2\delta})$-neighborhood of set $\hat{\chi}_q^0(k,V,\delta)$. Then for sufficiently large k, $k > k_0(V,\delta)$, there exists a unique eigenvalue of the operator $H(t)$ in the $(k^{-1-2\delta})$-neighborhood of the point $\lambda = p_j^2(t) + \Delta\hat{A}(t)_{jj}$. It is given by the absolutely converging series:*

$$\hat{\lambda}(t) = p_j^2(t) + \Delta\hat{A}(t)_{jj} + \sum_{r=2}^{\infty}\hat{g}'_r(k,t), \tag{4.6.18}$$

The spectral projection corresponding to $\hat{\lambda}(t)$ is determined by the series:

$$\hat{E}(t) = \hat{E}_j + \sum_{r=1}^{\infty}\hat{G}'_r(k,t), \tag{4.6.19}$$

which converges in the class $\mathbf{S}_1$. For the functions $\hat{g}'_r(k,t)$ and the operator-valued functions $\hat{G}'_r(k,t)$ the following estimates hold:

$$|\hat{g}'_2(k,t)| < k^{-8/5+7\delta}, \tag{4.6.20}$$

$$\|\hat{G}'_1(k,t)\|_1 < k^{-1+4\delta}, \tag{4.6.21}$$

$$\|\hat{G}'_2(k,t)\|_1 < k^{-7/5+6\delta}, \tag{4.6.22}$$

$$|\ \hat{g}'_r(k,t)\ | < k^{-1-\delta-\gamma_2 r}, \tag{4.6.23}$$

$$\|\hat{G}'_r(k,t)\|_1 < k^{-\gamma_2 r}, \tag{4.6.24}$$

$$\gamma_2 = 1/15 - 20\delta. \tag{4.6.25}$$

Corollary 4.8 .

$$|\hat{\lambda}(t) - p_j^2(t) - \Delta\hat{A}(t)_{jj}| < ck^{-1-2\gamma_2}, \tag{4.6.26}$$

$$\|\hat{E}(t) - E_j\| < ck^{-3\gamma_2},\ c \neq c(k). \tag{4.6.27}$$

Theorem 4.5 . *Under the conditions of Theorem 4.4 the functions $\hat{g}'_r(k,t)$ and the operator-valued functions $\hat{G}'_r(k,t)$ depend analytically on t in the complex $(k^{-2-2\delta})$-neighborhood of each simply connected component of $\hat{\chi}_q^0(k,V,\delta)$. They satisfy the estimates:*

$$| T(m)\hat{g}_2'(k,t) | < m!k^{-8/5+7\delta+2(1+\delta)|m|}, \quad (4.6.28)$$

$$\|T(m)\hat{G}_1'(k,t)\| < m!k^{-1+4\delta+2(1+\delta)|m|}, \quad (4.6.29)$$

$$\|T(m)\hat{G}_2'(k,t)\| < m!k^{-7/5+6\delta+2(1+\delta)|m|}, \quad (4.6.30)$$

$$| T(m)\hat{g}_r'(k,t) | < m!k^{-1-\delta-\gamma_2 r+2(1+\delta)|m|}, \quad (4.6.31)$$

$$\|T(m)\hat{G}_r'(k,t)\| < m!k^{-\gamma_2 r+2(1+\delta)|m|}, \quad (4.6.32)$$

Corollary 4.9 . *The function $\hat{\lambda}(t)$ and the operator-valued function $\hat{E}(t)$ analytically depend on t in the complex $(k^{-2-2\delta})$-neighborhood of each simply connected component of the set $\hat{\chi}_q^0(k,V,\delta)$. They satisfy the following estimates:*

$$| T(m)(\hat{\lambda}(t) - p_j^2(t) - \Delta\hat{A}(t)_{jj}) | \leq cm!k^{-1-2\gamma_2+2(1+\delta)|m|}, \quad (4.6.33)$$

$$\|T(m)(\hat{E}(t) - \hat{E}_j)\|_1 \leq cm!k^{-3\gamma_2+2(1+\delta)|m|}. \quad (4.6.34)$$

Theorems 4.4 and 4.5 enable us to conclude that the points, satisfying the equation $\hat{\lambda}(t) = k^2$, are situated near the isoenergetic surface of the operator $\hat{H}$.

The similar results hold for $\mu_q(k,\delta)$, which belongs to the isoenergetic surface of the free operator.

Lemma 4.17 . *Let $0 < \delta < 1/300$. Then for sufficiently large k, $k > k_0(V,\delta)$, there exists a set $\hat{\chi}_q'(k,V,\delta) \subset \mu_q(k,\delta)$, such that for any t in its $(k^{-2-2\delta})$-neighborhood there exists a unique $j \in Z^3$, such that*

$$|p_j^2(t) - k^2| < k^{-1-\delta} \quad (p_j^2(t) = k^2 \ \ \textit{for } t \textit{ of } \hat{\chi}_q'(k,V,\delta)), \quad (4.6.35)$$

inequality (4.6.4) and the conditions $1' - 5'$ are satisfied: The set $\hat{\chi}_q'(k,V,\delta)$ has an asymptotically full measure on $\mu_q(k,\delta)$; moreover the inequality

$$\frac{s(\mu_q(k,\delta) \setminus \hat{\chi}_q'(k,V,\delta))}{s(\mu_q(k,\delta))} < k^{-3\delta} \quad (4.6.36)$$

holds.

The proof of this lemma is similar to that of Lemma 4.16. Naturally, for $\hat{\chi}_q'(k,V,\delta)$ the theorems similar to Theorems 4.4 and 4.5 hold:

Theorem 4.6 . *Suppose $0 < \delta < 1/300$ and t is in the $(k^{-2-2\delta})$-neighborhood of the set $\hat{\chi}_q'(k,V,\delta)$ Then, for sufficiently large k, $k > k_0(V)$ there exists a unique eigenvalue $\hat{\lambda}(t)$ of operator $H(t)$ in the $(k^{-2-2\delta})$-neighborhood of the point $p_j^2(t)$. It is given by the absolutely converging series (4.6.18), j being uniquely determined by the relation (4.6.35). The corresponding spectral projection is given by formula (4.6.19). Estimates (4.6.20) - (4.6.24) hold.*

Corollary 4.10 . *If t is in the $(k^{-2-2\delta})$-neighborhood of $\hat{\chi}_q'(k,V,\delta)$, then estimates (4.6.26) - (4.6.27) hold.*

Theorem 4.7 . *Under the conditions of Theorem 4.6 the functions $\hat{g}'_r(k,t)$ and the operator-valued functions $\hat{G}'_r(k,t)$ depend analytically on t in the complex $(k^{-2-2\delta})$-neighborhood of each simply connected component of the set $\hat{\chi}'_q(k,V,\delta)$. They satisfy estimates (4.6.28) – (4.6.32).*

Corollary 4.11 . *The function $\hat{\lambda}(t)$ and the operator-valued function $\hat{E}(t)$ analytically depend on t in the complex $(k^{-2-2\delta})$-neighborhood of each simply connected component of the set $\hat{\chi}'_q(k,V,\delta)$. They satisfy the estimates (4.6.33) and (4.6.34).*

4.7 Proof of the Perturbation Formulae on the Singular Set

The formal construction of the perturbation series on the singular set is quite similar to that on the nonsingular set. The proof of convergence is based on the conditions $1'-5'$, which are analogous to the conditions 1^0-5^0 on the nonsingular set, and so it is similar to the proof of Theorem 4.1. In this section we prove Lemma 4.19, which, in fact, asserts the convergence of the perturbation series for the resolvent of the operator $H(t)$ with respect to the model operator $\hat{H}(t)$. The validity of estimates (4.6.20)–(4.6.22) is verified in Lemma 4.20. Basing on these two lemmas we prove Theorems 4.4 and 4.5.

Lemma 4.18 . *If t is in the $(k^{-2-2\delta})$-neighborhood of the set $\hat{\chi}^0_q(k,V,\delta)$, then the operator $\hat{H}(t)$ has a unique eigenvalue inside the circle C_1. For any $z \in C_1$ the following estimate holds:*

$$\|(\hat{H}(t)-z)^{-1/2}\| < k^{1/2+2\delta}. \tag{4.7.1}$$

Proof. We first prove estimate (4.7.1). According to Lemma 4.16 for any t of $\hat{\chi}^0_q(k,V,\delta)$ there exists a unique $j \in Z^3$ such that $p_j^2(t)+\Delta\hat{A}_{jj}(t)=k^2$. By the definition of the circle C_1 the point $k^2=p_j^2(t)+\Delta\hat{A}_{jj}(t)$ is at its center and

$$|p_j^2(t)+\Delta\hat{A}_{jj}(t)-z| = k^{-1-\delta}, \quad z \in C_1. \tag{4.7.2}$$

Let $P=\sum_{q\in\Gamma(R_0)}P_q$. Suppose $i: P_{ii}=0$. Than, using (4.6.12) and (4.7.2), we get

$$|p_i^2(t)-z| > k^{-1-\delta}. \tag{4.7.3}$$

Since $\hat{H}(t)$ acts as $H_0(t)$ in $(I-P)l_2^3$ and inequality (4.7.3) holds, we easily obtain

$$\|(I-P)(\hat{H}(t)-z)^{-1/2}\| < k^{1/2+\delta/2}. \tag{4.7.4}$$

Suppose $i: P_{ii}=1$, $i \neq j$. It is clear that

$$|p_i^2(t)+\Delta\hat{A}_{ii}(t)-z| >$$
$$|p_i^2(t)+\Delta\hat{A}_{ii}(t)-p_j^2(t)-\Delta\hat{A}_{jj}(t)| - |p_j^2(t)+\Delta\hat{A}_{jj}(t)-z|. \tag{4.7.5}$$

Using relations (4.6.13) and noting that $|(\mathbf{p}_i(0), \mathbf{p}_{q'}(0))| < k^{1/5}$ when $q' \in \Pi_{q'}(k^{1/5})$ (see (4.2.3), we get

$$|p_i^2(t) + \Delta\hat{\Lambda}_{ii}(t) - p_j^2(t) - \Delta\hat{\Lambda}_{jj}(t)| > 2k^{-1/5-6\delta}. \qquad (4.7.6)$$

Taking into account (4.7.2), we easily obtain:

$$|p_i^2(t) + \Delta\hat{\Lambda}_{ii}(t) - z| > k^{-1/5-6\delta} \quad i \neq j. \qquad (4.7.7)$$

Note that $p_i^2(t) + \Delta\hat{\Lambda}_{ii}(t)$ are the eigenvalues of $P\hat{H}P$. Therefore,

$$\|(P(\hat{H}(t) - z)^{-1/2}\| < k^{1/10+3\delta}. \qquad (4.7.8)$$

Inequalities (4.7.4) and (4.7.8) together give estimate (4.7.1).

Now we prove that $\hat{H}$ has a unique eigenvalue inside C_1. It follows from relations (4.6.12) and (4.7.7) that the points $p_i^2(t)$, $i : P_{ii} = 0$ and $p_i^2(t)+\Delta\hat{\Lambda}_{ii}(t)$, $i : P_{ii} = 1$, $i \neq j$ lie outside the circle C_1. Since they are all eigenvalues of operator $\hat{H}(t)$, this means that there are an unique eigenvalue of operator $\hat{H}(t)$ inside this circle. It is equal to $p_j^2(t) + \Delta\Lambda_{jj}(t)$. The lemma is proved.

Lemma 4.19 . *Suppose t is in the $(k^{-2-2\delta})$-neighbourhood of the set $\hat{\chi}_q^0(k, V, \delta)$, $z \in C_1$. Then for sufficiently large k, $k > k_0(V, \delta)$, the following estimates hold:*

$$\|\hat{A}(z,t)\| < k^{8\delta}, \qquad (4.7.9)$$

$$\|\hat{A}^3(z,t)\| < k^{-1/5+31\delta}. \qquad (4.7.10)$$

Proof. We consider the operator A_1:

$$A_1 = A_0 + A_1^{(1)} + A_1^{(2)} + A_1^{(3)}, \qquad (4.7.11)$$

where

$$A_0 = (H_0 - z)^{-1/2}B_0(H_0 - z)^{-1/2}, \qquad (4.7.12)$$

$$B_0 = (I - P)V(I - P), \qquad (4.7.13)$$

$$A_1^{(i)} = (H_0(t) + \Delta\hat{\Lambda} - z)^{-1/2}B_1^{(i)}(H_0(t) + \Delta\hat{\Lambda} - z)^{-1/2}, \qquad (4.7.14)$$

$$B_1^{(1)} = P_0\mathcal{E}_q(I - P_0), \qquad (4.7.15)$$

$$B_1^{(2)} = B_1^{(1)*} = (I - P_0)\mathcal{E}_qP_0, \qquad (4.7.16)$$

$$B_1^{(3)} = P_0\mathcal{E}_qP_0, \qquad (4.7.17)$$

P_0 being the diagonal one-dimensional projection: $(P_0)_{jj} = 1$. The operator $\mathcal{E}_q$ was introduced in the proof of Lemma 4.9. Its main properties are described by formulae (4.5.41) – (4.5.43).

To prove estimates (4.7.9) and (4.7.10) we verify the following relations:

1)

$$\|A_1\| < k^{7\delta}, \qquad (4.7.18)$$

2)

$$\|\hat{A} - A_1\| < 5k^{-1/5+8\delta},$$ (4.7.19)

3)

$$\|A_1^3\| < 8k^{-1/5+32\delta}.$$ (4.7.20)

First, we check that

$$A_1^{(3)} = B_1^{(3)} = 0.$$ (4.7.21)

Indeed, from relation (4.5.43), taking into account that $P_0 P_{k^2}(k^{1/5}) = P_0$, we obtain $B_1^{(3)} = 0$ and, therefore $A_1^{(3)} = 0$.

1) We prove inequality (4.7.18). Let us estimate $\|A_0\|$. It is clear that

$$(A_0)_{mi} = v_{m-i}(p_m^2(t) - z)^{-1/2}(p_i^2(t) - z)^{-1/2}.$$ (4.7.22)

We prove that

$$|(p_m^2(t) - z)(p_i^2(t) - z)| > k^{-4\delta}$$ (4.7.23)

$$i, m \in Z^3, \ i \neq m, \ |i - m| < k^\delta, P_{mm} = P_{ii} = 0.$$

for all $z \in C_1$. Inequality (4.7.23) is symmetric with respect to m, i, so we can assume that $|p_m^2(t) - z| \leq |p_i^2(t) - z|$. Now, if $|p_m^2(t) - z| > k^{-2\delta}$, then inequality (4.7.23) is obvious. Suppose

$$|p_m^2(t) - z| \leq k^{-2\delta}.$$ (4.7.24)

Let $m \neq j + n_0 q$, $n_0 \in Z$. Using inequalities (4.6.9) – (4.6.12) and considering as in the proof of Corollary 3.1 we verify that relation (4.7.23) holds. Next, let $m = j + n_0 q, n_0 \in Z \backslash \{0\}$. Suppose, $|n_0| > k^{1/5-\delta}$. Then $|p_m^2(t) - p_j^2(t)| > ck^{1/5-\delta}$, because $|(\mathbf{p}_j(t), \mathbf{p}_q(0))| < k^\delta$ (see (4.6.4). Considering relation (4.7.2), we obtain that

$$|p_m^2(t) - z| > k^{1/5-2\delta},$$ (4.7.25)

where $m = j + n_0 q$, $n_0 \in Z$, $|n_0| > k^{1/5-\delta}$. This inequality is in contradiction with the assumption (4.7.24). Therefore, it remains to consider the case $|n| < k^{1/5-\delta}$, i.e., $m \in \Pi_q(k^{1/5})$. It is easy to see that $m \notin \Pi_q(k^{1/5}) \backslash T(k, \delta)$. Otherwise $P_{mm} = 1$, but we consider inequality (4.7.23) in the case $i, m : P_{mm} = P_{ii} = 0$. Thus, $m \in T(k, \delta)$. Hence, inequality (4.6.14) holds. This inequality contradicts to (4.7.24). Thus, estimate (4.7.23) is valid. Taking into account that $R_0 < k^{\delta/8}$, we obtain

$$\|A_0\| < k^{7\delta}.$$ (4.7.26)

We estimate $A_1^{(1)}$. Considering (4.7.2), we get:

$$\|A_1^{(1)}\|^2 \leq k^{1+\delta} \sum_i (\mathcal{E}_q)_{ij}^2 |p_i^2 + \Delta\hat{\Lambda}_{ii} - z|^{-1}.$$ (4.7.27)

We prove the estimate

$$3|p_i^2 + \Delta\hat{\Lambda}_{ii} - z| > k^{1-11\delta}.$$ (4.7.28)

By (4.5.44) $(\mathcal{E}_q)_{ij}$ can differ from zero only if $i-j = l_0 q + l q'$, where $q' \in \Gamma(R_0)$, $q' \neq q$, $|l_0| < 15k^{1/5}$, $1 \le |l| < R_0$. First we estimate $|p_i^2(t) - p_j^2(t)|$. It is easy to see that

$$|p_i^2(t) - p_j^2(t)| >$$
$$\big||\mathbf{p}_j(t) + l\mathbf{p}_{q'}(0)|^2 - p_j^2(t)\big| - 2|(\mathbf{p}_j(t) + l\mathbf{p}_{q'}(0), l_0\mathbf{p}_q(0))| - l_0^2 p_q^2(0). \quad (4.7.29)$$

Taking into account inequality (4.6.9) and the relation $|(\mathbf{p}_j(0), \mathbf{p}_q(0))| < k^\delta$, which holds because of (4.6.4), we obtain

$$|p_i^2(t) - p_j^2(t)| > k^{1-11\delta}. \quad (4.7.30)$$

Considering (4.7.2) and (4.7.30) we obtain (4.7.28). From estimates (4.7.27) and (4.7.28), taking into account (4.5.41), we get

$$\|A_1^{(1)}\| < ck^{6\delta}. \quad (4.7.31)$$

Similarly we prove estimate

$$\|A_1^{(2)}\| < ck^{6\delta}. \quad (4.7.32)$$

Adding estimates (4.7.26), (4.7.31) and (4.7.32), and considering that $A_1^{(3)} = 0$, we get (4.7.18). 2) Estimate (4.7.19) is proved similarly to the analogous estimate (4.5.100). The difference is that we use conditions $3'$ and $4'$ instead of conditions 3^0 and 4^0 and we consider the operator $(I - P_0)\mathcal{E}_q(I - P_0)$ instead of $\mathcal{E}_q$ in the definition of ε_{1q}.

3) We prove estimate (4.7.20). The proof up to some modifications is similar to the proof of relation (4.5.101). In fact, first we verify that

$$P_{k^2}(k^{1/5-10\delta})A_1 P_{k^2}(k^{1/5-10\delta}) = 0. \quad (4.7.33)$$

We represent A_1 as the sum (4.7.11) and prove relation (4.7.33) for each of the operators A_0, $A_1^{(1)}$, $A_1^{(2)}$ $(A_1^{(3)} = 0)$. The relation

$$P_{k^2}(k^{1/5-10\delta})A_0 P_{k^2}(k^{1/5-10\delta}) = 0 \quad (4.7.34)$$

is proved just as the similar relation (4.5.113. (instead of condition 5^0 we use the condition $5'$). The relation

$$P_{k^2}(k^{1/5-10\delta})A_1^{(1)} P_{k^2}(k^{1/5-10\delta}) = 0. \quad (4.7.35)$$

easily follows from relation (4.5.43). Similarly,

$$P_{k^2}(k^{1/5-10\delta})A_1^{(2)} P_{k^2}(k^{1/5-10\delta}) = 0. \quad (4.7.36)$$

Summing equalities (4.7.34) – (4.7.36), we get formula (4.7.33). Further, we represent A_1 in the form:

$$A_1 = M_0 + M_+ + M_-, \quad (4.7.37)$$

$$M_0 = (I - P_{k^2}(k^{1/5-10\delta}))A_1(I - P_{k^2}(k^{1/5-10\delta})), \tag{4.7.38}$$

$$M_+ = P_{k^2}(k^{1/5-10\delta})A_1(I - P_{k^2}(k^{1/5-10\delta})) = P_{k^2}(k^{1/5-10\delta})A_1, \tag{4.7.39}$$

$$M_- = (I - P_{k^2}(k^{1/5-10\delta}))A_1 P_{k^2}(k^{1/5-10\delta}) = A_1 P_{k^2}(k^{1/5-10\delta}). \tag{4.7.40}$$

Obviously,

$$M_+^2 = M_-^2 = M_0 M_+ = M_- M_0 = 0. \tag{4.7.41}$$

From this,

$$A_0^3 = M_0^3 + M_0^2 M_- + M_+ M_0^2 + M_0 M_- M_+ + M_+ M_0 M_+ M_- M_+ M_0$$
$$+ M_+ M_- M_+ + M_- M_+ M_-. \tag{4.7.42}$$

Using the definition of $P_{k^2}(k^{1/5-10\delta})$ it is easy to see that

$$\|M_0\| < k^{-1/5+9\delta}, \quad \|M_-\| < k^{3\delta}, \quad \|M_+\| < k^{3\delta}. \tag{4.7.43}$$

From (4.7.43) we obtain

$$\|M_0^3\| + \|M_0^2 M_-\| + \|M_+ M_0^2\| + \|M_0 M_- M_+\| +$$
$$\|M_+ M_0 M_-\| + \|M_- M_+ M_0\| < ck^{-1/5+16\delta}. \tag{4.7.44}$$

Assume that we have proved the inequality

$$\|M_+ M_-\| < k^{-1/5+26\delta} \tag{4.7.45}$$

Then, we have

$$\|M_+ M_- M_+\|, \|M_- M_+ M_-\| < k^{-1/5+32\delta}, \tag{4.7.46}$$

and summing the estimates (4.7.44) and (4.7.46), we get (4.7.20). Thus, to prove the necessary relation (4.7.20) it remains to check the estimate (4.7.45).

First, we consider the diagonal part of the operator $M_+ M_-$:

$$(M_+ M_-)_{ii} = \sum_{|m|<R_0} \frac{|(B_1)_{i,i+m}|^2}{(p_i^2(t) + \Delta\hat{A}_{ii} - z)(p_{i+m}^2(t) + \Delta\hat{A}_{i+m,i+m} - z)}, \tag{4.7.47}$$

where

$$B_1 = P_{k^2}(k^{1/5-10\delta})(B_0 + B_1^{(1)} + B_1^{(2)})(I - P_{k^2}(k^{1/5-10\delta})). \tag{4.7.48}$$

Since $P_0(I - P_{k^2}(k^{1/5-10\delta})) = 0$,

$$B_1 = P_{k^2}(k^{1/5-10\delta})(B_0 + B_1^{(1)})(I - P_{k^2}(k^{1/5-10\delta})). \tag{4.7.49}$$

Let us consider different i in (4.7.47). If $i \neq j$, $i : P_{ii} = 1$, then taking into account the definitions of B_0 and $B_1^{(1)}$ (formulae (4.7.13) and (4.7.15), we get $(M_+ M_-)_{ii} = 0$. The same is valid if $i : P_{k^2}(k^{1/5-10\delta})_{ii} = 0$. It remains the case

$i : P_{ii} = 0,\ P_{k^2}(k^{1/5-10\delta})_{ii} = 1$. Considering as in Lemma 4.14 ($4^0 \to 4'$), it is not hard to show that formula

$$(M_+M_-)_{ii} = \sum_{|m|<R_0} \frac{|v_m|^2(p_i^2(t) - z + p_m^2(0))}{(p_i^2(t) - z)(p_{i+m}^2(t) - z)(p_{i-m}^2(t) - z)} \tag{4.7.50}$$

valid for such i (see (4.5.121)). We prove that

$$2|(M_+M_-)_{ii}| < k^{-1/5+25\delta}. \tag{4.7.51}$$

Let $i : |p_i^2(t) - k^2| > k^{-1/5-4\delta}$. Considering the definitions of M_+, M_- and $P_{k^2}(k^{1/5-10\delta})$ we get (4.7.51). Suppose $i : |p_i^2(t) - k^2| \le k^{-1/5-4\delta}$. Considering as in the proof of (4.7.18), we obtain that

$$36|(p_i^2(t) - z)(p_{i\pm m}^2(t) - z)| > k^{-4\delta}. \tag{4.7.52}$$

Using formula (4.7.50) and the last inequality, it is not hard to show that estimate (4.7.51) holds.

It remains only to consider the case $i = j$. It is clear that

$$(M_+M_-)_{jj} = \sum_{|m|<R_0, m\neq 0} \frac{|(\mathcal{E}_q)_{jj+m}|^2}{(p_j^2(t) + \Delta\hat{A}_{jj} - z)(p_{j+m}^2(t) + \Delta\hat{A}_{j+mj+m} - z)}. \tag{4.7.53}$$

According to (4.5.44) the matrix element $(\mathcal{E}_q)_{jj+m}$ can differ from zero only if $m = l_0q + lq'$, $|l_0| < 15k^{1/5}$, $q' \in \Gamma(R_0)$, $q' \neq q$, $1 \le |l| < R_0$. From this we obtain

$$p_{j+m}^2(t) + \Delta\hat{A}_{j+m,j+m} - z = 2l(\mathbf{p}_j(0), \mathbf{p}_{q'}(0)) + O(k^{2/5}). \tag{4.7.54}$$

From relation (4.6.9) it follows

$$8|(\mathbf{p}_j(0), \mathbf{p}_{q'}(0))| > k^{1-10\delta}. \tag{4.7.55}$$

Let

$$a_l = \sum_{m\in\Omega_{lq'}} |(\mathcal{E}_q)_{jj+m}|^2,$$

$$\Omega_{lq'} = \{m : \mathbf{p}_m(0) = l_0q + lq'\}.$$

Using relations (4.7.53) – (4.7.55), it is not hard to show that

$$(M_+M_-)_{jj} = \left(\sum_l a_l(2l)^{-1}\right)(p_j^2(t)+\Delta A_{jj}-z)^{-1}|(\mathbf{p}_j(0), \mathbf{p}_{q'}(0))|^{-1}+O(k^{-3/5+20\delta}). \tag{4.7.56}$$

Using again relations (4.7.2) and (4.7.55) we get

$$(M_+M_-)_{jj} = \left(\sum_{lq} a_l(2l)^{-1}\right) O(k^{11\delta}) + O(k^{-3/5+20\delta}). \tag{4.7.57}$$

We prove that $a_l = a_{-l}$. Indeed, let us recall the definition of $\mathcal{E}_q$ (see (4.5.62). Noting that $(R_q^0)_{j,j+m} = 0$, we verify that

$$a_l = \sum_{l_1,l_2} \left| U_{j,j+l_1 q} v_{lq'} U^*_{j+l_1 q+lq', j+l_1 q+lq'+l_2 q} \right|^2, \tag{4.7.58}$$

where U is the unitary operator reducing $\hat{H}$ to the diagonal form. Summing with respect to indexes l_1 and l_2, we get:

$$a_l = |v_{lq'}|^2 = a_{-l}. \tag{4.7.59}$$

Now we see that $\sum_l a_l (2l)^{-1} = 0$. Therefore,

$$|(M_+M_-)_{jj}| \le ck^{-3/5+20\delta}. \tag{4.7.60}$$

Next, we consider non-diagonal elements: $(M_+M_-)_{il}$, $i \ne l$,

$$(M_+M_-)_{il} = \tag{4.7.61}$$

$$\sum_m \frac{(B_1)_{im}(B_1)_{ml}}{(p_i^2(t) + \Delta\hat{\Lambda}_{ii}(t) - z)^{1/2}(p_l^2(t) + \Delta\hat{\Lambda}_{ll}(t) - z)^{1/2}(p_m^2(t) + \Delta\hat{\Lambda}_{mm}(t) - z)},$$

where

$$P_{k^2}(k^{1/5-10\delta})_{ii} = P_{k^2}(k^{1/5-10\delta})_{ll} = 1, \tag{4.7.62}$$

$$P_{k^2}(k^{1/5-10\delta})_{mm} = 0. \tag{4.7.63}$$

From the last relation it follows that $m \ne j$. Suppose $i = j$, then $l \ne j$ (because we consider non-diagonal elements). It is clear that $j - l = l_0 q + lq'$, $|l_0| < 15k^{1/5}$, $1 \le |l_1| < R_0$, $l_1 \ne 0$. Since $|(\mathbf{p}_j(0), \mathbf{p}_{q'}(0))| > k^{3/5}$, $|(\mathbf{p}_j(0), \mathbf{p}_q(0))| < 6k^{1/5}$, we readily obtain

$$|p_l^2(t) - k^2| = |p_l^2(t) - p_j^2(t)| > ck^{3/5}. \tag{4.7.64}$$

This inequality contradicts relation (4.7.62). Therefore $i \ne j$. Similarly, we obtain that $l \ne j$, otherwise $(M_+M_-)_{il} = 0$ (for $i \ne l$). Since

$$(B_1)_{im} = (B_0)_{im} \tag{4.7.65}$$

for $i \ne j$, $m \ne j$, we have

$$(M_+M_-)_{il} = \sum_m \frac{(B_0)_{im}(B_0)_{ml}}{(p_i^2(t) - z)^{1/2}(p_l^2(t) - z)^{1/2}(p_m^2(t) - z)}, \tag{4.7.66}$$

Using relation (4.7.23), which is valid for all $i, m : |i - m| < k^\delta$, we get

$$6|(p_i^2(t) - z)(p_l^2(t) - z)|^{1/2} > k^{-2\delta}. \tag{4.7.67}$$

Taking into account that $|p_m^2(t) - k^2| > k^{1/5}$, (see (4.7.63)) we arrive at the estimate

$$|(M_+M_-)_{il}| < ck^{-1/5+12\delta}, \quad i \ne l. \tag{4.7.68}$$

Since $(M_+M_-)_{il} = 0$ if $|i-l| > 2R_0$, it follows that

$$\|M_+M_-\| < k^{\delta} \max_{i,j} |(M_+M_-)_{ij}|.$$

Taking into account (4.7.51), (4.7.60) and (4.7.68), we get (4.7.45). Thus, estimate (4.7.10) is proved. The lemma is proved.

Lemma 4.20 . *If t is in the $(k^{-2-2\delta})$-neighborhood of the set $\hat{\chi}_q^0(k,V,\delta)$ and $z \in C_1$, then for sufficiently large k, $k > k_0(V,\delta)$, the following estimates hold:*

$$\|\hat{G}_1'(k,t)\|_1 < k^{-1+4\delta}, \tag{4.7.69}$$

$$\|\hat{G}_2'(k,t)\|_1 < k^{-7/5+6\delta}, \tag{4.7.70}$$

$$|\hat{g}_2'(k,t)| < k^{-8/5+8\delta}. \tag{4.7.71}$$

Proof. Evaluating the integral $U\hat{G}_1'(k,t)U^*$ (U is the unitary operator, reducing $\hat{H}(t)$ to the diagonal form) by a residue, we obtain $(U\hat{G}_1'(k,t)U^*)_{im} = 0$, if $i \neq j, m \neq j$ or $i = m = j$ and

$$(U\hat{G}_1'(k,t)U^*)_{ji} = \overline{(U\hat{G}_1'(k,t)U^*)_{ij})} =$$

$$B_{ji}\left(p_i^2(t) + \Delta\hat{\Lambda}_{ii}(t) - p_j^2(t) - \Delta\hat{\Lambda}_{jj}(t)\right)^{-1}, \tag{4.7.72}$$

where $B = B_0 + \sum_q \mathcal{E}_q + L + D$, operators L and D being described in Lemma 4.9. Here we only will use their properties. Taking into account that $(B_0)_{ji} = L_{ji} = 0$ $(j \in \Pi_q(k^{\delta}))$ and $(\mathcal{E}_{q'})_{ji} = 0$ if $q' \neq q$, we get

$$B_{ji} = (\mathcal{E}_q)_{ji} + D_{ji}. \tag{4.7.73}$$

From (4.7.73), using (4.7.72), we obtain

$$|(U\hat{G}_1'(k,t)U^*)_{ji}| < a_i + c_i,$$

where

$$a_i = |(\mathcal{E}_q)_{ji}||p_i^2(t) + \Delta\hat{\Lambda}_{ii}(t) - p_j^2(t) - \Delta\hat{\Lambda}_{jj}(t)|^{-1},$$

$$c_i = |D_{ji}||p_i^2(t) + \Delta\hat{\Lambda}_{ii}(t) - p_j^2(t) - \Delta\hat{\Lambda}_{jj}(t)|^{-1}.$$

Terms similar to a_i were estimated in the proof of Lemma 4.19 (see formulae (4.7.27) – (4.7.32)). Using estimate (4.7.28) and (4.7.2) we get

$$2a_i < |(\mathcal{E}_q)_{ji}|k^{-1+11\delta}. \tag{4.7.74}$$

Relations (4.7.1) and (4.7.2) yield

$$c_i < |D_{ji}|k^{1+\delta}. \tag{4.7.75}$$

It is clear that

$$\|\hat{G}_1'(k,t)\|_1^2 = \|U\hat{G}_1'(k,t)U^*\|_1^2 \leq \sum_i |(U\hat{G}_1'(k,t)U^*)_{ji}|^2. \tag{4.7.76}$$

Considering estimates (4.7.74) – (4.7.76) we get

$$\|\hat{G}_1'(k,t)\|_1 \leq \|\mathcal{E}_q\| k^{-1+3\delta} + \|D\| k^{1+\delta}.$$

Taking into account that

$$\|\mathcal{E}_q\| < \|V\|, \tag{4.7.77}$$

$$\|D\| < k^{-k^{1/5-8\delta}} \tag{4.7.78}$$

(see (4.5.49)), we arrive at estimate (4.7.69).

Next, we consider $\|\hat{G}_2'(k,t)\|_1$. Evaluating the integral by a residue, we get

$$(U\hat{G}_2'(k,t)U^*)_{im} = (U\hat{G}_1'(k,t)U^*)_{ij}(U\hat{G}_1'(k,t)U^*)_{jm}, \tag{4.7.79}$$

when $i \neq j$, $m \neq j$ and

$$(U\hat{G}_2'(k,t)U^*)_{jm} = \overline{(U\hat{G}_2'(k,t)U^*)_{mj})} = \tilde{a}_m + \tilde{c}_m, \tag{4.7.80}$$

$$\tilde{a}_m = -B_{jj}B_{jm}(p_m^2(t) + \Delta\hat{\Lambda}_{mm}(t) - p_j^2(t) - \Delta\hat{\Lambda}_{jj}(t))^{-2} +$$

$$\tilde{c}_m =$$

$$\sum_{i\neq j} B_{ji}B_{im}(p_i^2(t)+\Delta\hat{\Lambda}_{ii}(t)-p_j^2(t)-\Delta\hat{\Lambda}_{jj}(t))^{-1}(p_m^2(t)+\Delta\hat{\Lambda}_{mm}(t)-p_j^2(t)-\Delta\hat{\Lambda}_{jj}(t))^{-1}$$

when $m \neq j$ and

$$(U\hat{G}_2'(k,t)U^*)_{jj} = -\sum_{m\neq j} |B_{jm}|^2(p_m^2(t) + \Delta\hat{\Lambda}_{mm}(t) - p_j^2(t) - \Delta\hat{\Lambda}_{jj}(t))^{-2}. \tag{4.7.81}$$

From relations (4.7.76) – (4.7.81) we see

$$\|\hat{G}_2'(k,t)\|_1^2 \leq \|\hat{G}_1'(k,t)\|_1^2 + 2I,$$

$$I = \sum_m |(U\hat{G}_2'(k,t)U^*)_{jm}|^2. \tag{4.7.82}$$

To estimate I we represent it as a sum $I = I_1 + I_2 + I_3$, where I_k corresponds to the summation over the set Ω_k:

$$\Omega_1 = \{m : m \in \Pi_q(\frac{1}{2}k^{1/5})\},$$

$$\Omega_2 = \{m : m \in \Pi_q(\frac{3}{2}k^{1/5}) \setminus \Pi_q(\frac{1}{2}k^{1/5})\},$$

$$\Omega_3 = \{m : m \notin \Pi_q(\frac{3}{2}k^{1/5-8\delta})\}.$$

Suppose $m \in \Omega_1$. Then, $L_{im} = (B_0)_{im} = 0$ and, therefore,

$$B_{im} = (\mathcal{E}_q)_{im} + D_{im}. \tag{4.7.83}$$

Considering as in the proof of (4.7.69), we obtain

$$I_1 < ck^{-2+8\delta}. \tag{4.7.84}$$

Suppose $m \in \Omega_2$. Using (4.7.80) we get

$$\sum_{m\in\Omega_2} |(U\hat{G}_2'(k,t)U^*)_{jm}|^2 < \sum_{m\in\Omega_2} \tilde{a}_m + \sum_{m\in\Omega_2} \tilde{c}_m.$$

To estimate $\tilde{a}_m$ note that B_{jm} is given by formula (4.7.73). Considering as in the proof of (4.7.69) we get

$$\sum_{m\in\Omega_2} \tilde{a}_m < k^{-2+8\delta}.$$

To estimate $\tilde{c}_m$ we recall (see (4.5.73) that for any $m \in \Omega_2$

$$|p_m^2(t) + \Delta\hat{\Lambda}_{mm}(t) - p_j^2(t) - \Delta\hat{\Lambda}_{jj}(t)| > ck^{2/5}.$$

Using this inequality and (4.7.28) we get

$$\sum_{m\in\Omega_2} \tilde{c}_m < k^{-7/5+8\delta}.$$

Thus,

$$I_2 < ck^{-7/5+8\delta}. \tag{4.7.85}$$

If $m \in \Omega_3$, then in formula (4.7.80) $B_{im} = (B_0)_{im}$. This means that $|i - m| < R_0$. On the other hand B_{ji} is given by (4.7.73). We consider that $\mathbf{p}_j(0) - \mathbf{p}_i(0) = l_0\mathbf{p}_q(0) + l\mathbf{p}_{q'}(0)$, where $|l_0| < 15k^{1/5}$, $1 \leq |l| < R_0$, because otherwise $(\mathcal{E}_q)_{ji} = 0$ while D_{ji} is estimated by tiny value $k^{-k^{1/5}}$. Therefore, $\mathbf{p}_j(0) - \mathbf{p}_m(0) = l_1\mathbf{p}_q(0) + l_2\mathbf{p}_{q''}(0)$, where $|l_1| < k^{1/5}$, $|l_2| < 2R_0$, $l_2\mathbf{p}_{q''}(0) \neq 0$ (the last relation holds because otherwise $m \in \Pi_q(3k^{1/5}/2) \not\subset \Omega_3$). Considering as in the proof of (4.7.69) we show that

$$|p_i^2(t) + \Delta\hat{\Lambda}_{ii}(t) - p_j^2(t) - \Delta\hat{\Lambda}_{jj}(t)| > ck^{1-8\delta},$$

$$|p_m^2(t) + \Delta\hat{\Lambda}_{mm}(t) - p_j^2(t) - \Delta\hat{\Lambda}_{jj}(t)| > ck^{1-8\delta}.$$

Using this inequality and the estimate $\|B\| \leq \|V\|$, we get

$$I_3 < ck^{-2+8\delta}. \tag{4.7.86}$$

Adding estimates (4.7.84), (4.7.85) and (4.7.86), we get

$$I < ck^{-7/5+8\delta}. \tag{4.7.87}$$

Using formula (4.7.82) and estimates (4.7.69), (4.7.87), we obtain (4.7.70).

Inequality (4.7.71) we obtain, calculating $\hat{g}_2'(k,t)$ by a residue:

$$\hat{g}_2'(k,t) = \sum_{m\neq j} |B_{jm}|^2 (p_m^2(t) + \Delta\hat{\Lambda}_{mm}(t) - p_j^2(t) - \Delta\hat{\Lambda}_{jj}(t))^{-1}.$$

Using formula (4.7.73) and the inequalities

$$\|D\| < k^{-k^{1/5-8\delta}},$$

$$2|p_m^2(t) + \Delta\hat{\Lambda}_{mm}(t) - p_j^2(t) - \Delta\hat{\Lambda}_{jj}(t)| > k^{-1-\delta}, \quad m \neq j$$

(see inequalities (4.6.12) and (4.6.13)), we obtain

$$\hat{g}_2'(k,t) = \sum_{m\neq j} |(\mathcal{E}_q)_{jm}|^2 (p_m^2(t) + \Delta\hat{\Lambda}_{mm}(t) - p_j^2(t) - \Delta\hat{\Lambda}_{jj}(t))^{-1} + O(k^{-2}).$$

From the last relation, (4.7.53) and (4.7.2) it is easy to see that $|\hat{g}_2'(k,t)| = |(M_+M_-)_{jj}|k^{-1-\delta} + O(k^{-2})$. Using (4.7.60), we get estimate(4.7.71).

The lemma is proved.

The proofs of Theorems 4.4 and 4.5 are quite similar to the proofs of Theorems 4.1 and 4.2, Lemma 4.19 being using instead of Lemma 4.14.

4.8 Geometric Constructions on the Singular Set

We construct now the subset $\hat{\chi}_q^0(k,V,\delta)$ of $\hat{\mu}_q(k,\delta)$ on which conditions $1' - 5'$ hold. The analytical description of $1' - 5'$ is similar to that of $1^0 - 5^0$ for the nonsingular set. Howver, to prove that $1' - -5'$ in fact can be satisfied, we have to develop much more subtle reasoning than in Section 4.4. This is due to the fact that conditions $1^0 - -5^0$ were proved to be satisfied for some t on μ_q which is, as a matter of fact, only a small part of S_k. So more precise estimates and complicated reasoning have to be done. We proved that the nonsingular set has an asymptotically full measure on S_k. Here we will prove that $\hat{\chi}_q^0(k,V,\delta)$ has an asymptotically full measure on $\hat{\mu}_q(k,\delta)$.

We consider

$$\hat{\chi}_q^0(k,V,\delta) \equiv \hat{\mu}_q(k,\delta) \setminus \cup_{i=1}^5 \hat{\Upsilon}_q(k,V,\delta)_i, \qquad (4.8.1)$$

where the set $\hat{\Upsilon}_q(k,V,\delta)_i$ contains all points t, for which condition i' (see Lemma 4.16, formulae (4.6.9) – (4.6.14)) does not hold. These sets will be described below. The definition of the set $\hat{\chi}_q^0(k,V,\delta)$ is similar to the definition of the nonsingular set $\chi_3(k,V,\delta)$ in Section 4.4, where conditions $1^0 - 5^0$ are satisfied.

First, we introduce the notations. Let $\Pi_m(k,a)$ be the plane layer:

$$\Pi_m(k,a) = \{x : ||x|^2 - |x + \mathbf{p}_m(0)|^2| < 4k^{-a}\} \qquad (4.8.2)$$

Let $\hat{\Pi}_m(k,a)$ be the curve analog of the plane layer $\Pi_m(k,a)$:

$$\hat{\Pi}_m(k,a) = \{x : ||x|^2 + \varphi_0(x) - |x + \mathbf{p}_m(0)|^2| < 4k^{-a}\}, \qquad (4.8.3)$$

where the function $\varphi_0(x)$ is given by formula (4.6.7).

We determine $\hat{\Upsilon}_q(k,V,\delta)_1$ by the formula:

$$\hat{\Upsilon}_q(k,V,\delta)_1 = \mathcal{K}\left(\cup_{r\in\Omega_1}\hat{\Pi}_r(k,-1+10\delta)\cap\hat{S}_q(k,-\delta)\right), \tag{4.8.4}$$

Ω_1 is given by (4.6.10), $\mathcal{K}$ is defined by (4.6.3). Next,

$$\hat{\Upsilon}_q(k,V,\delta)_2 = \mathcal{K}\left(\cup_{m\in\Omega_2,i\in\Omega_3}\hat{\Upsilon}_q(k,V,\delta)_2^{mi}\cap\hat{S}_q(k,-\delta)\right), \tag{4.8.5}$$

$$\Omega_2 = \{m, m\in Z^3, m\neq n_0q, n_0\in Z\}, \tag{4.8.6}$$

$$\Omega_3 = \{i, i\in Z^3, i\neq 0, |i|<k^\delta\},$$

$$\hat{\Upsilon}_q(k,V,\delta)_2^{mi} = \cup_{n=0}^{N}\hat{\Upsilon}_q(k,V,\delta)_2^{min},\quad N=[1/\delta]-19, \tag{4.8.7}$$

$$\hat{\Upsilon}_q(k,V,\delta)_2^{min} = \hat{\Pi}_m(k,V,1-\delta n)\cap\hat{\Pi}_{m+i}(k,V,-1+\delta(n+11)). \tag{4.8.8}$$

Furthermore, let

$$\hat{\Upsilon}_q(k,V,\delta)_3 = \mathcal{K}\left(\cup_{m\in\Omega_2}\hat{\Pi}_m(k,-1+\delta)\cap\hat{S}_q(k,-\delta)\right), \tag{4.8.9}$$

$$\hat{\Upsilon}_q(k,V,\delta)_4 = \cup_{q'\in\Gamma(R_0)}\hat{\Upsilon}_q(k,V,\delta)_{4q'}, \tag{4.8.10}$$

where $\hat{\Upsilon}_q(k,V,\delta)_{4q'}$ is set of t in $\hat{\mu}_q(k,\delta)$ such that

$$|p_m^2(t)+\Delta\hat{\Lambda}_{mm}(t)-p_j^2(t)-\Delta\hat{\Lambda}_{jj}(t)|<k^{-1/5-5\delta} \tag{4.8.11}$$

for any m of $\Pi_{q'}(k^{1/5})$. This means that the eigenvalues $p_m^2(t)+\Delta\hat{\Lambda}_{mm}(t)$ are situated rather close to the point $p_j^2(t)+\Delta\hat{\Lambda}_{jj}(t)=k^2$. Finally,

$$\hat{\Upsilon}_q(k,\delta)_5 = \mathcal{K}\hat{\pi}(k,1/5,3/5,\delta), \tag{4.8.12}$$

where $\hat{\pi}$ is determined by the formula:

$$\hat{\pi} = \cup_{q,q'\in\Gamma(R_0),q\neq q'}\hat{\pi}_{qq'}, \tag{4.8.13}$$

$$\hat{\pi}_{qq'} = \Big\{x, x\in R^3:\ ||x|-k|<k^{1/5-10\delta},$$

$$|(x,\mathbf{p}_q(0))|<2k^{1/5},\ |(x,\mathbf{p}_{q'}(0))|<k^{3/5}\Big\}.$$

Lemma 4.21 . *If t belongs to $\hat{\chi}_q^0(k,V,\delta)$, then there exists a unique j, such that $p_j^2(t)+\Delta\hat{\Lambda}_{jj}(t)=k^2$ and conditions $1'-5'$ hold. For t in the $(k^{-2-2\delta})$-neighbourhood of $\hat{\chi}_q^0(k,V,\delta)$ there exists a unique j such that $|p_j^2(t)+\Delta\hat{\Lambda}_{jj}(t)-k^2|<k^{-1-2\delta}$ and conditions $1'-5'$ hold.*

Proof. Since $t \in \hat{\mu}_q(k,\delta)$, there exist at least one j such that $\mathbf{p}_j(t) \in \hat{S}_q(k,-\delta)$, i.e., $p_j^2(t) + \Delta\hat{A}_{jj}(t) = k^2$, and $|p_j^2(t) - p_{j+q}(t)^2| < k^\delta$. Suppose, inequality (4.6.9) does not hold. Then for $x = \mathbf{p}_j(t)$ and some $r \in \Omega_1$:

$$\left||x|^2 + \varphi_0(x) - |x + \mathbf{p}_m(0)|^2\right| < k^{1-10\delta}. \tag{4.8.14}$$

Thus, we have $x \in \hat{\Pi}_m(k,-1+10\delta) \cap \hat{S}_q(k,-\delta)$. Therefore, $t \in \mathcal{K}(\hat{\Pi}_m \cap \hat{S}_q) \subset \hat{\Upsilon}_q(k,V,\delta)_1$. However, $\hat{\chi}_q^0(k,V,\delta) \cap \hat{\Upsilon}_q(k,V,\delta)_1 = \emptyset$ (see (4.8.4)). This contradiction proves inequality (4.6.9).

Next, suppose condition 3′ does not hold. This means that there exists m such that $m \neq n_0 q$, $(n_0 \in Z)$ and

$$|p_{m+j}^2(t) - p_j^2(t) - \Delta\hat{A}(t)_{jj}| < 2k^{-1-\delta}. \tag{4.8.15}$$

The last inequality means that for $x = \mathbf{p}_j(t)$ the next relation is valid:

$$\left||x|^2 + \varphi_0(x) - |x + \mathbf{p}_m(t)|^2\right| < 2k^{-1-\delta}, \tag{4.8.16}$$

i.e., $x \in \hat{\Pi}_m(k,1+\delta) \cap \hat{S}_q(k,-\delta)$. Therefore $t \in \mathcal{K}(\hat{\Pi}_m(k,1+\delta) \cap \hat{S}_q(k,-\delta)) \subset \hat{\Upsilon}_q(k,V,\delta)_3$. However, $\hat{\chi}_q^0(k,V,\delta) \cap \hat{\Upsilon}_q(k,V,\delta)_3 = \emptyset$. This contradiction proves that condition 3′ holds.

Suppose condition 2′ does not hold. This means that there exists $m, m \neq n_0 q$ $(n_0 \in Z)$, such that

$$|p_{m+j}^2(t) - p_j^2(t) - \Delta\hat{A}(t)_{jj}| < k^{-20\delta}. \tag{4.8.17}$$

Additionally, there exist some i, $|i| < k^\delta$, $i \neq 0$, such that the inequality opposite to (4.6.11) is satisfied:

$$|p_{m+j+i}^2(t) - p_j^2(t) - \Delta\hat{A}(t)_{jj}| \leq k^{-12\delta}|p_{m+j}^2(t) - p_j^2(t) - \Delta\hat{A}(t)_{jj}|^{-1}. \tag{4.8.18}$$

Inequalities (4.8.17) and (4.6.12) mean that

$$\mathbf{p}_j(t) \in \left(\cup_{n=0}^N \hat{\Pi}_m(k,1-\delta n)\right) \setminus \hat{\Pi}_m(k,1+\delta), N = [1/\delta] - 19. \tag{4.8.19}$$

The sets $\hat{\Pi}_m(k,1-\delta n)$, $n = 0,...,N$, form a sequence of expanding sets. It is clear that

$$\mathbf{p}_j(t) \in \hat{\Pi}_m(k,1-\delta n) \setminus \hat{\Pi}_m(k,1-\delta(n-1)) \tag{4.8.20}$$

for some $n \in \{0,...,N\}$. Relation (4.8.18) means that

$$|p_{m+i+j}^2(t) - p_j^2(t) - \Delta\hat{A}(t)_{jj}| < k^{1-\delta n-11\delta}. \tag{4.8.21}$$

i.e., $\mathbf{p}_j(t) \in \Pi_{i+m}(k,-1+\delta n+11\delta)$. Using relation (4.8.20) and formula (4.8.8), we get

$$\mathbf{p}_j(t) \in \hat{\Upsilon}_q(k,V,\delta)_2^{min} \subset \hat{\Upsilon}_q(k,V,\delta)_2^{mi}. \tag{4.8.22}$$

Therefore, $t \in \hat{\Upsilon}_q(k,V,\delta)_2$. However, $\hat{\Upsilon}_q(k,V,\delta)_2 \cap \hat{\chi}_q^0(k,V,\delta) = \emptyset$. This contradiction proves that condition 2′ is satisfied.

Suppose, condition $4'$ does not hold. Therefore, there exists $m \in \Pi_0(k)$, $\Pi_0 = \cup_{q' \in \Gamma(R_0)} \Pi_{q'}(k^{1/5})$ such that

$$|p_m^2(t) + \Delta\hat{A}(t)_{mm} - p_j^2(t) - \Delta\hat{A}(t)_{jj}| < k^{-1/5-10\delta}, \tag{4.8.23}$$

i.e., $t \in \hat{\Upsilon}_q(k,V,\delta)_4$. Since $\hat{\chi}_q(k,V,\delta) \cap \hat{\Upsilon}_q(k,V,\delta)_4 = \emptyset$, condition $4'$ holds.

Next, suppose condition $5'$ is not satisfied. This means that there exists $m \in T(k,\delta)$ such that

$$|p_m^2(t) - p_j^2(t) - \Delta\hat{A}(t)_{jj}| < k^{1/5-10\delta}. \tag{4.8.24}$$

By the definition of $T(k,\delta)$ the relation $m \in T(k,\delta)$ means that there exists the pair q', q'', where $q', q'' \in \Gamma(R_0)$, $q' \neq q''$, such that

$$|(\mathbf{p}_m(t), \mathbf{p}_{q'}(0))| < k^{1/5}, \quad |(\mathbf{p}_m(t), \mathbf{p}_{q''}(0))| < k^{3/5}. \tag{4.8.25}$$

Let $x \equiv \mathbf{p}_m(t)$. From inequalities (4.8.24) and (4.8.25), considering that $|\Delta\hat{A}(t)_{jj}| < k^{\delta}$, we obtain the estimates:

$$\left||x|^2 - k^2\right| < k^{1/5-10\delta}, \ |(x, \mathbf{p}_{q'}(0))| < 2k^{1/5}, \ |(x, \mathbf{p}_{q''}(0))| < 2k^{3/5}. \tag{4.8.26}$$

This means that $x \in \hat{\pi}_{q'q''}$ (see (4.4.22)). Hence, $t \in \mathcal{K}\hat{\pi} \equiv \hat{\Upsilon}_q(k,V,\delta)_5$. However, $\hat{\chi}_q(k,V,\delta) \cap \hat{\Upsilon}_q(k,V,\delta)_5 = \emptyset$. This contradiction proves that condition $5'$ holds.

Finally, we prove that j is determined uniquely. Suppose, there exists $m \neq j$, such that $\mathbf{p}_m(t) \in \hat{S}_q(k,-\delta)$ and conditions $1' - 5'$ are satisfied. Since $\mathbf{p}_m(t) \in \hat{S}_q(k,-\delta)$, we see that $m \in \Pi_q(k^{1/5})$. Therefore, estimate (4.6.13) holds. But, this contradicts the assumption that $\mathbf{p}_m(t)$ and $\mathbf{p}_j(t)$ belong to $\hat{S}_q(k,-\delta)$ $\left(p_j^2(t) + \Delta\hat{A}(t)_{jj} = p_m^2(t) + \Delta\hat{A}(t)_{mm} = k^2\right)$. Estimates (4.8.14)–(4.8.26) are stable with respect to a perturbation of order $k^{-2-2\delta}$. Therefore conditions $1' - 5'$ hold in $(k^{-2-2\delta})$-neighbourhood of $\hat{\chi}_q^0(k,V,\delta)$, j being uniquely determined from the relation $|p_j^2(t) + \Delta\hat{A}(t)_{jj} - k^2| < k^{-2-2\delta}$. The lemma is proved.

Next, we prove that the set $\hat{\chi}_q^0(k,V,\delta)$ has an asymptotically full measure on the set $\hat{\mu}_q(k,\delta)$ (see Lemma 4.16). To prove this we verify that each of the sets $\hat{\Upsilon}_q(k,V,\delta)_i$, $i = 1,...,5$ has an asymptotically full measure on the set $\hat{\mu}_q(k,\delta)$. Lemmas 4.23 - 4.28 are devoted to the proofs of these assertions. Lemma 4.22 estimates the area of $\hat{S}_q(k,-\delta) \cap \hat{\Pi}_m(k,\xi)$. Technically complicated, but nevertheless important parts of the proofs, are carried out to Appendixes at the end of the paper.

First, we introduce some notations. Let T_q be the body of the torus with the radii equal to k and k', $k' \equiv \sqrt{k^2 - p_q^2(0)/4 - \varphi(\mathbf{p}_q(0)/2)}$, the main circle O_q of the radius k' being centered at the point $\mathbf{p}_q(0)/2$ and lying in the plane orthogonal to $\mathbf{p}_q(0)$. Thus,

$$T_q = \{x : |x - x_s| \leq k, \ x_s \in O_q\},$$

$$O_q = \{x : |x - \mathbf{p}_q(0)/2| = k',\ (x, \mathbf{p}_q(0)) = 0\}.$$

It is clear that the sphere $|x - \mathbf{p}_m(0)| = k$ intersects with the circle O_q if and only if $\mathbf{p}_m(0)$ belongs to T_q. Let ρ_{qm} be the distance from $\mathbf{p}_m(0)$ to the torus (ρ_{mq} is positive if $\mathbf{p}_m(0)$ is inside torus and negative otherwise).

Let us estimate the area of $\hat{S}_q(k,-\delta) \cap \hat{\Pi}_m(k,\xi)$. To understand the structure of $\hat{S}_q(k,-\delta) \cap \hat{\Pi}_m(k,\xi)$ note that in the case $\varphi_0(x) = 0$ this is the intersection of the sphere $|x| = k$ and two plane layers: $\Pi_q(k,-\delta)$ and $\Pi_m(k,\xi)$. So, this is a vicinity on the sphere of two points, which are the intersection of two circles. When $\varphi_0(x) \neq 0$, the picture is a little bit "curve". Let us give more precise description of these two points. Using the definitions of $\hat{S}_q(k,-\delta)$ and $\Pi_m(k,\xi)$ it is easy to show that

$$\hat{S}_q(k,-\delta) \cap \hat{\Pi}_m(k,\xi) = \hat{S}_q(k,-\delta) \cap \hat{S}_m(k,\xi) = \tag{4.8.27}$$

$$\{x :\ |x|^2 + \varphi_0(x) = k^2,\ \left||x|^2 - |x + \mathbf{p}_q(0)|^2\right| < k^\delta,$$
$$\left||x|^2 + \varphi_0(x) - |x + \mathbf{p}_m(0)|^2\right| < k^{-\xi}\}.$$

Thus, $\hat{S}_q(k,-\delta) \cap \hat{\Pi}_m(k,\xi)$ is a neighbourhood on the surface $|x|^2 + \varphi_0(x) = k^2$ of the points defined by the equations

$$|x|^2 + \varphi_0(x) = k^2,$$

$$|x|^2 - |x + \mathbf{p}_q(0)|^2 = 0,$$
$$|x + \mathbf{p}_m(0)|^2 = k^2.$$

It is easy to see that this equations define the intersection of the circle O_q with the sphere $|x + \mathbf{p}_m(0)| = k$, which is, obviously, two points. As it is noted above O_q intersects with the sphere if and only if ρ_{qm} is positive. Moreover, it turns out that the area of $\hat{S}_q(k,-\delta) \cap \hat{\Pi}_m(k,\xi)$ essentially depends on ρ_{qm}. Naturally, it depends on k, δ, ξ. The following lemma give estimates for the area of $\hat{S}_q(k,-\delta) \cap \hat{\Pi}_m(k,\xi)$.

Lemma 4.22 . *If* $8p_q(0) < k$, $0 < \delta < 1$, $-1 < \xi < 1$, *then the following estimates for* $s(\hat{S}_q(k,-\delta) \cap \hat{\Pi}_m(k,\xi))$ *hold:*

$$s(\hat{S}_q(k,-\delta) \cap \hat{\Pi}_m(k,\xi)) \leq ck^{\delta-\xi} p_q^{-1}(0) p_m(0)_\perp^{-1/2} \rho_{mq}^{-1/2}, \tag{4.8.28}$$

when $\rho_{mq} > k^{-2\delta}$;

$$s(\hat{S}_q(k,-\delta) \cap \hat{\Pi}_m(k,\xi)) \leq ck^{\delta/2-\xi+3/2} p_q^{-1/2}(0) p_m(0)^{-3/2} (4k^2 - p_m^2(0))^{-1/4}, \tag{4.8.29}$$

when $\rho_{mq} < k^\delta$; $k^{3\delta} < p_m(0) < 2k - k^{-1+3\delta}$ *and*

$$s(\hat{S}_q(k,-\delta) \cap \hat{\Pi}_m(k,\xi)) \leq ck^{\delta-\xi/2} p_q^{-1}(0), \tag{4.8.30}$$

when $\rho_{mq} < k^\delta$; $p_m(0) > 2k - k^\delta$.

The proof is only a slight modification of Lemma 2.10, where we consider the area of $\hat{S}_q(k,-\delta) \cap \hat{\Pi}_m(k,\xi)$ for $\varphi_0 = 0$, i.e., the case when $\hat{S}_q(k,-\delta)$ is a plane-spherical layer and $\hat{\Pi}_m(k,\xi)$ is a plane layer. We give it in Appendix 1.

Lemma 4.23 . *For $0 < \delta < 10^{-3}$ and sufficiently large k, $k > k_0(V,\delta)$, the following estimate holds:*

$$\frac{s(\hat{\mu}_q(k,\delta) \cap \hat{\Upsilon}_q(k,V,\delta)_1))}{s(\hat{\mu}_q(k,\delta))} < k^{-3\delta}. \tag{4.8.31}$$

Proof. By the definition of $\hat{\Upsilon}_q(k,V,\delta)_1$ (see (4.8.4)) we have

$$s(\hat{\Upsilon}_q(k,V,\delta)_1) \leq \sum_{m \in \Omega_1} s(\hat{S}_q(k,\delta) \cap \hat{\Pi}_m(k,-1+10\delta)). \tag{4.8.32}$$

Taking into account that Ω_1 contains less than $ck^{3\delta}$ elements, we obtain:

$$s(\hat{\Upsilon}_q(k,V,\delta)_1) \leq ck^{3\delta} \max_{m \in \Omega_1} s(\hat{S}_q(k,-\delta) \cap \hat{\Pi}_m(k,-1+10\delta)). \tag{4.8.33}$$

We prove now that

$$2\rho_{mq} > k^{-\delta}. \tag{4.8.34}$$

Indeed, by the definition of ρ_{mq}, we have

$$(k-\rho_{mq})^2 = (k'-r_m)^2 + z_m'^2, \quad z_m' = z_m - p_q(0)/2, \tag{4.8.35}$$

where $r_m \equiv p_m(0)_\perp$ is the absolute value of the projection of the vector $\mathbf{p}_m(0)$ onto the plane orthogonal to $\mathbf{p}_q(0)$ and z_m is the projection of $\mathbf{p}_m(0)$ onto $\mathbf{p}_q(0)$. It is clear that $r_m = (p_m^2(0) - z_m^2)^{1/2}$. Taking into account that $z_m^2 + r_m^2 = p_m^2(0) = O(k^{2\delta})$ we get

$$\rho_{mq}(2k - \rho_{mq}) = 2kr_m + O(k^{2\delta}). \tag{4.8.36}$$

From the relation

$$r_m = |[\mathbf{p}_m(0), \mathbf{p}_q(0)]| p_q^{-1}(0), \tag{4.8.37}$$

considering that the vectors $\mathbf{p}_m(0)$ and $\mathbf{p}_q(0)$ are linearly independent we obtain $r_m > cp_q^{-1}(0) > c_1 k^{-\delta}$. Estimate (4.8.34) follows from the last inequality and (4.8.36). Using formula (4.8.36) and taking into account that $r_m \equiv p_m(0)_\perp$ and $\rho_{mq} > k^{-2\delta}$, we obtain from (4.8.28) that

$$s(\hat{S}_q(k,-\delta) \cap \hat{\Pi}_m(k,-1+10\delta)) < ck^{1-8\delta} p_q(0)^{-1}. \tag{4.8.38}$$

Noting that

$$s(\hat{\mu}_q(k,\delta)) \approx k^{1+\delta} p_q(0)^{-1}, \tag{4.8.39}$$

we get

$$s(\hat{S}_q(k,-\delta) \cap \Pi_m(k,-1+10\delta)) \leq k^{-9\delta} s(\hat{\mu}_q(k,\delta)). \tag{4.8.40}$$

Using estimate (4.8.40) in inequality (4.8.33) yields (4.8.31). The lemma is proved.

Lemma 4.24 . *For* $0 < \delta < 10^{-3}$ *and sufficiently large* k, $k > k_0(V,\delta)$, *the following estimate holds:*

$$\frac{s(\hat{\mu}_q(k,\delta) \cap \hat{\Upsilon}_q(k,V,\delta)_3))}{s(\hat{\mu}_q(k,\delta))} < k^{-3\delta}. \tag{4.8.41}$$

Proof. By the definition of $\hat{\Upsilon}_q(k,V,\delta)_3$ we have

$$s(\hat{\mu}_q(k,\delta) \cap \hat{\Upsilon}_q(k,V,\delta)_3) \leq s(\hat{\Upsilon}_q(k,V,\delta)_3)) \leq \sum_m s(\hat{S}_q(k,-\delta) \cap \hat{\Pi}_m(k,1+\delta)). \tag{4.8.42}$$

Arguing further as in Lemma 2.11 and using inequalities (4.8.28) – (4.8.30) we verify relation (4.8.41). The lemma is proved.

Lemma 4.25 . *For* $0 < \delta < 10^{-3}$ *and sufficiently large* k, $k > k_0(V,\delta)$ *the following estimate holds:*

$$\frac{s(\hat{\mu}_q(k,\delta) \cap \hat{\Upsilon}_q(k,V,\delta)_2))}{s(\hat{\mu}_q(k,\delta))} < k^{-4\delta}. \tag{4.8.43}$$

The proof of this lemma is very technical, so in this consideration we describe the principal steps. Technically complicated parts we send to Appendixes.

Using the definition of $\hat{\Upsilon}_q(k,V,\delta)_2$ (see (4.8.5)– (4.8.8)) we obtain

$$s(\hat{\mu}_q(k,\delta) \cap \hat{\Upsilon}_q(k,V,\delta)_2) \leq$$

$$N \max_{n=1,\dots,N} \sum_{|i|<k^\delta} \sum_{m \in Q_{ni}} s(\hat{S}_q(k,-\delta) \cap \hat{\Pi}_m(k,1-\delta n)), \tag{4.8.44}$$

where Q_{ni} is the set of the indeces m, such that $m \neq n_0 q$, $n_0 \in Z$, and

$$\hat{S}_q(k,-\delta) \cap \hat{\Pi}_m(k,1-\delta n) \cap \hat{\Pi}_{m+i}(k,-1+\delta n+10\delta) \neq \emptyset. \tag{4.8.45}$$

Let us estimate the terms of the sum (4.8.44).

Using estimates (4.8.28)–(4.8.30), we obtain that

$$s(\hat{S}_q(k,-\delta) \cap \hat{\Pi}_m(k,1-\delta n)) \leq k^{-1+\delta(n+1)} p_q^{-1}(0) \rho_{mq}^{-1/2} p_m(0)_\perp^{-1/2}, \tag{4.8.46}$$

when $\rho_{mq} > k^{-2\delta}$;

$$s(\hat{S}_q(k,-\delta) \cap \hat{\Pi}_m(k,1-\delta n)) \leq$$

$$k^{1/2+\delta(n+1/2)} p_q^{-1/2}(0) p_m(0)^{-3/2} (4k^2 - p_m^2(0))^{-1/4}, \tag{4.8.47}$$

when $\rho_{mq} < k^{2\delta}$; $k^{3\delta} < p_m(0) < 2k - k^{-1+3\delta}$ and

$$s(\hat{S}_q(k,-\delta) \cap \hat{\Pi}_m(k,1-\delta n)) \leq k^{-1/2+\delta(n/2+1)} p_q^{-1}(0), \tag{4.8.48}$$

when $\rho_{mq} < k^\delta$; $p_m(0) > 2k - k^\delta$.

We represent Q_{ni} in the form $Q_{ni} = \cup_{k=1}^7 Q_{ni}^k$,

$$Q^1_{ni} = \{m : m \in Q_{ni}, \rho_{mq} > k^{2\delta}\}, \tag{4.8.49}$$

$$Q^2_{ni} = \{m : m \in Q_{ni}, \rho_{mq} < k^{2\delta}, r_m > k^{1-6\delta}, 2k - r_m > k^{1-3\delta}\}, \tag{4.8.50}$$

$$Q^3_{ni} = \{m : m \in Q_{ni}, \rho_{mq} < k^{2\delta}, r_m \le k^{1-6\delta}, p_m(0) > k^{2\delta}\}, \tag{4.8.51}$$

$$Q^4_{ni} = \{m : m \in Q_{ni}, \rho_{mq} < k^{2\delta}, 2k - r_m < k^{1-3\delta}, 2k - p_m(0) > k^{3\delta}\}, \tag{4.8.52}$$

$$Q^5_{ni} = \{m : m \in Q_{ni}, \rho_{mq} < k^{2\delta}, 2k - p_m(0) < k^{3\delta}, |z_m| \ge k^{2\delta}\}, \tag{4.8.53}$$

$$Q^6_{ni} = \{m : m \in Q_{ni}, \rho_{mq} < k^{2\delta}, 2k - p_m(0) < k^{2\delta}, |z_m| < k^{2\delta}\}, \tag{4.8.54}$$

$$Q^7_{ni} = \{m : m \in Q_{ni}, \rho_{mq} < k^{2\delta}, p_m(0) < k^{2\delta}\}, \tag{4.8.55}$$

where r_m as before is the projection of $\mathbf{p}_m(0)$ onto the plane orthogonal to $\mathbf{p}_q(0)$ and z_m is the projection of $\mathbf{p}_m(0)$ onto $\mathbf{p}_q0)$. We break the sum over Q_{ni} in (4.8.44) into six sums corresponding to Q^k_{ni}, $k = 1, ..., 6$. Taking into account that the summation with respect to i contains less than $ck^{3\delta}$ terms we get:

$$s(\hat{\mu}_q(k,\delta) \cap \hat{\Upsilon}_q(k, V, \delta)_2) \le ck^{3\delta} \max_{n=1,\dots,N, |i|<k^\delta} \left(\sum_{k=1}^{7} \Sigma^k_{in}\right), \tag{4.8.56}$$

where

$$\Sigma^k_{in} = \sum_{m \in Q^k_{ni}} s(\hat{S}_q(k, -\delta) \cap \hat{\Pi}_m(k, 1 - \delta n)), \tag{4.8.57}$$

We estimate each sum Σ^k_{in} separately.

1) The estimate of Σ^1_{in}.

Using inequality (4.8.46), we obtain

$$\Sigma^1_{in} < ck^{-1+\delta(n+1)} p_q^{-1}(0) \sum_{m \in Q^1_{ni}} (\rho_{mq} r_m)^{-1/2}, \tag{4.8.58}$$

Let us estimate the sum by an integral. Let us define $\rho(y)$, $r(y)$ and $z(y)$ for y just as ρ_{mq}, r_m and z_m for $\mathbf{p}_m(0)$. Suppose a point y belongs to a cell of the dual lattice, including the point $\mathbf{p}_m(0)$. It is clear that $|y - \mathbf{p}_m(0)| < A$, $A = 2\pi(a_1^{-2} + a_2^{-2} + a_3^{-2})^{1/2}$ and $\rho(y) > \rho_{mq} - A$. From the relation $\rho_{mq} > 2k^\delta$ it follows that

$$2\rho(y) > \rho_{mq} > k^{2\delta}. \tag{4.8.59}$$

Similarly, since

$$r_m > \rho_{mq}, \tag{4.8.60}$$

we have $4r(y) > r_m$. Relations (4.8.59) and (4.8.60) enable us to estimate the sum Σ^1_{in} by the integral I_1:

$$\Sigma^1_{in} \le c\frac{a_1 a_2 a_3}{(2\pi)^3} k^{-1+\delta(n+1)} p_q^{-1}(0) I_1, \tag{4.8.61}$$

$$I_1 = \int_{\tilde{Q}^1_{ni}} (\rho(y) r(y))^{-1/2} dy, \tag{4.8.62}$$

where $\tilde{Q}^1_{ni}$ is the A-neighbourhood of Q^1_{ni}. In Appendix 2 we obtain the formula for $\tilde{Q}^1_{ni}$:

$$\tilde{Q}^1_{ni} \subset \{y : 2\rho(y) > k^{2\delta}, \exists x_0 \in S_0 : |y - x_0| = k,$$
$$|(y - x_0, \mathbf{p}_i(0))| < 4k^{1-\delta n-10\delta} + Ak(r(y)\rho(y))^{-1/2}\}, \tag{4.8.63}$$

where S_0 the middle circle of the layer $\hat{S}_q(k, -\delta)$:

$$S_0 = \{x, x \in \hat{S}_q(k, -\delta), |x|^2 - |x + \mathbf{p}_q(0)|^2 = 0\}. \tag{4.8.64}$$

In Appendix 3 we prove the estimate

$$I_1 < ck^{2-n\delta-9\delta} \tag{4.8.65}$$

Using estimates (4.8.61) and (4.8.65) we obtain

$$\Sigma^1_{in} < ck^{1-4\delta}p_q^{-1}(0). \tag{4.8.66}$$

2) The estimate of the sum Σ^2_{in}. For $m \in Q^2_{ni}$ (see (4.8.50)), estimate (4.8.47) holds. Note that

$$p_m(0) = \sqrt{r_m^2 + z_m^2} \geq r_m. \tag{4.8.67}$$

We estimate $4k^2 - p_m^2(0)$. It follows from relation (4.8.35) that

$$4k^2 - p_m^2(0) = 2k(2k - r_m k'/k) + 2\rho_{mq}k - \rho_{mq}^2 + O(k^{1+\delta}) > k(2k - r_m). \tag{4.8.68}$$

Since $m \in Q^2_{ni}$, we have $2k - r_m > k^{1-6\delta}$, $r_m > k^{1-6\delta}$. Therefore,

$$p_m(0) > k^{1-6\delta}, \tag{4.8.69}$$

$$4k^2 - p_m^2(0) > k^{2-6\delta}. \tag{4.8.70}$$

Using the last pair of the estimates in inequality (4.8.47), we obtain

$$s(\hat{S}_q(k, -\delta) \cap \hat{\Pi}_m(k, V, 1 - \delta n)) \leq cp_q^{-1}(0)k^{-3/2+\delta(n+12)}. \tag{4.8.71}$$

Using this estimate in formula (4.8.57) for Σ^2_{in}, we get:

$$\Sigma^2_{in} \leq cp_q^{-1}(0)k^{-3/2+\delta(n+12)} \sum_{m \in Q^2_{ni}} 1. \tag{4.8.72}$$

The series on the right of estimate (4.8.72) can be estimated by the volume of the A-neighbourhood $\tilde{Q}^2_{ni}$ of the set Q^2_{in}:

$$\Sigma^2_{in} \leq cp_q^{-1}(0)k^{-3/2+\delta(n+12)} V(\tilde{Q}^2_{ni}). \tag{4.8.73}$$

In Appendix 4 we obtain the formula for $\tilde{Q}^k_{ni}$, $k \geq 2$ in a cylindrical coordinates. In Appendix 5 we show that

$$V(\tilde{Q}^2_{ni}) < k^{1+3\delta}(k^{1-n\delta-10\delta} + k^{1/2+3\delta}). \tag{4.8.74}$$

It follows from estimates (4.8.73) and (4.8.74) that

$$\Sigma^2_{in} \leq ck^{1-4\delta}p_q^{-1}(0). \tag{4.8.75}$$

3) The estimate of the sum Σ^3_{in}.

For $m \in Q^3_{ni}$ the estimate (4.8.47) is valid. It follows from relation (4.8.35) that

$$r_m^2 + z_m^2 = 2r_m k' - 2\rho_{mq}k + \rho^2_{mq} + O(k^{1+\delta}). \tag{4.8.76}$$

Taking into account that $\rho_{mq} < k^{2\delta}$, we obtain

$$p_m^2(0) = r_m^2 + z_m^2 = 2k'r_m + O(k^{1+\delta}). \tag{4.8.77}$$

Using this inequality and considering that $r_m < k^{1-6\delta}$, we arrive at the relation $4k^2 - p_m^2(0) > k^2$. From the last estimate and inequality (4.8.47), we get:

$$s\left(\hat{S}_q(k,-\delta) \cap \hat{\Pi}_m(k, 1-\delta n)\right) < ck^{\delta(n+1/2)}p_q^{-1}(0)p_m^{-3/2}(0). \tag{4.8.78}$$

Substituting (4.8.78) in formula (4.8.57) for Σ^3_{in}, we obtain:

$$\Sigma^3_{in} \leq ck^{\delta(n+1/2)}p_q^{-1}(0) \sum_{m \in Q^3_{ni}} p_m(0)^{-3/2}. \tag{4.8.79}$$

Since $p_m(0) > k^{2\delta}$, when $m \in Q^3_{ni}$, the series on the right of (4.8.79) can be estimated by the integral:

$$\Sigma^3_{in} \leq ck^{\delta(n+1/2)}p_q^{-1}(0)I_3, \tag{4.8.80}$$

$$I_3 = \int_{\tilde{Q}^3_{ni}} (r^2 + z^2)^{-3/4} r\,dr\,dz\,d\vartheta. \tag{4.8.81}$$

In Appendix 6 we show that

$$I_3 \leq 3k^{1/2-n\delta-9\delta} + 4k^{2\delta}. \tag{4.8.82}$$

Substituting the last estimate in the inequality (4.8.80) yields:

$$\Sigma^3_{in} < cp_q^{-1}(0)k^{1-4\delta}. \tag{4.8.83}$$

Next, we consider Σ^4_{in}. Since $m \in Q^4_{ni}$, then estimate (4.8.47) holds. Using relations $p_m(0) > r_m > k$ and $p_q(0) < k^\delta$, we rewrite (4.8.47) in the form:

$$s(\hat{S}_q(k,-\delta) \cap \hat{\Pi}_m(k, 1-\delta n) < cp_q^{-1}(0)k^{-1+\delta(n+1)}(4k^2 - r_m^2 - z_m^2)^{-1/4}. \tag{4.8.84}$$

Substituting estimate (4.8.84) in formula (4.8.57) for Σ^4_{in}, we obtain:

$$\Sigma^4_{in} \leq cp_q^{-1}(0)k^{-1+\delta(n+1)} \sum_{m \in Q^4_{ni}} (4k^2 - r_m^2 - z_m^2)^{-1/4}. \tag{4.8.85}$$

It is clear that $\tilde{Q}^4_{ni}$ can be described by the formula similar to that for Q^4_{ni}. Since $4k^2 - r_m^2 - z_m^2 > k^{1+2\delta}$, when $m \in Q^4_{ni}$, the series on the right of (4.8.85) can be estimated by the integral

$$\Sigma^4_{in} \leq cp_q^{-1}(0)k^{-1+\delta(n+1)}I_4, \tag{4.8.86}$$

$$I_4 = \int_{\tilde{Q}^4_{ni}} \frac{rdrdzd\vartheta}{(4k^2 - r^2 - z^2)^{1/4}}. \tag{4.8.87}$$

In the Appendix 7 we prove that

$$I_4 k^{7/4-\delta n-6\delta} + k^{1+2\delta}. \tag{4.8.88}$$

Using this estimate we (4.8.86), we get

$$\Sigma^4_{in} \leq cp_q^{-1}(0)k^{1-4\delta} \tag{4.8.89}$$

for any $\mathbf{p}_i(0)$, $p_i(0) < k^\delta$.

5) Let us consider Σ^5_{in}. We prove that $4k^2 - p_m^2(0) > k^{-1+3\delta}$. Indeed, using (4.8.35), we obtain

$$(2k' - r_m)r_m = z_m'^2 + 2\rho_{mq}k - \rho_{mq}^2 + O(k^{2\delta}). \tag{4.8.90}$$

Therefore,

$$4k^2 - r_m^2 - z_m^2 =$$
$$(z_m'^2 + 2\rho_{mq}k - \rho_{mq}^2 + O(k^{2\delta}))(2k + r_m)r_m^{-1} - z_m^2 > z_m^2 > k^{4\delta}. \tag{4.8.91}$$

From this $2k - p_m(0) > k^{-1+3\delta}$. Thus, estimate (4.8.47) holds. Using relations $p_m(0) > r_m > k$, $p_q(0) < k^\delta$, we rewrite (4.8.47) in the form (4.8.84). Taking into account (4.8.91), we obtain

$$\Sigma^5_{in} \leq k^{-1+\delta(n+1)}p_q^{-1}(0)\sum_{Q^5_{ni}} 1. \tag{4.8.92}$$

Obviously,

$$\Sigma^5_{in} \leq k^{-1+\delta(n+1)}p_q^{-1}(0)V(\tilde{Q}^5_{ni}), \tag{4.8.93}$$

where $\tilde{Q}^5_{ni}$ is the A-neighbourhood of Q^5_{ni}. In Appendix 8 we show that

$$V(\tilde{Q}^5_{ni} < k^{7/4-\delta n-6\delta} + k^{1+2\delta}. \tag{4.8.94}$$

Using this estimate we get

$$\Sigma^5_{in} \leq cp_q^{-1}(0)k^{1-4\delta} \tag{4.8.95}$$

for any $\mathbf{p}_i(0)$, $p_i(0) < k^\delta$.

6) Next, we consider Σ^6_{in}. Using estimate (4.8.48), we get:

$$\Sigma^6_{in} \leq k^{-1/2+n\delta/2+\delta}p_q^{-1}(0)\sum_{Q^6_{ni}} 1. \tag{4.8.96}$$

It is clear that the sum on the right can be estimated by the volume of the A-neighbourhood of Q^6_{ni}:

$$\Sigma^6_{in} \le k^{-1/2+n\delta/2+\delta} V(\tilde{Q}^6_{ni}).$$

It is not hard to show that $\tilde{Q}^6_{ni}$ can be described by the formula similar to the formula for Q^6_{ni}:

$$\tilde{Q}^6_{ni} = \{y : |4k^2 - r^2 - z^2| < 2k^{1+2\delta}, \rho < 2k^{\delta},\ |z| < k^{2\delta}\}.$$

It is not hard to show that $\tilde{Q}^6_{ni}$ belongs to the $(2k^{2\delta})$-neighbourhood of the circle: $r = 2k,\ z = 0$. Therefore, $V(\tilde{Q}^6_{ni}) < ck^{1+2\delta}$. Using the last estimate in the inequality (4.8.96), we get:

$$\Sigma^6_{in} < cp_q^{-1}(0)k^{1-4\delta}. \tag{4.8.97}$$

7) Finally, we consider Σ^7_{in}. We proved in Lemma 4.23 that $\rho_{mq} > k^{-2\delta}$ (see (4.8.34)), when $r_m^2 + z_m^2 < k^{2\delta}$. Arguing as in the proof of Lemma 4.23 and using estimate (4.8.46), we obtain the inequality:

$$\Sigma^7_{in} < cp_q^{-1}(0)k^{1-4\delta}. \tag{4.8.98}$$

It follows from (4.8.56) and estimates (4.8.66), (4.8.75), (4.8.83), (4.8.89), (4.8.95), (4.8.97) and (4.8.98) that

$$s(\hat{\mu}_q(k,\delta) \cap \hat{\Upsilon}_q(k,V,\delta)_2) < cp_q^{-1}(0)k^{1-4\delta}.$$

Taking into account (4.8.39), we get (4.8.43). The lemma is proved.

Lemma 4.26 . *For $0 < \delta < 10^{-2}$ and sufficiently large k, $k > k_0(V,\delta)$, the following estimate holds:*

$$\frac{s(\hat{\mu}_q(k,\delta) \cap \hat{\Upsilon}(k,V,\delta)_4)}{s(\hat{\mu}_q(k,\delta))} < ck^{-2\delta}, c \ne c(k,\delta). \tag{4.8.99}$$

Proof. Using formula (4.8.10) for $\hat{\Upsilon}(k,V,\delta)_4$ and the inequality $|q'| < k^{\delta}$ we obtain

$$s(\hat{\Upsilon}(k,V,\delta)_4 \cap \hat{\mu}_q(k,\delta)) \le ck^{3\delta} \max_{q' \in \Gamma(R_0)} s(\hat{\Upsilon}_q(k,V,\delta)_{4q'} \cap \hat{\mu}_q(k,\delta)). \tag{4.8.100}$$

We prove that

$$\frac{s(\hat{\mu}_q(k,\delta) \cap \hat{\Upsilon}_q(k,V,\delta)_{4q'})}{s(\hat{\mu}_q(k,\delta))} < k^{-5\delta}. \tag{4.8.101}$$

Then, estimate (4.8.99) easily follows from inequalities (4.8.100) and (4.8.101).

In the case of linearly independent q and q' the proof of estimate (4.8.101) is similar to that of (4.4.3) in Lemma 4.2. Indeed, the layer $\hat{S}_q(k,-\delta)$ consists of the pieces $S_{kj} \cap \hat{S}_q(k,-\delta)$, where

$$S_{kj} = \{t, t \in K, p_j^2(t) + \hat{\Delta}\Lambda_{jj}(t) = k^2\}. \tag{4.8.102}$$

Using estimate (4.8.28), it is not hard to show that the sum area of the pieces S_{kj}, such that $j \in \Pi_{q'}(k^{1-12\delta})$, $q' \neq q$, $|q'| < k^{\delta}$, is infinitely small with respect to $s(\hat{S}_q(k, -\delta))$, namely

$$s\left(\cup_{j \in \Pi_{q'}(k^{1-12\delta})} S_{kj}\right) < ck^{-6\delta} s(\hat{S}_q(k, -\delta)). \tag{4.8.103}$$

Considering relation (4.8.103), we obtain:

$$s(\hat{\mu}_q(k,\delta) \cap \hat{\Upsilon}_q(k, V, \delta)_{4q'}) <$$

$$\sum_{j \notin \Pi_q(k^{1-12\delta})} s(S_{kj} \cap \hat{\Upsilon}_q(k, V, \delta)_{4q'}) + O(k^{-6\delta}) s(\hat{S}_q(k, -\delta))$$

$$< cs(\hat{S}_q(k, -\delta))(\sup_{j \notin \Pi_q(k^{1-12\delta})} s(S_{kj} \cap \hat{\Upsilon}_q(k, V, \delta)_{4q'}) + O(k^{-6\delta})). \tag{4.8.104}$$

To prove (4.8.101) it suffices to show that

$$s(S_{kj} \cap \hat{\Upsilon}_q(k, V, \delta)_{4q'}) < k^{-5\delta}. \tag{4.8.105}$$

The proof of this relation is quite similar to that of estimate (ref3.4.4), where instead of the inequality $|\nabla \Lambda_{mm}(t)| < k^{\eta}$ we use the following one: $|\nabla \hat{\Lambda}_{mm}(t) - \nabla \hat{\Lambda}_{jj}(t)| < k^{\eta}, \quad \eta = 1/5$. Thus,

$$s(\hat{\mu}_q(k,\delta) \cap \hat{\Upsilon}_q(k, V, \delta)_{4q'}) < ck^{1-4\delta}, \tag{4.8.106}$$

when $q' \neq \alpha q$, $\alpha \in R$. Next, suppose $q' = \alpha q$, $\alpha \in R$. It is clear that

$$\hat{\Upsilon}_q(k, V, \delta)_{4q'} = \hat{\Upsilon}_q(k, V, \delta)'_{4q'} \cap \hat{\Upsilon}_q(k, V, \delta)''_{4q'}, \tag{4.8.107}$$

where

$$\hat{\Upsilon}_q(k, V, \delta)'_{4q'} = \{t : |p_m^2(t) + \Delta\hat{\Lambda}^q_{mm}(t) - p_j^2(t) - \Delta\hat{\Lambda}_{jj}(t)| < k^{-1/5-8\delta},$$

$$m \in \Pi_q(k^{1/5}), \mathbf{p}_m(0) - \mathbf{p}_j(0) = n\mathbf{p}_q(0), n \in Z\}. \tag{4.8.108}$$

Naturally,

$$\hat{\Upsilon}_q(k, V, \delta)''_{4q'} = \{t : |p_m^2(t) + \Delta\hat{\Lambda}^q_{mm}(t) - p_j^2(t) - \Delta\hat{\Lambda}_{jj}(t)| < k^{-1/5-8\delta},$$

$$m \in \Pi_q(k^{1/5}), \mathbf{p}_m(0) - \mathbf{p}_j(0) \neq n\mathbf{p}_q(0), n \in Z\}. \tag{4.8.109}$$

Using estimate (4.5.89), which follows from the definition of $\hat{H}$, we obtain:

$$p_m^2(t) + \Delta\hat{\Lambda}^q_{mm}(t) - p_j^2(t) - \Delta\hat{\Lambda}_{jj}(t) = p_m^2(t) + \Delta\Lambda^q_{mm}(t) - p_j^2(t) - \Delta\Lambda_{jj}(t) + O(k^{-k^{2/5}}).$$

Obviously that

$$p_m^2(t) + \Delta\Lambda^q_{mm}(t) - p_j^2(t) - \Delta\Lambda_{jj}(t) = (\lambda_{l(m)} - \lambda_{l(j)})(\tau(t)),$$

where $l(m), l(j), \tau(t)$ can be uniquely determined from the relations (4.3.25), (4.3.26), λ_l being eigenvalues of a one-dimensional Schrödinger equation. It is well-known that for eigenvalues of a one-dimensional periodic Schrödinger equation with a smooth potential the relation

$$|(\lambda_{l(m)} - \lambda_{l(j)})(\tau)| < k^{-1/5-8\delta} \tag{4.8.110}$$

can be satisfied only when $\min\{\tau^2, (\tau-\pi)^2, (\tau-2\pi)^2\} < ck^{-1/5-8\delta}$; therefore

$$\hat{\Upsilon}_q(k, V, \delta)'_{4q'} \subset \tilde{M}, \tag{4.8.111}$$

$$\tilde{M} = \{x, x \in \hat{S}_q(k, -\delta), 2\pi|(x, q)||q|^{-2} - l/2| < k^{-1/5-8\delta}, l \in Z\}.$$

It is not hard to show, that $\tilde{M}$ is the union of the layers of the width $\pi^{-1}|q|k^{-1/5-8\delta}$ and that all of them belong to $\hat{S}_q(k, -\delta)$ and, so there are no more than k^δ of them, we obtain the estimate

$$s(\hat{S}_q(k, -\delta) \cap \tilde{M}) < ck^{4/5-7\delta}. \tag{4.8.112}$$

Using relations (4.8.111), (4.8.112), we obtain:

$$s(\hat{\Upsilon}_q(k, V, \delta)'_{4q'}) < cp_q^{-1}(0)k^{1-4\delta}, \tag{4.8.113}$$

It remains to consider $\hat{\Upsilon}_q(k, V, \delta)''_{4q'}$. It is clear that

$$\hat{\Upsilon}_q(k, V, \delta)''_{4q'} \subset \cup_{m \in Q} \left(\hat{S}_q(k, -\delta) \cap \hat{\Pi}'_m(k, 1/5 + 8\delta)\right), \tag{4.8.114}$$

where

$$\hat{\Pi}'_m(k, 1/5 + 8\delta) = \tag{4.8.115}$$

$$\{x : ||x|^2 + \varphi(x) - |x + \mathbf{p}_m(0)|^2 - \varphi(x + \mathbf{p}_m(0))| < k^{-1/5-8\delta}\}.$$

$$Q = \{m : |(\mathbf{p}_m(0), \mathbf{p}_q(0))| < k^{1/5}\}. \tag{4.8.116}$$

It is not hard to show that the set $\hat{S}_q(k, -\delta) \cap \hat{\Pi}'_m(k, 1/5 + 8\delta)$ satisfies the estimates similar to (4.8.28), (4.8.30) and inequality (4.8.29) when $p_m(0) > k^{1/5+\delta}$. Further considerations will be quite similar to those in Lemma 4.25. Indeed, we estimate $s(\hat{\Upsilon}_q(k, V, \delta)''_{4q'})$ as follows:

$$s(\hat{\Upsilon}_q(k, V, \delta)''_{4q'}) \leq \sum_{k=1}^{6} \Sigma^k, \tag{4.8.117}$$

where

$$\Sigma^k = \sum_{m \in Q^k} s(\hat{S}_q(k, -\delta) \cap \hat{\Pi}'_m(k, 1/5 + 8\delta)), \tag{4.8.118}$$

the sets Q^k being given by the formula

$$Q^1 = \{m : m \in Q, \rho_{mq} > k^{2\delta}\}, \tag{4.8.119}$$

$$Q^2 = \{m : m \in Q, \rho_{mq} < k^{2\delta}, r_m > k^{1-\delta}, 2k - r_m > k^{1-6\delta}\}, \tag{4.8.120}$$

$$Q^3 = \{m : m \in Q, \rho_{mq} < k^{2\delta}, r_m \le k^{1-\delta}, p_m(0) > k^{2\delta}\}, \tag{4.8.121}$$

$$Q^4 = \{m : m \in Q, \rho_{mq} < k^{2\delta}, 2k - r_m < k^{1-6\delta}, 2k - p_m(0) > k^{2\delta}\}, \tag{4.8.122}$$

$$Q^5 = \{m : m \in Q, \rho_{mq} < k^{2\delta}, 2k - p_m(0) < k^{2\delta}, |z| \ge k^{2\delta}\}, \tag{4.8.123}$$

$$Q^6 = \{m : m \in Q, \rho_{mq} < k^{2\delta}, 2k - p_m(0) < k^{2\delta}, |z| < k^{2\delta}\}, \tag{4.8.124}$$

$$Q^7 = \{m : m \in Q, \rho_{mq} < k^{2\delta}, p_m(0) < k^{2\delta}\}, \tag{4.8.125}$$

We estimate the sums Σ^k in Appendix 9. Adding the inequalities for Σ^k we get

$$s(\hat{\Upsilon}_q(k, V, \delta)''_{4q'}) < cp_q^{-1}(0)k^{1-4\delta}. \tag{4.8.126}$$

Adding (4.8.113) and (4.8.126), we get

$$s(\hat{\Upsilon}_q(k, V, \delta)_{4q'}) < cp_q^{-1}(0)k^{1-4\delta}. \tag{4.8.127}$$

Taking into account (4.8.39) we get (4.8.99). The lemma is proved.

Suppose $\mu'(k,\delta) \subset \hat{\mu}_q(k,\delta)$. Let $\Gamma(\mu'(k,\delta), k^{-r})$ be the (k^{-r})-neighbourhood in K of the surface $\mu'(k,\delta)$.

Lemma 4.27 . *Suppose* $0 < \delta < 10^{-2}$ *and* $\mu'(k,\delta) \subset \hat{\mu}_q(k,\delta)$, *and* $s(\mu'(k,\delta)) > k^{1-2\delta}$. *Then for sufficiently large* k, $k > k_0(V,\delta)$ *the volume of the* $(k^{-1-3\delta})$-*neighbourhood of* $\mu'(k,\delta)$ *is greater than* $k^{-5\delta}$:

$$V(\Gamma(\mu'(k,\delta), k^{-1-3\delta})) > k^{-5\delta}. \tag{4.8.128}$$

Proof. The proof is similar to that of Lemma 3.5. Indeed, we consider the set

$$\chi = \mathcal{K}\left(\chi_0 \cap \hat{S}_q(k, -\delta)\right),$$

$$\chi_0 = \cup_{m \in \Pi_q(k^\delta)} \hat{\Pi}'_m(k, 2\delta). \tag{4.8.129}$$

It is obvious that

$$\Gamma(\mu', k^{-1-3\delta}) \supset \Gamma(\mu' \setminus \chi, k^{-1-3\delta}). \tag{4.8.130}$$

The surface $\hat{\mu}_q(k,\delta)$ consists of the pieces S_{kj} ($t \in S_{kj}$, if $x \in \hat{S}_q(k,-\delta)$, $x = \mathbf{p}_j(t)$). We denote the intersection of the j-th piece with μ' by μ'_j. Thus,

$$\Gamma(\mu', k^{-1-3\delta}) = \cup_j \Gamma(\mu'_j \setminus \chi, k^{-1-3\delta}). \tag{4.8.131}$$

Using the definition of χ we readily show that

$$\Gamma(\mu'_{j_1} \setminus \chi, k^{-1-3\delta}) \cap \Gamma(\mu'_{j_2} \setminus \chi, k^{-1-3\delta}) = \emptyset, \quad \text{if } j_1 \ne j_2. \tag{4.8.132}$$

From this we obtain:

$$V(\Gamma(\mu', k^{1-3\delta})) = \sum_j V\left(\Gamma(\mu'_j \setminus \chi, k^{1-3\delta})\right). \tag{4.8.133}$$

It is not hard to show that

$$V(\Gamma(\mu_j' \setminus \chi, k^{1-3\delta})) = s(\mu_j' \setminus \chi)k^{-1-3\delta}(1+o(1)). \tag{4.8.134}$$

Therefore,

$$V(\Gamma(\mu', k^{1-3\delta})) > cs(\mu' \setminus \chi)k^{-1-3\delta}. \tag{4.8.135}$$

Set χ coincides with $\hat{\Upsilon}_q(k, V, \delta)_4$ up to replacement of $\Pi_q(k^{1/5})$ by $\Pi_q(k^\delta)$ and $(1 + |(\mathbf{p}_m(0), \mathbf{p}_{q'}(0))|)^{-1}$ by $k^{-2\delta}$. Arguing as in Lemma 4.26, we obtain that

$$s(\chi) < cp_q^{-1}(0)k^{1-4\delta}. \tag{4.8.136}$$

Taking into account the hypothesis of the lemma on $\mu'(k,\delta)$, we get

$$2s(\mu' \setminus \chi) > k^{1-2\delta}. \tag{4.8.137}$$

Using the last inequality in relation (4.8.135) we obtain estimate (4.8.128).
The lemma is proved.

Lemma 4.28 . *For* $0 < \delta < 10^{-2}$ *and sufficiently large* k, $k > k_0(V,\delta)$ *the following estimate holds:*

$$\frac{s(\hat{\mu}_q(k,\delta) \cap \hat{\Upsilon}_q(k,\delta)_5)}{s(\hat{\mu}_q(k,\delta))} < k^{-3\delta}. \tag{4.8.138}$$

Proof. Suppose inequality (4.8.138) does not hold. Then,

$$s(\hat{\mu}_q(k,\delta) \cap \hat{\Upsilon}_q(k,\delta)_5) > k^{1-2\delta}.$$

By the previous lemma

$$V(\Gamma(\hat{\mu}_q(k,\delta) \cap \hat{\Upsilon}_q(k,\delta)_5, k^{-1-3\delta})) > k^{-5\delta}. \tag{4.8.139}$$

On the other hand

$$\hat{\mu}_q(k,\delta) \cap \hat{\Upsilon}_q(k,\delta)_5 \subset \hat{\Upsilon}_q(k,\delta)_5 = \mathcal{K}\hat{\pi}.$$

(see (4.4.22)). In Lemma 4.3 we proved that

$$V(\Gamma(\mathcal{K}\hat{\pi}, k^{-1-3\delta})) < k^{-6\delta}. \tag{4.8.140}$$

Thus, inequalities (4.8.139) and (4.8.140) are in contradiction. This means that estimate (4.8.138) holds. The lemma is proved.

Lemma 4.29 . *For* $0 < \delta < 10^{-2}$ *and sufficiently large* k, $k > k_0(V,\delta)$, *the following estimate holds:*

$$\frac{s(\hat{\mu}_q(k,\delta) \setminus \chi_q^0(k,V,\delta))}{s(\hat{\mu}_q(k,\delta))} < k^{-4\delta}. \tag{4.8.141}$$

Proof. From (4.8.1) it follows that

$$s(\hat{\mu}_q(k,\delta)\cap \chi_q^0(k,V,\delta)) \le \sum_{r=1}^{5} s(\hat{\mu}_q(k,\delta)\cap \hat{\Upsilon}_q(k,V,\delta)_i).$$

Adding estimates (4.8.31), (4.8.41), (4.8.43), (4.8.99) and (4.8.138) for $s(\hat{\mu}_q(k,\delta)\cap \hat{\Upsilon}_q(k,V,\delta)_i)$, $i = 1,2,3,4,5$, we obtain inequality (4.8.141). The lemma is proved.

Geometric Lemma 4.16 is the union of Lemma 4.21 and Lemma 4.29.

Let us consider the set

$$\hat{\chi}_q'(k,V,\delta) = \mu_q(k,-\delta) \setminus \cup_{i=1}^{6} \Upsilon_q(k,V,\delta)_i, \tag{4.8.142}$$

where $\Upsilon_q(k,V,\delta)_i$, $i = 1,2,3,4,5$ coincide with $\hat{\Upsilon}_q(k,V,\delta)_i$ given by formulae (4.8.4)–(4.4.22) up to replacement of $\hat{S}_q(k,-\delta)$ for $S_q(k,-\delta)$ and

$$\Upsilon_6(k,\delta) = \mathcal{K}\left(\cup_{m\in Z^3} \Pi_m(k,1+\delta)\cap S_q(k,-\delta)\right). \tag{4.8.143}$$

Lemma 4.30 . *If t belongs to $\hat{\chi}_q'(k,V,\delta)$, then there exists j, such that $p_j^2(t) = k^2$ and conditions $1'-5'$ hold. For t in the $(k^{-2-2\delta})$-neighbourhood of $\hat{\chi}_q'(k,V,\delta)$ there exists j such that $|p_j^2(t) - k^2| < k^{-1-2\delta}$, and conditions $1'-5'$ hold. The set $\hat{\chi}_q'(k,V,\delta)$ has an asymptotically full measure on $S_q(k,-\delta)$ and the following estimates holds:*

$$\frac{s(\hat{\chi}_q'(k,V,\delta))}{s\left(S_q(k,-\delta)\right)} < ck^{-\delta}. \tag{4.8.144}$$

Proof. It is easy to see that j is uniquely determined from the relation $|p_j^2(t) - k^2| < k^{-1-6\delta}$ for any t in $k^{-2-\delta}$-neighbourhood of $\hat{\chi}_q'(k,V,\delta)$. Indeed, suppose it is not so. Then, there exists $m \ne j$ such that $|p_m^2(t) - k^2| < k^{-1-2\delta}$, i.e., $|p_m^2(t) - p_j^2(t)| < 2k^{-1-2\delta}$. From the last relation it follows that t is in the $(k^{-2-2\delta})$- neighbourhood of $\Upsilon_6(k,\delta)$. But this contradicts the initial assumption that t is in the neighbourhood of $\hat{\chi}_q'(k,V,\delta)$. Considering just as in Lemma 4.21, we check that conditions $1'-5'$ are satisfied. To obtain estimate (4.8.144) one has to repeat the considerations of Lemma 2.11 which asserts that $\Upsilon_6(k,\delta)$ has an asymptotically full measure on $\mu_q(k,\delta)$. The lemma is proved.

4.9 Appendixes

4.9.1 Appendix 1 (The Proof of Lemma 4.22)

First, we prove estimate (4.8.28). We split the layer $\hat{S}_q(k,-\delta)$ into parallel layers $M_n(k,\xi+\delta)$, $|n| < k^{\xi+2\delta}$ of the width $k^{-\xi-\delta}$. It is clear that

$$s(\hat{S}_q(k,\delta)\cap \hat{\Pi}_m(k,\xi)) \le k^{\xi+2\delta} \max_n s(M_n(k,\xi+\delta)\cap \hat{\Pi}_m(k,\xi)). \tag{4.9.1}$$

Inequality (4.8.28) immediately follows from the estimate:

$$s(M_n(k,\xi+\delta)\cap \hat{\Pi}_m(k,\xi)) \le ck^{-2\xi-\delta}p_q^{-1}(0)p_m^{-1/2}(0)_\perp \rho_{mq}^{-1/2}. \tag{4.9.2}$$

We verify the last inequality. For each layer $M_n(k,\xi+\delta)$ the relation

$$\varphi_0(x) = \varphi_n + O(k^{-\xi}), \quad \varphi_n \neq \varphi_n(x), \quad |\varphi_n| < k^{\delta}. \tag{4.9.3}$$

holds. Therefore, from the inequality

$$||x|^2 + \varphi_0(x) - |x+\mathbf{p}_m(0)|^2| < k^{-\xi-\delta} \tag{4.9.4}$$

it follows that

$$||x|^2 + \varphi_n - |x+\mathbf{p}_m(0)|^2| < 2k^{-\xi}, \quad m \in Z^n. \tag{4.9.5}$$

This means that

$$M_n(k,\xi+\delta)\cap \hat{\Pi}_m(k,\xi) \subset$$
$$\{x : x \in M_n(k,\xi+\delta), ||x|^2 + \varphi_n - |x+\mathbf{p}_m(0)|^2| < 2k^{-\xi}\}. \tag{4.9.6}$$

Now we have the intersection of a plane layer and a "quasi"-spherical shell. Arguing as in the proof of Lemma 2.10 and taking into account that $|\varphi_n| < k^{\delta}$, we obtain estimate (4.9.2), and it gives (4.8.28). Estimate (4.8.30) is proved similarly.

To prove inequality (4.8.29) we direct axis $\tilde{x}_1$ along $\mathbf{p}_m(0)$. We consider axis $\tilde{x}_2$ to be situated in the plane of the vectors $\mathbf{p}_q(0)$ and $\mathbf{p}_m(0)$. We introduce the spherical coordinates $\tilde{R},\tilde{\vartheta},\tilde{\varphi}$ such that $\tilde{x}_1 = \tilde{R}\cos\tilde{\vartheta}$, $\tilde{x}_2 = \tilde{R}\sin\tilde{\vartheta}\cos\tilde{\varphi}$, $\tilde{x}_3 = \tilde{R}\sin\tilde{\vartheta}\sin\tilde{\varphi}$. Arguing again as in Lemma 2.10 and taking into account that $|\varphi_0(x)| < k^{\delta}$, we get:

$$\hat{S}_q(k,-\delta)\cap \hat{\Pi}_m(k,1+\xi) \subset \tag{4.9.7}$$
$$\Big\{x : |\cos\tilde{\vartheta} - \cos\tilde{\vartheta}_1 + \tilde{\Phi}(\tilde{R},\tilde{\vartheta},\tilde{\varphi})| < \tilde{\varepsilon}_1,$$
$$|\sin\tilde{\vartheta}(\cos\tilde{\varphi} - \cos\tilde{\varphi}_1)| \le \tilde{\varepsilon}_2, \quad \tilde{R} = k + O(k^{-1})\Big\},$$

where $\cos\tilde{\vartheta}_1 = p_m(0)/2\tilde{R}$, $\cos\tilde{\varphi}_1 = (p_q^2(0) - 2p_q(0)_{||}k\cos\tilde{\vartheta})(2kp_q(0)_\perp \sin\tilde{\vartheta})^{-1}$, $\Phi(\tilde{R},\tilde{\vartheta},\tilde{\varphi}) = (2\tilde{R})^{-1}p_m^{-1}(0)\varphi_0(x)$, $\tilde{\varepsilon}_1 = k^{-\xi-1}p_m^{-1}(0)/2$, and $\tilde{\varepsilon}_2 = k^{-\delta-1}p_q(0)_\perp^{-1}$. We will prove that from the inequality

$$|\cos\tilde{\vartheta} - \cos\tilde{\vartheta}_1 + \tilde{\Phi}(\tilde{R},\tilde{\vartheta},\tilde{\varphi})| < \tilde{\varepsilon}_1 \tag{4.9.8}$$

it follows that

$$|\cos\tilde{\vartheta} - \cos\tilde{\vartheta}_1| < \tilde{\varepsilon}_1. \tag{4.9.9}$$

Indeed, it is easy to see that

$$\left|\frac{\partial\Phi}{\partial\cos\tilde{\vartheta}}\right| \le |p_m(0)\sin\tilde{\vartheta}|^{-1}k^{\delta}. \tag{4.9.10}$$

Suppose that for all $\tilde{\vartheta}$ satisfying inequality (4.9.8) the following estimate holds:

$$|p_m(0)\sin\tilde{\vartheta}| > 2k^{\delta}. \tag{4.9.11}$$

Then (4.9.9) follows from (4.9.8). If for some $\tilde{\vartheta}$ we have the opposite inequality $|p_m(0)\sin\tilde{\vartheta}| \le 2k^{\delta}$, then (4.9.8) gives

$$\cos\tilde{\vartheta}_1 = 1 + O(p_m^{-2}(0)k^{2\delta}). \tag{4.9.12}$$

On the other hand

$$\cos\tilde{\vartheta}_1 = p_m(0)/2k. \tag{4.9.13}$$

by the definition of $\tilde{\vartheta}_1$. By hypothesis $p_m(0) > k^{3\delta}$, $2k - p_m(0) > k^{-1+3\delta}$. It is not hard to show, using the last two inequalities that the inequalities (4.9.12) and (4.9.13) are in contradiction. Thus, (4.9.11) is proved, and therefore estimate (4.9.9) holds. Using inequality (4.9.8) instead of (4.9.9) in relation (4.9.7), we get:

$$M_n(k,\xi+\delta)\cap\hat{\Pi}_m(k,\xi)\subset \tag{4.9.14}$$
$$\{x :|\cos\tilde{\vartheta}-\cos\tilde{\vartheta}_1|<\tilde{\varepsilon}_1, |\sin\tilde{\vartheta}(\cos\tilde{\varphi}-\cos\tilde{\varphi}_1)|\,\tilde{\vartheta}_1\le\tilde{\varepsilon}_2, \tilde{R}=k+O(k^{-1})\}.$$

Thus, we arrive to the case of a plane layer. This was considered before in the proof of Lemma 2.10. Thus, we obtain estimate (4.8.29). The lemma is proved.

4.9.2 Appendix 2

Considering formulae (4.8.45) and (4.8.49) we get:

$$Q_{ni}^1 \subset \Big\{m : \rho_{mq} > k^{2\delta}, \exists x\in\hat{S}_q(k,-\delta) : ||x|^2 - |x+\mathbf{p}_m(0)|^2 + \varphi_0(x)| < k^{-1+\delta n},$$
$$|(x+\mathbf{p}_m(0),\mathbf{p}_i(0))| < 2k^{1-\delta n-10\delta}\Big\}. \tag{4.9.15}$$

Let us prove that

$$\tilde{Q}_{ni}^1 \subset \{y : 2\rho(y) > k^{2\delta}, \exists x_0\in S_0 : |y-x_0| = k,$$
$$|(y-x_0,\mathbf{p}_i(0))| < 4k^{1-\delta n-10\delta} + Ak(r(y)\rho(y))^{-1/2}\}, \tag{4.9.16}$$

where S_0 the middle circle of the layer $\hat{S}_q(k,-\delta)$:

$$S_0 = \{x, x\in\hat{S}_q(k,-\delta), |x|^2 - |x+\mathbf{p}_q(0)|^2 = 0\}. \tag{4.9.17}$$

Indeed, let $y\in\tilde{Q}_{ni}^1$. Hence, there exists $\mathbf{p}_m(0) : |\mathbf{p}_m(0)-y| < A$, $\mathbf{p}_m(0)\in Q_{ni}^1$. Estimate (4.8.59) means that the distance from the point y to the circle S_0 is less than $k - k^{2\delta}/2$. Therefore the sphere of the radius k centered at the point y intersects the circle S_0 at two points $x_\pm$:

$$|y-x_\pm| = k, \quad x_\pm\in S_0. \tag{4.9.18}$$

It remains to prove that

$$|(y-x_\pm,\mathbf{p}_i(0))| < 4k^{1-\delta n-10\delta} + k^{1+2\delta}(r(y)\rho(y))^{-1/2}. \tag{4.9.19}$$

Since $m \in Q^1_{ni}$, by the definition of Q^1_{ni} there exists $x \in \hat{S}_q(k,\delta)$ such that

$$|k^2 + \varphi_0(x) - |x + \mathbf{p}_m(0)|^2| < k^{-1+\delta n} \tag{4.9.20}$$

and

$$|(x + \mathbf{p}_m(0), \mathbf{p}_i(0))| < k^{1-\delta n - 10\delta}. \tag{4.9.21}$$

From estimate (4.9.20), using the obvious inequalities $|\mathbf{p}_m(0) - y| < A$ and $|\varphi(x) - \varphi(x_0)| \le \|V\|$, we get that the point x is in the $(2A + \|V\|)$-neighbourhood of the sphere centered at y with radius k, i.e., in the spherical shell $\{x : ||y - x| - k| < 2A + 2\|V\|\}$. We show (see Appendix 2A) that the intersection of this spherical shell and the quasi-spherical layer $\hat{S}_q(k, -\delta)$ lies in the $(k^{1+\delta}(r(y)\rho(y))^{-1/2})$-neighbourhood of the points $x_\pm$. Thus, x lies in this neighbourhood, i.e.,

$$|x - x_0| < k^{1+\delta}(r(y)\rho(y))^{-1/2}, \tag{4.9.22}$$

where x_0 one of the points $x_\pm$. Considering (4.9.21) and (4.9.22), we arrive at (4.9.19) and, therefore, at (4.9.17).

4.9.3 Appendix 2A

. Suppose x is in the intersection of the $2(A + \|V\|)$-neighbourhood of the sphere of radius k centered at the point y and the layer $\hat{S}_q(k, -\delta)$. We prove that x is in the $(k^{1+\delta}(\rho(y)r(y))^{-1/2})$-neighbourhood of the points $x_\pm$ defined by (4.9.17) and (4.9.18); i.e.,

$$|x - x_0| \le k^{1+\delta}(\rho(y)r(y))^{-1/2}, \tag{4.9.23}$$

where x_0 is one of the points $x_\pm$.

We direct axis x_1 along $\mathbf{p}_q(0)$; axis x_2 is situated in the plane of the vectors $\mathbf{p}_q(0)$, y and directed so that the projection of y on this axis is positive (the vectors $\mathbf{p}_q(0)$ and y are linearly independent because $\rho(y) > k^{2\delta}$). The axis x_3 is orthogonal to the axes x_1, x_2. In this coordinate system vector y has the form: $(z, r, 0)$. It is clear that vector x satisfies the relations

$$\begin{gathered} x_1 = a, \\ x_2^2 + x_3^2 = k^2 + \varphi_0(a) - a^2, \\ (z-a)^2 + (r - x_2)^2 + x_3^2 = (k - l)^2 \end{gathered} \tag{4.9.24}$$

for some $a, l : |a| < k^\delta, |l| < 2(\|V\| + A)$. Note that vector x_0 satisfies the same relations with $a = p_q(0)/2$ and $l = 0$:

$$\begin{gathered} x_{01} = p_q(0)/2, \\ x_{02}^2 + x_{03}^2 = k'^2, \quad k'^2 = k^2 + \varphi_0(p_q(0)) - p_q(0)^2/4, \\ (z - p_q(0)/2)^2 + (r - x_{02})^2 + x_{03}^2 = k^2. \end{gathered} \tag{4.9.25}$$

We prove that

$$\left|\frac{\partial x}{\partial a}\right| + \left|\frac{\partial x}{\partial l}\right| < \frac{12k}{\sqrt{\rho(y)r(y)}}. \tag{4.9.26}$$

Then, inequality (4.9.23) easily follows from relations (4.9.24) - (4.9.26). Let us now prove (4.9.26). Thus, we calculate $\partial x/\partial l$. By differentiation of equations (4.9.24) we obtain

$$\frac{\partial x_1}{\partial l} = 0,$$

$$x_2 \frac{\partial x_2}{\partial l} + x_3 \frac{\partial x_3}{\partial l} = 0, \tag{4.9.27}$$

$$(x_2 - r)\frac{\partial x_2}{\partial l} + x_3 \frac{\partial x_3}{\partial l} = k - l. \tag{4.9.28}$$

We consider this as a linear system with respect to $\partial x_2/\partial l$, $\partial x_3/\partial l$. Its determinant is:

$$\Delta = \left| \begin{pmatrix} x_2 & x_3 \\ x_2 - r & x_3 \end{pmatrix} \right| = r x_3. \tag{4.9.29}$$

Therefore,

$$\frac{\partial x_2}{\partial l} = -(k - l) r^{-1}, \tag{4.9.30}$$

$$\frac{\partial x_3}{\partial l} = x_2 (k - l) r^{-1} x_3^{-1}. \tag{4.9.31}$$

From the relations (4.9.30) and (4.9.31), taking into account that $\partial x_1/\partial l = 0$, we get

$$\sqrt{\left(\frac{\partial x_2}{\partial l}\right)^2 + \left(\frac{\partial x_3}{\partial l}\right)^2} = (k - l) r^{-1} x_3^{-1} \sqrt{k^2 - \varphi_0(a) - a^2} \le 2k^2 r^{-1} x_3^{-1}. \tag{4.9.32}$$

Similarly,

$$\sqrt{\left(\frac{\partial x_1}{\partial a}\right)^2 + \left(\frac{\partial x_2}{\partial a}\right)^2 + \left(\frac{\partial x_3}{\partial a}\right)^2} < 2k^2 r^{-1} x_3^{-1}. \tag{4.9.33}$$

Next, we prove that $r x_3 > k \rho^{1/2} r^{1/2}$. From (4.9.24) we easily obtain

$$x_3^2 + x_2^2 = R_1^2, \tag{4.9.34}$$

$$x_3^2 + (r - x_2)^2 = R_2^2, \tag{4.9.35}$$

where

$$R_1^2 = k^2 - \varphi_0(a) - a^2,$$

$$R_2^2 = (k - l)^2 - (z - a)^2.$$

Hence,

$$(r - 2x_2) r = R_2^2 - R_1^2, \tag{4.9.36}$$

i.e.,

$$-2x_2 = \left((R_2^2 - R_1^2) - r^2\right) r^{-1}. \tag{4.9.37}$$

From relations (4.9.34) and (4.9.37) we get

$$x_3^2 = R_1^2 - (R_2^2 - R_1^2 - r^2)^2/4r^2. \tag{4.9.38}$$

Thus,

$$4x_3^2r^2 = 4R_1^2r^2 - (R_2^2 - R_1^2 - r^2)^2 = -r^4 + 2(R_2^2 + R_1^2)r^2 - (R_2^2 - R_1^2)^2. \tag{4.9.39}$$

Considering the right as a polynomial with respect to r, we obtain

$$4x_3^2r^2 = -(r - y_{1+})(r - y_{1-})(r - y_{2+})(r - y_{2-}), \tag{4.9.40}$$

where $y_{1\pm}$, $y_{2\pm}$ are the roots of the polynomial

$$y_{1+} = R_1 + R_2, \quad y_{1-} = -(R_1 + R_2), \quad y_{2+} = R_1 - R_2, \quad y_{2-} = R_2 - R_1. \tag{4.9.41}$$

We define ρ_a so that $k - l - \rho_a$ is the distance between the point y and the circle $\{x : x_1 = a, x_2^2 + x_3^2 = R_1^2\}$. Then ρ_a satisfies the equation

$$(k - l - \rho_a)^2 = (R_1 - r)^2 + (z - a)^2. \tag{4.9.42}$$

It is easily follows from the last relation and the formula for R_2 that

$$2(k - l)\rho_a - \rho_a^2 = (k - l)^2 - (R_1 - r)^2 - (z - a)^2 = (R_2 - R_1 + r)(R_2 + R_1 - r) =$$
$$= -(r - y_{2+})(r - y_{1+}). \tag{4.9.43}$$

Note that $|k - l - \rho_a - k + \rho| < a + O(ak^{-1})$, since the circles S_0 and S_a are separated the distance $a + O(ak^{-1})$. Hence, $|\rho - \rho_a| < 2a + l$. From the relation $\rho > k^\delta$, it follows that $2\rho_a > \rho$. Considering the inequality $\rho_a < k$, we obtain

$$2(k - l)\rho_a - \rho_a^2 > \rho k/2. \tag{4.9.44}$$

Relations (4.9.43) and (4.9.44) together give:

$$\rho k < -2(r - y_{2+})(r - y_{1+}). \tag{4.9.45}$$

Note that

$$r - y_{1-} = r + R_1 + R_2 > R_1 > k/2. \tag{4.9.46}$$

Suppose $z > k^{2\delta}$. Then, $R_2 < R_1$, because $a < k^\delta$. Therefore, $y_{2-} < 0$ and, naturally,

$$r - y_{2-} > r. \tag{4.9.47}$$

If $z \le k^{2\delta}$, then $R_2 - R_1 = O(k^{4\delta - 1})$ and using that $2r > \rho > k^{2\delta}$, we obtain $r - y_{2-} > r/2$. Therefore,

$$r - y_{2-} > r/2. \tag{4.9.48}$$

Substituting inequalities (4.9.45), (4.9.46) and (4.9.48) in the relation (4.9.40), we get

$$4x_3^2r^2 > \rho k^2 r/8.$$

Hence,

$$6x_3 r > k\sqrt{\rho r}.$$

Using the last relation in inequalities (4.9.32) and (4.9.33) we obtain (4.9.26). Thus, estimate (4.9.23) is proved.

4.9.4 Appendix 3

We prove the estimate

$$I_1 < ck^{2-n\delta-9\delta}. \tag{4.9.49}$$

First, suppose $\mathbf{p}_i(0)$ and $\mathbf{p}_q(0)$ are linearly independent. We introduce the cylindrical coordinates z, r, ϑ, where z is directed alone $\mathbf{p}_q(0)$, and angle ϑ is assumed to be zero on the projection of the vector $\mathbf{p}_i(0)$ on the plane $z = 0$. In these coordinates

$$\begin{gathered}\mathbf{p}_i(0) = (\gamma_2, \gamma_1, 0),\\ x_0 = (p_q(0)/2, k', \vartheta_0),\ x_0 \in S_0,\end{gathered} \tag{4.9.50}$$

where

$$k' = \sqrt{k^2 - \varphi_0(x_0) - p_q^2(0)/4}, \tag{4.9.51}$$

γ_2 is the projection of the vector $\mathbf{p}_i(0)$ onto $\mathbf{p}_q(0)$ and $\gamma_1 = \sqrt{p_i^2(0) - \gamma_2^2}$. Note that in the cylindrical coordinates $\tilde{Q}_{ni}^1$ is described as follows:

$$\begin{aligned}\tilde{Q}_{ni}^1 = \{y = (r, z, \vartheta) : \rho(y) > k^{2\delta}, \exists \vartheta_0 : r^2 - 2rk' \cos(\vartheta - \vartheta_0) + z'^2 = b,\\ |\gamma_1(r\cos\vartheta - k'\cos\vartheta_0) + \gamma_2 z'| < 4k^{1-\delta n} + k^{1+\delta}(r\rho)^{-1/2}\}\end{aligned} \tag{4.9.52}$$

$$b \equiv k^2 - k'^2 = \varphi_0(x_0) + p_q^2(0)/4.$$

Taking into account that

$$(\rho - k)^2 = (r - k')^2 + z'^2, \tag{4.9.53}$$

we get that the relation $r^2 - 2rk' \cos(\vartheta - \vartheta_0) + z'^2 = b$ can be rewritten in the form:

$$r(1 - \cos(\vartheta - \vartheta_0)) = \rho\left(\frac{k - \rho/2}{k'}\right). \tag{4.9.54}$$

Thus,

$$\begin{aligned}\tilde{Q}_{ni}^1 = \{y = (r, z, \vartheta) : \rho > k^{\delta}, \exists \vartheta_0 : r(1 - \cos(\vartheta - \vartheta_0)) = \frac{\rho(k - \rho/2)}{k'},\\ |\gamma_1(r\cos\vartheta - k'\cos\vartheta_0) + \gamma_2 z'| < 4k^{1-\delta n-2\delta} + k^{1+\delta}(r\rho)^{-1/2}\}.\end{aligned} \tag{4.9.55}$$

To calculate I_1 we change the variables: $r, z, \vartheta \to \vartheta_0, l, \vartheta$

$$l = \gamma_1 r \cos\vartheta + \gamma_2 z', \tag{4.9.56}$$

$$\cos(\vartheta - \vartheta_0) = \frac{r^2 + z'^2 - b}{2rk'}, \quad -\pi/2 \le \vartheta - \vartheta_0 < \pi/2. \tag{4.9.57}$$

We describe the set $\tilde{Q}_{ni}^1$ in the new coordinates. It follows from formula (4.9.54) and relations $k^{2\delta} < \rho(y) < k$ and $r(y) < 2k$ that

$$2(1 - \cos(\vartheta - \vartheta_0)) > k^{-1+2\delta}. \tag{4.9.58}$$

Thus,

$$\tilde{Q}^1_{ni} \subset \{y = (\vartheta_0, l, \vartheta) : 0 \le \vartheta < 2\pi, \tag{4.9.59}$$
$$|l - \gamma_1 k' \cos\vartheta_0| < 4k^{1-\delta n-10\delta} + k^{1+2\delta}(r(y)\rho(y))^{-1/2}, 2(1-\cos(\vartheta-\vartheta_0)) > k^{-1+2\delta}\}.$$

The determinant corresponding to this change of variables is given by the following calculation:

$$D(y) = \left|\begin{pmatrix} \frac{\partial\vartheta_0}{\partial r} & \frac{\partial\vartheta_0}{\partial z} \\ \frac{\partial l}{\partial r} & \frac{\partial l}{\partial z} \end{pmatrix}\right| = \frac{1}{2k'\sin(\vartheta-\vartheta_0)}\left|\begin{pmatrix} 1-\frac{z'^2-b}{r^2} & \frac{2z'}{r} \\ \gamma_1\cos\vartheta & \gamma_2 \end{pmatrix}\right| =$$

$$\frac{1}{k'r\sin(\vartheta-\vartheta_0)}|z'\gamma_1\cos\vartheta - \gamma_2 r + \frac{\gamma_2(r^2+z'^2-b)}{2r}| =$$

$$\frac{|z'\gamma_1\cos\vartheta - \gamma_2 r + \gamma_2 k'\cos(\vartheta-\vartheta_0)|}{k'r\sin(\vartheta-\vartheta_0)}. \tag{4.9.60}$$

Introducing the new coordinate η orthogonal to l in the plane of variables z' and r:

$$\eta = z'\gamma_1\cos\vartheta - \gamma_2 r, \tag{4.9.61}$$

and defining $\hat{a}$, α_1 and α_2 by the formulae: $\hat{a} = \sqrt{\gamma_1^2\cos^2\vartheta + \gamma_2^2}$, $\alpha_1 = \hat{a}^{-1}\gamma_1\cos\vartheta$, and $\alpha_2 = \hat{a}^{-1}\gamma_2$, we obtain

$$D(y) = \frac{|\hat{a}(\eta\hat{a}^{-1} + \alpha_2 k'\cos(\vartheta-\vartheta_0))|}{k'r|\sin(\vartheta-\vartheta_0)|}. \tag{4.9.62}$$

To estimate $D(y)$ we rewrite (4.9.57) in the form

$$(r - k'\cos(\vartheta-\vartheta_0))^2 + z'^2 = k'^2\cos^2(\vartheta-\vartheta_0) + b.$$

Using formulae (4.9.56) and (4.9.61) for l and η, we obtain that

$$(\hat{a}^{-1}l - \alpha_1 k'\cos(\vartheta-\vartheta_0))^2 + (\hat{a}^{-1}\eta + \alpha_2 k'\cos(\vartheta-\vartheta_0))^2 =$$

$$k'^2\cos^2(\vartheta-\vartheta_0) + b. \tag{4.9.63}$$

From this, using the relation (4.9.59) and taking into account that

$$l = \gamma_1 k'\cos\vartheta_0 + \tilde{l}, \tag{4.9.64}$$

$$|\tilde{l}| < 2k^{1-\delta n-10\delta} + k^{1+\delta}(r\rho)^{-1/2},$$

we obtain the estimate for the determinant:

$$D \ge (r\sin(\vartheta-\vartheta_0))^{-1}\sqrt{f(\vartheta,\vartheta_0,l)}, \tag{4.9.65}$$

where

$$f(\vartheta,\vartheta_0,l) =$$
$$\hat{a}^2\cos^2(\vartheta-\vartheta_0) - (\gamma_1\cos\vartheta_0 - \gamma_1\cos\vartheta\cos(\vartheta-\vartheta_0) + \tilde{l}/k')^2 + \hat{a}^2 bk'^{-2}. \tag{4.9.66}$$

Using relation (4.9.66), we get the estimate for the integral:

$$I_1 \le \int_{\tilde{Q}^1_{rm}} \frac{r^2 \sin(\vartheta - \vartheta_0)}{\sqrt{f(\vartheta, \vartheta_0, l) r\rho}} d\vartheta d\vartheta_0 dl. \tag{4.9.67}$$

It follows from relations (4.9.54), (4.8.60) and the estimate $\rho > k^{2\delta}$ that

$$k^{-1+2\delta} < r/\rho < 2(1 - \cos(\vartheta - \vartheta_0))^{-1}. \tag{4.9.68}$$

Using this relation and producing the change of the variable $\vartheta_0 \to \tau$, $\tau = \cos(\vartheta - \vartheta_0)$, we get

$$I_1 \le \int_{\tilde{Q}^1_{rm}} \frac{r}{\sqrt{f(\vartheta, \tau, l)(1-\tau)}} d\tau d\vartheta dl. \tag{4.9.69}$$

Taking into account formula (4.9.59) for $\tilde{Q}^1_{rm}$, it is not hard to show that $I_1 \le I_2 + I_3$, where I_2 and I_3 are the integrals over ω_2 and ω_3 given by:

$$\omega_2 = \{y : 0 < \vartheta \le 2\pi, |l - \gamma_1 k' \cos\vartheta_0| < 8k^{1-\delta n - 10\delta},\ 2(1-\tau) > k^{-1+2\delta}\},$$

$$\omega_3 = \{y : 0 < \vartheta \le 2\pi, |l - \gamma_1 k' \cos\vartheta_0| < k^{1+\delta}(r\rho)^{-1/2},\ 2(1-\tau) > k^{-1+2\delta}\}.$$

Integrating with respect to l and taking into account (4.9.68), we obtain

$$I_2 \le k^{2-\delta n - 10\delta} \max_l \int_0^{1-\tau_0} \frac{d\tau}{(1-\tau)^{1/2}} \int_0^{2\pi} \frac{d\vartheta}{\sqrt{f(\vartheta, \tau, l)}}, \tag{4.9.70}$$

$$I_3 \le k^{1+\delta} \max_l \int_0^{1-\tau_0} \frac{d\tau}{1-\tau} \int_0^{2\pi} \frac{d\vartheta}{\sqrt{f(\vartheta, \tau, l)}}, \tag{4.9.71}$$

where $\tau_0 = k^{-1+2\delta}/2$. It is not hard to show that

$$f(\vartheta, \tau, l) = \gamma_2^2 \tau^2 + \gamma_1^2 \tau^2 \cos^2\vartheta - \gamma_1^2 (1 - \tau^2) \sin^2\vartheta$$
$$+ 2\gamma_1 k'^{-1} \tilde{l} \sqrt{1-\tau^2} \sin\vartheta + \tilde{l}^2 k'^{-2} + k'^{-2} b (\gamma_2^2 + \gamma_1^2 \cos^2\vartheta) =$$
$$-(\mu_1 \sin^2\vartheta + 2\alpha \sin\vartheta + \beta), \tag{4.9.72}$$

where

$$\alpha = -\gamma_1 k'^{-1} \tilde{l} \sqrt{1-\tau^2}, \tag{4.9.73}$$

$$\beta = -(\gamma_1^2 + \gamma_2^2)\tau^2 - k'^{-2}\tilde{l}^2 - (\gamma_1^2 + \gamma_2^2) k'^{-2} b, \tag{4.9.74}$$

$$\mu_1 = \gamma_1^2 (1 + k'^{-2} b). \tag{4.9.75}$$

We show in Appendix 3A that integral with respect to ϑ has only logarithmic singularities in some points $\tau_n = \tau_n(l)$, $l = 1, 2, 3, 4$. Therefore,

$$\max_l \int_0^{1-\tau_0} \frac{d\tau}{\sqrt{1-\tau}} \int_0^{2\pi} \frac{d\vartheta}{\sqrt{f}} < k^{2\delta}. \tag{4.9.76}$$

Hence, $I_2 \le k^{2-\delta n - 8\delta}$. Similarly, we prove that $I_3 \le k^{1+2\delta}$. Thus,

$$I_1 < 2k^{2-\delta n-8\delta}, \quad \text{when} \quad \gamma_1 \neq 0. \tag{4.9.77}$$

If $\gamma_1 = 0$, then we direct the semiaxis $\vartheta = 0$ arbitrary. It is not hard to show $f(\vartheta, \vartheta_0, l) = |\gamma|^2\tau^2 + k'^{-2}\tilde{l}^2 + k'^{-2}|\gamma|^2 b^2$. Therefore,

$$I_2 \le k \int_{|l|<k^{1-\delta n-10\delta}} dl \int_0^{1-\tau_0} \frac{d\tau}{(1-\tau)^{1/2}(|\gamma|^2\tau^2 + k'^{-2}l^2 + k'^{-2}|\gamma|^2 b^2)^{1/2}}$$

$$\le k \int_{|l|<k^{1-\delta n-10\delta}} \ln \frac{l}{k} dl \le k^{2-\delta n-9\delta}. \tag{4.9.78}$$

Similarly, $I_3 \le k^{1+2\delta}$. Therefore, $I_1 \le k^{2-\delta n-9\delta}$ for any $\mathbf{p}_i(0)$, $p_i(0) < k^\delta$. The estimate (4.9.49) is proved.

4.9.5 Appendix 3A

We prove that for $\gamma_1 \neq 0$

$$\tilde{I} \equiv \int_0^{2\pi} f(\vartheta, \tau, l)^{-1/2} d\vartheta = \sum_{i=1}^{4} c_i \ln(\tau - \tau_i(l)) + c_0, \quad c_i \neq c_i(k), \ i = 0,1,2,3,4. \tag{4.9.79}$$

It is clear that

$$\tilde{I} = -\frac{1}{\sqrt{-\mu_1}} \int_0^{2\pi} \frac{d\varphi}{\sqrt{(\sin\varphi - b_1)(\sin\varphi - b_2)}} =$$

$$-\frac{1}{\sqrt{-\mu_1}} \int_{-1}^{1} \frac{d\xi}{\sqrt{(1-\xi^2)(\xi - b_1)(\xi - b_2)}}, \tag{4.9.80}$$

$$b_i = b_i(l, \tau), \ i = 1, 2.$$

It is easy to see from the linear independence of $\mathbf{p}_i(0)$ and $\mathbf{p}_q(0)$ that $|\gamma_1| > k^{-2\delta}$. Suppose $|\gamma||\tau||\gamma_1|^{-1} > 1/2$. Then, in formulae (4.9.73)–(4.9.75)) $\alpha\mu_1^{-1} = o(1)$, $\beta\mu_1^{-1} = \gamma_1^{-2}|\gamma|^2\tau^2 + o(1)$; therefore $b_1 \approx \gamma_1^{-1}|\gamma|\tau > 1/2$, $b_2 \approx -\gamma_1^{-1}|\gamma|\tau < -1/2$. Considering the last two equations, it is not hard to show that

$$\tilde{I} \le c \ln|1 - b_1(l,\tau)||1 + b_2(l,\tau)|.$$

We see that $\tilde{I}$ has singularities on the curves $b_1(l,\tau) = 1$ and $b_2(l,\tau) = -1$. It is not hard to show that these equation are satisfied by roots τ_{1+}, τ_{1-}, τ_{2+} and τ_{2-} of a polynomial of the fourth power and $\tau_{1,2\pm}^2 = \gamma_1^2|\gamma|^{-2} + o(1)$, The function $(1 - b_1(l,\tau))(1 + b_2(l,\tau))$ is infinitely differentiable in these points, because

$$(1 - b_1(l,\tau))(1 + b_2(l,\tau)) = (1 - \sqrt{\tilde{\alpha}^2 - \tilde{\beta}})^2 - \tilde{\alpha}^2,$$

$$\tilde{\alpha}^2 = -\mu_1^{-1}\alpha^2 = \tilde{l}^2 k'^{-2}(1 - \tau^2) = o(1),$$
$$-\tilde{\beta} = \mu_1^{-1}\beta = \gamma_1^{-2}|\gamma|^2\tau^2 + o(1) \approx 1.$$

Taking into account that $(1 - b_1(l,\tau))(1 + b_2(l,\tau))$ is differentiable, we obtain estimate (4.9.79).

If $\gamma|\gamma_1|^{-1}\tau \le 1/2$, then $|1 - b_1| > 1/4$, $|1 + b_1| > 1/4$. It follows from relation (4.9.80) that

$$\tilde{I} \le |\gamma_1|^{-2} \ln |b_1 - b_2| = \ln |\alpha^2 - \beta|.$$

Taking into account that $\alpha^2 - \beta$ is a quadratic polynomial with respect to τ, it is not hard to prove estimate (4.9.79).

4.9.6 Appendix 4 (formulae in the cylindrical coordinates for Q^k_{ni} and $\tilde{Q}^k_{ni}$, k ¿ 1)

Suppose $\mathbf{p}_i(0)$ and $\mathbf{p}_q(0)$ are linearly independent. We introduce the cylindrical coordinates z, r, ϑ, where z is directed alone $\mathbf{p}_q(0)$, and angle ϑ is assumed to be zero on the projection of the vector $\mathbf{p}_i(0)$ on the plane $z = 0$. In these coordinates

$$\begin{gathered} \mathbf{p}_i(0) = (\gamma_2, \gamma_1, 0), \\ x_0 = (p_q(0)/2, k', \vartheta_0), \ x_0 \in S_0, \end{gathered} \tag{4.9.81}$$

where k' is given by (4.8.35), γ_2 is the projection of the vector $\mathbf{p}_i(0)$ onto $\mathbf{p}_q(0)$ and $\gamma_1 = \sqrt{p_i^2(0) - \gamma_2^2}$.

We prove that Q^k_{ni} and $\tilde{Q}^k_{ni}$, $k = 2,3,4,5,6,7$ are described by similar formulae:

$$\begin{gathered} Q^k_{ni} \subset \{m : |\sqrt{(r_m - k')^2 + z'^2_m} - k| < k^\delta, \\ |\gamma_1(r_m - k)\cos\vartheta_m + \gamma_2 \hat{z}_m| < 2k^{1-\delta n - 10\delta} + k^{1+2\delta} r^{-1/2}\}; \end{gathered} \tag{4.9.82}$$

$$\begin{gathered} \tilde{Q}^k_{ni} \subset \{y = (z, r, \vartheta) : |\sqrt{(r - k')^2 + z'^2} - k| < 2k^{2\delta}, \\ |\gamma_1(r - k)\cos\vartheta + \gamma_2 z| < 2k^{1-\delta n - 10\delta} + k^{1+\delta} r^{-1/2}\}. \end{gathered} \tag{4.9.83}$$

Indeed, the polar coordinates of a point x in $\hat{S}_q(k, -\delta)$ are

$$\begin{gathered} x = (a, \hat{k}, \vartheta_0), \ \ \hat{k} \equiv \sqrt{k^2 - a^2 - \varphi_0(a)}, \\ |a| < k^\delta, 0 \le \vartheta_0 < 2\pi. \end{gathered} \tag{4.9.84}$$

If $m \in Q_{ni}$, then relation (4.8.45) holds. This means that for some x in $\hat{S}_q(k, -\delta)$

$$(r_m \cos\vartheta_m - \hat{k}\cos\vartheta_0)^2 + (r_m \sin\vartheta_m - \hat{k}\sin\vartheta_0)^2 + (z_m - a)^2$$

$$= k^2 - \varphi_0(a) + O(k^{-1+\delta n}), \tag{4.9.85}$$

$$|\gamma_1(r_m \cos\vartheta_m - \hat{k}\cos\vartheta_0) + \gamma_2(z_m - a)| < k^{1-\delta n - 10\delta}, \tag{4.9.86}$$

r_m, z_m, φ_m being the polar coordinates of the vector $-\mathbf{p}_m(0)$. If $m \in \cup_{k=2}^7 Q^k_{ni}$, then $\rho_{mq} < k^{2\delta}$, i.e., (see (4.9.53))

$$|\sqrt{(r_m - k')^2 + z'^2_m} - k| < k^{2\delta}. \tag{4.9.87}$$

Using relations (4.9.85) – (4.9.87) and the inequalities $|a| < k^\delta$, $1-\delta n - 10\delta > \delta$, we easily show that for $k = 2,3,4,5,6,7$

$$Q^k_{ni} \subset \{m : \exists\vartheta_0 : |\sqrt{(r_m - k')^2 + z'^2_m} - k| < k^{2\delta},$$

$$|\gamma_1(r_m \cos\vartheta_m - \hat{k}\cos\vartheta_0) + \gamma_2 \hat{z}_m| < 2k^{1-\delta n-10\delta},$$

$$|r^2_m - 2r_m\hat{k}\cos(\vartheta_m - \vartheta_0) + \hat{z}^2_m - a^2| < 4k^{-1+\delta n}\}, \tag{4.9.88}$$

$$\hat{z}_m = z_m - a.$$

From the last equality in the right part of (4.9.88) and (4.9.53) we obtain

$$2r_m(k' - \hat{k}\cos(\vartheta_m - \vartheta_0)) = \rho_{mq}(2k - \rho_{mq}) + O(k^{2\delta}). \tag{4.9.89}$$

From this relation we easily get:

$$r_m(1 - \cos(\vartheta_m - \vartheta_0)) < k^{3\delta}. \tag{4.9.90}$$

Thus, for $k = 2,3,4,5,6,7$

$$Q^k_{ni} \subset \{m : |\sqrt{(r_m - k')^2 + z'^2_m} - k| < k^{2\delta}, \exists\vartheta_0 : r_m(1 - \cos(\vartheta_m - \vartheta_0)) < k^{3\delta},$$

$$|\gamma_1(r_m \cos\vartheta_m - \hat{k}\cos\vartheta_0) + \gamma_2\hat{z}_m| < 2k^{1-\delta n-10\delta}\}. \tag{4.9.91}$$

From inequality (4.9.90) we obtain that $|\cos\vartheta_m - \cos\vartheta_0| < 2k^{2\delta}r^{-1/2}$. Thus, we arrive at (4.9.82). If $y = (z, r, \vartheta)$ is in the A-neighborhood of $\mathbf{p}_m(0)$, $m \in Q^k_{ni}$, $k = 2,3,4,5,6,7$, then

$$\begin{aligned} |r_m - r| &\le A, \\ |z_m - z| &\le A, \\ |\cos\vartheta_m - \cos\vartheta| &< 2A/r. \end{aligned} \tag{4.9.92}$$

It is clear now that formula for $\tilde{Q}^k_{ni}$ (A-neighborhood of Q^k_{ni}) coincides with formula (4.9.82) with respect to constant factors, namely, (4.9.83) holds.

4.9.7 Appendix 5

We prove that

$$V(\tilde{Q}^2_{ni}) < k^{1+3\delta}(k^{1-n\delta-10\delta} + k^{1/2+3\delta}). \tag{4.9.93}$$

Indeed, from (4.9.83), taking into account that $2r > k^{1-6\delta}$ when $y \in \tilde{Q}^2_{ni}$, we obtain

$$\tilde{Q}^2_{ni} \subset \{y : 0 \le \vartheta \le 2\pi, (r, z) \in M_\vartheta\}, \tag{4.9.94}$$

where M_ϑ is the intersection of a plane layer and a ring in the plane of the variables r and z:

$$M_\vartheta = \{r, z : |\sqrt{(k' - r)^2 + z'^2} - k| < 2k^{2\delta}, |\gamma_1(r - k)\cos\vartheta + \gamma_2 z| < h\}, \tag{4.9.95}$$

$$h \equiv 2k^{1-n\delta-10\delta} + 2k^{1/2+4\delta}.$$

It is clear that

$$V(\tilde{Q}_{ni}^2) \leq \int_0^{2\pi} d\vartheta \int_{M_\vartheta} r dr dz \leq k \int_0^{2\pi} s(M_\vartheta) d\vartheta, \tag{4.9.96}$$

where $s(M_\vartheta) < ck^{2\delta} \min\{h\hat{a}^{-1}, 3k\}$. Taking into account the obvious inequality

$$\min\{h\hat{a}^{-1}, 3k\} < 2h(\hat{a} + k^{-1})^{-1},$$

we obtain

$$s(M_\vartheta) < \frac{ck^{2\delta}(k^{1-n\delta-10\delta} + k^{1/2+4\delta})}{\sqrt{\gamma_1^2 \cos^2\vartheta + \gamma_2^2 + k^{-1}}}. \tag{4.9.97}$$

Using estimate (4.9.97) in the inequality (4.9.96) and integrating with respect to ϑ yields formula (4.9.93).

4.9.8 Appendix 6

We prove the estimate:

$$I_3 \leq 3k^{1/2-n\delta-9\delta} + 4k^{2\delta}. \tag{4.9.98}$$

Suppose $\mathbf{p}_i(0)$ and $\mathbf{p}_q(0)$ are linearly independent. As in Appendix 4 we use the cylindrical coordinates z, r, ϑ, where z is directed alone $\mathbf{p}_q(0)$, and angle ϑ is assumed to be zero on the projection of the vector $\mathbf{p}_i(0)$ on the plane $z = 0$. In these coordinates

$$\begin{gathered} \mathbf{p}_i(0) = (\gamma_2, \gamma_1, 0), \\ x_0 = (p_q(0)/2, k', \vartheta_0), \;\; x_0 \in S_0, \end{gathered} \tag{4.9.99}$$

where k' is given by (4.9.51), γ_2 is the projection of the vector $\mathbf{p}_i(0)$ onto $\mathbf{p}_q(0)$ and $\gamma_1 = \sqrt{p_i^2(0) - \gamma_2^2}$. Using formula (4.8.37) for the linearly independent vectors $\mathbf{p}_i(0)$, $\mathbf{p}_q(0)$, it is easy to show that $|\gamma_1| > k^{-\delta}$. We represent I_3 in the form $I_3 = I_3' + I_3''$, I_3' being the integral over the region $r > k^{3\delta}$. Let us consider I_3'. Taking into account that $|r| < 2k^{1-6\delta}$ and $\rho < 2k^{2\delta}$, from the relation (4.8.76) we obtain the estimates:

$$z = \sqrt{(2k' - r)r} + O(k^{1/2+\delta} r^{-1/2}), \tag{4.9.100}$$

$$2z > \sqrt{(2k' - r)r} > k^{1/2} r^{1/2}. \tag{4.9.101}$$

Considering (4.9.100) and the last estimate in (4.9.83) and taking into account that $|\gamma_2/\gamma_1^{-1}| < k^\delta$, we verify that

$$|\cos\vartheta - \xi_0(r)| < \xi_1(r), \tag{4.9.102}$$

$$\xi_0(r) = \gamma_2\gamma_1^{-1}\sqrt{(2k - r)r}(k - r)^{-1} = O(k^{-2\delta}),$$

$$\gamma_1\xi_1(r) = 2k^{-n\delta-10\delta} + 4k^\delta r^{-1/2}.$$

Producing in the integral I_3 the change of the variables $z \to \rho$ (see (4.9.53)), we get

$$I_3' \le k \int_{\tilde{Q}^3_{ni}} \frac{r dr d\rho d\vartheta}{z(r,\rho,\varphi)(r^2+z^2)^{3/4}}. \tag{4.9.103}$$

Taking into account estimate (4.9.101) we show that

$$I_3' \le k^{-1/4} \int_{\tilde{Q}^3_{ni}} r^{-1/4} dr d\rho d\vartheta. \tag{4.9.104}$$

Integrating with respect to ρ and ϑ (see (4.9.102)), we obtain:

$$I_3' \le k^{-1/4+\delta} \int_0^k r^{-1/4} \xi_1(r) dr. \tag{4.9.105}$$

Evaluating the integral we get:

$$I_3' \le 2k^{1/2-n\delta-9\delta} + 4k^{2\delta}. \tag{4.9.106}$$

Next, we consider I_3''. It is not hard to show that $z^2 < 2k^{1+3\delta}$ (see (4.9.53)), when $r < k^{3\delta}$. Taking into account that $2\sqrt{r^2+z^2} > k^{2\delta}$, when $(r,z,\vartheta) \in \tilde{Q}^3_{ni}$, we obtain $I_3'' < k^{1/2+7\delta}$. Therefore, (4.9.98) holds when $\mathbf{p}_i(0)$, $\mathbf{p}_q(0)$ are linearly independent, i.e., $\gamma_1 \ne 0$.

If $\gamma_1 = 0$, then it follows from formula (4.9.88) that

$$Q^3_{ni} \subset \{m : |\sqrt{(r_m - k')^2 + z_m'^2} - k| < k^{2\delta}, |z_m| < k^{1-\delta n-10\delta}\}. \tag{4.9.107}$$

From this we obtain:

$$\tilde{Q}^3_{ni} \subset \{y : |\sqrt{(r - k')^2 + z^2} - k| < 2k^{2\delta}, |z| < 2k^{1-\delta n-10\delta}\}. \tag{4.9.108}$$

It is not hard to show that

$$\tilde{Q}^3_{ni} \subset \{y : |z| < 2k^{1-\delta n-10\delta}, |r - r_0(z)| < k^{2\delta}\}, \tag{4.9.109}$$

$r_0(z)$ being uniquely determined from the relations $z^2 = r_0(2k - r_0)$, $r_0 < k^{1-3\delta}$. We use again formula (4.8.81). To estimate I_3 we again represent it as the sum $I_3' + I_3''$. Using (4.9.101), we get

$$I_3' \le k^{-3/4} \int r^{1/4} dr dz.$$

Taking into account (4.9.108), it is not hard to check estimate (4.9.106). The integral I_3'' we estimate as before. Thus, inequality (4.9.98) holds for linearly dependent $\mathbf{p}_i(0)$, $\mathbf{p}_q(0)$ too.

4.9.9 Appendix 7

We prove that

$$I_4 < k^{7/4-\delta n-6\delta} + k^{1+2\delta}. \tag{4.9.110}$$

From the relations $4k^2 - r_m^2 - z_m^2 > k^{1+3\delta}$, $\rho < k^{2\delta}$ and (4.9.53) it easily follows that $2k - r > (1/4)k^{3\delta}$. Taking into account that $2k - r < k^{1-3\delta}$ we obtain from (4.8.76):

$$|z| = \sqrt{(2k-r)r} + O(k^{1/2+2\delta}(2k-r)^{-1/2}), \tag{4.9.111}$$

$$2|z| > k^{1/2}(2k-r)^{1/2}, \tag{4.9.112}$$

$$4k^2 - r^2 - z^2 > k(2k-r). \tag{4.9.113}$$

Suppose $\mathbf{p}_i(0)$ and $\mathbf{p}_q(0)$ are linearly independent. As in Appendix 4 we used the cylindrical coordinates z, r, ϑ, where z is directed alone $\mathbf{p}_q(0)$, and angle ϑ is assumed to be zero on the projection of the vector $\mathbf{p}_i(0)$ on the plane $z = 0$. In these coordinates

$$\begin{aligned} \mathbf{p}_i(0) &= (\gamma_2, \gamma_1, 0), \\ x_0 &= (p_q(0)/2, k', \vartheta_0), \ x_0 \in S_0, \end{aligned} \tag{4.9.114}$$

where k' is given by (4.9.51), γ_2 is the projection of the vector $\mathbf{p}_i(0)$ onto $\mathbf{p}_q(0)$ and $\gamma_1 = \sqrt{p_i^2(0) - \gamma_2^2}$. Using formula (4.8.37) for the linearly independent vectors $\mathbf{p}_i(0)$, $\mathbf{p}_q(0)$, it is easy to show that $|\gamma_1| > k^{-\delta}$ Considering estimate (4.9.111), the second inequality on the right of (4.9.83) and the relation $2k - r > k^{2\delta}$, we verify that

$$|\cos\vartheta - \xi_0(r)| < \xi_2, \tag{4.9.115}$$

$$\gamma_1 \xi_2 = 2k^{-n\delta-2\delta} + 2k^{-1/2+\delta}, \tag{4.9.116}$$

ξ_0 being given by (4.9.102). Producing in the integral I_4 the change of the variables $z \to \rho$, and using estimate (4.9.113) we obtain:

$$I_4 \le k^{5/4+\delta} \int_{\tilde{Q}^4_{ni}} (2k-r)^{-3/4} dr d\rho d\vartheta. \tag{4.9.117}$$

Integrating with respect to ρ $(0 < \rho < k^\delta)$, ϑ (see (4.9.115)) and r, $0 \le r \le 2k$, we arrive at the inequality: $I_4 < k^{3/2+\delta}\xi_2$. Using (4.9.116), we get

$$I_4 < k^{3/2-n\delta-\delta} + k^{1+2\delta}, \ \ ,\gamma_1 \ne 0. \tag{4.9.118}$$

Suppose $\gamma_1 = 0$. Using (4.9.111), from relation (4.9.108) we obtain

$$\tilde{Q}^k_{ni} \subset \{y : |z| < 2k^{1-\delta n-10\delta}, |r - r_0(z)| < k^\delta\},$$

where r_0 is uniquely determined from the relations $k^{3\delta} < 2k - r_0 < k^{1-3\delta}$, $r_0(2k - r_0) = z^2$. Now, from relations (4.8.87) and (4.9.113), we get:

$$I_4 < k^{3/4} \int_{\tilde{Q}^4_{ni}} (2k-r)^{-1/4} dr dz d\vartheta < k^{7/4-\delta n-6\delta}, \ \gamma_1 = 0. \tag{4.9.119}$$

Estimates (4.9.118), (4.9.119) together give (4.9.110). subsectionAppendix 8 We prove that

$$V(\tilde{Q}_{ni}^5) < k^{7/4-\delta n-6\delta} + k^{1+2\delta}. \qquad (4.9.120)$$

It is not hard to show that the formula for $\tilde{Q}_{ni}^5$ is similar to that of $\tilde{Q}_{ni}^5$. It is clear that

$$V(\tilde{Q}_{ni}^5) = \int_{\tilde{Q}_{ni}^5} r dr dz d\vartheta.$$

We use formula (4.9.83) for $\tilde{Q}_{ni}^5$. Considering as above (see (4.9.115)), we show that $|\cos\vartheta - \xi_0| < \xi_2$. From the relation $4k^2 - r^2 - z^2 < k^{1+3\delta}$ and (4.8.91) it follows that $z^2 < k^{1+3\delta}$. Using (4.8.90), we obtain $2k - r_m = O(k^{3\delta})$. Considering this estimate in the formula (4.9.102) for ξ_0 we show that , $\xi_0 = o(1)$. Therefore, $V(\tilde{Q}_{ni}^5) < ck^{3/2+5\delta}\xi_2$ and (4.9.120) holds.

4.9.10 Appendix 9

We estimate each term of the sum (4.8.117) and obtain the estimate

$$s(\hat{\Upsilon}_q(k, V, \delta)_{4q'}) < cp_q^{-1}(0)k^{1-4\delta}, \qquad (4.9.121)$$

Using the estimate similar to (4.8.28) and considering as in the proof of Lemma 4.25, we verify that

$$\Sigma^1 < ck^{-1/5-8\delta}p_q^{-1}(0)I_1, \qquad (4.9.122)$$

$$I_1 = \int_{\tilde{Q}} (\rho(y)r(y))^{-1/2} r dr dz d\vartheta,$$

$\tilde{Q}$ being A-neighbourhood of Q. Note that

$$2k\rho - \rho^2 = 2k'r - r^2 - z^2 + O(k^{4\delta}). \qquad (4.9.123)$$

Taking into account that $r \geq \rho > k^{2\delta}$, $|z| < 2k^{1/5}$ (see (4.8.116)), we obtain $\rho > r/2$ for $y \in \tilde{Q}$. Therefore,

$$I_1 < \int_{\tilde{Q}} dr dz d\vartheta < ck^{6/5}.$$

Using this estimate in (4.9.122), we obtain

$$\Sigma^1 < cp_q^{-1}(0)k^{1-4\delta}. \qquad (4.9.124)$$

It is not hard to show, using (4.8.116) and (4.9.123) that $Q^2 = \emptyset$. Therefore, $\Sigma^2 = 0$.

We consider Σ^3. Since $m \in Q^3$, then estimate (4.8.47) is valid. From relation (4.9.123), taking into account that $\rho_{mq} < k^{2\delta}$, we obtain

$$p_m^2(0) = r_m^2 + z_m^2 = 2r_m k + O(k^{1+2\delta}). \qquad (4.9.125)$$

Taking into account this inequality, and considering that $r_m < k^{1-\delta}$, $|z| < 2k^{1/5}$, we obtain $4k^2 - p_m^2(0) > k^2$. Using the last estimate in inequality (4.8.47), we get

$$s\left(\hat{S}_q(k,-\delta) \cap \Pi'_m(k, 1/5 + 8\delta)\right) < ck^{4/5-7\delta} p_q^{-1}(0) p_m^{-3/2}(0). \tag{4.9.126}$$

Substituting estimate (4.9.126) in formula for Σ^3, we obtain:

$$\Sigma^3 \leq ck^{4/5-7\delta} p_q^{-1}(0) \sum_{m \in Q^3} p_m(0)^{-3/2}. \tag{4.9.127}$$

Since $p_m(0) > k^{2\delta}$, when $m \in Q^3$, the series on the right of estimate (4.9.127) can be estimated by the integral

$$\Sigma^3 \leq ck^{4/5-7\delta} p_q^{-1}(0) I_3. \tag{4.9.128}$$

$$I_3 = \int_{\tilde{Q}^3_{ni}} r(r^2 + z^2)^{-3/4} dr dz d\vartheta. \tag{4.9.129}$$

From the relations $\rho < k^{2\delta}$, $|z| < k^{1/5}$ and (4.9.123) it follows that $r < ck^{2\delta}$. Using that $p_m(0) > k^{2\delta}$, we obtain $I_3 < k^{1/5+2\delta}$. Therefore,

$$\Sigma^3 < cp_q^{-1}(0) k^{1-4\delta}, \tag{4.9.130}$$

We estimate Σ^4. Since $m \in Q^4$, estimate (4.8.29) is valid. Using relations $p_m(0) > r_m > k$, $p_q(0) < k^{\delta}$, we rewrite estimate (4.8.29) in the form:

$$s\left(\hat{S}_q(k,-\delta) \cap \hat{\Pi}_m(k, 1/5 + 8\delta)\right) < cp_q^{-1}(0) k^{-1/5-7\delta} (4k^2 - r_m^2 - z_m^2)^{-1/4}. \tag{4.9.131}$$

Using estimate (4.9.131) in formula for Σ^4, we obtain

$$\Sigma^4 \leq cp_q^{-1}(0) k^{-1/5-7\delta} \sum_{m \in Q^4} (4k^2 - r_m^2 - z_m^2)^{-1/4}. \tag{4.9.132}$$

It is clear that $\tilde{Q}^4$ can be described by a formula similar to the formula for Q^4. Since $4k^2 - r_m^2 - z_m^2 > k^{1+2\delta}$, when $m \in Q^4$, the series on the right of (4.9.132) can be estimated by the integral

$$\Sigma^4 \leq cp_q^{-1}(0) k^{-1/5-7\delta} I_4, \tag{4.9.133}$$

$$I_4 = \int_{\tilde{Q}^4} \frac{r dr dz d\vartheta}{(4k^2 - r^2 - z^2)^{1/4}}. \tag{4.9.134}$$

From the relations $|z| < k^{1/5}$, $\rho < k^{2\delta}$ and (4.9.123) it is easily follows that $2k - r < k^{3\delta}$. Therefore,

$$I_4 < ck$$

and

$$\Sigma^4 \le c p_q^{-1}(0) k^{1-4\delta}. \tag{4.9.135}$$

Let us consider Σ^5. We prove that $4k^2 - p_m^2(0) > k^{-1+3\delta}$. Indeed, using (4.9.53), we obtain

$$(2k' - r_m) r_m = z_m'^2 + 2\rho_{mq} k - \rho_{mq}^2 + O(k^{2\delta}).$$

Therefore,

$$4k^2 - r_m^2 - z_m^2 = (z_m'^2 + 2\rho_{mq} k - \rho_{mq}^2 + O(k^{2\delta}))(2k + r_m) r_m^{-1} - z_m^2 > z_m^2 > k^{4\delta}. \tag{4.9.136}$$

From this $2k - p_m(0) > k^{-1+3\delta}$. Thus, estimate (4.8.29) is valid. Using relations $p_m(0) > r_m > k$, $p_q(0) < k^{\delta}$, (4.8.91) and (4.9.135) we obtain

$$\Sigma^5 \le k^{-1/5-7\delta} p_q^{-1}(0) \sum_{Q^5} 1. \tag{4.9.137}$$

It is clear that

$$\Sigma^5 \le k^{-1/5-7\delta} p_q^{-1}(0) V(\tilde{Q}^5), \tag{4.9.138}$$

$\tilde{Q}^5$ being the A-neighbourhood of Q^5. Obviously, the formula for $\tilde{Q}^5$ is similar to that for $\tilde{Q}^5$ and

$$V(\tilde{Q}^5) = \int_{\tilde{Q}^5} r dr dz d\vartheta.$$

From the relation $4k^2 - r^2 - z^2 < k^{1+3\delta}$ (see (4.8.123)) and (4.9.53) it follows that

$$2k - r = -\rho + \rho^2/2k + O(k^{3\delta}) = O(k^{3\delta}).$$

Taking into account that $|z| < k^{1/5}$, we obtain

$$V(\tilde{Q}^5) < c k^{6/5+3\delta}.$$

Using this estimate in (4.9.138), we get

$$\Sigma^5 < c p_q^{-1}(0) k^{1-4\delta}. \tag{4.9.139}$$

The sums Σ_6 and Σ_7 can be estimated just as Σ_{in}^6, Σ_{in}^7 in Lemma 4.25. Adding the inequalities for Σ^k we get (4.9.121).

4.10 The Main Results for a Nonsmooth Potential.

In this section we consider $V(x)$ all of whose smoothness requirements are contained in the condition of convergence of the series:

$$\sum_{m \in Z^3, m \neq 0} |v_m|^2 |m|^{-1+\beta} \tag{4.10.1}$$

for some positive β. This class contains not only smooth functions. For example, a function, with a singularity of the form x_0 as $| x - x_0 |^{-\zeta}$, $\zeta < 2$, in particular

, the Coulomb potential, is contained in this class. As in the case of a smooth potential we construct the perturbation series with respect to the model operator $\hat{H}(t)$. We prove that the corresponding perturbation series converge. Moreover, initial terms of the series with respect to $\hat{H}(t)$ can be replaced by the analogous terms of the series with respect to $H_0(t)$, the accuracy of this replacement being determined by the smoothness of potential, i.e., by the value of β for which the series (4.10.1) converges.

Geometric considerations play an important role in justifying the convergence of the series. As before they consist in describing the nonsingular set of quasimomenta on which the perturbation series converge. Without going into details, this nonsingular set is the intersection of the nonsingular set $\chi_2(k, V, \delta)$ for the polyharmonic operator in the case $n = 3$ (see Section 3.6) and the nonsingular set for the Schrödinger operator with a potential being a trigonometric polynomial (Section 4.4). In Section 4.10 we describe a set $\hat{\chi}_4(k, \rho, \delta, V)$, of an asymptotically full measure on which perturbation series converge for a nonsmooth potential.

Now, we formulate the main results precisely. Let

$$V_\rho = \sum_{|m|<k^\rho} v_m \exp(i(\mathbf{p}_m(0), x)), \tag{4.10.2}$$

where $0 < \rho < 10^{-2}$. The trigonometric polynomial V_ρ is close to V in the sense that

$$\sum_{|m|\geq k^\rho} | v_m |^2 | m |^{-1+\beta} \to_{k\to\infty} 0.$$

Let $V_{\rho q}$ is the trigonometric polynomial (4.2.2) for $q \in \Gamma(R_0)$, $R_0 = k^\rho$. It is clear that

$$V_\rho = \sum_{q\in\Gamma(k^\rho)} V_{\rho q}.$$

We define $\hat{H}(t)$ by the formula:

$$\hat{H}(t) = H_0(t) + \sum_{q\in\Gamma(k^\rho)} P_q(1/5 - \delta\nu) V_{\rho q} P_q(1/5 - \delta\nu), \tag{4.10.3}$$

where $0 < \rho < 10^{-2}$, $\nu = \max\{1, \beta^{-1}\}$, $0 < \delta\nu < 1/200$. We use the notation (4.2.7), i.e.,

$$W = V_\rho - \sum_{q\in\Gamma(k^\rho)} P_q(1/5 - \delta\nu) V_{\rho q} P_q(1/5 - \delta\nu), \tag{4.10.4}$$

Theorem 4.8 . *Suppose* $0 < \rho < 1/100$, $0 < \delta\nu < 1/200$ *and* t *is in the* $(k^{-2-2\delta})$*-neighbourhood of the nonsingular set* $\chi_4(k, \rho, \delta, V)$. *Then for sufficiently large* k, $k > k_0(V, \delta)$, *in the interval* $\varepsilon(k, \delta) \equiv [k^2 - k^{-1-\delta}, k^2 + k^{-1-\delta}]$ *there exists a unique eigenvalue of the operator* H. *It is given by the series*

$$\lambda(t) = p_j^2(t) + \sum_{r=2}^{\infty} \hat{g}_r(k, t), \tag{4.10.5}$$

where the index j is uniquely determined from the relation $p_j^2(t) \in \varepsilon(k,\delta)$. The spectral projection corresponding to $\lambda(t)$ is determined by the series

$$E(t) = E_j + \sum_{r=1}^{\infty} \hat{G}_r(k,t), \tag{4.10.6}$$

which converges in the class $\mathbf{S}_1$.

The following estimates hold for the functions $\hat{g}_r(k,t)$ and the operator-valued functions $\hat{G}_r(k,t)$:

$$| T(m)\hat{g}_r(k,t) | < m!k^{-1-\gamma_6 r+2|m|(1+\delta)} \tag{4.10.7}$$

$$\|T(m)\hat{G}_r(k,t)\|_1 < m!k^{-\gamma_6 r+2|m|(1+\delta)} \tag{4.10.8}$$

$$6\gamma_6 = \min\{\rho\beta - 170\delta\nu, 1/5 - 15\rho - 20\delta\nu\}. \tag{4.10.9}$$

Corollary 4.12 . *The perturbed eigenvalue and the corresponding spectral projection satisfy the following estimates:*

$$| T(m)(\lambda(t) - p_j^2(t)) | \le cm!k^{-1-2\gamma_6+2|m|(1+\delta)}, \tag{4.10.10}$$

$$\|E(t) - E_j\|_1 \le cm!k^{-\gamma_6+2|m|(1+\delta)}. \tag{4.10.11}$$

Theorem 4.8 is inconvenient in that the asymptotic terms $\hat{g}_r$, $\hat{G}_r$ depend on V in a rather complicated manner. It turns out, however, that some number of first terms of the series can be replaced with sufficient accuracy by g_r, G_r (see estimates (4.12.88) and (4.12.87)). This makes it possible to replace initial segments of the series for $\hat{H}(t)$ by the analogous segments of the series for $H_0(t)$. The accuracy of this replacement increases together with the smoothness of potential, i.e., with β.

Theorem 4.9 . *Under the conditions of Theorem 4.8 the eigenvalue $\lambda(t)$ and its spectral projection $E(t)$ can be represented in the form*

$$\lambda(t) = p_j^{2l}(t) + \sum_{r=2}^{2m_0} g_r(k,t) + \varphi_1(k,t), \tag{4.10.12}$$

$$E(t) = E_j + \sum_{r=1}^{m_0} G_r(k,t) + \psi_1(k,t), \tag{4.10.13}$$

$$m_0 = [(\beta/2 - \alpha_0)(\gamma_6 - 2\delta)^{-1}] + 1, \quad \alpha_0 = \beta(3\rho + \delta)/2.$$

For the functions g_r and the operator-valued functions G_r the estimates

$$| T(m)g_r(k,t) | < m!k^{-1-\delta-\gamma_7 r+2|m|(1+\delta)}, \tag{4.10.14}$$

$$\|T(m)G_r(k,t)\| < m!k^{-\gamma_6 r+2|m|(1+\delta)}, \tag{4.10.15}$$

$$\gamma_7 = \beta/(5\beta + 100),$$

are fulfilled, while $\varphi(k,t)$, $\psi(k,t)$ satisfy:

$$| T(m)\varphi(k,t) | < m!k^{-1-\beta+\nu_0\delta+2\alpha_0}k^{2(1+\delta)|m|} \tag{4.10.16}$$

$$\|T(m)\psi(k,t)\|_1 < m!k^{-\beta/2+\nu_0\delta+\alpha_0}k^{2(1+\delta)|m|} \tag{4.10.17}$$

$$\nu_0 = 300\nu\beta\gamma_7^{-1}.$$

We obtain estimates (4.10.15) and (4.10.16) adding estimates (4.10.7) and (4.10.8) for $\hat{g}_r(k,t)$, $\hat{G}_r(k,t)$, when $r > 2m_0$ and $r > m_0$, correspondingly.

4.11 Geometric Constructions for a Nonsmooth Potential.

Geometric constructions play an important role in justifying the convergence of perturbation series. They consist in describing the nonsingular set on which the convergence conditions are satisfied. Here such constructions will be carried out for the case of a nonsmooth potential.

We recall that the operator $(-\Delta)^l + V$ in $L_2(R^n)$, $4l > n+1$, $n > 1$ with a nonsmooth potential was studied in in Sections 4.5-4.7. In particular, for the case when $n = 3$, $l > 1$ we have constructed a set $\chi_2(k,V,\delta)$ of an asymptotically full measure on $\mathcal{K}S_k$, such that for all t in the $(k^{-2-2\delta})$-neighbourhood of it we have:

$$\max_{z\in C_0} \|A(z,t,V)\|_2^2 < k^{4(l-1+\delta)}\|V\|_*, \tag{4.11.1}$$

$$\|V\|_* = \sum_{m\in Z^3} |v_m|^2|m|^{-4l+3}. \tag{4.11.2}$$

It is essential that $\chi_2(k,V,\delta)$ is one and the same for all $l > 1$. We consider this set in the case $l = 1$. In this case inequality (4.11.1) takes the form

$$\max_{z\in C_0} \|A(z,t,V)\|_2^2 < k^{4\delta}\|V\|_* \tag{4.11.3}$$

and convergence is not ensured. We consider

$$\tilde{\chi}_2(k,V,\delta) = \chi_2(k,V,\delta) \cap \left(\cap_{p=1}^{P}\chi_2(k,V-V_{p\delta\nu},\delta)\right), \quad P = [(\delta\nu)^{-1}], \tag{4.11.4}$$

where $V_{p\delta\nu}$ is defined by (4.10.2) for $\rho = p\delta\nu$. It is clear that $\tilde{\chi}_2(k,V,\delta)$ possesses an asymptotically full measure on $\mathcal{K}S_k$. In the $(k^{-2-2\delta})$-neighbourhood of $\tilde{\chi}_2(k,V,\delta)$ the inequalities (4.11.3) are satisfied, and

$$\max_{z\in C_0} \|A(z,t,V-V_{p\delta\nu})\|_2^2 < k^{-p\delta\nu\beta+4\delta}, \quad 1 \le p \le P. \tag{4.11.5}$$

These relations follow immediately from (4.11.3) and the obvious inequality

$$\|V - V_{p\delta\nu}\|_* < ck^{-p\delta\nu\beta}. \tag{4.11.6}$$

The nonsingular set $\chi_3(k,V,\delta)$ for the Schrödinger operator was described in Section 4.4 for the case of a trigonometric polynomial. In constructing $\chi_3(k,V,\delta)$

there arise the new objects $\Upsilon_4'(k,\xi,\eta,V_q)$ (see (4.4.1)) and $\Upsilon_5(k,\eta,\nu,\varepsilon,\delta)$ (see (4.4.21)). We consider the set $\tilde{\Upsilon}_4(k,V,\delta)$ given by

$$\tilde{\Upsilon}_4(k,V,\delta) = \cap_{q\in\Gamma(k^{5\delta})} \cap_{r=1}^{[1/5\delta]} \Upsilon_4'(k,(r+42)\delta, r\delta, V_q). \tag{4.11.7}$$

Lemma 4.31 . *If t is in the (k^{-2})-neighbourhood of $\tilde{\Upsilon}_4(k,V,\delta)$ then for all q in $\Gamma(k^{5\delta})$ the condition*

$$|p_m^2(t) + \Delta\hat{\Lambda}_{mm}(t) - k^2| > (1 + |(\mathbf{p}_m(0), \mathbf{p}_q(0))|)^{-1} k^{-43\delta}. \tag{4.11.8}$$

is satisfied when $m \in \Pi_q(k^{1/5})$. The set $\tilde{\Upsilon}_4(k,V,\delta)$ has an asymptotically full measure on $\mathcal{K}S_k$. Moreover,

$$s(\mathcal{K}S_k \setminus \tilde{\Upsilon}_4(k,V,\delta))/s(\mathcal{K}S_k) < k^{-\delta}. \tag{4.11.9}$$

Proof. It is obvious that the set $\Pi_q(k^{1/5})$ can be represented as the union of expanding sets:

$$\Pi_q(k^{1/5}) \subset \cap_{r=1}^{[1/(5\delta)]+1} \Pi_q(k^{r\delta}), \quad \Pi_q(k^{(r+1)\delta}) \supset \Pi_q(k^{r\delta}). \tag{4.11.10}$$

Let $m \in \Pi_q(k^{(r+1)\delta}) \setminus \Pi_q(k^{r\delta})$. Since t is in the (k^{-2})-neighbourhood of $\Upsilon_4'(k,(r+42)\delta, r\delta, V_q)$, we have

$$|p_m^2(t) + \Delta\hat{\Lambda}_{mm}(t) - k^2| > k^{-(42+r)\delta}, \tag{4.11.11}$$

here we use the notation (4.5.68). Noting that $|(\mathbf{p}_m(0),\mathbf{p}_q(0))| > k^{(r-1)\delta}$ ($m \notin \Pi_q(k^{(r-1)\delta})$), we obtain (4.11.8). Substituting the concrete values $\xi = (r+42)\delta$ and $\eta = r\delta$ into (4.4.3), we see that

$$s(\mathcal{K}S_k \setminus \Upsilon_4'(k,(r+42)\delta, r\delta, V_q))/s(\mathcal{K}S_k) < k^{-16\delta}. \tag{4.11.12}$$

Noting that

$$s(\mathcal{K}S_k \setminus \tilde{\Upsilon}_4(k,V,\delta) \le \sum_{r,q} s(\mathcal{K}S_k \setminus \Upsilon_4'(k,(r+42)\delta, r\delta, V_q))$$

$$\le ck^{15\delta} \max s(\mathcal{K}S_k \setminus \Upsilon_4'(k,(r+42)\delta, r\delta, V_q)), \tag{4.11.13}$$

we obtain (4.11.9). The lemma is proved.

A point t belongs to $\Upsilon_5(k,\eta,\nu,\varepsilon,\delta)$ if and only if the inequality

$$|p_m^2(t) - k^2| > k^{\varepsilon - 10\rho} \tag{4.11.14}$$

is satisfied for any m in $T(k,\varepsilon,\nu,\rho)$:

$$T(k,\varepsilon,\nu,\rho) = \tag{4.11.15}$$

$$\{j : \exists q,q' \in \Gamma(k^\rho), q \neq q' : |(\mathbf{p}_j(0),\mathbf{p}_q(0))| < k^\varepsilon, |(\mathbf{p}_j(0),\mathbf{p}_q(0))| < k^\nu\}.$$

According to Lemma 4.3 this set has an asymptotically full measure on $\mathcal{K}S_k$.

Let us define a set $\Upsilon_6(k,\rho,\delta) \subset \chi_2(k,V,\delta)$. Note that for every $t \in \chi_2(k,V,\delta)$ there is a unique j such that $\mathbf{p}_j(t) = k$. We define $\Upsilon_6(k,\rho,\delta)$ as follows: $t \in \Upsilon_6(k,\rho,\delta)$ if and only if for any q in $\Gamma(k^\rho)$

$$|(\mathbf{p}_j(0), \mathbf{p}_q(0))| > k^{1-2\rho-\delta}. \tag{4.11.16}$$

It easily follows from Lemma 3.4 that $\Upsilon_6(k,\rho,\delta)$ is a set of an asymptotically full measure on $\mathcal{K}S_k$.

For the nonsingular set for the Schrödinger operator with a nonsmooth potential we propose

$$\chi_4(k,\rho,\delta,V) =$$

$$\tilde{\chi}_2(k,\delta,V) \cap \tilde{\Upsilon}_4(k,\delta\nu,V) \cap \Upsilon_5(k,1/5,3/5\nu,1/5-10\rho,\rho) \cap \Upsilon_6(k,\rho,\delta), \tag{4.11.17}$$

here $0 < \rho < 10^{-2}$, $0 < \delta < 10^{-2}\rho$. From the foregoing arguments we easily obtain

Lemma 4.32 . *If t is in the $(k^{-2-2\delta})$-neighbourhood of $\chi_4(k,\rho,V,\delta)$ then inequalities (4.11.3) and (4.11.5) hold. Inequality (4.11.14) holds for $\varepsilon = 1/5$ and m in $T(k,\rho,1/5,3/5)$; (4.11.8) holds for all q in $\Gamma(k^{5\delta})$ and $m \in \Pi_q(k^{1/5})$ and (4.11.16) holds for $q \in \Gamma(k^\rho)$. The set $\chi_4(k,\rho,V,\delta)$ has an asymptotically full measure on $\mathcal{K}S_k$:*

$$s(\mathcal{K}S_k \setminus \chi_4(k,\rho,V,\delta))/s(\mathcal{K}S_k) < k^{-\delta}. \tag{4.11.18}$$

4.12 Proof of Convergence of the Perturbation Series.

To construct the series we use the formula:

$$(H(t)-z)^{-1} = (\hat{H}(t)-z)^{-1/2}(I-\hat{A}(z,W+V_2,t))^{-1}(\hat{H}(t)-z)^{-1/2}, \tag{4.12.1}$$

where the operators $\hat{H}(t)$, W are given by formulae (4.10.3), (4.10.4),

$$\hat{A}(z,t,W+V_2) = (\hat{H}(t)-z)^{-1/2}(W+V_2)(\hat{H}(t)-z)^{-1/2}, \tag{4.12.2}$$

$$V_2 = V - V_\rho.$$

Formally expanding the resolvent in a series in powers of $\hat{A}(z,W+V_2,t)$ and arguing as in the proof of Theorem 4.1, we arrive at (4.10.5) and (4.10.6). To justify the relations it suffices to verify the inequalities

$$\|\hat{A}^3(z,W+V_2,t)\| < ck^{-6\gamma_6}, \tag{4.12.3}$$

$$\|\hat{A}(z,W+V_2,t)\| < ck^{2\delta}, \quad z \in C_0 \tag{4.12.4}$$

in the $(k^{-2-2\delta})$-neighbourhood of the nonsingular set. We represent $\hat{A}(z,W+V_2,t)$ in the form:

$$\hat{A}(z,W+V_2,t) = \hat{A}(z,W,t) + U^*B(z)A(z,t,V_2)B(\overline{z})^*U, \tag{4.12.5}$$

$$A(z,t,V_2) = (H_0(t) - z)^{-1/2}V_2(H_0(t) - z)^{-1/2}, \quad (4.12.6)$$

$$B(z) = (H_0(t) + \Delta\hat{A}(t) - z)^{-1/2}U(H_0(t) - z)^{1/2}, \quad (4.12.7)$$

where U is the unitary operator reducing $\hat{H}(t)$ to diagonal form. It differs from I only on the subspace $\sum_{q\in\Gamma(k^\rho)} P_q l_2^3$. Estimates (4.12.3) and (4.12.4) follow easily from (4.12.2) and the inequalities:

$$\|A(z,V_2,t)\| < ck^{-6\gamma_6-164\delta\nu}, \quad (4.12.8)$$

$$\|\hat{A}^3(z,W,t)\| < ck^{-6\gamma_6}, \quad (4.12.9)$$

$$\|\hat{A}(z,W,t)\| < ck^{2\delta}, \quad (4.12.10)$$

$$\|B(z)\| < ck^{80\delta\nu}. \quad (4.12.11)$$

Lemma 4.33 is devoted to the proof of (4.12.8), Lemma 4.34 to the proof of (4.12.9) and (4.12.10), and Lemma 4.35 to the proof of (4.12.11). Rigorous proofs of Theorems 4 and 5 then follow.

Lemma 4.33 . *If t is in the $(k^{-2-2\delta})$-neighbourhood of $\chi_4(k,\rho,\delta,V)$, then*

$$\|A(z,t,V)\| < k^{2\delta}, \quad (4.12.12)$$

$$\|A(z,t,V_2)\| < k^{-\rho\beta+5\delta}, \quad z \in C_0. \quad (4.12.13)$$

This assertion is a simple consequence of (4.11.5) and the relation

$$\|A(z,t,V-V_\rho)\|_2 \le \|A(z,t,V-V_{p\delta\nu})\|_2, \quad p = [\rho(\delta\nu)^{-1}].$$

Note that (4.12.8) follows from (4.12.13) and the definition of γ_6 (see (4.10.9).

Lemma 4.34 . *If t is in the $(k^{-2-2\delta})$-neighbourhood of $\chi_4(k,\rho,\delta,V)$, then*

$$\|\hat{A}^3(z,W,t)\| < ck^{-1/5+15\rho+15\delta\nu}, \quad (4.12.14)$$

$$\|\hat{A}(z,W,t)\| < ck^{2\delta}. \quad (4.12.15)$$

The proof is based on (4.11.3), (4.11.8) and (4.11.14) and is a repetition of the proof of Lemma 4.14 up to the replacement of $\Gamma(k^\delta)$ by $\Gamma(k^\rho)$.

The operator B differs from I only on the subspace $\sum_{q\in\Gamma(k^\rho)} P_q l_2^3$. Therefore in the next lemma in estimating $\|B\|$, we assume that B acts in this subspace.

Lemma 4.35 . *If t is in the $(k^{-2-2\delta})$ neighbourhood of $\chi_4(k,\rho,\delta,V)$, then*

$$\|B(z)\| < ck^{80\delta\nu}. \quad (4.12.16)$$

The proof of the lemma reduces essentially to verification of the inequalities

$$|U_{ij}| \ll 1. \quad (4.12.17)$$

in the case where $|p_i^2(t) + \Delta\hat{A}(t)_{ii} - z| \ll |p_j^2(t) - z|$. Obviously, to estimate $\|B\|$ it is required to sum inequalities of the type (4.12.17). In order that the sum not be too large in a number of cases we must prove the estimates (4.12.17) in a rather strong version. Using the definition of the circle $|k^2 - z| = k^{-1-\delta}$ it is not hard to show that (4.12.16) easily follows from $\|B(k^2)\| < ck^{80\delta\nu}$. So, it suffices to prove (4.12.16) for $z = k^2$.

From the construction of $\hat{H}(t)$ it follows that B is the orthogonal sum of the operators $B_i^q = P_i^q B P_i^q$, here P_i^q is the diagonal projection defined in Section 4.2 (see (4.3.24)). It is obvious that $\|B\| < \max_{i,q} \|B_i^q\|$. We shall estimate $\|B_i^q\|$. Below we drop the indices i and q on P_i^q and B_i^q to simplify notation. However, for us it is essential that in this point we have actually gone over from l_2^3 to l_2^1, because we have already established the isomorphism between $P_i^q l_2^3$ and l_2^1 by the mapping described in Section 4.2 (see (4.3.25)). For this reason we will write l_2^1 instead of $P_i^q l_2^3$. Thus, we represent B in the form:

$$B = B_1 + B_2 + B_3, \tag{4.12.18}$$

$$B_1 = P(k^{10\delta\nu})B,$$

$$B_2 = P_0(I - P(k^{10\delta\nu}))B,$$

$$B_3 = (I - P_0)(I - P(k^{10\delta\nu}))B,$$

where projection $P \equiv P_i^q(k^{1/5})$, $P_i^q(k^{1/5})$ being determined by formula (4.5.13), P_0 is a diagonal projection:

$$(P_0)_{jj} = \begin{cases} 1 & \text{when } |p_j^2(t) - k^2| \le |(\mathbf{p}_j, \mathbf{p}_q(0))|^{1-\delta} \\ 0 & \text{otherwise.} \end{cases}$$

We first prove the lemma under the assumption that $|q| < k^{5\delta\nu}$.

We consider B_1. It is obvious that

$$\|B_1(k^2)\|_2^2 = \sum_{m,j\in\sigma} |U_{jm}|^2 |p_m^2(t) - k^2||p_j^2(t) + \Delta\hat{\Lambda}_{jj} - k^2|^{-1},$$

$$\sigma = \{m, j : \delta_m \in P(k^{1/5})l_2^1, \delta_j \in P(k^{10\delta\nu})l_2^1\}.$$

Noting that $|U_{jm}|^2 = (\hat{E}_j)_{mm}$ (here $\hat{E}_j$ is the spectral projection of the operator $\hat{H}_{qi} \equiv P_i^q \hat{H} P_i^q$), we obtain:

$$\|B_1\|_2^2 = \sum_{m,j\in\sigma} (\hat{E}_j)_{mm} |p_m^2(t) - k^2||p_j^2(t) + \Delta\hat{\Lambda}_{jj} - k^2|^{-1}. \tag{4.12.19}$$

It is obvious that

$$\sum_{j:\delta_j \in P(k^{10\delta\nu})l_2^1} (\hat{E}_j)_{mm} \le (2\pi i)^{-1} \oint_{C_3} (\hat{H}_{qi} - z)_{mm}^{-1} dz, \tag{4.12.20}$$

where C_3 is a circle in the complex plane of radius $R \approx k^{20\delta\nu}$, containing in its interior all points $p_j^2(t)$, $\delta_j \in P(k^{10\delta\nu})l_2^1$. We may assume, that its center is located at some point $p_b^2(t)$, $\delta_b \in P(k^{\delta\nu})l_2^1$, i.e.,

$$|(\mathbf{p}_b(0), \mathbf{p}_q(0))| < k^{\delta\nu}. \tag{4.12.21}$$

We choose R so that all points $p_l^2(t)$, $\delta_l \in P(k^{10\delta\nu})l_2^1$ lie sufficiently far from contour: $\min_{z\in C_3} |p_l^2(t) - z| > R^{(1-\delta)/2}$. Such a contour can always be chosen, since the situation, in fact, is one-dimensional. We expand the resolvent on the contour in the perturbation series. For this it is necessary to estimate the norm of the operator

$$A_1(z,t) = P(k^{1/5})(H_0 - z)^{-1/2} V (H_0 - z)^{-1/2} P(k^{1/5}). \tag{4.12.22}$$

It turns out that

$$\|A_1(z,t)\|_2^2 < c\min\{R^{-(1-\delta)/2}, R^{-\beta+\delta}\}. \tag{4.12.23}$$

Indeed,

$$\|A_1(z,t)\|_2^2 = \sum_{m,l:\delta_m,\delta_l \in P(k^{1/5})l_2^1} a_{lm}(z,t), \tag{4.12.24}$$

$$a_{lm}(z,t) = \frac{|v_{l-m}|^2}{|p_l^2(t) - z||p_m^2(t) - z|}.$$

It is obvious that

$$\|A_1(z,t)\|_2^2 = \sum_{m,l:\delta_m,\delta_l \in P(k^{1/5})l_2^1} \frac{|v_{l-m}|^2}{|l-m|^{1-\beta}} b_{m,l-m}; \tag{4.12.25}$$

$$b_{m,l-m} = \frac{|l-m|^{1-\beta}}{|p_l^2(t) - z||p_m^2(t) - z|}.$$

Introducing a new summation variable and considering the condition on the potential (Section 4.9), we obtain

$$\|A_1(z,t)\|_2^2 < c\max_r \sum_m b_{m,r}. \tag{4.12.26}$$

We break the region of summation into two:

$$\Xi_1 = \{m : |p_m^2(t) - p_{m+r}^2(t)| > r^{1-\beta}\},$$

$$\Xi_2 = \{m : |p_m^2(t) - p_{m+r}^2(t)| \le r^{1-\beta}\}.$$

In the first case we represent $b_{m,r}$ in the form:

$$b_{m,r} = \left| \frac{|r|^{1-\beta}}{(p_m^2(t) - z)(p_m^2(t) - p_{m+r}^2(t))} - \frac{|r|^{1-\beta}}{(p_{m+r}^2(t) - z)(p_m^2(t) - p_{m+r}^2(t))} \right|.$$

Considering the definition of Ξ_1, we obtain

$$\sum_{m\in\Xi_1} b_{mr} \le 2 \sum_{\delta_m \in P(k^{1/5})l_2^1} |p_m^2(t) - z|^{-1}. \tag{4.12.27}$$

Since

$$\min_{\delta_m \in P(k^{1/5})l_2^1} |p_m^2(t) - z|^2 > R^{1-\delta}, \tag{4.12.28}$$

the sum on the right can be estimated in terms of an integral:

$$\sum_{\delta_m \in P(k^{1/5})l_2^1} |p_m^2(t) - z|^{-1} <$$

$$c \int_{|x^2 - x_0^2| > R^{1-\delta}} \frac{dx}{x^2 - x_0^2} + cR^{-(1-\delta)/2} < 2cR^{-1/2+\delta}, \quad x_0^2 \approx R. \tag{4.12.29}$$

We consider Ξ_2. We shall prove that the number of points contained in it does not exceed some finite N not depending on r or q. Indeed, noting that $\mathbf{p}_r(0) = n\mathbf{p}_q(0)$, $n \in Z$, we obtain $r = nq$. The inequality defining Ξ_2 can be written in the form:

$$2|(\mathbf{p}_m(t), \mathbf{p}_q(0)) + np_q(0)^2| < c|n|^{-\beta} p_q(0)^{1-\beta}, \quad c \neq c(n, q).$$

Since the situation is one-dimensional ($\delta_m \in P_i^q l_2^3$),

$$\mathbf{p}_m - \mathbf{p}_b = n_1 \mathbf{p}_q(0), \quad n_1 \in Z, \tag{4.12.30}$$

$p_b^2(t)$ being the center of C_3. From this it follows that the number of points m satisfying the last inequality does not exceed some finite N, $N \neq N(n, q)$, and $(\mathbf{p}_m(t), \mathbf{p}_q(0)) \approx np_q(0)^2 \approx rp_q(0)$. Considering this, we obtain

$$\sum_{m \in \Xi_2} b_{mr} < c \max_{m \in \Xi_2} \frac{|(\mathbf{p}_m(t), \mathbf{p}_q(0))|^{1-\beta} p_q(0)^{-1+\beta}}{|(p_m^2(t) - z)(p_{m+r}^2(t) - z)|}, \quad c \neq c(r, q, k). \tag{4.12.31}$$

If $|(\mathbf{p}_m, \mathbf{p}_q(0))| < 8Rp_q(0)$, then, using inequality (4.12.28), we verify that the right side of (4.12.31) is bounded by $cR^{-\beta+\delta}$. If $|(\mathbf{p}_m(t), \mathbf{p}_q(0))| \geq 8Rp_q(0)$, then, taking into account (4.12.30) and (4.12.21), we obtain:

$$2|p_m^2(t) - z| > 2|p_m^2(t) - p_b^2(t)| - 2R = |2(\mathbf{p}_m, \mathbf{p}_q(0))^2 p_q(0)^{-2} - 2(\mathbf{p}_b, \mathbf{p}_q(0))^2 p_q(0)^{-2}| - 2I$$

$$2(\mathbf{p}_m, \mathbf{p}_q(0))^2 p_q(0)^{-2} - 4R^2 > (\mathbf{p}_m, \mathbf{p}_q(0))^2 p_q(0)^{-2}.$$

Considering the last inequality and (4.12.28), we see that the right side of (4.12.31) does not exceed R^{-1}. From this we obtain (4.12.23). From the regularity of the resolvent on the contour C_3 uniformly with respect to V, we easily get:

$$|\Delta \hat{A}_q(t)_{jj}| < k^{20\delta\nu}. \tag{4.12.32}$$

We represent $\|B_1\|_2^2$ (see (4.12.19)) in the form

$$\|B_1\|_2^2 = \Sigma_1 + \Sigma_2, \tag{4.12.33}$$

$$\Sigma_1 = \sum_{j: \delta_j \in P(k^{10\delta\nu})l_2^1} (p_j^2(t) - k^2)|p_j^2(t) + \Delta\hat{A}(t)_{jj} - k^2|^{-1} Sp(\hat{E}_j \chi),$$

$$\Sigma_2 = \sum_{j,m\in\sigma} |p_j^2(t) + \Delta\hat{\Lambda}(t)_{jj} - k^2|^{-1} \sum_m (p_m^2(t) - p_j^2(t))(\hat{E}_j)_{mm}\chi_{mm},$$

where χ is the diagonal matrix with elements equal to ± 1:

$$\chi_{mm} = \operatorname{sgn}(p_m^2(t) - k^2).$$

Since $t \in \chi_4(k,\rho,\delta,V) \subset \tilde{\Upsilon}_4(k,\delta\nu,V)$ (see (4.11.17), the relation

$$|p_m^2(t) + \Delta\hat{\Lambda}_{mm}(t) - k^2| > (1 + |(\mathbf{p}_m(0),\mathbf{p}_q(0))|)^{-1}k^{-43\delta\nu}. \tag{4.12.34}$$

holds. Setting in this inequality $m = j$ and taking into account that $|(\mathbf{p}_j(t),\mathbf{p}_q(0))| < k^{10\delta\nu}$, we obtain:

$$2|p_j^2(t) + \Delta\hat{\Lambda}(t)_{jj} - k^2| > k^{-53\delta\nu}. \tag{4.12.35}$$

From (4.12.32), (4.12.35) it follows that $|p_j^2(t) - k^2| < 2k^{20\delta\nu}$ and

$$|p_j^2(t) - k^2||p_j^2(t) + \Delta\hat{\Lambda}(t)_{jj} - k^2|^{-1} < k^{74\delta\nu}.$$

Noting that the summation over j contains the order of $k^{10\delta\nu}$ terms, and using the inequality $|\mathrm{Tr}(\hat{E}_j\chi)| < 1$, we obtain $\Sigma_1 < ck^{84\delta\nu}$.

In Σ_2 we break the region of summation σ into two (σ_1 and σ_2):

$$\sigma_1 = \{m,j : \delta_m \in P(k^{30\delta\nu})l_2^1, \delta_j \in P(k^{10\delta\nu})l_2^1\}.$$

$$\sigma_2 = \{m,j : \delta_m \notin P(k^{30\delta\nu})l_2^1, \delta_j \in P(k^{10\delta\nu})l_2^1\}.$$

Correspondingly, $\Sigma_2 = \Sigma_2^{(1)} + \Sigma_2^{(2)}$. If $m,j \in \sigma_1$, then $|p_m^2(t) - p_j^2(t)| < k^{61\delta\nu}$. Hence, using again (4.12.35) and noting that summation over j contains the order of $k^{10\delta\nu}$ terms, we obtain $\Sigma_2^{(1)} < k^{155\delta\nu}$. We consider $\Sigma_2^{(2)}$. Since $p_m^2(t)$ does not lie inside the contour C_3 ($\delta_m \notin P(k^{30\delta\nu})l_2^1$), we have

$$\left|(\hat{E}_j)_{mm}(p_m^2(t) - p_j^2(t))\chi_{mm}\right| \le \left|(2\pi i)^{-1}\oint_{C_3}((\hat{H}_{qi}(t) - z)^{-1}(H_0(t) - z)\chi)_{mm}dz\right|. \tag{4.12.36}$$

Expanding the resolvent of the operator $\hat{H}_{qi}(t)$ in powers of A_1 (see (4.12.22) and (4.12.23)) and noting that $(A_1)_{mm} = 0$, we obtain

$$|\sum_m (\hat{E}_j)_{mm}(p_m^2(t) - p_j^2(t))\chi_{mm}| < R\|A_1\|_2^2 < k^{20\delta\nu}. \tag{4.12.37}$$

Taking (4.12.35) and (4.12.37) into account, we prove that $|\Sigma_2^{(2)}| < ck^{83\delta\nu}$. Summing the estimates for Σ_1, $\Sigma_2^{(1)}$ and $\Sigma_2^{(2)}$, we see that

$$\|B_1\|_2^2 < ck^{155\delta\nu}. \tag{4.12.38}$$

To estimate B_2 we again use the relation similar to (4.12.19):

$$\|B_2\|_2^2 = \sum_{m,j\in\sigma_0} (\hat{E}_j)_{mm}|p_m^2(t) - k^2||p_j^2(t) + \Delta\hat{\Lambda}_{jj} - k^2|^{-1}, \tag{4.12.39}$$

$$\sigma_0 = \{m, j : \delta_m \in P(k^{1/5})l_2^1, \delta_j \in (I - P(k^{10\delta\nu})P_0 l_2^1)\}.$$

To evaluate $(\hat{E}_j)_{mm}$ we draw a circle C_4 with center at the point $p_j^2(t)$ and of radius r, with $r \approx |p_j^2(t) - k^2|/2$, if $|p_j^2(t) - k^2||(\mathbf{p}_j(t), \mathbf{p}_q(0))| > k^{6\delta\nu}$ and $r \approx |(\mathbf{p}_j(t), \mathbf{p}_q(0))|^{-1} k^{6\delta\nu}$ otherwise. We assume that the distance from the circle to any point $p_l^2(t)$, $\delta_l \in P(k^{1/5})l_2^1$ more that $r/2$. This can always be achieved by the choice of r, the interior of the circle containing no more than two points. In order to estimate $\hat{E}_j$, $\Delta\hat{A}_{jj}$, we expand the resolvent in the perturbation series on this contour. For this it suffices to prove the estimate:

$$\|A_1(z,t)\|_2^2 < ck^{-\delta\nu}, \quad c \neq c(k). \tag{4.12.40}$$

We use again formula (4.12.24). We represent $\|A_1(z,t)\|_2^2$ as the sums S_1 and S_2, where

$$S_1 = \sum_{l,m:|l-m| \geq k^{5\delta\nu}} a_{lm}(z_0, t)|c_l||c_m|,$$

$$S_2 = \sum_{l,m:|l-m| < k^{5\delta\nu}} a_{lm}(z, t).$$

$$c_l = (p_l^2(t) - z_0)(p_l^2(t) - z)^{-1},$$

z_0 being some point of the circle C_0, i.e., $|z_0 - k^2| = k^{-1-\delta}$. Let us consider S_1. We see that $|c_l| < 30$. Indeed, $c_l = c_j + (1 - c_j)(1 - (p_j^2(t) - z)(p_l^2(t) - z)^{-1})$. From the definition of the circle C_4 it easily follows that $|c_j| < 4$ and $|(p_j^2(t) - z)(p_l^2(t) - z)^{-1}| < 2$. Hence, $|c_l| < 30$ and

$$S_1 \leq c \sum_{l,m:|l-m| \geq k^{5\delta\nu}, l,m \in \sigma_0} a_{lm}(z_0, t).$$

Considering estimate (4.11.5) for $p = 5$ and $z = z_0$, we immediately obtain

$$|S_1| < k^{-5\delta\nu\beta + 4\delta} \leq k^{-\delta}. \tag{4.12.41}$$

We represent S_2 in the form

$$S_2 = S_3 + S_4, \tag{4.12.42}$$

$$S_3 = \sum_{l,m \in \sigma_3} a_{lm}(z, t),$$

$$S_4 = \sum_{l,m \in \sigma_4} a_{lm}(z, t),$$

where

$$\sigma_3 = \{m, l : |l - m| < k^{5\delta\nu}, \delta_l \in P((3/4)k^{10\delta\nu})l_2^1\},$$
$$\sigma_4 = \{m, l : |l - m| < k^{5\delta\nu}, \delta_l \notin P((3/4)k^{10\delta\nu})l_2^1\}.$$

We consider S_3. It is clear that

$$4|p_l^2(t) - z| > 4|p_l^2(t) - p_j^2(t)| - 4|p_j^2(t) - z|. \tag{4.12.43}$$

Noting that $\delta_l \in P((3/4)k^{10\delta\nu})l_2^1$, $\delta_j \notin P(k^{10\delta\nu})l_2^1$, and using the relation

$$|p_l^2(t) - p_j^2(t)| = ((\mathbf{p}_j(t), \mathbf{p}_q(0))^2 - (\mathbf{p}_l(t), \mathbf{p}_q(0))^2)p_q(0)^{-2}, \tag{4.12.44}$$

we obtain

$$4|p_l^2(t) - p_j^2(t)| \geq (\mathbf{p}_j(t), \mathbf{p}_q(0))^2 p_q(0)^{-2}. \tag{4.12.45}$$

By the definition of the circle C_4 we have

$$|p_j^2(t) - z| = r. \tag{4.12.46}$$

It is not hard to verify that

$$r < c|(\mathbf{p}_j(t), \mathbf{p}_q(0))|^{1-\delta\nu}. \tag{4.12.47}$$

Indeed, by the definition of C_4 $r \approx |p_j^2(t) - k^2|/2$ or $r \approx |(\mathbf{p}_j(0), \mathbf{p}_q(0))|^{-1}k^{6\delta\nu}$. Suppose the first relation holds. Note that $|p_j^2(t) - k^2| \leq |(\mathbf{p}_j(t), \mathbf{p}_q(0))|^{1-\delta\nu}$ by the definition of P_0. Therefore, estimate (4.12.47) is valid. When the second relation holds, we remark that $\delta_j \notin P(k^{10\delta\nu})l_2^1$, i.e., $|(\mathbf{p}_j(t), \mathbf{p}_q(0))|p_q(0)^{-2} > k^{10\delta\nu}$ (see (4.5.13). Using the last inequality, we get $r < cp_q(0)^2k^{-4\delta\nu} < c|(\mathbf{p}_j(t), \mathbf{p}_q(0))|^{1-\delta\nu}$. Thus, (4.12.47) holds. Note that

$$|(\mathbf{p}_j(t), \mathbf{p}_q(0))|^{1-\delta\nu} < ck^{-\delta\nu}(\mathbf{p}_j(t), \mathbf{p}_q(0))^2 p_q(0)^{-2}, \tag{4.12.48}$$

because $\delta_j \notin P(k^{10\delta\nu})l_2^1$, $|q| < k^{5\delta\nu}$. Using relations (4.12.43) and (4.12.45)–(4.12.48), we get:

$$8|p_l^2(t) - z| > (\mathbf{p}_j(t), \mathbf{p}_q(0))^2 p_q(0)^{-2}. \tag{4.12.49}$$

Hence,

$$8|p_l^2(t) - z| > k^{10\delta\nu}. \tag{4.12.50}$$

Note that $\delta_m \in P((7/8)k^{10\delta\nu})l_2^1$, because $|l-m| < k^{5\delta\nu}$. Arguing as in the proof of inequality (4.12.50), we show that

$$8|p_m^2(t) - z| > k^{10\delta\nu}. \tag{4.12.51}$$

Using estimate (4.12.50), we get

$$S_3 \leq \sum_{l,m \in \sigma_3} \frac{k^{-5\delta\nu}}{|p_m^2(t) - z|} \sum_{l:|l-m|<k^{5\delta\nu}} |v_{l-m}|^2 k^{-5\delta\nu}.$$

Taking into account the condition on the potential, we easily show that

$$\sum_{l:|l-m|<k^{5\delta\nu}} |v_{l-m}|^2 k^{-5\delta\nu} < c, \quad c \neq c(k).$$

Now, using (4.12.51) and (4.12.29), it is not hard to verify the estimate:

$$S_3 < k^{-5\delta\nu}. \tag{4.12.52}$$

We next consider S_4. It is obvious that $S_4 \le S_4' + S_4''$, where

$$S_4' = \sum_{\sigma_4} |d_{lm}|, \quad S_4'' = \sum_{\sigma_4} |d_{ml}|,$$

$$d_{lm} = |v_{l-m}|^2 (p_l^2(t) - z)(p_l^2(t) - p_m^2(t))^{-1}. \tag{4.12.53}$$

It is clear that $p_l^2(t) - p_m^2(t) = -p_{l-m}^2(0) + 2(\mathbf{p}_l(t), \mathbf{p}_{l-m}(0))$ and $\mathbf{p}_{l-m}(0) = n\mathbf{p}_q(0), n \in Z, n \neq 0$. Therefore, $|(\mathbf{p}_l(t), \mathbf{p}_{l-m}(0))| = (\mathbf{p}_l(t), \mathbf{p}_q(0))p_q(0)^{-1}p_{l-m}(0)$. Since $|l-m| < k^{5\delta\nu}$ and $|(\mathbf{p}_l(t), \mathbf{p}_q(0))|p_q(0)^{-2} > (3/4)k^{10\delta\nu}$, it follows that

$$|p_l^2(t) - p_m^2(t)| > |(\mathbf{p}_l(t), \mathbf{p}_q(0))|p_q(0)^{-1}p_{l-m}(t).$$

Considering the last relation , we obtain

$$S_4' < c \sum_{m,l:|m-l|<k^{5\delta\nu}} \frac{|v_{l-m}|^2 p_q(0)}{|(p_l^2(t) - z)||(\mathbf{p}_l(t), \mathbf{p}_q(0))||l-m|}$$

$$< c\sum_l \frac{p_q(0)}{|(p_l^2(t) - z)||(\mathbf{p}_l(t), \mathbf{p}_q(0))|}.$$

We break the region of summation into two: $\Xi_3 = \{l : |p_l^2(t) - z| \ge k^{5\delta\nu}\}$ and $\Xi_4 = \{l : |p_l^2(t) - z| < k^{5\delta\nu}\}$. Correspondingly,

$$S_4' \le S_4''' + S_4'''',$$

$$S_4''' = \sum_{l\in\Xi_3} \frac{p_q(0)}{|(p_l^2(t) - z)||(\mathbf{p}_l(t), \mathbf{p}_q(0))|},$$

$$S_4'''' = \sum_{l\in\Xi_4} \frac{p_q(0)}{|(p_l^2(t) - z)||(\mathbf{p}_l(t), \mathbf{p}_q(0))|}.$$

It is clear that

$$S_4''' \le \sum_{l\in\Xi_3} |(\mathbf{p}_l(t), \mathbf{p}_q(0))|^{-1},$$

Considering the relation $\delta_l \in (P(k^{1/5}) - P((3/4)k^{10\delta\nu}))l_2^3$, it is not hard to show that

$$S_4''' < ck^{-5\delta\nu}.$$

We estimate S_4''''. Since $|p_l^2(t) - z| < k^{5\delta\nu}$, $\delta_l \notin P((3/4)k^{10\delta\nu})l_2^3$, the set Ξ_4 contains no more than two points. From (4.12.46) and (4.12.47) it follows that $|(\mathbf{p}_l(t), \mathbf{p}_q(0))| \approx |(\mathbf{p}_j(t), \mathbf{p}_q(0))|$. Thus,

$$S_4'''' \le 2\max_{l\in\Xi_4} p_q(0)|(p_l^2(t) - z)|^{-1}|(\mathbf{p}_l(t), \mathbf{p}_q(0))|^{-1} < cr^{-1}|(\mathbf{p}_j(t), \mathbf{p}_q(0))|^{-1}p_q(0).$$

Suppose $|p_j^2(t) - k^2||(\mathbf{p}_j(t), \mathbf{p}_q(0))| > k^{6\delta\nu}$. Then, $2r \approx |p_j^2(t) - k^2|$ and $S_4'''' < ck^{-\delta\nu}$. If $|p_j^2(t) - k^2||(\mathbf{p}_j(t), \mathbf{p}_q(0))| \le k^{6\delta\nu}$ then $r \approx k^{6\delta\nu}|(\mathbf{p}_j(t), \mathbf{p}_q(0))|^{-1}$ and, therefore, $S_4'''' < ck^{-\delta\nu}$. Thus, $S_4' < ck^{-\delta\nu}$. We estimate S_4'' similarly. Summing the estimates for S_4' and S_4'', we obtain $|S_4| < ck^{-\delta}$. Using the last estimate and (4.12.52), (4.12.42), we get $|S_2| < ck^{-\delta\nu}$. Considering this estimate and (4.12.41), we obtain inequality (4.12.40).

In order to estimate $\|B_2\|_2^2$ we use the representation similar to (4.12.33):

$$\|B_2\|_2^2 = \Sigma_1 + \Sigma_2, \tag{4.12.54}$$

$$\Sigma_1 = \sum_{j:\delta_j \in P_0(I-P(k^{10\delta\nu}))l_2^1} (p_j^2(t) - k^2)|p_j^2(t) + \Delta\hat{A}(t)_{jj} - k^2|^{-1} Sp(\hat{E}_j \chi),$$

$$\Sigma_2 = \sum_{j,m\in\sigma'} |p_j^2(t) + \Delta\hat{A}(t)_{jj} - k^2|^{-1} \sum_m (p_m^2(t) - p_j^2(t))(\hat{E}_j)_{mm}\chi_{mm},$$

$$\sigma' = \{m, j : \delta_m \in P(k^{1/5})l_2^1, \delta_j \in P_0(I - P(k^{10\delta_1}))l_2^1\}.$$

We estimate Σ_1. From the foregoing arguments we have

$$2|p_j^2(t) + \Delta A_{jj} - k^2| > rk^{-49\delta\nu}. \tag{4.12.55}$$

Indeed, since the resolvent can be expanded in the series on the contour, it follows that $|\Delta A_{jj}| < r$. In the case where $|p_j^2(t) - k^2||(\mathbf{p}_j(t), \mathbf{p}_q(0))| > k^{6\delta\nu}$, we have chosen $r \approx |p_j^2(t) - k^2|/2$ and therefore (4.12.55) holds. If $|p_j^2(t) - k^2||(\mathbf{p}_j(t), \mathbf{p}_q(0))| \le k^{6\delta\nu}$, then according to (4.11.8)

$$|p_j^2(t) + \Delta A_{jj} - k^2| > k^{-43\delta\nu}|(\mathbf{p}_j(t), \mathbf{p}_q(0))|^{-1} > rk^{-49\delta\nu}.$$

On the other hand by the definition of the circle C_4

$$|p_j^2(t) - k^2| \le cr, \tag{4.12.56}$$

The summation on j contains no more than two terms, since $\delta_j \notin P(k^{10\delta\nu})l_2^3$ and $|p_j^2(t) - k^2| < |(\mathbf{p}_j(t), \mathbf{p}_q(0))|^{1-\delta}$ by the definition of P_0. Hence, noting that $\mathrm{Tr}(\hat{E}_j\chi) < 1$, we obtain $\Sigma_1 < k^{60\delta\nu}$.

In Σ_2 we break the set of summation on m into two: $\tilde{\Xi}_1$ and $\tilde{\Xi}_2$: $\tilde{\Xi}_1$ contains those m for which $p_m^2(t)$ lies inside C_4, and $\tilde{\Xi}_2$ corresponds to points $p_m^2(t)$ lying outside the disk. Correspondingly, $\Sigma_2 = \Sigma_2^{(1)} + \Sigma_2^{(2)}$. We note that $\Sigma_2^{(1)}$ contains no more than four terms, and $|p_m^2(t) - p_j^2(t)| < 2r$, $|(\hat{E}_j)_{mm}\chi_{mm}| < 1$. Using (4.12.55) and (4.12.56), we obtain $|\Sigma_2^{(1)}| < 10k^{50\delta\nu}$. To estimate $\Sigma_2^{(2)}$ we use (4.12.36). Expanding the resolvent in a series (see (4.12.40)), we obtain

$$\sum_m \oint_{C_4} ((\hat{H}_q(t) - z)^{-1}(H_0(t) - z)\chi)_{mm} dz < ck^{-\delta\nu}r.$$

Using (4.12.55) and (4.12.37) and noting that the summation over j contains no more than two terms, we obtain $\Sigma_2^{(2)} < k^{50\delta\nu}$. Thus,

$$\|B_2\|_2^2 < ck^{60\delta\nu}. \tag{4.12.57}$$

We estimate $\|B_3\|$ using formula similar to (4.12.33):

$$\|B_3\|_2^2 = \Sigma_1 + \Sigma_2, \tag{4.12.58}$$

$$\Sigma_1 = \sum_{j:\delta_j \in (I-P_0)(I-P(k^{10\delta_\nu})l_2^1} (p_j^2(t) - k^2)|p_j^2(t) + \Delta\hat{A}(t)_{jj} - k^2|^{-1} Sp(\hat{E}_j\chi),$$

$$\Sigma_2 = \sum_{j,m\in\sigma''} |p_j^2(t) + \Delta\hat{A}(t)_{jj} - k^2|^{-1} \sum_m (p_m^2(t) - p_j^2(t))(\hat{E}_j)_{mm}\chi_{mm},$$

$$\sigma'' = \{m, j : \delta_m \in P(k^{1/5})l_2^1, \delta_j \in (I - P_0)(I - P(k^{10\delta_1}))l_2^1\}.$$

To estimate $\hat{E}_j$, for the contour C_j we take a circle with center at the point p_j^2 and of radius r,

$$r \approx |(\mathbf{p}_j(t), \mathbf{p}_q(0))|^{1-2\delta}. \tag{4.12.59}$$

We note that inside this circle there are no more than two points $p_l^2(t)$: $\delta_l \in P(k^{1/5})l_2^3$, and $\delta_l \notin P(k^{10\delta_\nu})l_2^3$, since $\delta_j \notin P(k^{10\delta_\nu})l_2^3$. It is possible to choose the circle C_j so that it is at a distance greater than $r/2$ from the nearest point $p_l^2(t)$: $\delta_l \in P(k^{1/5})l_2^3$. In order to expand the resolvent in a series we estimate $\|A_1\|_2^2$, using (4.12.24).

We break the region of summation into two:

$$\sigma_5 = \{l, m : |p_l^2 - p_m^2| \geq |l - m|(\ln k)^{-1}\},$$

$$\sigma_6 = \{l, m : |p_l^2 - p_m^2| < |l - m|(\ln k)^{-1}\}.$$

It is obvious that $a_{lm} = d_{lm} + d_{ml}$ (see (4.12.53)). Hence,

$$\sum_{l,m\in\sigma_5} a_{lm} < 2\ln k(\sum_r |v_r|^2|r|^{-1})(\sum_l |p_l^2(t) - z|^{-1}) <$$

$$c\ln k|(\mathbf{p}_j(t), \mathbf{p}_q(0))|^{-1+3\delta} < |(\mathbf{p}_j(t), \mathbf{p}_q(0))|^{-1+4\delta}. \tag{4.12.60}$$

If $l, m \in \sigma_6$, then, taking into account that $l - m = n_1 q$, $n_1 \in Z$ and $p_{l-m}(0) < c|l - m|$, we obtain

$$|(\mathbf{p}_l(t) + \mathbf{p}_m(t), \mathbf{p}_q(0))| < cp_q(0)(\ln k)^{-1}. \tag{4.12.61}$$

It is easy to verify that for each l there exists no more than one $m(l) \neq l$ satisfying the last inequality. Therefore

$$\sum_{\sigma_6} a_{lm} < \sum_l |v_{l-m(l)}|^2 |p_l^2(t) - z|^{-1} |p_{m(l)}^2(t) - z|^{-1}. \tag{4.12.62}$$

Noting that $2|p_{m(l)}^2(t) - z| > r \approx |(\mathbf{p}_j(t), \mathbf{p}_q(0))|^{1-2\delta}$, we obtain

$$\sum_{\sigma_6} a_{lm} < c|(\mathbf{p}_j(t), \mathbf{p}_q(0))|^{-1+2\delta} \sum_l |v_{l-m(l)}|^2 |p_l^2(t) - z|^{-1}. \tag{4.12.63}$$

We shall show that

$$20|p_l^2(t) - z| > |(\mathbf{p}_l(t), \mathbf{p}_q(0))|^{1-3\delta} p_q(0)^{-1+3\delta}. \tag{4.12.64}$$

If $|(\mathbf{p}_l(t), \mathbf{p}_q(0))|^{1-\delta} p_q(0)^{-1+\delta} < |(\mathbf{p}_j(t), \mathbf{p}_q(0))|$, then this is so by the definition of the circle (see (4.12.59)). For other l, taking into account (4.12.43) and (4.12.44) we obtain

$$2|p_l^2(t) - z| > 2|p_l^2(t) - p_j^2(t)| - 2|p_j^2(t) - z| >$$

$$(\mathbf{p}_l(t), \mathbf{p}_q(0))^2 p_q(0)^{-2} - (\mathbf{p}_j(t), \mathbf{p}_q(0))^2 p_q(0)^{-2} - 2r.$$

Considering (4.12.59), we get

$$2|p_l^2(t) - z| > |(\mathbf{p}_l(t), \mathbf{p}_q(0))|^{1-3\delta} p_q(0)^{-1+3\delta}. \tag{4.12.65}$$

Thus, relation (4.12.64) is proved. Further,

$$2|(\mathbf{p}_l(t), \mathbf{p}_q(0))| = |(\mathbf{p}_{l-m}(t), \mathbf{p}_q(0)) + (\mathbf{p}_{l+m}(t), \mathbf{p}_q(0))| >$$

$$A^{-1}|l - m| p_q(0) - 2p_q(0)(\ln k)^{-1} > |l - m| p_q(0)(2A)^{-1}. \tag{4.12.66}$$

From relations (4.12.65) and (4.12.66) it follows that $|p_l^2(t) - z| > c|l - m|^{1-3\delta}$. Using the last estimate in inequality (4.12.62), we get:

$$\sum_{l,m \in \sigma_6} a_{lm} < c|(\mathbf{p}_j(t), \mathbf{p}_q(0))|^{-1+3\delta} \sum_l \frac{|v_{l-m(l)}|^2}{|l - m(l)|^{1-3\delta}}.$$

Let us introduce the new variable $\tilde{r} = l - m(l)$. It is clear that $\delta_{\tilde{r}} \in P_i^q l_2^3$, because δ_m, $\delta_{l(m)} \in P_i^q l_2^3$. Using this fact it is not hard to show that there is no more than two l corresponding to a given r. Therefore,

$$\sum_l \frac{|v_{l-m(l)}|^2}{|l - m(l)|^{1-3\delta}} < 2 \sum_r \frac{|v_r|^2}{|r|^{1-3\delta}} < c.$$

Using this inequality we easily obtain

$$\sum_{l,m \in \sigma_6} a_{lm} < c|(\mathbf{p}_j(t), \mathbf{p}_q(0))|^{-1+3\delta}. \tag{4.12.67}$$

Thus, summing (4.12.60) and (4.12.67), we obtain

$$\|A_1\|_2^2 < c|(\mathbf{p}_j(t), \mathbf{p}_q(0))|^{-1+4\delta}. \tag{4.12.68}$$

Therefore, the perturbation series for the resolvent converges on the circle C_j (see (4.12.59)). Hence,

$$|\Delta \hat{A}_{jj}(t)| < c|(\mathbf{p}_j(t), \mathbf{p}_q(0))|^{1-2\delta}. \tag{4.12.69}$$

Note that

$$|p_j^2(t) - k^2| > |(\mathbf{p}_j(t), \mathbf{p}_q(0))|^{1-\delta}, \tag{4.12.70}$$

since $\delta_j \in (I - P_0)l_2^3$. We use the representation (4.12.33). Considering (4.12.69), (4.12.70), we obtain

$$2|p_j^2(t) + \Delta \Lambda_{jj}(t) - k^2| > |p_j^2(t) - k^2|. \tag{4.12.71}$$

Hence,

$$\Sigma_1 \leq \sum_{jm}(\hat{E}_j)_{mm}. \tag{4.12.72}$$

It is clear that

$$\sum_m(\hat{E}_j)_{mm} < \left|\sum_m \oint_{C_j} (\hat{H}_{qi} - z)_{mm}^{-1}dz\right|.$$

Expanding the resolvent in the series in powers of A, taking into account that

$$\oint_{C_j} (\hat{H}_0 - z)_{mm}^{-1}dz = 0,$$

because p_m^2 lie outside the circle by the definition of B_3', and noting that $(A_1)_{mm} = 0$, we obtain

$$\sum_m(E_j)_{mm} \leq c\|A_1\|_2^2.$$

Considering inequality (4.12.68), we verify that

$$|\sum_m(E_j)_{mm}| < c|(\mathbf{p}_j(t), \mathbf{p}_q(0))|^{-1+4\delta}.$$

Substituting the last estimate into (4.12.72), we get

$$\Sigma_1 \leq c\sum_j |(\mathbf{p}_j(t), \mathbf{p}_q(0))|^{-1+4\delta}. \tag{4.12.73}$$

Noting that the summation goes only over those j for which $\delta_j \in (P(k^{1/5}) - P(k^{10\delta\nu}))l_2^3$, we obtain

$$|\Sigma_1| < k^{5\delta\nu}. \tag{4.12.74}$$

In Σ_2 we break the region of summation into two: $\sigma'' = \sigma_7 \cup \sigma_8$,

$$\sigma_7 = \{j, m \in \sigma : 2|p_m^2(t) - p_j^2(t)| > |(\mathbf{p}_j(t), \mathbf{p}_q(0))|^{1-\delta}\},$$

$$\sigma_8 = \{j, m \in \sigma : 2|p_m^2(t) - p_j^2(t)| \leq |(\mathbf{p}_j(t), \mathbf{p}_q(0))|^{1-\delta}\}.$$

Correspondingly, $\Sigma_2 = \Sigma_2' + \Sigma_2''$. Let us estimate Σ_2'. It is clear that for $m, j \in \sigma_7$ formula (4.12.36) holds, because $p_m^2(t)$ lies outside the circle C_4. Expanding $(\hat{H}_{qi} - z)^{-1}$ in the series in powers of A_1, we obtain

$$\left|\sum_m(E_j)_{mm}(p_j^2(t) - p_m^2(t))\chi_{mm}\right| < \|A\|_2^2 r < |(\mathbf{p}_j(t), \mathbf{p}_q(0))|^{2\delta}. \tag{4.12.75}$$

Considering estimates (4.12.69) and (4.12.70), we arrive at the inequality

$$2|p_j^2(t) + \Delta\Lambda_{jj}(t) - k^2| > |(\mathbf{p}_j(t), \mathbf{p}_q(0))|^{1-\delta}. \tag{4.12.76}$$

Estimates (4.12.75) and (4.12.76) give:

$$\Sigma_2' \leq c \sum_j |(\mathbf{p}_j(t), \mathbf{p}_q(0))|^{-1+3\delta}. \tag{4.12.77}$$

Taking into account that $\delta_j \in (P(k^{1/5}) - P(k^{10\delta\nu}))l_2^3$, we arrive at the inequality

$$|\Sigma_2'| < k^{3\delta}. \tag{4.12.78}$$

It remains to consider Σ_2''. Since $\delta_j \notin P(k^{10\delta\nu})$ and $2|p_m^2(t) - p_j^2(t)| < |(\mathbf{p}_j(t), \mathbf{p}_q(0))|^{1-\delta}$, for each j there are no more than two m satisfying this condition and, similarly, for each m there are no more than two j satisfying the indicated inequality. Therefore,

$$\|\Sigma_2''\| < 4 \max_{m,j}(|p_m^2(t) - p_j^2(t)||p_j^2(t) + \Delta\Lambda_{jj} - k^2|^{-1}). \tag{4.12.79}$$

By the definition of σ_8

$$|p_m^2(t) - p_j^2(t)| < |(\mathbf{p}_j(t), \mathbf{p}_q(0))|^{1-\delta}. \tag{4.12.80}$$

Taking into account inequality (4.12.70), which holds because $\delta_j \in (I - P_0)l_2^3$, we obtain

$$|p_m^2(t) - k^2| < 2|p_j^2(t) - k^2|. \tag{4.12.81}$$

Using estimates (4.12.69) and (4.12.79), we obtain the inequality

$$\|\Sigma_2''\| < 8. \tag{4.12.82}$$

The estimates for (4.12.78) and (4.12.82) for $\|\Sigma_2'\|$ and $\|\Sigma_2''\|$ together give

$$\|\Xi_2\| < ck^{60\delta\nu}. \tag{4.12.83}$$

Using (4.12.74), (4.12.83), we get

$$\|B_3\| < ck^{60\delta\nu}. \tag{4.12.84}$$

Summing inequalities (4.12.38), (4.12.57) and (4.12.83), we get:

$$\|B\| < ck^{80\delta\nu}. \tag{4.12.85}$$

We have proved this estimate for the case where $|q| < k^{5\delta\nu}$. Suppose $|q| \geq k^{5\delta\nu}$. The proof is the simplified version of that for the case $|q| < k^{5\delta\nu}$. Indeed, we represent B in the form: $B = C_1 + C_2$, $C_1 = P_0'B$, $C_2 = (I - P_0')B$. Here P_0' is a diagonal projection $(P_0')_{jj} = 1$, when $|p_j^2(t) - k^2| < (|(\mathbf{p}_j(t), \mathbf{p}_q(0))| + p_q(0)^2)^{1-\delta}$, and $(P_0')_{jj} = 0$ otherwise. The operator C_1 can be estimated in the same way as B_2. Here we choose a circle of radius $r \approx |p_j^2(t) - k^2|/2$. We remark that in the proof of (4.12.40) the sum S_2 vanishes, since $|l-m| > cp_{l-m}(0) \geq |q| > k^{5\delta\nu}$. The operator C_2 can be estimated just as B_3. It is represented as a sum of two operators C_2' and C_2'', with $(C_2')_{mj} = (C_2)_{mj}$, when $|p_m^2(t) - p_j^2(t)| \geq (|(\mathbf{p}_j(t), \mathbf{p}_q(0))| + p_q(0)^2)^{1-\delta}$ and $(C_2')_{mj} = 0$ otherwise. The operators C_2' and C_2'' can be estimated in a manner similar to B_3' and B_3'' respectively. In the estimate of C_2' for C_j it is necessary to take a contour with center at the point $p_j^2(t)$ and of the radius $r \approx (|(\mathbf{p}_j(t), \mathbf{p}_q(0))| + p_q(0)^2)^{1-2\delta}$. The lemma is proved.

Lemma 4.36 . *If t is in the $(k^{-2-2\delta})$ neighbourhood of $\chi_4(k,\rho,\delta,V)$, then*

$$\hat{g}_1(k,t) = 0 \tag{4.12.86}$$

and the following inequalities hold:

$$|g_r(k,t) - \hat{g}_r(k,t)| < k^{-1-\beta+2r\delta+2\alpha_0}, \tag{4.12.87}$$

$$\alpha_0 = \beta(3\rho + 5\delta)/2,$$

when $2 < r < 2r_0$, $r_0 \equiv [(\beta/2 - \alpha_0)(\gamma_6 + 2\delta)^{-1}] + 1$, *and*

$$\|G_r(k,t) - \hat{G}_r(k,t)\| < k^{-\beta/2+2r\delta+\alpha_0}, \tag{4.12.88}$$

when $1 < r < r_0$.

Proof. Let us consider $\hat{g}_1(k,t)$. It is clear that

$$\hat{g}_1(k,t) = \frac{(-1)^r}{2\pi i r} \sum_m \oint_{C_0} \hat{A}_{mm}.$$

Note that $\hat{A}_{mm}$ is holomorphic inside C_0, when $m \neq j$ and $\hat{A}_{jj} = A_{jj} = 0$. Thus, (4.12.86) is proved.

Suppose $r \geq 2$. We write $g_r(k,t) - \hat{g}_r(k,t)$ in the form:

$$g_r(k,t) - \hat{g}_r(k,t) = -\frac{(-1)^r}{2\pi i r} \sum_{i_1,\ldots,i_r=1,2} \operatorname{Tr} \oint_C Q_{i_1} \ldots Q_{i_r} dz + \frac{(-1)^r}{2\pi i r} \operatorname{Tr} \oint_C Q_2^r dz,$$

$$Q_1 = \hat{A} - A, \quad Q_2 = A.$$

From this representation we obtain

$$|g_r(k,t) - \hat{g}_r(k,t)| \leq 2^r \max_{i_2,\ldots,i_r=1,2} |\operatorname{Tr} \oint_C Q_1 Q_{i_2} \ldots Q_{i_r} dz|. \tag{4.12.89}$$

We remark that $(I-P)Q_1(I-P) = 0$. Therefore the integrand on the right side of (4.12.89) can be represented in the form

$$\operatorname{Tr} \oint_C P \Pi_r(Q) dz, \tag{4.12.90}$$

where $\Pi_r(Q)$ is the product of r operators each of them being equal to Q_1 or Q_2 or any of these operators multiplied by P or $I-P$. Noting that $E_jP_q(k^{1/5})=0$ when $q\in\Gamma(k^\rho)$ (see (4.5.79)), we can easily verify the relation: $E_jQE_j=0$. Thus Q can be represented in the form $Q=Q'+Q^-+Q^+$,where

$$Q'=(I-E_j)Q(I-E_j),\quad Q^+=E_jQ(I-E_j).\quad Q^-=(I-E_j)QE_j.$$

The integral on the right side of (4.12.90) can be written in the form of sum of 3^r expressions of the type

$$\operatorname{Tr}\oint_C P\Pi_r(Q',Q^+,Q^-)dz, \tag{4.12.91}$$

where $\Pi_r(Q',Q^+,Q^-)$ is a product of operators Q', Q^+ and Q^- containing r factors. We note that

$$\oint_C P(Q')^{r-1}dz=0,$$

since the integrand is regular inside the contour. Thus, the integral (4.12.91) can be nonzero only when it can be represented in the form

$$\operatorname{Tr}\oint_C P\Pi_{r'}(Q',Q^+,Q^-)Q^-Q^+\Pi_{r-2-r'}(Q',Q^+,Q^-)dz,$$

$0\le r'\le r-2$, since otherwise the integrand is holomorphic. We rewrite the last integral in the form

$$\operatorname{Tr}\oint_C S^-_{r'}S^+_{r-2-r'}dz,$$

$$S^-_r=P\Pi_rQ^-,\quad S^+_r=Q^+\Pi_rP,$$

It is obvious that

$$\|S^-_r\|_1=\|S^-_rE_j\|_1\le\|S^-_r\|\|E_j\|_1=\|S^-_r\|.$$

Similarly,

$$\|S^+_r\|_1\le\|S^+_r\|.$$

Since P is the orthogonal sum of the operators P_q, it follows that

$$\|S^\pm_r\|\le\max_{q\in\Gamma(k^\rho)}\|S^\pm_{r,q}\|,$$

$$S^-_{r,q}=P_q\Pi_rQ^-,\quad S^+_{r,q}=Q^+\Pi_rP_q.$$

It is easy now to see that

$$|g_r(k,t)-\hat g_r(k,t)|\le 6^rk^{-1-\delta}\max_{q\in\Gamma(k^\rho),0\le r'\le r-2}\|S^-_{r',q}\|\|S^+_{r-2-r',q}\|. \tag{4.12.92}$$

We shall estimate $\|S^\pm_{r,q}\|$. We understand by $\hat Q$ any one of the operators Q', Q^+, or Q^-. We represent $\hat Q$ in the form

$$\hat{Q} = \sum_{l=0}^{L} \hat{Q}_l, \quad L = [(1 - 3\rho - 4\delta\nu)\delta^{-1}]. \tag{4.12.93}$$

The operators $\hat{Q}_l$ for $l < L$ are defined as follows: $(\hat{Q}_l)_{ij} = \hat{Q}_{ij}$, if $k^{l\delta} \le |i-j| < k^{(l+1)\delta}$, and $(\hat{Q}_l)_{ij} = 0$ otherwise. It is natural that $(\hat{Q}_L)_{ij} = \hat{Q}_{ij}$, if $|i-j| \ge k^{L\delta}$, and $(\hat{Q}_L)_{ij} = 0$ otherwise. Obviously,

$$\|S_r\| \le (L)^r \max_{l_1,\dots,l_r} \|T_{l_1\dots l_r}\|, \tag{4.12.94}$$

$$T_{l_1\dots l_r} = P_q \hat{Q}_{l_1} \dots \hat{Q}_{l_r} E_j.$$

It is not hard to verify that an operator $T_{l_1\dots l_r}$ vanishes if all its indeces differ from L. Indeed, an element of the matrix $(T_{l_1\dots l_r})_{mn}$ can be nonzero only if $n = j$ and $\delta_m \in P_q l_2^3$. Using relation (4.11.16) and the definition of P_q (see (4.2.3)–(4.2.5)), we obtain

$$4|(\mathbf{p}_j(t) - \mathbf{p}_m(t), \mathbf{p}_q(0))| > k^{1-2\rho-\delta}.$$

From this it follows that

$$|j - m| > k^{1-3\rho-\delta}.$$

The last inequality can be valid only if at last one of indeces $l_1, \dots, l_r$ is equal L. We prove that

$$\|\hat{Q}_L\|_2^2 < ck^{-L\delta\beta+4\delta}. \tag{4.12.95}$$

In fact, by the definition of $\hat{Q}_L$ one of the equalities

$$(\hat{Q}_L)_{mn} = A_{mn}, \quad (\hat{Q}_L)_{mn} = (A - \hat{A})_{mn} \tag{4.12.96}$$

hold. For the first case we use the estimate

$$\|Q_L\|_2^2 \le \max_{z \in C} \|A(z, t, V - V_{L\delta})\|_2^2.$$

Considering inequality (4.11.5), we obtain (4.12.95). For the second case it suffices to estimate the operator $\tilde{Q}$:

$$(\tilde{Q}_L)_{mn} = \begin{cases} \hat{A}_{mn} & \text{when } |m-n| > k^{L\delta}, \\ 0 & \text{otherwise.} \end{cases}$$

We use formula (4.12.5). Note that $U_{im} = B_{im} = 0$ when $|i - m| > k^{1/5}$. Thus the estimate of $\tilde{Q}$ is easily reduces to the estimate of $\hat{Q}$ in the first case. Thus, we get (4.12.95). Considering relations (4.12.12) and (4.12.15) we verify that

$$\|Q_l\| < k^{2\delta}, l = 1, \dots, L. \tag{4.12.97}$$

Inequalities (4.12.95) and (4.12.97) together give:

$$\|T_{l_1\dots l_r}\| < k^{-\beta/2+2\delta r+\alpha_0}. \tag{4.12.98}$$

Using (4.12.92), (4.12.94) and (4.12.98) together give (4.12.87). Inequality (4.12.88) is proved similarly. The lemma is proved.

Proof of Theorem 4.8. To construct the series we use (4.12.1). Taking into account estimates (4.12.12) – (4.12.16), we obtain (4.12.3) and (4.12.4), where $6\gamma_6 = \min\{1/5 - 15\rho - 20\delta, \rho\beta - 170\delta\nu\}$. The estimates (4.12.3) and (4.12.4) make it possible to expand the resolvent in the perturbation series. Arguing as in the proof of Theorem 4.1, we arrive at (4.10.5) and (4.10.6). Here we have

$$\|\hat{G}_r(k,t)\|_1 < k^{-6\gamma_6[r/3]+4\delta},$$

$$|\hat{g}_r(k,t)| < k^{-1-6\gamma_6[r/3]+3\delta}.$$

Relations (4.10.7) and (4.10.8) for $m \geq 3$ follow easily from these inequalities.

It remains to estimate $\hat{g}_2(k,t)$, $\hat{G}_1(k,t)$ and $\hat{G}_2(k,t)$. We first consider $g_2(k,t)$, $G_1(k,t)$ and $G_2(k,t)$. It is obvious that

$$g_2(k,t) = \frac{1}{2\pi i}\mathrm{Tr}\oint_C E_j A^2 E_j dz = \frac{1}{2\pi i}\mathrm{Tr}\oint_C E_j(A' + A'')^2 E_j dz,$$

where $A'_{mn} = A_{mn}$ if $|n-m| < k^{\rho_1}$, $0 < \rho_1 < 1$ and $A'_{mn} = 0$ otherwise; $A'' \equiv A - A'$. Since t is in the $(k^{-2-2\delta})$-neighbourhood of $\chi_4(k,\rho,\delta,V)$, it follows that

$$\|A''\|_2^2 < k^{-\rho_1\beta+5\delta\nu}.$$

Noting that $\|E_j\| = 1$, we obtain

$$\left|\mathrm{Tr}\oint_C E_j A''^2 E_j dz\right| < k^{-1+4\delta-\rho_1\beta}, \tag{4.12.99}$$

$$\left|\mathrm{Tr}\oint_C E_j A' A'' E_j dz\right| < k^{-1+4\delta-\rho_1\beta/2}, \tag{4.12.100}$$

$$\left|\mathrm{Tr}\oint_C E_j A'' A' E_j dz\right| < k^{-1+4\delta-\rho_1\beta/2}. \tag{4.12.101}$$

We shall prove that

$$\left|\mathrm{Tr}\oint_C E_j A'^2 E_j dz\right| < k^{-2+8\delta+7\rho_1}. \tag{4.12.102}$$

Indeed, evaluating the integral by the residue it is not hard to show that it is equal to $(M_+M_-)_{jj}$ (see (4.5.121)) for $R_0 = k^{\rho_1}$. Arguing as in the proof of estimate (4.5.122) for $i = j$ we obtain inequality (4.12.102). Taking $\rho_1 = 2(\beta+14)^{-1}$, we get

$$|g_2(k,t)| < k^{-1-\beta/(\beta+14)+8\delta}. \tag{4.12.103}$$

We estimate $G_1(k,t)$ and $G_2(k,t)$ in a similar way. It turns out that

$$\|G_1(k,t)\|_1 < k^{-\beta/(\beta+5)+8\delta}, \tag{4.12.104}$$

$$\|G_2(k,t)\|_1 < k^{-2\beta/(\beta+10)+8\delta}. \tag{4.12.105}$$

Considering Lemma 4.36, we obtain (4.10.7) and (4.10.8) for $r = 1, 2$, $m = 0$. From (4.5.160) and estimates (4.10.8) for $m = 0$ it follows that inside C there is a unique eigenvalue.

The functions $\hat{g}_r$ and the operator-valued functions $\hat{G}_r(k,t)$ depend analytically on t in each simply connected component of the $(k^{-2-2\delta})$-neighbourhood of $\chi_4(k,\rho,\delta,V)$; the estimates (4.10.7) and (4.10.8) hold for $r = 0$. Estimating the derivative of the analytic function at the point t in terms of its value on the boundary of the disk of radius $k^{-2-2\delta}$ in the complex plane, we obtain inequalities (4.10.7) and (4.10.8) for $m > 0$. The theorem is proved.

Lemma 4.37 . *If t is in the $(k^{-2-2\delta})$-neighbourhood of $\chi_4(k,\rho,\delta,V)$, then the following inequalities hold:*

$$|T(m)(g_r(k,t) - \hat{g}_r(k,t))| < k^{-1-\beta+2r\delta+2\alpha_0+(2+\delta)|m|}, \tag{4.12.106}$$

when $r < 2r_0$, and

$$\|T(m)(G_r(k,t) - \hat{G}_r(k,t))\| < k^{-\beta/2+2r\delta+\alpha_0+(2+\delta)|m|}, \tag{4.12.107}$$

when $m < m_0$.

Proof. Estimates (4.12.87), (4.12.88) are satisfied in the $(k^{-2-2\delta})$-neighbourhood of $\chi_4(k,\rho,\delta,V)$. Considering that $g_r(k,t)$, $\hat{g}_r(k,t)$, $G_r(k,t)$, $\hat{G}_r(k,t)$ depend analytically on t in this neighbourhood, we obtain (4.12.106) and (4.12.107). The lemma is proved.

Proof of the Theorem 4.9. We represent $\lambda(t)$ and $E(t)$ in the forms (4.10.12) and (4.10.13), setting $\rho = 1/100$,

$$\varphi_1 = \sum_{2r_0+1}^{\infty} \hat{g}_r(k,t) + \sum_{r=2}^{2r_0}(\hat{g}_r(k,t) - g_r(k,t)),$$

$$\psi_1 = \sum_{r_0+1}^{\infty} \hat{G}_r(k,t) + \sum_{r=1}^{r_0}(\hat{G}_r(k,t) - G_r(k,t)).$$

Using estimates (4.10.7), (4.10.8) and (4.12.106), (4.12.107) and taking into account that $\gamma_6 r > \beta/2 - \alpha_0$ for $r > r_0$ and also that $\gamma_7 < \gamma_6$ for the given ρ we obtain (4.10.14) – (4.10.17). The theorem is proved.

4.13 The Description of the Isoenergetic Surface.

4.13.1 Proof of the Bethe-Sommerfeld Conjecture.

Theorem 4.10 . *Suppose a potential $V(x)$ satisfies the condition*

$$\sum_{m \in Z^3} |v_m|^2 |m|^{-1+\beta} < \infty \tag{4.13.1}$$

for some positive β. Then, there exists only a finite number of gaps in the spectrum of operator H.

The proof is quite similar to that in the case when $2l > n$ (Theorem 2.5), it being based on Theorem 4.8 instead of Theorem 2.1.

This is the first proof valid for such a general class of potentials. The Bethe-Sommerfeld conjecture in the three dimensional situation for a smooth potential was proved in [Sk6],[Sk7]and [Ve3]. In [Ve3], for the first time, an asymptotic formula for an eigenvalue was used. The formula contained first few terms of the asymptotic. Here, we have constructed an infinite series for the eigenvalue. The terms of asymptotics have a simpler form (in [Ve3] a recursive procedure and an additional asymptotic expansion are needed). Moreover, termwise infinite differentiability of the asymptotic series with respect to t is proved. It is here that for the first time an asymptotic formula for the spectral projection is constructed (see Theorems 4.1 – 4.9). Theorem 4.8, being valid for a potential satisfying condition (4.13.1), enables us to obtain even a stronger result – to describe the behavior of the perturbed isoenergetic surface near such planes. Using asymptotic formulae near planes of diffraction, we describe the behavior of the perturbed isoenergetic surface near the planes of diffraction.

4.13.2 The behavior of the isoenergetic surface near the nonsingular set.

Theorem 4.11 . *There exists a single piece $S'(k)$ of the perturbed isoenergetic surface in the $(k^{-2-2\delta})$-neighbourhood of each simply connected component of the nonsingular set $\chi_4(k,\rho,V,\delta)$, $0<\rho<1/100$, $0<\delta<1/100$. For any given i, $i=1,2,3$ the points t of $S'(k)$ can be represented in the form:*

$$t = t_0 + \varphi_i(k^2, t_0), \quad t_0 \in \chi_4(k,\rho,\delta,V), \tag{4.13.2}$$

where $\varphi_i(k^2,t_0)$ is a continuously differentiable vector-valued function with a single nonzero component $(\varphi_i)_i$. The following asymptotic estimates are fulfilled for $(\varphi_i)_i$:

$$| \varphi_i(k^2, t_0)_i | =_{k\to\infty} O(k^{-2-2\gamma_6}), \tag{4.13.3}$$

$$| \nabla\varphi_i(k^2, t_0)_i | =_{k\to\infty} O(k^{-2\gamma_6}), \tag{4.13.4}$$

Estimates (4.13.3) and (4.13.4) are uniform in $t_0, t_0 \in \chi_4(k,\rho,\delta,V)$.

Remark 4.3. The piece of the isoenergetic surface $S'(k)$ is a unique one in the $(k^{-2-2\delta})$-neighbourhood of $\chi_4(k,\rho,\delta,V)$. But as a matter of fact $S'(k)$ lies in a smaller $(k^{-2\gamma_6})$-neighbourhood of $\chi_4(k,\rho,\delta,V)$.

Remark 4.4. For a smooth potential, estimates (4.13.3) and (4.13.4) can be improved. For example, for a trigonometric polynomial they have the form

$$| \varphi_i(k^2, t_0)_i | =_{k\to\infty} O(k^{-3+8\delta}), \tag{4.13.5}$$

$$| \nabla\varphi_i(k^2, t_0)_i | =_{k\to\infty} O(k^{-3+8\delta}). \tag{4.13.6}$$

Proof.The proof is similar to that of Theorem 2.6. Indeed, the eigenvalue $\lambda(t)$ is determined by formula (4.10.5) in the $(2k^{-2-2\delta})$-neighbourhood of the nonsingular set $\chi_4(k,\rho,\delta,V)$. The function $\lambda(t)$ is continuously differentiable with respect to t in this neighbourhood of each simply connected component of the nonsingular set, the estimates being valid:

$$\mid \lambda(t) - p_j^2(t) \mid =_{k\to\infty} O(k^{-1-2\gamma_6}), \tag{4.13.7}$$

$$\left| \frac{\partial\lambda(t)}{\partial t_i} - \frac{\partial p_j^2(t)}{\partial t_i} \right| =_{k\to\infty} O(k^{1-2\gamma_6+\delta}). \tag{4.13.8}$$

Taking into account that t is in the $(k^{-2-2\delta})$-neighbourhood of the nonsingular set, it is not hard to check the inequality:

$$\left| \frac{\partial p_j^2(t)}{\partial t_i} \right| > k^{1-2\delta}.$$

Considering in the same way as in the proof of Theorem 2.6, we verify estimates (4.13.3) and (4.13.4) and prove the uniqueness of $S'(k)$. If $V(x)$ is a trigonometric polynomial, then, using the estimate (4.2.23) and (4.2.33) instead of (4.13.7) and (4.13.8), we obtain (4.13.5) and (4.13.6). The theorem is proved.

As in the previous chapters, we denote by $S_H(k)_0$ the part of the perturbed isoenergetic surface situated in the $(k^{-2-2\delta})$-neighbourhood of $\chi_4(k,\rho,\delta,V)$. By Theorem 4.11, it is the union of smooth pieces. We denote by $e(t)$ a normal to $S_H(k)_0$ at a point t, and by $e_0(t_0), t_0 \in \chi_4(k,\rho,\delta,V)$, a normal to the isoenergetic surface of the free operator at a point t_0. It is clear that $e_0(t_0) = \mathbf{p}_j(t_0)/p_j(t_0)$, j being determined uniquely from the relation $p_j^2(t_0) = k^2$. If t is in the $(k^{-2-2\delta})$-neighbourhood of $\chi_4(k,\rho,\delta,V)$, then j is uniquely determined from the relation $p_j^{2l}(t_0) \in \varepsilon(k,\delta)$. Hence, the vector $\mathbf{p}_j(t_0)/p_j(t_0)$ is correctly defined in the $(k^{-2-2\delta})$- neighbourhood of $\chi_4(k,\rho,\delta,V)$.

Theorem 4.12 . *Suppose $t \in S_H(k)_0$. Then,*

$$e(t) =_{k\to\infty} e_0(t) + O(k^{-2\gamma_6+2\delta}). \tag{4.13.9}$$

The measure of surface $S_H(k)_0$ is asymptotically close to that of $S_0(k)$:

$$\frac{s(S_H(k)_0)}{s(S_0(k)} \to_{k\to\infty} 1. \tag{4.13.10}$$

Corollary 4.13 . *For the area of the perturbed isoenergetic surface $S_H(k)$ the following estimate holds:*

$$\lim_{k\to\infty} \frac{s(S_H(k))}{s(S_0(k))} \geq 1. \tag{4.13.11}$$

The estimate (4.13.11) is fulfilled because $S_H(k)_0 \subset S_H(k)$ and relation (4.13.10) holds.

The proof of the theorem is based on formula (4.10.5) and completely similar to that of Theorem 2.7

4.13.3 The behavior of the isoenergetic surface in a vicinity of the singular set.

Since the eigenvalues and the spectral projections of operator H are close to those of $\hat{H}$ on the essential part of the singular set, the isoenergetic surface of H is close to that of $\hat{H}$.

Let V be a trigonometric polynomial. Let us consider the surface $\hat{\mu}_q(k,\delta)$ in K (see (4.6.5)). We denote by $\hat{\mu}_q(k,\delta)_j$ the part of $\hat{\mu}_q(k,\delta)$ corresponding to some given j. A point t belongs to $\hat{\mu}_q(k,\delta)_j$ if and only if

$$p_j^2(t) + \Delta\Lambda(t)_{jj} = k^2. \tag{4.13.12}$$

Let $\mathbf{p}_{j\perp}(t)$ be the component of the vector $\mathbf{p}_j(t)$ orthogonal to $\mathbf{p}_q(0)$. Equation (4.13.12) is resolvable with respect to $|\mathbf{p}_{j\perp}(t)|$, since $\Delta\Lambda(t)_{jj}$ depends only on $(\mathbf{p}_j(t), \mathbf{p}_q(0))$:

$$|\mathbf{p}_{j\perp}(t)| = f_0((\mathbf{p}_j(t), \mathbf{p}_q(0))),$$

$$f_0((\mathbf{p}_j(t), \mathbf{p}_q(0))) = (k^2 - \Delta\Lambda(t)_{jj} - (\mathbf{p}_j(t), \mathbf{p}_q(0))^2 p_q(0)^{-2})^{1/2}. \tag{4.13.13}$$

We recall that $\hat{\mu}_q(k,\delta) = \mathcal{K}\hat{S}_q(k,-\delta)$ (see (4.6.6)). It is clear now that $\hat{S}_q(k,-\delta)$ can be represented by the formula for a curved cylinder:

$$x_\perp = f_0(x_\parallel), \quad x_\parallel = (x, \mathbf{p}_q(0))p_q^{-1}(0).$$

To obtain the surface $\mathcal{K}\hat{\mu}(k,\delta)$ in K one has to break this curved cylinder into pieces by the dual lattice and to translate all pieces into the elementary cell K. According to Theorem 4.4, there exists the subset $\hat{\chi}_q^0(k,V,\delta)$ of an asymptotically full measure on $\hat{\mu}_q(k,\delta)$, on which the perturbation series converge. We denote by $\hat{\chi}_q^0(k,V,\delta)_j$ the intersection of $\hat{\chi}_q^0(k,V,\delta)$ with $\hat{\mu}_q(k,\delta)_j$.

Lemma 4.38 . *In the $(k^{-2-2\delta})$-neighbourhood of every piece $\hat{\chi}_q^0(k,V,\delta)_j$, there are no other pieces $\hat{\chi}_q^0(k,V,\delta)_i$, $i \neq j$.*

Proof. Suppose that in the $(k^{-2-2\delta})$-neighbourhood of point t, $t \in \hat{\chi}_q^0(k,V,\delta)_j$, there exists a point t_0, $t_0 \in \hat{\chi}_q^0(k,V,\delta)_i$, $i \neq j$. Then,

$$p_j^2(t) + \hat{\Delta}\Lambda(t)_{jj} = k^2, \quad p_i^2(t_0) + \hat{\Delta}\Lambda(t_0)_{ii} = k^2, \tag{4.13.14}$$

$$|t - t_0| < k^{-2-2\delta}.$$

It is obvious that

$$|p_j^2(t) + \hat{\Delta}\Lambda(t)_{jj} - p_i^2(t) + \hat{\Delta}\Lambda(t)_{ii}| < k^{-1-2\delta}, \tag{4.13.15}$$

but this contradicts the definition of $\hat{\chi}_q^0(k,V,\delta)$ (see inequalities (4.6.12) and (4.6.13)). The lemma is proved.

Let $\hat{e}(t)$ be a unit vector orthogonal to $\hat{\mu}_q(k,\delta)$ at a point t.

Theorem 4.13 . *For $0 < \delta < 1/300$ and sufficiently large k, $k > k_0(V,\delta)$, in the $(k^{-2-2\delta})$-neighbourhood of each simply connected component of $\hat{\chi}_q^0(k,V,\delta)$, there exists a unique piece $\hat{S}_H^q(k)$ of the isoenergetic surface of H. In fact, it is in the smaller $(k^{-2-\gamma_4})$-neighbourhood of $\hat{\chi}_q^0(k,V,\delta)$ and can be described by the equation*

$$\hat{\lambda}(t) = k^2, \tag{4.13.16}$$

where the function $\hat{\lambda}(t)$ is determined by the series (4.6.18). The corresponding spectral projection is given by formula (4.6.19). Terms of these series satisfy estimates (4.6.20) – (4.6.24) and (4.6.28) – (4.6.32). The unit vector $e(t)$, $t \in S'_H(k)$,, orthogonal to $S'_H(k)$ at a point t, can be represented in the form:

$$e(t) = \hat{e}(t) + O(k^{-2\gamma_4}), \quad \gamma_4 = 1/15 - 20\delta. \tag{4.13.17}$$

Proof. Suppose t belongs to the $(k^{-2-2\delta})$-neighbourhood of $\hat{\chi}_q^0(k,V,\delta)$. According to Corollaries 4.8 and 4.9 the following estimates hold:

$$|\hat{\lambda}(t) - p_j^2(t) - \Delta A(t)_{jj}| < ck^{-1-2\gamma_4}, \tag{4.13.18}$$

$$|\nabla\hat{\lambda}(t) - \nabla(p_j^2(t) - \Delta A(t)_{jj})| < ck^{1-2\gamma_4+\delta}. \tag{4.13.19}$$

From the last relation and inequality (4.3.33), it follows that

$$|\nabla\hat{\lambda}(t)| > ck. \tag{4.13.20}$$

Using estimates (4.13.18) and (4.13.19), we obtain that equation (4.13.16) has a solution in $(k^{-2-2\gamma_4})$-neighbourhood of any point t_0 belonging to $\hat{\chi}_q^0(k,V,\delta)$; and these solutions form a surface in $(k^{-2-\gamma_4})$-neighbourhood of $\hat{\chi}_q^0(k,V,\delta)$, because (4.13.19) holds.

Next, we prove that equation (4.13.16) has no other solutions in $(k^{-2-2\delta})$-neighbourhood of $\hat{\chi}_q^0(k,V,\delta)$. Indeed, let $\tilde{t}$ be in the $(k^{-2-2\delta})$-neighbourhood of $\hat{\chi}_q^0(k,V,\delta)$ and let there exist an eigenvalue $\tilde{\lambda}(\tilde{t})$ of operator H, which is not represented as the series (4.6.18), i.e., $\tilde{\lambda}(\tilde{t}) \neq \hat{\lambda}(\tilde{t})$. According to Theorem 4.4, $\hat{\lambda}(\tilde{t})$ is a unique eigenvalue in the interval $(k^2 - k^{-1-\delta}, k^2 + k^{-1-\delta})$. However, $\tilde{\lambda}(\tilde{t}) = k^2$. From this it follows that $\tilde{\lambda}(\tilde{t}) = \hat{\lambda}(\tilde{t})$. The theorem is proved.

Let $q = q_0 \equiv (1,0,0)$. We consider $S_H^{q_0}(k)$, i.e., the part of the perturbed isoenergetic surface lying in the $(k^{-2-2\gamma_4})$-neighbourhood of $\hat{\chi}_q^0(k,V,\delta)$ (see the foregoing theorem). In solving the semicrystal problem, we shall use the following lemma.

Lemma 4.39 . *If $t \in S_H^{q_0}$, then for sufficiently large k, $k > k_0(V,\delta)$, there exists a unique $t_0 \in \hat{\mu}_{q_0}(k,\delta)$, such that*

$$t_{\|} = t_{0\|}, \quad |t_1 - t_{01}| < k^{-8/5+12\delta} \tag{4.13.21}$$

and

$$|\sin(a_1 t_{01})| > k^{3\delta}. \tag{4.13.22}$$

Proof. If $t \in S_H^{q_0}$, then $\hat{\lambda}(t) = k^2$ and

$$\hat{\lambda}(t) = p_j^2(t) + \Delta\hat{A}(t)_{jj} + O(k^{-8/5+10\delta}). \tag{4.13.23}$$

To prove (4.13.21) it suffices to show that

$$\left|\frac{\partial(p_j^2(t) + \Delta\hat{A}(t)_{jj})}{\partial t_1}\right| > ck^{-2\delta}. \tag{4.13.24}$$

From the last two estimates, (4.13.21) easily follows. Note that $p_j^2(t) + \Delta\hat{A}(t)_{jj}$ is an eigenvalue of the operator $P_q V_q P_q$. Using that

$$|(\mathbf{p}_j(0), \mathbf{p}_q(0))| < k^\delta << k^{1/5}, \tag{4.13.25}$$

we easily show (see Proposition 4.1) that $p_j^2(t) + \Delta\hat{A}(t)_{jj} = p_j^2(t) + \Delta A_{q_0}(t)_{jj} + O(k^{-k^{1/5}})$, where $p_j^2(t) + \Delta A_{q_0}(t)_{jj}$ is an eigenvalue of the operator $H_0 + V_{q_0}$, with the potential V_{q_0} depending only on x_1. From condition 4′ (see (4.6.13)), it follows that the point $p_j^2(t) + \Delta\hat{A}(t)_{jj}$ is situated at a distance greater than $k^{-3\delta}$ from the nearest one $p_m^2(t) + \Delta\hat{A}(t)_{mm}$, $m \neq j$, $\mathbf{p}_m(0) = \mathbf{p}_j(0) + l\mathbf{p}_{q_0}(0)$, $l \in Z$, $|(\mathbf{p}_m(0), \mathbf{p}_q(0))| < 2k^\delta$. For this nearest point we also have $p_m^2(t) + \Delta\hat{A}(t)_{mm} = p_m^2(t) + \Delta A_{q_0}(t)_{mm} + O(k^{-k^{1/5}})$. Now, considering as in Proposition 4.1, we easily get:

$$\left|\frac{\partial(p_j^2(t) + \Delta\hat{A}(t)_{jj})}{\partial t_1} - \frac{\partial(p_j^2(t) + \Delta A_{q_0}(t)_{jj})}{\partial t_1}\right| < k^{-k^{2/5}}k^{4\delta}. \tag{4.13.26}$$

We remark that the study of the operator $H_0 + V_q$ can be reduced to that of the Schrödinger operator in $L_2(R)$. It is known that for such an operator the derivative of an eigenvalue with respect to a quasimomentum t is separated from zero when this eigenvalue is far enough from the others eigenvalues. Noting that $p_j^2(t) + \Delta A_q(t)_{jj}$ is far enough from the nearest point $p_m^2(t) + \Delta A_q(t)_{mm}$, $m \neq j$, and putting the foregoing arguments on a rigorous basis, we obtain that

$$\left|\frac{\partial(p_j^2(t) + \Delta A_{q_0}(t)_{jj})}{\partial t_1}\right| > ck^{-3\delta}. \tag{4.13.27}$$

From (4.13.27) and (4.13.26), inequality (4.13.24) and, therefore, (4.13.21) easily follows.

Let us prove (4.13.22). Using notations (4.3.29) – (4.3.31)), we obtain that

$$\left|\frac{\partial(p_j^2(t) + \Delta\hat{A}(t)_{jj})}{\partial t_1}\right| = \frac{\partial\lambda_{j_1}^{q_0}(t_1)}{\partial t_1},$$

where $\lambda_{j_1}^{q_0}(t_1)$ is an eigenvalue of the one-dimensional Schrödinger operator (see (4.3.27)) for $q = q_0$. From (4.13.27) it follows

$$\left|\frac{\partial \lambda_{j_1}^{q_0}(t_1)}{\partial t_1}\right| > k^{-3\delta}.$$

Taking into account that the derivative is equal to zero only when $\sin(a_1 t_1) = 0$, we get

$$\left|\frac{\partial \lambda_{j_1}^{q_0}(t_1)}{\partial t_1}\right| > c_j |\sin(a_1 t_1)|, c_j > 0.$$

Taking into account that

$$\left|\frac{\partial^2 \lambda_{j_1}^{q_0}(t_1)}{\partial t_1^2}\right| \to_{|j_1| \to \infty} \infty,$$

when $t_1 = 0, \pi$, we get (4.13.22). The lemma is proved.

Let $\chi_{q_0}^2(k, V, \delta)$ be a subset of $\hat{\mu}_{q_0}(k, \delta)$:

$$\chi_{q_0}^2(k, V, \delta) = \{t_0 : \exists t \in S_H^{q_0}, t_{\|} = t_{0\|}, |t_1 - t_{01}| < k^{-8/5+12\delta}\}, \tag{4.13.28}$$

i.e., $\chi_{q_0}^2(k, V, \delta)$ is the "trace" of the set $S_H^{q_0}$ on $\hat{\mu}_q(k, \delta)$, when $S_H^{q_0}$ is projected on the plane $t_1 = 0$.

Lemma 4.40 . *The set $\chi_{q_0}^2(k, V, \delta)$ has an asymptotically full measure on $\hat{\mu}(k, \delta)$. Moreover,*

$$\frac{s(\hat{\mu}(k, \delta) \setminus \chi_{q_0}^2(k, V, \delta))}{s(\hat{\mu}(k, \delta)} < k^{-\delta}. \tag{4.13.29}$$

Proof. From the proof of the previous lemma, we see that χ_2 lies in the $(k^{-8/5+12\delta})$-neighbourhood of $S_H^{q_0}$ and their normals are close, because the formula for $\hat{\lambda}(t)$ can be differentiated. From this it follows that

$$s(\chi_2) = s(S_H^{q_0})(1 + o(k)).$$

Taking into account that $s(S_H^{q_0}) = s(\hat{\chi}_{q_0}^0)(1 + o(k))$, and using (4.6.15), we obtain estimate (4.13.29). The lemma is proved.

4.14 Formulae for Eigenfunctions on the Isoenergetic Surface.

Suppose V is a trigonometric polynomial. Following the scheme described in Section 2.7, we construct the formulae for the eigenfunctions for t being on $S_H(k)_0$.

Theorem 4.14 . *Suppose t belongs to $\chi_3(k, V, \delta)$. Then at the point $t(t_0)$, determined by the formula (4.13.2), the following formulae for $\nabla\lambda(t)$ and $E(t)$ hold:*

$$\nabla\lambda(t) = \mathbf{p}_j(t_0)(1 + O(k^{-3+10\delta})), \tag{4.14.1}$$

$$E(t) = E_j + G_1(k^2, t_0) + G_2(k^2, t_0) + O(k^{-3+12\delta}). \tag{4.14.2}$$

The proof is similar to that of Theorem 2.9. Indeed, according to Corollary 4.3, the following estimate hold, for any point t' in the $(k^{-2-2\delta})$-neighbourhood of $\chi_3(k,V,\delta)$:

$$| T(m)\lambda(t') - T(m)p_j^2(t) |< ck^{-2+12\delta}, \quad |m| = 2. \tag{4.14.3}$$

Therefore,

$$| \nabla\lambda(t) - \nabla\lambda(t_0) |< c \,|\, t - t_0 \,|\, k^{-2+12\delta} = O(k^{-4+10\delta}). \tag{4.14.4}$$

On the other hand,

$$\nabla\lambda(t_0) = \mathbf{p}_j(t_0)(1 + O(k^{-3+9\delta})). \tag{4.14.5}$$

Formula (4.14.1) immediately follows from (4.14.4) and (4.14.5). According to Corollary 4.3, we have:

$$\|\partial E(t)/\partial t_1\|_1 < ck^{-1+6\delta}. \tag{4.14.6}$$

Taking into account this estimate we prove relation (4.14.2) similarly as (4.14.1). The theorem is proved.

We can calculate $T(m)\lambda(t)$ and $T(m)E(t)$ in a similar way. However, the accuracy of the formulae is restricted by that of the approximation $t_1 \approx t_{01}$. To write out the next terms of the asymptotic expansion one has to solve more precisely the equation $\lambda(t) = k^2$. But it seems not to be an effective way. In Section 2.7 we described another way for constructing of a formula for the eigenfunction. That way is not connected with solving the equation $\lambda(t) = k^2$. The scheme turns out to be valid for the case of the Schrödinger equation too. Let us describe its main points.

Suppose $t \in S_H(k)_0$. According to Theorem 4.11, t can be represented in the form (4.13.2). We set $i = 1$ and consider the integral

$$I(k^2, t_0) = \frac{1}{2\pi i}\int_{C_1} (H(t) - k^2)^{-1} dt_1, \tag{4.14.7}$$

where $t = (t_1, t_{02}, t_{03})$ and C_1 is the circle of radius $k^{-2-2\delta}$ centered at point t_{01}.

Lemma 4.41 . *Suppose t_0 belongs to $\chi_3(k,V,\delta)$. Then the operator $(H(t) - k^2)^{-1}$ has a unique pole inside C_1 at the point $t(t_0)$, given by relation (4.13.2), i being equal to 1. The following formula holds for $I(k^2, t_0)$:*

$$I(k^2, t_0) = \left.\frac{E(t)}{(\partial\lambda(t)/\partial t_1)}\right|_{t=t(t_0)}. \tag{4.14.8}$$

The proof is similar to that of Lemma 2.14 . We use Theorem 4.9 and formula (4.13.2) instead of Theorem 2.6 and formula (2.6.1).

Further, expanding formally $(H(t) - k^2)^{-1}$ in the series (4.1.2), (4.1.3), we obtain

$$I(k^2, t_0) = \sum_{r=0}^{\infty} \hat{D}_r(k, t_0), \tag{4.14.9}$$

where

$$\hat{D}_r(k, t_0) = \frac{(-1)^r}{2\pi i} \oint_{C_1} (\hat{H}(t) - k^2)^{-1} (W(\hat{H}(t) - k^2)^{-1})^r dt. \tag{4.14.10}$$

It is clear that $\hat{D}_0(k, t_0)$ is the integral $I(k^2, t_0)$ for the operator $\hat{H}(t)$. According to Lemma 4.13, the operator $\hat{H}(t)$ has a unique eigenvalue inside C_0. It is equal to k^2. Note that $k^2 = p_j^2(t_0)$, j being uniquely determined by the equation. Now we see that, in fact, $\hat{D}_0(k, t_0)$ is equal to the integral $I(k^2, t_0)$ for the free operator:

$$D_0(k, t_0) = \frac{E_j}{p_{j_1}(t_1)}, \tag{4.14.11}$$

$$p_{j_1}(t_1) = t_1 + 2\pi j_1 a_1^{-1}.$$

To justify the asymptotic formula for $I(k^2, t_0)$,, it suffices to prove the power estimates similar to (4.2.16) for $\hat{D}_r(k, t_0)$. In formula (4.14.9) the asymptotic terms explicitly depend on k^2 and t_0. It is not necessary to solve the equation $\lambda(t) = k^2$. Now it remains to prove the estimates for $\hat{D}_r(k, t_0)$. Let us prove that (4.14.9) can be simplified, because $\hat{D}_r(k, t_0) = D_r(k, t_0)$ for r small enough, $D_r(k^2, t_0)$ being given by formula (2.7.8).

Lemma 4.42 . *Suppose t_0 belongs to $\chi_3(k, V, \delta)$. Then at the point $t(t_0)$, defined by formula (4.13.2), the following relations hold:*

$$\hat{D}_r(k, t) = D_r(k, t), \quad r < R_1, \quad R_1 = k^{1-4\delta} R_0^{-1}. \tag{4.14.12}$$

The proof is quite similar to the proof of relation (4.5.125), and is based on the fact that V is a trigonometric polynomial.

Theorem 4.15 . *Suppose t_0 belongs to $\chi_3(k, V, \delta)$. Then at the point $t(t_0)$, defined by formula (4.13.2), the following formula holds:*

$$\left. \frac{E(t)}{(\partial \lambda(t) / \partial t_1)} \right|_{t=t(t_0)} = \sum_{r=0}^{\infty} \hat{D}_r(k, t_0), \tag{4.14.13}$$

where the series converges in the class $\mathbf{S}_1$. The operator-valued functions $\hat{D}_r(k, t_0)$ satisfy the estimates:

$$\| \hat{D}_r(k, t_0) \|_1 < k^{-1-r/20}. \tag{4.14.14}$$

Proof. It suffices to prove estimate (4.14.14). Suppose we have verified the following inequalities:

$$\max_{t \in C_1} \| \hat{A}(k^2, W, t) \|_1 < k^{4\delta}, \tag{4.14.15}$$

$$\max_{t\in C_1}\|\hat{A}^3(k^2,W,t)\|_1 < k^{-1/5+16\delta}, \tag{4.14.16}$$

$$\max_{t\in C_1}\|(\hat{H}(t)-k^2)^{-1/2}\| < k^{1/2+2\delta} \tag{4.14.17}$$

(see formulae (4.2.7) and (4.2.8). Then, estimating $\hat{D}_r(k,t_0)$ by the norm of the integrand and the length $2\pi k^{-2-2\delta}$ of the circle C_1, we obtain:

$$\|\hat{D}_r(k,t_0)\|_1 < k^{-1-r/20}. \tag{4.14.18}$$

Thus, it remains to verify estimates (4.2.9) and (4.5.84). Considering as in the proof of Theorem 2.10, we obtain them from estimates (4.5.97), (4.5.98), (4.5.84). Let us prove (4.14.15). Note that in the formulae for $\hat{g}_r$ and $\hat{G}_r$, the contour C_0 is the same for all t in the $(k^{-2-2\delta})$-neighbourhood of the nonsingular set. Let us take for each t its special contour $C(t)$, centered at the point $z=p_j^2(t)$ and with the radius $r(t)k^{-1-\delta}$, $r(t)$ satisfying the inequality $k^{-2\delta} < r(t) < 1$. The results of the integrations preserve, when replacing C_0 by $C(t)$, because the integrands are holomorphic between C_0 and $C(t)$. Moreover, the estimates for the integrands are stable relative to such a perturbation of the contour. Indeed, after replacing δ by 3δ, repeating all the considerations of Theorem 4.1, we obtain, instead of estimates (4.2.9), the similar ones:

$$\max_{z\in C(t)}\|\hat{A}(k^2,W,t)\|_1 < k^{4\delta}, \tag{4.14.19}$$

$$\max_{z\in C(t)}\|\hat{A}^3(k^2,W,t)\|_1 < k^{-1/5+16\delta}, \tag{4.14.20}$$

$$\max_{z\in C(t)}\|(\hat{H}(t)-k^2)^{-1/2}\| < k^{1/2+2\delta}. \tag{4.14.21}$$

Suppose $t_0 \in \chi_3(k,V,\delta)$, $t\in C_1(t_0)$. It is clear that t belongs to the complex $(k^{-2-2\delta})$-neighbourhood of $\chi_3(k,V,\delta)$. Hence, estimates (4.14.19) – (4.14.21) hold for t. It is not hard to show that

$$|\,k^2-p_j^2(t)\,|=|\,p_j^2(t)-p_j^2(t_0)\,|=r(t)k^{-1-\delta},$$

where $r(t)\approx k^{-\delta}$. Therefore, $k^2\in C(t)$ for this $r(t)$. Estimates (4.14.19)–(4.14.21) for $z=k^2$ imply inequalities (4.14.15) – (4.14.17).
The theorem is proved.

Theorem 4.16 . *Under the conditions of Theorem 4.15, the following formula holds:*

$$\left.\frac{E(t)}{\partial\lambda(t)/\partial t_1}\right|_{t=t(t_0)} = \sum_{r=0}^{R_1} D_r(k,t_0)+\hat{\Psi}(k,t_0), \tag{4.14.22}$$

$$\hat{\Psi}(k,\delta)=\sum_{r=R_1+1}^{\infty}\hat{D}_r(k,t_0). \tag{4.14.23}$$

The operator-valued functions $D_r(k,t_0)$ and $\hat{\Psi}(k,\delta)$ satisfy the estimates:

$$\|D_r(k,t_0)\|_1 < k^{-1-r/20}, \tag{4.14.24}$$

$$\|\hat{\Psi}(k,\delta)\|_1 < k^{-1-(R_1+1)/20}. \tag{4.14.25}$$

This theorem immediately follows from Theorem 4.15 and Lemma 4.42.

We consider the following function:

$$\Psi(k^2,t_0,x)=\sum_{m\in Z^3}\oint_{C_1}(H(t)-k^2)^{-1}_{mj}\exp(i(\mathbf{p}_m(t),x))dt_1, \tag{4.14.26}$$

where $t_0\in\chi_3(k,V,\delta)$, and j is uniquely determined from the relation $p_j^2(t_0)=k^2$. It is easy to see that $\Psi(k^2,t_0,x)$ satisfies the equation $(-\Delta+V)\Psi=k^2\Psi$ and the quasiperiodic conditions with the real quasimomentum $t=(t_1,t_{02},t_{03})$, t_1 being given by formula (4.13.2). Thus, $\Psi(k^2,t_0,x)$ is an eigenfunction of the operator $H(t)$, $t\in S_H(k)_0$. Formally expanding the resolvent into series (4.1.2), we get:

$$\Psi(k^2,t_0,x)=\sum_{r=0}^{\infty}\hat{B}_r(k^2,t_0,x), \tag{4.14.27}$$

where the functions $\hat{B}_r(k^2,t_0,x)$ are defined by the formula:

$$\hat{B}_r(k^2,t_0,x)=$$

$$\frac{(-1)^r}{2\pi i}\sum_{m\in Z^3}\oint_{C_1}[(\hat{H}(t)-k^2)^{-1}(W(\hat{H}(t)-k^2)^{-1})^r]_{mj}\exp(i(\mathbf{p}_m(t),x))dt_1, \tag{4.14.28}$$

$$t=(t_1,t_{02},t_{03}).$$

It is not hard to show that the functions $\hat{B}_r(k^2,t_0,x)$ satisfy quasiperiodic conditions in the directions orthogonal to x_1. The corresponding components of the quasimomentum are t_{02},t_{03}.

Let us describe the results we shall need for solving a semicrystal problem (Chapter 5). We introduce the notation:

$$\|\hat{B}_r(k^2,t_0,x)\|_{2,M}\equiv\max_{|x_1|<M}\int_{K_\parallel}|\hat{B}_r|^2\,dx_2dx_3, \tag{4.14.29}$$

where $K_\parallel=[0,a_2)\times[0,a_3)$. We recall that B_r is given by the formula:

$$B_r(k^2,t_0,x)=$$

$$\frac{(-1)^r}{2\pi i}\sum_{m\in Z^3}\oint_{C_1}[(H_0(t)-k^2)^{-1}(V(H_0(t)-k^2)^{-1})^r]_{mj}\exp(i(\mathbf{p}_m(t),x))dt_1, \tag{4.14.30}$$

$$t=(t_1,t_{02},t_{03}).$$

(see also (2.7.19)). Let us consider the space $L_2(K_\parallel)$. We denote by $x_\parallel$ the independent variable $x_\parallel\in K_\parallel$, $x_\parallel=(x_2,x_3)$. Let $m_\parallel$ be a two-dimensional integer, i.e., $m_\parallel\in Z^2$, $m_\parallel=(m_2,m_3)$, $m_2,m_3\in Z$. We use the notations:

$$p_{m_\parallel}(0)=\left(\frac{2\pi m_2}{a_2},\ \frac{2\pi m_3}{a_3}\right),\quad t_\parallel=(t_2,t_3),\quad t_\parallel\in\left[0,\frac{2\pi}{a_2}\right)\times\left[0,\frac{2\pi}{a_3}\right).$$

It is clear that the functions $\left\{\exp i(p_{m_\|}(0)+t_\|,x_\|)\right\}_{m_\| \in Z^2}$ form a basis in $L_2(K_\|)$. We call this basis $t_\|$*-basis.*

Let us consider the function $\hat{B}_r(k^2,t_0,x)$ as a function of $x_\|$, $x_\| \in K_\|$ for a fixed x_1. This function satisfies the quasiperiodic conditions in the directions a_2, a_3 with the quasimomentum $t_{0\|}=(t_{02},t_{03})$. Let $\left\{\hat{B}_r(k^2,t_0,x_1)_{m_\|}\right\}_{m_\| \in Z^2}$ be the Fourier coefficients of this function in $t_{0\|}$-basis:

$$\hat{B}_r(k^2,t_0,x_1)_{m_\|}=\frac{1}{|K_\||}\int_{K_\|}\hat{B}_r(k^2,t_0,x)exp-i(p_{m_\|}(0)+t_\|,x_\|)dx_\|,$$

$$|K_\||=a_2a_3.$$

Let us recall that j is uniquely determined from the equation $p_j^2(t_0)=k^2$ for a given $t_0 \in \chi_3(k,V,\delta)$ and k^2 large enough. Let $j_\|$ be the $\|$-component of j: $j_\|=(j_2,j_3)$.

Lemma 4.43 . *Under conditions of Theorem 4.15, for any x_1 and r*

$$B_r(k^2,t_0,x_1)_{m_\|}=0 \tag{4.14.31}$$

when $|j_\|-m_\||>rR_0$, and

$$\hat{B}_r(k^2,t_0,x_1)_{m_\|}=0, \tag{4.14.32}$$

when $|j_\|-m_\||>rk^{1/5}$.

When $r<R_1$, $R_1=k^{1-4\delta}R_0^{-1}$, the following relation holds:

$$\hat{B}_r(k^2,t_0,x)=B_r(k^2,t_0,x). \tag{4.14.33}$$

Proof The proofs of (4.14.31), (4.14.32) are similar to the proof of (2.7.24). In proving (4.14.32) we take into account that $W_{im}=0$, when $|i-m|>k^{1/5}$. The proof of (4.14.33) is similar to that of (4.5.125) and based on the fact that V is a trigonometric polynomial. The lemma is proved.

Lemma 4.44 . *Under conditions of Theorem 4.15, the functions $B_r(k^2,t_0,x)$ and $\hat{B}_r(k^2,t_0,x)$ can be represented as follows::*

$$B_r(k^2,t_0,x)=\sum_{0<l<r,|n-j|<rR_0}a_{nl}^{(r)}\exp i(\mathbf{p}_n(t_0),x)\frac{x_1^l}{l!}, \tag{4.14.34}$$

$$\hat{B}_r(k^2,t_0,x)=\sum_{0<l<r,|n-j|<rk^{1/5}}\hat{a}_{nl}^{(r)}\exp i(\mathbf{p}_n(t_0),x)\frac{x_1^l}{l!}. \tag{4.14.35}$$

For all r, the coefficients $a_{nl}^{(r)}$ and $\hat{a}_{nl}^{(r)}$ satisfy the estimates:

$$|a_{nl}^{(r)}|<k^{-1-r/20+(2+\delta)l}, \tag{4.14.36}$$

$$|\hat{a}_{nl}^{(r)}| < k^{-1-r/20+(2+\delta)l}. \tag{4.14.37}$$

The estimates for $a_{nl}^{(r)}$ can be improved when $r < k^{3\delta}$:

$$|a_{nl}^{(r)}| < k^{-1-(1-8\delta)r+(2+\delta)l}. \tag{4.14.38}$$

Proof. We use formula (4.14.27). The proof is quite similar to that of Theorem 2.11. For proving (4.14.38), we take into account that the integrand vanishes when $|j_{\|} - m_{\|}| > rR_0$. Then, considering that $|p_i^2(t) - p_j^2(t)| > k^{1-8\delta}$, for $t \in \chi_3(k, V, \delta)$, we obtain (4.14.38). The lemma is proved.

Theorem 4.17 . *Suppose t_0 belongs to $\chi_3(k, V, \delta)$. Then for sufficiently large k, $k > k_0(k, V, \delta)$, the function $\psi(k^2, t_0, x)$ can be represented by the series:*

$$\psi(k^2, t_0, x) = \sum_{r=0}^{R_1} B_r(k^2, t_0, x) + \sum_{r=R_1+1}^{\infty} \hat{B}_r(k^2, t_0, x), \tag{4.14.39}$$

$$B_0(k^2, t_0, x) = \frac{\exp(i(\mathbf{p}_j(t), x))}{2p_{j_1}(t_1)}, \tag{4.14.40}$$

$$p_{j_1}(t_1) = 2\pi j_1 a_1^{-1} + t_1,$$

$$t = (t_1, t_{02}, t_{03}).$$

The functions $B_r(k^2, t_0 x)$ and $\hat{B}_r(k^2, t_0, x)$ satisfy the estimates:

$$\left\|B_r(k^2, t_0, x)\right\|_{2,M} < c(1 + M^r)k^{-1-r/20}, \tag{4.14.41}$$

$$\left\|\frac{\partial B_r(k^2, t_0, x)}{\partial x_1}\right\|_{2,M} < c(1 + M^r)k^{-r/20}, \tag{4.14.42}$$

$$\left\|\hat{B}_r(k^2, t_0, x)\right\|_{2,M} < c(1 + M^r)k^{-1-r/20}, \tag{4.14.43}$$

$$\left\|\frac{\partial \hat{B}_r(k^2, t_{20}, t_{20}, x)}{\partial x_1}\right\|_{2,M} < c(1 + M^r)k^{-r/20}. \tag{4.14.44}$$

The estimates for $B_r(k^2, t_0, x)$ can be improved when $r < k^{3\delta}$:

$$\left\|B_r(k^2, t_0, x)\right\|_{2,M} < c(1 + M^r)k^{-1-(1-8\delta)r}, \tag{4.14.45}$$

$$\left\|\frac{\partial B_r(k^2, t_0, x)}{\partial x_1}\right\|_{2,M} < c(1 + M^r)k^{-(1-8\delta)r}. \tag{4.14.46}$$

Proof. The proof of estimates (4.14.41) – (4.14.46) is similar to the proof of estimates (2.7.27) and (2.7.28) in Theorem 2.12. We use estimates (4.14.36) – (4.14.38). The theorem is proved.

In Chapter 5 we shall use this result in the following form:

Theorem 4.18 . *Suppose t_0 belongs to $\chi_3(k,V,\delta)$. Then for sufficiently large k, $k > k_0(k,V,\delta)$, the function $\psi(k^2,t_0,x)$ can be represented by the formula:*

$$\psi(k^2,t_0,x) = \sum_{r=0}^{R_2} B_r(k^2,t_0,x) + C(k^2,t_0,x), \tag{4.14.47}$$

$$R_2 = k^{3\delta}/R_0].$$

The functions $B_r(k^2,t_0,x)$, $C(k^2,t_0,x)$ satisfy the estimates:

$$\|B_r(k^2,t_0,x)\|_{2,M} < c(1+M^r)k^{-1-(1-8\delta)r}, \tag{4.14.48}$$

$$\|C(k^2,t_0,x)\|_{2,M} < c(1+M^{R_2+1})k^{-1-(R_2+1)(1-8\delta)}, \tag{4.14.49}$$

$$\left\|\frac{\partial B_r(k^2,t_0,x)}{\partial x_1}\right\|_{2,M} < c(1+M^r)k^{-(1-8\delta)r}, \tag{4.14.50}$$

$$\left\|\frac{\partial C(k^2,t_0,x)}{\partial x_1}\right\|_{2,M} < c(1+M^{R_2+1})k^{-(R_2+1)(1-8\delta)}. \tag{4.14.51}$$

Proof. Using Theorem 4.17, we obtain formula (4.14.47), where

$$C(k^2,t_0,x) = \sum_{r=R_2+1}^{\infty} \hat{B}_r(k^2,t_0,x). \tag{4.14.52}$$

Estimates (4.14.48)) – (4.14.51) follow from (4.14.41) – (4.14.46).
The theorem is proved.

Theorem 4.19 . *Suppose t belongs to $\hat{\chi}_q^0(k,V,\delta)$. Then, at the point $t(t_0)$ of $S_H^q(k)$, being in the $(k^{-2-\gamma_4})$ of t_0, the following formulae for $\nabla\lambda(t)$ and $E(t)$ are valid:*

$$\nabla\lambda(t) = \nabla(p_j^2(t_0) + \Delta\Lambda(t_0)_{jj})(1 + O(k^{-2/5}), \tag{4.14.53}$$

$$E(t) = \hat{E}_j^0 + \hat{G}_1'(k,t_0) + \hat{G}_2'(k,t_0) + \hat{G}_3'(k,t_0) + O(k^{-4\gamma_4}). \tag{4.14.54}$$

The proof is similar to that of Theorem 2.9. Indeed, according to Corollary 4.9, the following estimate holds for any point t' in the $(k^{-2-2\delta})$-neighbourhood of $\hat{\chi}_q^0(k,V,\delta)$:

$$| T(m)\nabla\hat{\lambda}(t') | < ck^{1+2\delta-2\gamma_4}, \quad |m| = 1. \tag{4.14.55}$$

Therefore,

$$| \nabla\hat{\lambda}(t) - \nabla\hat{\lambda}(t_0) | < c \,| t - t_0 |\, k^{1+2\delta-2\gamma_4} = O(k^{-3\gamma_4+4\delta}). \tag{4.14.56}$$

On the other hand,

$$\nabla\hat{\lambda}(t_0) = \nabla(p_j^2(t_0) + \Delta\hat{\Lambda}(t_0)_{jj})(1 + O(k^{-2/5+10\delta})). \tag{4.14.57}$$

It follows from (4.14.56) and (4.14.57) that

$$\nabla\lambda(t) = \nabla(p_j^2(t_0) + \Delta\hat{\Lambda}(t_0)_{jj})(1 + O(k^{-2/5}). \tag{4.14.58}$$

Taking into account (4.13.26)and that

$$\frac{\partial(p_j^2(t) + \Delta\hat{\Lambda}(t)_{jj})}{\partial t_i} = \frac{\partial(p_j^2(t) + \Delta\Lambda_{q_0}(t)_{jj})}{\partial t_i}, \quad i = 2, 3,$$

we get (4.14.53).

According to Corollary 4.9, we have:

$$\|\partial\hat{E}(t)/\partial t_1\|_1 < ck^{2-3\gamma_4+2\delta} \tag{4.14.59}$$

Taking into account this estimate we prove relation (4.14.54) similarly as (4.14.2). The theorem is proved.

Let $q_0 = (1, 0, 0)$, and V_{q_0} is the part of potential V depending only on x_1:

$$V_{q_0}(x) = \sum_{n\in Z} v_{nq_0} e^{(p_{nq_0}(0),x)}.$$

Let H_1 be the operator in $L_2(R^3)$ corresponding to such V_1. It is clear that the study of this operator can be reduced to that of the operator $\tilde{H}_1$ in $L_2(R)$:

$$\tilde{H}_1 = -\frac{d^2}{dx_1^2} + V_1. \tag{4.14.60}$$

Let $\hat{\chi}^2_{q_0}(k, V, \delta)$ be defined by (4.13.28).

Theorem 4.20 . *Suppose $t \in S_H^{q_0}$, $t_0 \in \hat{\chi}^2_{q_0}(k, V, \delta)$ and $t_{0\|} = t_\|$, $|t_{01} - t_1| < k^{-8/5+12\delta}$. Let j satisfy the equation*

$$p_j^2(t_0) + \Delta\Lambda q_0(t_0)_{jj} = k^2 \tag{4.14.61}$$

and $|j_1| < c$, $c \neq c(k)$. Then the eigenfunction $\psi(t, x)$ of $H(t)$ satisfies the asymptotic relations:

$$\psi(t, x)|_{x_1=0} = \psi_1(t_0, x)|_{x_1=0} + o(k^{-1}), \tag{4.14.62}$$

$$\left.\frac{\partial\psi(t, x)}{\partial x_1}\right|_{x_1=0} = \left.\frac{\partial\psi_1(t_0, x)}{\partial x_1}\right|_{x_1=0} + o(k^{-1}); \tag{4.14.63}$$

here $\psi_1(t_0, x)$ is the eigenfunction of the operator $H_1(t_0)$, corresponding to the eigenvalue (4.14.61) with the unit norm in $L_2(K)$, and $o(k^{-1})$ is infinitely small in the class $L_2(K_\|)$. The asymptotic hold in the $(k^{-2-2\delta})$-neighbourhood of $S_H^{q_0}$.

Proof. Since $S_H^{q_0}$ is in the $(k^{-2-2\delta})$-neighbourhood of $\hat{\chi}_{q_0}^0(k,V,\delta)$, for $t \in S_H^{q_0}$, the following formula holds:

$$E(t) = \hat{E}_j(t) + \sum_{r=1}^{\infty} G'_r(k,t).$$

We consider the functions

$$\hat{\psi}_1(t,x) = b\sum_n (\hat{E}_j)_{nm} \exp i(\mathbf{p}_n(t),x), \qquad (4.14.64)$$

$$\psi(t,x) = b\sum_n (E)_{nm} \exp i(\mathbf{p}_n(t),x), \qquad (4.14.65)$$

$$b = (\hat{E}_j)_{mm}^{-1/2},$$

where $m = m(j)$ is chosen to satisfy the condition $(\hat{E}_j)_{mm} = \max_n(\hat{E}_j)_{nn}$. Since $|j_1| < c$, we have $(\hat{E}_j)_{mm} > c_0$, $c_0 \neq c_0(k)$. It is clear that

$$\psi(t,x) = \psi_1(t,x) + \sum_{r=1}^{\infty} \tilde{B}'_r, \qquad (4.14.66)$$

$$\tilde{B}'_r = b\sum_n (G'_r)_{nm} \exp i(\mathbf{p}_n(t),x).$$

Let us estimate $\tilde{B}'_r|_{x_1=0}$ and $\frac{\partial}{\partial x_1}\tilde{B}'_r|_{x_1=0}$. First, we introduce the diagonal operator L: $L_{nn} = n_1 + t_1$. Noting that $(\hat{H})_{nm} = W_{nm} = 0$, when $|n-m| > k^{1/5+\delta}$, we obtain

$$(G'_r)_{nm} = (LG'_r)_{nm} = 0 \text{ when } |n-j| > rk^{1/5+\delta}, \qquad (4.14.67)$$

and, taking into account that $|j_1| < c$, we get

$$|L_{nn}| < rk^{1/5}. \qquad (4.14.68)$$

Using (4.14.67) and (4.14.68), it is easy to obtain that

$$\left\|\tilde{B}'_r(k^2,t_0,x)\right\|_{2,M} < (rk^{1/5})^{1/2}\|G'_r\|_2,$$

$$\left\|\frac{\partial\tilde{B}'_r(k^2,t_0,x)}{\partial x_1}\right\|_{2,M} < (rk^{1/5})^{3/2}\|G'_r\|_2. \qquad (4.14.69)$$

Let us estimate $\|G'_r\|_2$. When r is large enough, say, $r > 100$, we use estimate (4.6.24). When $r=1$ and $r=2$, we use (4.6.21) and (4.6.22). If $3 \le r \le 100$, let us prove that

$$\|G'_r(k,t)\|_2 \le k^{-7/5+6\delta r}, \quad r \le 100. \qquad (4.14.70)$$

We use similar considerations as in the proof of (4.6.21) and (4.6.22). Indeed, it is obvious that

$$G_r(k,t) = \sum_{i_1,\dots,i_{r+1}=0,1} \oint_C P_{i_1} A' A' \dots P_{i_r} A' P_{i_{r+1}},$$

where $P_0 = E_j$, $P_1 = I - E_j$. We consider a term of the sum. It vanishes if all indices $i_1, \dots, 1_{r+1}$ are equal to 1, because the integrand is holomorphic inside the contour. Suppose $i_1 = 0$. Taking into account that $P_0 A' P_0 = 0$, we obtain $i_2 \neq 0$. Suppose $i_3 = 0$. Then, noting that

$$2P_0 A P_1 A P_0 = (p_j^2(t) + \Delta\Lambda(t)_{jj} - z)^{-1} \sum_{i \neq j} |B_{ij}|^2 (p_i^2(t) + \Delta\Lambda(t)_{ii} - z)^{-1},$$

and usin the definition of the circle C_1:

$$|p_j^2(t) + \Delta\Lambda(t)_{jj} - z| = k^{-1-\delta},$$

and arguing as in the proof of estimate (4.6.20), we obtain that

$$|P_0 A' P_1 A' P_0| < k^{-3/5+7\delta}.$$

If $i_3 = 1$, then considering as in the proof of estimate (4.6.22), we get

$$|P_0 A' P_1 A' P_1| < k^{-2/5+6\delta}.$$

Taking into account that $\|A'\| < k^{3\delta}$, and the radius of the circle is equal to $k^{-1-\delta}$, we conclude that the integral does not exceed the value $k^{-7/5+6\delta(r-2)}$, when $i_1 = 0$. In the case where $i_k = 0$, $k \neq 1$, the integral can be estimated similarly. Thus, we obtain (4.14.70). Using relations (4.14.69), (4.6.24), (4.6.21), (4.6.22) and (4.14.70), we show that

$$\|\tilde{B}_r'(k^2, t_{20}, t_{20}, x)\|_{2,0} < ck^{-1-\gamma_4(r-100)},$$

$$\left\| \frac{\partial \tilde{B}_r'(k^2, t_{20}, t_{20}, x)}{\partial x_1} \right\|_{2,0} < k^{-1-\gamma_4(r-100)}.$$

From relation (4.14.66) it follows that

$$\psi(t,x)|_{x_1=0} = \hat{\psi}_1(t,x)\Big|_{x_1=0} + o(k^{-1}), \tag{4.14.71}$$

$$\left.\frac{\partial \psi(t,x)}{\partial x_1}\right|_{x_1=0} = \left.\frac{\partial \hat{\psi}_1(t,x)}{\partial x_1}\right|_{x_1=0} + o(k^{-1}); \tag{4.14.72}$$

here $o(k^{-1})$ is infinitely small in the class $L_2(K_{\|})$. As it was mentioned before the eigenvalue $p_j^2(t) + \Delta\hat{\Lambda}(t)_{jj}$ of $\hat{H}(t_0)$ coincides with the eigenvalue $p_j^2(t) + \Delta\Lambda_{q_0}(t)_{jj}$ of the operator $H_0(t) + V_{q_0}$ up to the value of order $k^{-k^{2/5}}$. Using the fact that both of these eigenvalues are on a distance greater than $k^{-\delta}$ (see (4.6.13)) from their neighbors, and considering as in Proposition 4.1, we easily get that the eigenfunctions, corresponding to these eigenvalues, are also close:

$$\psi_1(t,x) = \hat{\psi}_1(t,x) + O(k^{-k^{2/5}+\delta}).$$

It is clear that

$$\psi_1(t,x) = \tilde{\psi}_{t_1}(x) \exp i(\mathbf{p}_{j_\parallel}(t_\parallel), x), \tag{4.14.73}$$

where $\tilde{\psi}$ is the eigenfunction of the operator $\tilde{H}_1$ (see (4.14.60)), corresponding to the eigenvalue $p_{j_1}^2(t_1) + \Delta_{q_0}\Lambda(t_1)_{jj}$. It is well known that for $|j_1| < c_1$, and t_1, t_{01} such that $\sin^2 t_1 > k^{-2\delta}$, $\sin^2 t_{01} > k^{-2\delta}$:

$$|\psi_1(t_1,0) - \psi_1(t_{01},0)| < ck^{-7/5+9\delta},$$

$$|\frac{\partial}{\partial x_1}\psi_1(t_1,0) - \frac{\partial}{\partial x_1}\psi_1(t_{01},0)| < ck^{-6/5+9\delta}, \tag{4.14.74}$$

because $|t_1 - t_{01}| < k^{-8/5+6\delta}$ (see (4.13.28)). Using formulae (4.14.71) – (4.14.74), we obtain estimates (4.14.62) and (4.14.63). They hold also in the $(k^{-2-2\delta})$-neighbourhood of $S_H^{q_0}$, because all estimates used in the proof are stable with respect to such a perturbation.

The theorem is proved.

5. The Interaction of a Free Wave with a Semi-bounded Crystal.

5.1 Introduction.

The Schrödinger operator with a semiperiodic potential describes a motion of a particle in a solid body and the influence of a surface on this motion. Many books and papers are devoted to this problem [1]. It is proved by E.B. Davies and B. Simon [DavSi] that there are three components of the spectrum corresponding to a semiperiodic potential. They are: the component coinciding with the spectrum of the free operator, the component coinciding with the spectrum of the whole crystal, and the component corresponding to surface states.

In this chapter we consider the operator

$$H_+ = -\Delta + V_+ \tag{5.1.1}$$

in $L_2(R^n)$, $n = 2, 3$, where V_+ is the operation of multiplication by a semiperiodic potential:

$$V_+(x) = \begin{cases} V(x) & \text{if } x_1 \geq 0; \\ 0 & \text{if } x_1 < 0; \end{cases} \tag{5.1.2}$$

V being a periodic potential. The potential $V_+(x)$ corresponds to the case when there is a crystal in one half of the space while the other is empty. We call this model a semibounded crystal or semicrystal.

In this chapter we consider a free wave $\exp(i(\mathbf{k}, x))$ $x_1 \leq 0$, which is incident upon the crystal $x_1 > 0$. Interacting with the crystal, it engenders a reflected and a refracted waves. We construct asymptotic formulae for them when momentum $\mathbf{k}$ belongs to a rich set on the sphere $|\mathbf{k}| = k$ and $k \to \infty$. It is proved that there is no essential reflection inside the crystal and on the surface when $\mathbf{k}$ belongs to this set. Hence, the reflected wave is small and the refracted one is close to the incident wave $\exp(i(\mathbf{k}, x))$. Constructing more precisely the asymptotic formula for the reflected wave, we describe in the explicit form the connection of its asymptotic terms with the potential. This description enable us to determine the potential from the asymptotic expansion of the reflection coefficients in a high

[1] see f.e. [Ae, AgMi, DaLe, DavSi, GroHøMe, JaMolPas, K1, K3, Ki, Mad, Lo, Pas, PavPol, Zi].

energy region, when it is known beforehand that the potential is a trigonometric polynomial.

We suppose $V(x)$ to be a trigonometric polynomial, [2] its period a_1 lies on the axis x_1 and is orthogonal with the others, namely with a_2 when $n = 2$ and with a_2, a_3 when $n = 3$. For the sake of simplicity we assume that in the three-dimensional case the periods a_2, a_3 are orthogonal; however all the results are valid also for the case of non-orthogonal periods. Let $x_{\|}$ be the projection of vector x on the plane $x_1 = 0$, namely, $x_{\|} = (x_2, x_3)$ for $n = 3$; $x_{\|} = x_2$ for $n = 2$. Let $Q_{\|}$ be the elementary cell of the periods in the plane $x_1 = 0$:

$$Q_{\|} = [0, a_2) \times [0, a_3), \text{ when } n = 3, \tag{5.1.3}$$

$$Q_{\|} = [0, a_2), \text{ when } n = 2.$$

The potential is periodic in the direction(s) $x_{\|}$. Following [Ge, Ti, OdKe, Ea1, Ea2, Th, Wil], we consider the family of operators $H_+(t_{\|})$, described by formula (5.1.1), and the quasiperiodic boundary condition(s) in the strip Q:

$$Q = Q_{\|} \times (-\infty, \infty).$$

For the case $n = 3$ (and orthogonal periods a_2, a_3) the quasiperiodic conditions have the form:

$$\psi(x + a_2, t_{\|}) = \exp(it_2a_2)\psi(x, t_{\|}),$$

$$\frac{\partial\psi(x + a_2, t_{\|})}{\partial x_2} = \exp(it_2a_2)\frac{\partial\psi(x, t_{\|})}{\partial x_2};$$

$$\psi(x + a_3, t_{\|}) = \exp(it_3a_3)\psi(x, t_{\|}),$$

$$\frac{\partial\psi(x + a_3, t_{\|})}{\partial x_3} = \exp(it_3a_3)\frac{\partial\psi(x, t_{\|})}{\partial x_3}. \tag{5.1.4}$$

Quasimomentum $t_{\|}$, $t_{\|} = (t_2, t_3)$, parameterizing the conditions, varies over the elementary cell of the dual lattice:

$$K_{\|} = [0, 2\pi a_2^{-1}) \times [0, 2\pi a_3^{-1}). \tag{5.1.5}$$

In the case $n = 2$ only the first pair of the relations in (5.1.4) have to be satisfied, and $t_{\|} = t_2$, $K_{\|} = [0, 2\pi a_2^{-1})$.

The spectrum of H_+ is the union of the spectra of the operators $H_+(t_{\|})$. The eigenfunctions of H_+ are obtained by the quasiperiodic extensions of the eigenfunctions of all the operators $H_+(t_{\|})$.

We are looking for a solution $\Psi \in W_2^2(Q)$ of the equation $H_+\psi = k^2\psi$ in the form:

$$\Psi(\mathbf{k}, x) = \begin{cases} \exp(i(\mathbf{k}, x)) + \Psi_{refl}(\mathbf{k}, x), & x_1 \le 0, \\ \Psi_{refr}(\mathbf{k}, x), & x_1 \ge 0, \end{cases} \tag{5.1.6}$$

where $\exp(i(\mathbf{k}, x))$ is an incident wave; $k_1 > 0$, $|\mathbf{k}|^2 = k^2$; Ψ_{refl} is a reflected wave; Ψ_{refr} is a refracted wave. We call a function Ψ_{refl} a reflected wave if it can be

[2] We don't assume that $v_0 = 0$ in this chapter.

represented as a $\lim_{\varepsilon\downarrow 0}\psi_-(k^2+i\varepsilon,t_{||},x)$, where $\psi_-(k^2+i\varepsilon,t_{||},x)$ is described as follows. The function $\psi_-(k^2+i\varepsilon,t_{||},x)$ satisfies the equation $-\Delta\psi_-=(k^2+i\varepsilon)\psi_-$ and the quasiperiodic conditions (5.1.4), when $x_1\le 0$ for all $i\varepsilon$ in a closed upper neighborhood of zero ($0\le\Re\varepsilon\le\varepsilon_0$), depending analytically on $i\varepsilon$ in that neighborhood for any fixed $x_1\le 0$. It decays exponentially when $x_1\to-\infty$ for all ε with a positive real part.

We call a function Ψ_{refr} a refracted wave if it can be represented as a $\lim_{\varepsilon\downarrow 0}\psi_+(k^2+i\varepsilon,t_{||},x)$, where $\psi_+(k^2+i\varepsilon,t_{||},x)$ is described as follows. The function $\psi_+(k^2+i\varepsilon,t_{||},x)$ satisfies the equation $(-\Delta+V)\psi_+=(k^2+i\varepsilon)\psi_+$ and the quasiperiodic conditions (5.1.4), when $x_1\ge 0$ for all ε in a closed upper neighborhood of zero ($0\le\Re\varepsilon\le\varepsilon_0$) ; depending analytically on $i\varepsilon$ in that neighborhood for any fixed $x_1\ge 0$. It decays exponentially when $x_1\to\infty$ for all ε with a positive real part.

These definitions mean that the reflected and refracted waves decay simultaneously under a dissipation,[3] while the incident wave does not.

The incident wave $\exp(i(\mathbf{k},x))$ satisfies the quasiperiodic conditions (5.1.4) with $t_{||}$:

$$t_i=k_i-2\pi a_i^{-1}[k_ia_i/2\pi], \tag{5.1.7}$$

where $i=2,3$ when $n=3$ and $i=2$ when $n=2$. Naturally, we look for refracted and reflected waves satisfying the quasiperiodic conditions with the same $t_{||}$. Suppose $\mathbf{k}$ is such that $t=\mathcal{K}\mathbf{k}$ belongs to the nonsingular set for the periodic potential V. Then, a wave close to $\exp(i(\mathbf{k},x))$ can propagate inside the crystal (see Theorem 3.1 for $n=2$ and Theorem 4.1 for $n=3$). It is not hard to show that this wave satisfies the definition of a refracted wave. Taking the reflected wave equal to zero, we satisfy the continuity conditions on the surface with the accuracy to $o(1)$ ($k\to\infty$). Thus, we obtain an approximate solution. The question arises: is this approximate solution close to the accurate one (5.1.6), satisfying the continuity conditions precisely? To answer this question, first of all let us see whether reflected and refracted waves are always defined uniquely. Suppose the equation $H_+\Psi=k^2\Psi$ has a smooth solution Ψ_{surf}, which is a reflected wave for $x_1\le 0$ and a refracted wave for $x_1\ge 0$. This solution is called a surface state. The surface states were discovered by Rayleigh [Ray] at the end of the last century and are of great importance in modern physics [4]. Obviously, in the case of a surface state the reflected and refracted waves in (5.1.6) are not uniquely determined. However, the nondecaying component of the reflected and refracted waves are uniquely determined, because it can be shown that surface states exponentially decay as $x_1\to\pm\infty$.

It turns out that the surface of the crystal can essentially influence the nondecaying part of the reflected wave too. This happens when there exists a solution of the equation $H_+\psi=k^2\psi$, which can be approximated with good

---

[3]Of course, the relations $k_1>0$, $\varepsilon>0$ are not of principal. It is important only that k_1, ε have the same sign. The case $k_1<0$, $\varepsilon<0$ is complexly conjugate for the case of positive k_1, ε.

[4]see f.e. [Ae, AgMi, DaLe, DavSi, GroHøMe, JaMolPas, K1, K3, Ki, Mad, Lo, Pas, PavPol, Zi].

accuracy by a reflected and refracted wave in the sense that the error in the continuity conditions on the surface is small. We call such a solution a quasisurface state. The quasisurface state can have strong influence on the asymptotics of the reflected and refracted waves, even far from the surface. Unlike the surface state, it can also influence the nondecaying component of the reflected wave.

The operator $(H_+(t_\|) - z)^{-1}$ has a pole at the point $z = k^2$ in the case of a surface state. In the two and three-dimensional situations, surface states can exist in a high energy region, while in the one-dimensional situation they can exist only for sufficiently low energies. We suppose that there corresponds a pole of the resolvent on the non-physical sheet in a vicinity of the point $z = k^2$ to a quasiperiodic state. It is easy to see that all the points close to surface states are quasisurface states. In the one-dimensional situation there are only these trivial cases: there are no quasisurface states which are not in a vicinity of surface states. The similar situation is in the case of separable variables in two and three dimensional spaces. However, it seems that quasisurface states can exist separately from surface states in the case of nonseparable variables.

We are not going to describe surface and quasisurface states here. Our aim is to describe a nonsingular set of $\mathbf{k}$ on the sphere S_k ($S_k = \{\mathbf{k} : |\mathbf{k}| = k\}$) for which the influence of the surface and quasisurface states is weak, i.e., such $\mathbf{k}$ that the corresponding reflected and refracted waves have regular asymptotics determined by a segment of perturbation series for the resolvent. We can show that this takes place when there are no surface and quasisurface states with quasimomentum (5.1.6) and the energy being in a close vicinity of the point k^2.

How can surface and quasisurface states be excluded? The situation in the two-dimensional case turns out to be relatively simple. In order to exclude all the surface and quasisurface states, it suffices to delete from S_k only the singular set for the periodic part of the potential V_+. The situation is more complicated in the three-dimensional case. To exclude the surface and quasisurface states in the three dimensional situation, one has to delete not only the singular set of the periodic part of V_+, but also some additional set corresponding, namely, to surface and quasisurface states. We prove that the reflected wave is asymptotically small and the refracted wave is close to the incident one when $\mathbf{k}$ belongs to the nonsingular set for the semicrystal.

However, this weak asymptotic is not able to give any information about the potential. To obtain this information we describe the reflected and refracted waves more precisely. First, we consider the functions satisfying the Helmholtz equation $-\Delta\psi = k^2\psi$ and the quasiperiodic conditions in $\|$-directions with the quasimomentum $t_\|$. The easy calculation shows that this set consists of the functions:

$$\Psi^0_\pm(k^2, t_\| + p_{q_\|}(0), x) = \exp(i(t_\| + p_{q_\|}(0), x_\|) \mp \sqrt{|t_\| + p_{q_\|}(0)|^2 - k^2}x_1),$$

$$q_\| \in Z^{(n-1)}, \qquad (5.1.8)$$

where $Re\sqrt{\ } > 0$, $p_{q_\|}(0) \in R^{n-1}$,

$$p_{q_\parallel}(0)=\left(\frac{2\pi q_2}{a_2},\frac{2\pi q_3}{a_3}\right)$$

in the three-dimensional situation and

$$p_{q_\parallel}(0)=\frac{2\pi q_2}{a_2}$$

in the two-dimensional case. Obviously, the function $\Psi_+^0(k^2,t_\parallel+p_{q_\parallel}(0),x)$ depends analytically on k^2 in the complex plane with the cut, along the semiaxis $k^2>|t_\parallel+p_{q_\parallel}(0)|^2$. Note that

$$\exp(i(\mathbf{k},x))=\begin{cases}\Psi_+^0(k^2+i0,k_\parallel,x), & \text{if } k_1>0,\\ \Psi_-^0(k^2+i0,k_\parallel,x), & \text{if } k_1<0.\end{cases} \tag{5.1.9}$$

Recall that we assume k_1 to be positive for the incident wave. Hence,

$$\exp(i(\mathbf{k},x))=\Psi_+^0(k^2+i0,k_\parallel,x). \tag{5.1.10}$$

It is easy to see that the function $\Psi_+^0(k^2,k_\parallel,x)$, $\Im k^2>0$, increases exponentially when $x_1\to-\infty$ and decays exponentially when $x_1\to+\infty$. Conversely, the function $\Psi_-^0(k^2,k_\parallel,x)$ increases when $x_1\to+\infty$ and decreases when $x_1\to-\infty$. When $\Im k^2=0$, and $k^2<|k_\parallel|^2$

$$\Psi_\pm^0(k^2,k_\parallel,x)=\exp(i(\mathbf{k}_\parallel,x_\parallel)\pm\sqrt{|k_\parallel|^2-k^2}\,x_1).$$

Therefore, in this case the functions $\Psi_\pm^0(k^2,k_\parallel,x)$ have the same behavior at infinity as in the case $\Im k^2>0$.

The functions $\Psi_-^0(k^2+i0,k_\parallel+p_{q_\parallel}(0),x)$, $q_\parallel\in Z^{n-1}$, form the complete set of reflected waves. This means that any reflected wave can be represented as a linear combination of $\Psi_-^0(k^2+i0,k_\parallel+p_{q_\parallel}(0),x)$:

$$\Psi_{refl}=\sum_{q_\parallel\in Z^{n-1}}\beta_{q_\parallel}\Psi_-^0(k^2+i0,k_\parallel+p_{q_\parallel}(0),x). \tag{5.1.11}$$

Coefficients $\beta_{q_\parallel}$ are called reflection coefficients. We want to obtain an asymptotic expansion of the reflection coefficients as $k^2\to\infty$, giving an information about the potential. To make this, we determine the reflection coefficients more precisely, we describe a part, $S^{(n)}$, of the nonsingular set for the semicrystal. This part is proved to have an asymptotically full measure on S_k. If $\mathbf{k}\in S^{(n)}$, then there exists not only a wave close to $\Psi_+^0(k^2,k_\parallel,x)$ inside the crystal, but also waves close to $\Psi_+^0(k^2,k_\parallel+p_{q_\parallel}(0),x)$ when $q_\parallel\in Z^{(n-1)}$, $|q_\parallel|<k^{3\delta}$ (in the sense of Theorems 1.1, 1.3). If $\mathbf{k}\in S^{(n)}$, then, taking as a refracted wave a linear combination of the waves close to $\Psi_+^0(k^2,k_\parallel+p_{q_\parallel}(0),x)$, $|q_\parallel|<k^{3\delta}$, and as a reflected wave a linear combination of $\Psi_-^0(k^2,k_\parallel+p_{q_\parallel}(0),x)$, $|q_\parallel|<k^{3\delta}$, we satisfy the continuity conditions with the accuracy to $O(k^{-k^\delta R_0^{-1}})$ for V being a trigonometric polynomial. The reflection coefficients are determined with this accuracy too.

We show that the asymptotic expansion of the reflection coefficients contains rich information about the potential. This information is sufficient to determine the potential, if it is known beforehand that the potential is a trigonometric polynomial.

Now we describe the structure of the chapter. Section 5.2 contains rigorous definitions of surface and quasisurface states and describes a boundary operator. In Sections 5.3 and 5.4 we clear up the conditions of the absence of surface and quasisurface states for given k^2, $k_{\|}$. The conditions are obtained in the form of inequalities there. Section 5.3 is devoted to the two-dimensional case, Section 5.4 – to the three-dimensional case. In Section 5.5 we geometrically describe the subset of S_k, where these conditions hold and prove that this subset has an asymptotically full measure on S_k when $k \to \infty$. In Section 5.6 we obtain a high order asymptotic expansion for the reflection and refraction coefficients when $k \to \infty$. In Section 5.7 we solve the inverse problem, i.e., we obtain the potential $V(x)$ from the asymptotic expansion of the reflection coefficients. This method makes it possible to describe surface and quasisurface states, which are of great interest (see f.e. [Ae, AgMi, DaLe, DavSi, GroHøMe, JaMolPas, K1, K3, Ki, Mad, Lo, Pas, PavPol, Zi].) . This will be done in following papers. The results proved in this chapter were announced in [K16] and [K17].

5.2 A Boundary Operator.

Here we introduce a boundary operator T and give the definitions of the surface and quasisurface states in the terms of this operator. Let Q_+ be a semistrip corresponding to positive x_1 and $x_{\|} \in Q_{\|}$:

$$K_+ = \{x : x_{\|} \in Q_{\|}, x_1 \geq 0\},$$

$Q_{\|}$ being given by (5.1.3). Suppose $u \in W_2^2(Q_+)$ and satisfies the equation

$$(-\Delta + V(x) - z)u = 0, \tag{5.2.1}$$

quasiperiodic conditions (5.1.4) and the boundary condition: [5]

$$u|_{x_1=+0} = \varphi(x_{\|}), \quad \varphi \in W_2^1(Q_{\|}). \tag{5.2.2}$$

The function u is uniquely determined by φ when $\Im z \neq 0$. Thus, we can define the operator $A_+^V(z, k_{\|}) : L_2(Q_{\|}) \to L_2(Q_{\|})$:

$$A_+^V(z, k_{\|})\varphi = \psi, \qquad \psi = u_{x_1}|_{x_1=+0}. \tag{5.2.3}$$

We shall show that the operator $A_+^V(z, k_{\|})^{-1}$ is compact and depends analytically on z in the upper halfplane.

[5] here and below we use the notation $u|_{x_1=\pm 0}(x_{\|}) = \lim_{x_1 \to \pm 0} u(x_{\|}, x_1)$, $x_{\|} \in Q_{\|}$. We pass to the limits in $L_2(Q_{\|})$ in the sense of imbedding theorems. In the case when limits are equal, we use the notation $u|_{x_1=\pm 0} \equiv u|_{x_1=0}$.

Let $A_-^V(z, k_\parallel)$ be the analogous operator defined for the semistrip Q_-:

$$Q_- = \{x : x_\parallel \in Q_\parallel, x_1 \le 0\}.$$

The operators A_+^V and A_-^0 (the last corresponds to $V = 0$) play an important role in the study of the semicrystal. To show this, let us consider the continuity conditions for $\psi(\mathbf{k}, x)$ on the surface $x_1 = 0$ (see (5.1.6)). First we consider $\psi(\mathbf{k}, x)$ as an analytic function of $k^2 + i\varepsilon$ in some upper semineighborhood ρ of k^2. In fact, from the representation $\exp(i(\mathbf{k}, x)) = \exp(ik(\,^\circ, x))$, $|\,^\circ| = 1$ for the incident wave, we see that it depends analytically on k for a fixed $^\circ$. By definition, the functions Ψ_{refl}, Ψ_{refr} depend analytically on $k^2 + i\varepsilon$ in a closed upper semineighborhood ρ of the point k^2. Thus, we define $\psi(\mathbf{k}, x)$ in ρ. We claim the continuity conditions for the function Ψ to be satisfied for all $k^2 + i\varepsilon$ in ρ:

$$\Psi|_{x_1=+0} = \Psi|_{x_1=-0}; \qquad \Psi_{x_1}|_{x_1=+0} = \Psi_{x_1}|_{x_1=-0}. \tag{5.2.4}$$

Using formula (5.1.6) and the definitions of the functions Ψ_{refl}, Ψ_{refr}, we obtain

$$\Psi_+|_{x_1=+0}(k, x_\parallel) = \Psi_-|_{x_1=-0}(k, x_\parallel) + \exp i(k_\parallel, x_\parallel);$$

$$A_+^V\left(\Psi_+|_{x_1=+0}(k, x_\parallel)\right) = A_-^0\left(\Psi_-|_{x_1=-0}(k, x_\parallel)\right) + ik_1 \exp i(k_\parallel, x_\parallel). \tag{5.2.5}$$

Excluding from the first equation the function Ψ_-, we get

$$T\left(\Psi_+|_{x_1=+0}(k, x_\parallel)\right) = f, \tag{5.2.6}$$

$$T \equiv A_+^V - A_-^0, \quad T = T(k^2 + i\varepsilon, k_\parallel);$$

$$f = ik_1 f_0 - A_-^0 f_0$$

$$f_0 = \exp i(k_\parallel, x_\parallel).$$

We call T a boundary operator. We shall show that $T(k^2 + i\varepsilon, k_\parallel)$ is invertible when $\varepsilon \ne 0$ and T^{-1} is compact. Thus,

$$\Psi_+|_{x_1=+0}(k, x_\parallel) = T^{-1}f; \qquad \Psi_-|_{x_1=-0}(k, x_\parallel) = T^{-1}f - f_0, \tag{5.2.7}$$

$$T^{-1} = T^{-1}(k^2 + i\varepsilon, k_\parallel).$$

Formally passing to the limit when $\varepsilon \to 0$ we get:

$$\Psi_{refr}|_{x_1=+0}(k, x_\parallel) = T^{-1}f; \qquad \Psi_{refl}|_{x_1=-0}(k, x_\parallel) = T^{-1}f - f_0, \tag{5.2.8}$$

$$T^{-1} = T^{-1}(k^2 + i0, k_\parallel).$$

Thus, to construct (5.1.6) we have to consider the operator $T^{-1}(k^2 + i0, k_\parallel)$.

We shall use the following definitions of the absence of surface and quasisurface states. There is no a surface state or a quasisurface state if $T^{-1}(k^2 + i0, k_\parallel)$ exists and

$$\|T^{-1}(k^2 + i0, t_\parallel)\| < k^\delta, \tag{5.2.9}$$

δ being small positive and fixed.

We study analytic properties of $T^{-1}(z,k_{\|})$, $\Im z \geq 0$. Lemma 5.1 is primary here. Under its conditions there exists $\lim_{\varepsilon\downarrow 0} T(k^2+i\varepsilon, t_{\|})$.

We introduce in $L_2(Q_{\|})$ a new basis: this is the set of the exponents satisfying the quasiperiodic conditions (5.1.4):

$$\{\exp\left(i(t_{\|}+p_{q_{\|}}(0),x_{\|})\right)\}_{q_{\|}\in Z^{n-1}}. \tag{5.2.10}$$

We call this basis $t_{\|}$-basis. We denote by $v_{q_{\|}}$ the coordinates of function v in $t_{\|}$-basis .

Let us calculate the matrices $A^0_{\pm}$ in $t_{\|}$-basis.

Proposition 5.1 . *The matrices $A^0_{\pm}$ are diagonal and given by the following formula:*

$$(A^0_{\pm})_{q_{\|}q_{\|}} = \mp\sqrt{|t_{\|}+p_{q_{\|}}(0)|^2 - z}, \qquad \Re\sqrt{\ } \geq 0. \tag{5.2.11}$$

Proof. To construct A^0_{+}, we consider function u of $L_2(Q_+)$, which satisfies the equation (5.2.1) with $V=0$, the boundary condition (5.2.2) and the quasiperiodic condition (5.1.4). It is easy to see that this function is uniquely determined and can be represented in the form:

$$u = \sum_{q_{\|}\in Z^{n-1}} \varphi_{q_{\|}}\Psi^0_{+}(k^2, t_{\|}+p_{q_{\|}}(0), x). \tag{5.2.12}$$

It is not hard to show that $u \in W_2^2(Q_+)$ and

$$u|_{x_1=+0} = \sum_{q_{\|}\in Z^{n-1}} \varphi_{q_{\|}} \exp\left(i(t_{\|}+p_{q_{\|}}(0),x_{\|})\right).$$

$$u_{x_1}|_{x_1=+0} = -\sum_{q_{\|}\in Z^{n-1}} \varphi_{q_{\|}}\sqrt{|t_{\|}+p_{q_{\|}}(0)|^2-z}\,\exp\left(i(t_{\|}+p_{q_{\|}}(0),x_{\|})\right). \tag{5.2.13}$$

This means that matrix A^0_{+} has a diagonal form and is given by the formula (5.2.11). Similar arguments are valid for A^0_{-}. The proposition is proved.

Now, we describe some properties of the operator $A^V_{+}(z,t_{\|})$, $\Im z \geq 0$. We express A^V_{+} by the resolvent of the Schrödinger operator with periodic potential V.

Proposition 5.2 . *If $\phi \in L_2(Q_{\|})$, $t_{\|} \in K_{\|}$, and $\varepsilon \neq 0$, then there exists a unique function $U_V(t_{\|},x)$ in $L_2(Q_+ \cup Q_-)$, satisfying the equation*

$$(-\Delta - k^2 - i\varepsilon + V)U_V(t_{\|},x) = \phi(x_{\|})\delta(x_1) \tag{5.2.14}$$

and the quasiperiodic conditions (5.1.4). In the case $V=0$

$$U_0(t_{\|},x) = \begin{cases} \sum_{q\in Z^{n-1}} c_{q_{\|}}\Psi^0_{+}(k^2+i\varepsilon, t_{\|}+p_{q_{\|}}(0),x) & \text{when } x_1 \geq 0; \\ \sum_{q\in Z^{n-1}} c_{q_{\|}}\Psi^0_{-}(k^2+i\varepsilon, t_{\|}+p_{q_{\|}}(0),x) & \text{when } x_1 \leq 0; \end{cases} \tag{5.2.15}$$

$$c_{q_\parallel} = \phi_{q_\parallel}(2\sqrt{|t_\parallel + p_{q_\parallel}(0)|^2 - k^2 - i\varepsilon})^{-1},$$

$\Im\sqrt{\ } \geq 0$. *When $V \not\equiv 0$, the function U_V admits the representation:*

$$U_V = U_0 + w, \tag{5.2.16}$$

where $w \in W_2^3(Q_+ \cup Q_-)$ and

$$|w|_{W_2^3} \leq c|\phi|_{L_2(Q_\parallel)}, \qquad c = c(k^2 + i\varepsilon). \tag{5.2.17}$$

The functions $U_V|_{x_1=\pm 0}$ belong to $W_2^1(Q_\parallel)$ and

$$U_V|_{x_1=+0} = U_V|_{x_1=-0}. \tag{5.2.18}$$

Proof. Suppose there exist two solutions of equation (5.2.14). Therefore, the equation

$$(-\Delta - k^2 - i\varepsilon + V)U_V(t_\parallel, x) = 0$$

has a nonzero solution in $L_2(Q_- \cup Q_+)$. Thus, the operator $H(t_\parallel)$, described by the differential expression $H(t_\parallel) = -\Delta + V$ and the quasiperiodic conditions (5.1.4), has a nonreal eigenvalue $k^2 + i\varepsilon$. But this is not the case, because $H(t_\parallel)$ is selfadjoint. The contradiction proves the uniqueness of the solution of (5.2.14). It is easy to show that the function $U_0(k_\parallel, x)$ (see (5.2.15)) belongs to $W_2^1(Q_+ \cup Q_-)$, satisfies equation (5.2.14) with $V = 0$, and

$$U_0|_{x_1=+0} = U_0|_{x_1=-0}, \tag{5.2.19}$$

$$\|U_0\|_{W_2^1(Q_+ \cup Q_-)} < c_1|\phi|_{L_2(Q_\parallel)}, \qquad c_1 = c_1(k^2 + i\varepsilon). \tag{5.2.20}$$

In the case $V \neq 0$, we are looking for U_V in the form (5.2.16), where w belongs to $W_2^3(Q_+ \cup Q_-)$, satisfies the quasiperiodic conditions (5.1.4) and the equation

$$(-\Delta - k^2 - i\varepsilon + V)w = -VU_0.$$

Since $u_0 \in W_2^1(Q_+ \cup Q_-)$ and V is a smooth potential, there exists $w \in W_2^3(Q_+ \cup Q_-)$ and

$$\|w\|_{W_2^3(Q_+ \cup Q_-)} \leq c\|U_0\|_{W_2^1(Q_+ \cup Q_-)}. \tag{5.2.21}$$

Using estimate (5.2.20) we get (5.2.17). According to Imbedding Theorem (see f.e.[Ad]), the function w is continuous. Therefore,

$$w|_{x_1=+0} = w|_{x_1=-0}. \tag{5.2.22}$$

Adding (5.2.19) and (5.2.22), we obtain (5.2.18). The proposition is proved.

We introduce the operators $\Gamma_1^V(k^2 + i\varepsilon, t_\parallel)$ and $\Delta\Gamma_1^V(k^2 + i\varepsilon, t_\parallel)$, acting in $L_2(Q_\parallel)$ as follows:

$$\Gamma_1^V \phi = U_V|_{x_1=0}, \tag{5.2.23}$$

$$\Delta\Gamma_1^V = \Gamma_1^V - \Gamma_1^0 \tag{5.2.24}$$

Using formula (5.2.15), it is not hard to show that the matrix Γ_1^0 in $t_{\|}$-basis has the form:

$$(\Gamma_1^0)_{q_{\|}q_{\|}} = (2\sqrt{|t_{\|} + p_{q_{\|}}(0)|^2 - k^2 - i\varepsilon})^{-1}, \quad \Re\sqrt{\ } > 0. \tag{5.2.25}$$

It is obvious that Γ_1^0 belongs to the class $\mathbf{S}_n$ (n is the dimension of the space) and is invertible.

Proposition 5.3 . *The operator $\Gamma_1^V(k^2 + i\varepsilon, t_{\|})$, $\varepsilon \neq 0$, belongs to the class $\mathbf{S}_n$ and is invertible.*

Proof. Considering formula (5.2.16) and the definition of $\Delta\Gamma_1^V$, we see that $\Delta\Gamma_1^V\phi = w|_{x_1=0}$. It is obvious that $\Delta\Gamma_1^V = M_1M_2$, where $M_2 : L_2(Q_{\|}) \to W_2^3(Q_+ \cup Q_-)$, $M_1 : W_2^3(Q_+ \cup Q_-) \to L_2(Q_{\|})$,

$$M_2\phi = w, \qquad M_1 w = w|_{x_1=0}. \tag{5.2.26}$$

It follows from relation (5.2.17) that the operator M_2 is bounded. Since the imbedding from $W_2^3(Q_+ \cup Q_-)$ to $L_2(Q_{\|})$ is compact, the operator M_1 is compact. Thus, $\Delta\Gamma_1^V$ is compact as a product of bounded and compact operators. Taking into account that $\Gamma_1^0 \in \mathbf{S}_n$, we get $\Gamma_1^V \in \mathbf{S}_n$. To prove the invertibility of Γ_1^V, it suffices to show that the equation $\Gamma_1^V\phi = 0$ has only the zero solution. Suppose $\Gamma_1^V\phi = 0$, $\phi \neq 0$. Then there is a nonzero function U_V, which satisfies the equation

$$(-\Delta - k^2 - i\varepsilon + V)U_V(t_{\|}, x) = 0, \quad x_1 > 0,$$

the quasiperiodic condition (5.1.4) and the boundary condition

$$U_V|_{x_1=+0} = 0. \tag{5.2.27}$$

This means that the point $z = k^2 + i\varepsilon$ is an eigenvalue of the selfadjoint operator, described by the differential expression $-\Delta + V$, the quasiperiodic conditions (5.1.4) and the boundary condition (5.2.27). But this is not the case, because z is not real. Therefore $\phi = 0$. This contradiction proves that the equation $\Gamma_1\phi = 0$ has only the zero solution, i.e., Γ_1^V is invertible. The proposition is proved.

We introduce the operator $\Gamma_2^V(k^2 + i\varepsilon, t_{\|}) : L_2(Q_{\|}) \to L_2(Q_{\|})$ by the formula: $(-I/2 + \Gamma_2^V)\phi = (U_V)_{x_1}|_{x_1=+0}$. Using formula (5.2.15), it is easy to show that $\Gamma_2^0 = 0$.

Proposition 5.4 . *If $\varepsilon \neq 0$, then the operator $\Gamma_2^V(k^2 + i\varepsilon, t_{\|})$ is compact, and the operator $(-I/2 + \Gamma_2^V)^{-1}(k^2 + i\varepsilon, t_{\|})$ is bounded.*

Proof. It is easy to see that $\Gamma_2^V\phi = w_{x_1}|_{x_1=+0}$. It is obvious that $\Gamma_2^V = M_3M_4$, where $M_3 : L_2(Q_{\|}) \to W_2^2(Q_+ \cup Q_-)$, $Q_4 : W_2^2(Q_+ \cup Q_-) \to L_2(Q_{\|})$,

$$M_3\phi = w_{x_1}, \qquad M_4 w_{x_1} = w_{x_1}|_{x_1=+0}. \tag{5.2.28}$$

It follows from relation (5.2.17) that the operator M_3 is bounded. The operator M_4 is compact, since the imbedding from $W_2^2(Q_+ \cup Q_-)$ to $L_2(Q_{\|})$ is compact.

Thus, Γ_2^V is compact as a product of bounded and compact operators. The operator $-I/2+\Gamma_2^V$ is invertible, when $\varepsilon \neq 0$, because otherwise the selfadjoint operator described by the differential expression $-\Delta+V$, the quasiperiodic conditions and the boundary condition $u_{x_1}|_{x_1=+0} = 0$ has a nonreal eigenvalue. The proposition is proved.

Proposition 5.5 . *The operator $A_+^V(k^2+i\varepsilon, t_{\|})$, $\varepsilon \neq 0$, can be represented in the form:*

$$A_+^V = (-I/2+\Gamma_2^V)(\Gamma_1^V)^{-1}. \tag{5.2.29}$$

It is invertible and $(A_+^V)^{-1}$ belongs to class $\mathbf{S}_n$.

Proof. Let $u_+ \in L_2(Q_{\|})$, $u \in L_2(Q_+)$ and satisfies the equation $(-\Delta + V - k^2 - i\varepsilon)u = 0$, $x_1 \geq 0$, the boundary condition $u|_{x_1=+0} = u_+$ and the quasiperiodic conditions (5.1.4). By the definition of A_+^V: $u_{x_1}|_{x_1=+0} = A_+^V u_+$. We define u in Q_- by the equation $(-\Delta+V-k^2-i\varepsilon)u = 0$, $x_1 \leq 0$, the boundary condition $u|_{x_1=-0} = u_+$ and the quasiperiodic conditions (5.1.4). By the definition of A_-^V, we have $u_{x_1}|_{x_1=-0} = A_-^V u_+$. It is easy to see now that u belongs to $L_2(Q_+\cup Q_-)$ and satisfies equation (5.2.14), where

$$\phi = u_{x_1}|_{x_1=-0} - u_{x_1}|_{x_1=+0}.$$

From the definitions of Γ_1^V and Γ_2^V, we get: $u_+ = \Gamma_1^V\phi$, $A_+^V u_+ = (-I/2+\Gamma_2^V)\phi$. From the last relations we obtain $A_+^V = (-I/2+\Gamma_2^V)(\Gamma_1^V)^{-1}$. The operator $(-I/2+\Gamma_2^V)^{-1}$ is bounded. Noting that $\Gamma_1^V \in \mathbf{S}_n$, we obtain $(A_+^V)^{-1} \in \mathbf{S}_n$. The proposition is proved.

Next, we express the operators $\Delta\Gamma_1^V$, Γ_2^V by the resolvent of the operator H with the periodic potential $V(x)$.

Proposition 5.6 . *The matrices of the operators $\Delta\Gamma_1^V$, Γ_2^V in $t_{\|}$-basis can be represented in the form:*

$$(\Delta\Gamma_1^V)_{n_{\|}m_{\|}} = \sum_{n_1 m_1} \int_0^{2\pi a_1^{-1}} \left((H(t)-z)^{-1} - (H_0(t)-z)^{-1}\right)_{nm} dt_1, \tag{5.2.30}$$

$$(\Gamma_2^V)_{n_{\|}m_{\|}} = \sum_{n_1 m_1} \int_0^{2\pi a_1^{-1}} \left((H(t)-z)^{-1} - (H_0(t)-z)^{-1}\right)_{nm} i(2\pi a_1^{-1} n_1 + t_1)\, dt_1, \tag{5.2.31}$$

where $n=(n_1,n_{\|})$, $m=(m_1,m_{\|})$, $t=(t_1,t_{\|})$. The series (5.2.30) and (5.2.31) converge absolutely when $\Im z > 0$ and uniformly in z, when $\Im z > \varepsilon_0 > 0$.

Corollary 5.1 . *The matrix of the operator Γ_1^V can be represented in the form:*

$$(\Gamma_1^V)_{n_{\|}m_{\|}} = \sum_{n_1 m_1} \int_0^{2\pi a_1^{-1}} (H(t)-z)^{-1}_{nm} dt_1. \tag{5.2.32}$$

Proof. We consider the operator $H(t_{\|})$, defined in $L_2(Q_+ \cup Q_-)$ by the differential expression $H(t_{\|}) = -\Delta + V$ and the quasiperiodic condition (5.1.4). It is clear that its resolvent in $t_{\|}$-basis can be represented in the form:

$$(H(t_{\|}) - z)^{-1}_{n_{\|} m_{\|}}(x_1, \tilde{x}_1) = \tag{5.2.33}$$

$$= \sum_{n_1 m_1} \int_0^{2\pi a_1^{-1}} (H(t) - z)^{-1}_{nm} \exp it_1(x_1 - \tilde{x}_1) \exp(2\pi a_1^{-1} i(n_1 x_1 - m_1 \tilde{x}_1)) dt_1.$$

We consider the solution of the equation

$$(H(t_{\|}) - k^2 - i\varepsilon)v(x, y) = \phi(x_{\|})\delta(x_1 - y_1), \tag{5.2.34}$$

where H acts on variable x,

$$y = (y_1, 0).$$

It is clear that the function $v(x, y)$ belongs to $L_2(Q_+ \cup Q_-)$ as a function of x for a fixed y. By analogy with (5.2.16), we represent v in the form

$$v(x, y) = v_0(x, y) + \tilde{w}(x, y), \tag{5.2.35}$$

where v_0 is the solution of the equation (5.2.34) for $V = 0$. Combining formulae (5.2.33) and (5.2.35), we obtain:

$$\tilde{w}(x, y) = \tag{5.2.36}$$

$$\sum_{n,m} \int_0^{2\pi a_1^{-1}} ((H(t) - z)^{-1} - (H_0(t) - z)^{-1})_{nm} \phi_{m_{\|}}$$

$$\exp\left(i(\mathbf{p}_n(t), x) - i(t_1 + 2\pi a_1^{-1} m_1) y_1\right) dt_1.$$

From (5.2.35) and (5.2.16) we see that $\tilde{w}(x, 0) = w(x)$ and, therefore,

$$\Delta\Gamma_1^V \phi = \tilde{w}(x, 0)|_{x_1 = +0}, \qquad \Gamma_2^V \phi = \tilde{w}_{x_1}(x, 0)|_{x_1 = +0}. \tag{5.2.37}$$

Using relation (5.2.37), we get (5.2.30) and (5.2.31). The convergence of the series can easily be obtained from the boundedness of the operator $H_0(t)((H(t)-z)^{-1} - (H_0(t) - z)^{-1})H_0(t)$. The proposition is proved.

We consider the operator $(H(t_{\|}) - k^2 - i\varepsilon)^{-1}$. It analytically depends on $i\varepsilon$, when $\Re\varepsilon > 0$. Now the problem is whether it can be analytically extended on the non-physical sheet across the real axis.

Proposition 5.7 . *For any real k^2 and some natural q, $q = q(k)$, the operator $(H(t_{\|}) - k^2 - i\varepsilon)^{-1}$ can be analytically extended as an operator-valued function* [6] *of $\xi = (\varepsilon)^{1/q}$ in some neighborhood of zero. In this neighborhood the operator can have only a unique pole. The pole is at the point k^2 and has an order not greater than $q - 1$ (there is no pole in the case q=1).*

[6] According to [GoKr] an operator $T(z)$ analytically depends on z in ρ, if $T(z)$ is bounded for any z in ρ and its bilinear form $(T(z)f, g)$ is an analytical function of z for any f and g.

Proof. We represent the bilinear form of the resolvent of the operator $H(t_{\|})$ in the form:

$$((H(t_{\|})-k^2-i\varepsilon)^{-1}f,g)=\sum_{m=1}^{\infty}I_m,\quad I_m=\int_0^{2\pi a_1^{-1}}\frac{(E_m(t)f,g)}{\lambda_m(t)-k^2-i\varepsilon}dt_1,\quad (5.2.38)$$

where $\lambda_m(t)$ and $E_m(t)$ are the eigenvalues and the spectral projections of the operator $H(t)$, $t=(t_1,t_{\|})$. We can label $\lambda_m(t)$ and $E_m(t)$ so that they depend analytically on t_1 in a neighborhood of the interval $[0,2\pi a_1^{-1}]$ for a fixed $t_{\|}$.

In (5.2.38) we pick out from the region of summation the set S of m such that the function $\lambda_m(t)-k^2$ has a root t_1, $t_1\in[0,2\pi a_1^{-1})$. First, we consider the series $I=\sum_{m\notin S}I_m$. If $m\notin S$ then the function $\lambda_m(t)-k^2-i\varepsilon$ has no roots in some neighborhood of $\varepsilon=0$. We can choose this neighborhood to be the same for all $m\notin S$, because $\lambda_m(t)\to\infty$ when $m\to\infty$. From this it follows that I_m, $m\notin S$, depend analytically on ε in some neighborhood ρ of zero. Moreover, the series $\tilde{I}(f,g)=\sum_{m\notin S}I_m$ of analytical functions I_m converges to an analytic function uniformly with respect to ε. The operator, corresponding to the bilinear form $\tilde{I}(f,g)$, is bounded for all $z\in\rho$. Thus, it is an analytic operator-valued function of z. Next, we consider I_m, $m\in S$. The set S contains only a finite number of m. Note that $\lambda_m(t)$ and $E_m(t)$ are analytic functions (operator-valued functions) of t_1 in the complex plane with cuts that do not begin on the real axis. We consider the contour $\tilde{C}$, with the upper part of the contour being lower than the cuts. The zeros of each function $\lambda_m(t)-k^2$ and $\dot{\lambda}_m(t)\equiv\partial\lambda_m(t)/\partial t_1$ are separated from each other. Hence, we can choose so narrow a contour, that the zeros are absent strictly inside the contour and on its nonreal sides. With no loss of generality it may be assumed, that the real zeros of the functions $\lambda_m(t)-k^2$ are situated on $(0,2\pi a_1^{-1})$, because the interval of the integration in (5.2.38) is chosen up to a shift on arbitrary vector. We consider

$$\tilde{I}_m=\oint_{\tilde{C}}\frac{(E_m(t),f,g)}{\lambda_m(t)-k^2-i\varepsilon}dt_1.\qquad (5.2.39)$$

It is clear that for sufficiently small positive ε, there exist only such zeros of the functions $\lambda_m(t)-k^2-i\varepsilon$ inside $\tilde{C}$, which tend to the real axis as ε goes to zero. The zeros are simple strictly inside the contour, because $\dot{\lambda}_m(t)\neq 0$ there. By determining the residue we obtain:

$$\tilde{I}_m=\sum_i\frac{(E_m(t_i)(\varepsilon)),f,g)}{\dot{\lambda}_m(t_i(\varepsilon))},\qquad (5.2.40)$$

$t_i(\varepsilon)=(t_{1i}(\varepsilon),t_{\|})$, $t_{1i}(\varepsilon)$ being the zeros of the function $\lambda_m(t)-k^2-i\varepsilon$ inside the contour, and $\dot{\lambda}_m$ is the partial derivative of $\lambda_m(t)$ with respect to t_1. We consider i-th term of the sum.

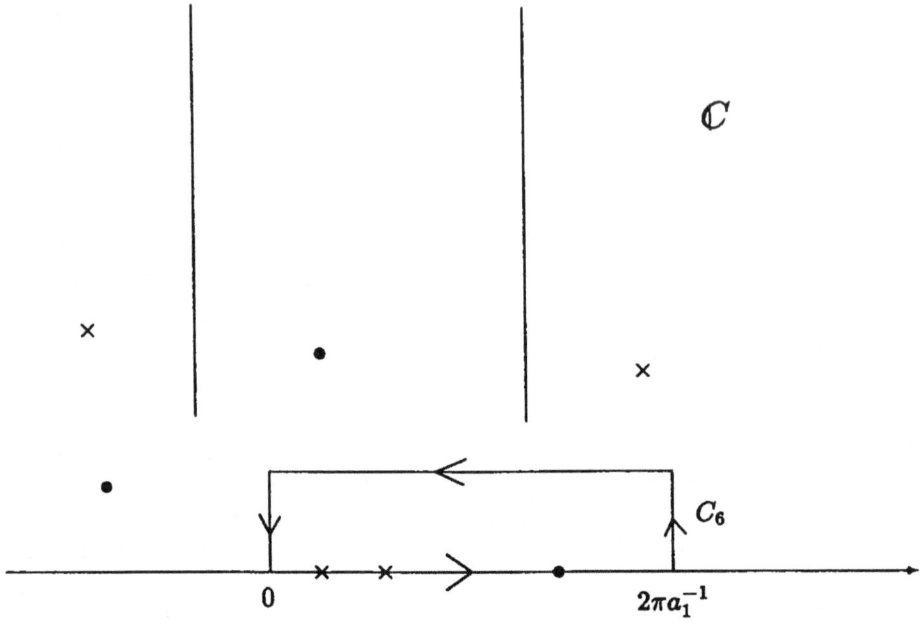

Fig 3. The contour C_6. Zero of the functions $\lambda_m(t)$-κ^2 and $\dot{\lambda}_m(t)$ are shown by × and •.

Suppose $\dot{\lambda}_m(t_i(0)) \neq 0$. Then $t_{1i}(\varepsilon)$ depends analytically on ε in some neighborhood of zero where $\dot{\lambda}_m(t_i(\varepsilon)) \neq 0$. Therefore, i-th term of the sum is a holomorphic operator-valued function. Next, suppose $\dot{\lambda}_m(t_i(0))$ has a zero of order p_i when $\varepsilon = 0$. Then $t_{1i}(\varepsilon)$ is an analytic function of $\xi_i = \varepsilon^{1/(p_i+1)}$ in some neighborhood of zero. Since $E_m(t)$ and $\lambda_m(t)$ analytically depend on t_1 in a neighborhood of the real axis, the i-th term of the sum analytically depends on ξ_i in some neighborhood ρ' of zero, with a single pole of the order p_i at zero. The sum $\tilde{I}_m$ contains only a finite number of terms. Therefore, it is an analytic function of $\xi = \varepsilon^{1/q}$, where $q = \prod_i (p_i + 1)$ (if $\dot{\lambda}_m(t_i(0)) \neq 0$ for all i, then $q = 1$). We represent I_m in the form $I_m = \tilde{I}_m - \hat{I}_m$, where $\hat{I}_m$ is the integral over the nonreal part $\tilde{C}'$ of the contour $\tilde{C}$. Note that $\hat{I}_m$ analytically depend on ε in some neighborhood of zero, because the function $\lambda_m(t) - k^2$ has no roots on $\tilde{C}$. Thus, I_m and, therefore, the resolvent of $H(t_{\|})$ analytically depends on $\xi = \varepsilon^{1/q}$ in some neighborhood of zero, having a single pole of the order not higher then $q - 1$ at zero. The proposition is proved.

From the previous proposition we see that the study of the analytic extension of the resolvent of the operator $H(t_{\|})$ is rather simple. The same problem for the whole operator H is complicated. This problem is studied in [Gér]. It is proved that $(H - \lambda)^{-1}$ extends across the spectrum of H to the complementary of a discrete set of points.

Using formulae (5.2.30) – (5.2.33) it is not hard to represent the operators Γ_1^V, Γ_2^V in the form:

$$\Gamma_1^V(z,t_{\|}) = (H(t_{\|})-z)^{-1}(0,0), \tag{5.2.41}$$

$$\Gamma_2^V(z,t_{\|}) = \frac{\partial}{\partial x_1}[(H(t_{\|})-z)^{-1} - (H_0(t_{\|})-z)^{-1}](0,0). \tag{5.2.42}$$

Obviously, their analytic properties are similar to those of the resolvent. Arguing as in the proof of the previous lemma, we obtain the following result :

Proposition 5.8 . *For any real k^2 and some natural q, $q=q(k)$, the operators $\Gamma_1^V(k^2+i\varepsilon,t_{\|})$, $\Gamma_2^V(k^2+i\varepsilon,t_{\|})$ can be extended as analytic operator-valued functions of $\xi=(\varepsilon)^{1/q}$ in some neighborhood of zero. In this neighborhood, each of the operators can have only a unique pole. This pole is at the point k^2 and has the order not higher than $q-1$ (there is no pole in the case q=1).*

Proposition 5.9 . *If $\varepsilon \neq 0$, then for the operators $A_\pm^V(k^2+i\varepsilon,t_{\|})$, the relation*

$$A_\pm^V(k^2+i\varepsilon,t_{\|})^* = A_\pm^V(k^2-i\varepsilon,t_{\|}) \tag{5.2.43}$$

holds. The operators $(A_+^V(k^2+i\varepsilon,t_{\|})-A_+^V(k^2-i\varepsilon,t_{\|}))/i\varepsilon$ and $(A_-^V(k^2-i\varepsilon,t_{\|})-A_-^V(k^2+i\varepsilon,t_{\|}))/i\varepsilon$ are positively determined.

Corollary 5.2 . *The operator $(T(k^2+i\varepsilon)-T^*(k^2+i\varepsilon))/i\varepsilon$ is positively determined, when $\varepsilon \neq 0$.*

Proof. Suppose functions u_+, u_- belonging to $L_2(Q_+)$ satisfy the equation

$$(-\Delta + V - k^2 \mp i\varepsilon)u_\pm = 0$$

and the quasiperiodic condition (5.1.4). Integrating by parts we show that

$$[u'_+|_{x_1=+0}, u_-|_{x_1=+0}] - [u_+|_{x_1=+0}, u'_-|_{x_1=+0}] = 0, \tag{5.2.44}$$

where $[\cdot,\cdot]$ is the scalar product in $L_2(Q_{\|})$. Since (5.2.44) is valid for any u_+, u_-, relation (5.2.43) holds for A_+^V.

Note that

$$[u'_+|_{x_1=+0}, u_+|_{x_1=+0}] - [u_+|_{x_1=+0}, u'_+|_{x_1=+0}] = 2i\varepsilon \int_{Q_+} |u_+|^2 dx_{\|} dx_1. \tag{5.2.45}$$

Therefore,

$$((A_+^V(k^2+i\varepsilon,t_{\|}) - A_+^V(k^2-i\varepsilon,t_{\|}))u_+, u_+)/i\varepsilon > 0.$$

The operator A_-^V is considered similarly. The proposition is proved.

Lemma 5.1 . *Suppose $T^{-1}(k^2+i\varepsilon, t_{\|})$ satisfies the estimate:*

$$\|T^{-1}(k^2+i\varepsilon, t_{\|})\| < c, \tag{5.2.46}$$

which is uniform in ε when $\varepsilon \in (0, \varepsilon_0)$, $\varepsilon_0 > 0$. Then, there exists the limit:

$$\lim_{\varepsilon \to 0}[T^{-1}(k^2+i\varepsilon, t_{\|})f, g]. \tag{5.2.47}$$

for any $f, g \in L_2(Q_{\|})$

Proof. Considering that $T^{-1}(k^2+i\varepsilon, t_{\|})$ is a function of Γ_1^V, Γ_2^V and relation (5.2.46) holds, we obtain that the operator $T^{-1}(k^2+i\varepsilon, t_{\|})$ depends on $\xi = \varepsilon^{1/q}$ analytically in a neighborhood of zero for some natural q. Obviously, ξ depends on ε continuously in the upper halfplane (the cut lies in the lower halfplane and begins on the real axis). Therefore, the function $[T^{-1}(k^2+i\varepsilon, t_{\|})f, g]$ is continuous in a closed upper semineighborhood of zero. Therefore, the limit (5.2.47) exists. The lemma is proved.

5.3 Elimination of Surface and Quasisurface States in the Plane Case.

Now we are interested in the case when there are no surface and quasisurface states. We will describe the conditions on $t_{\|}$ which provide the estimate:

$$\|T^{-1}(k^2 \pm i0, t_{\|})\| < 1. \tag{5.3.1}$$

It turns out that in the present plane case estimate (5.3.1) holds for all $t_{\|}$ being projections (on $K_{\|} = [0, 2\pi a_2^{-1})$ of vectors t, which are in the set $\chi_1(k, \delta, \delta)$ [7] and its small neighborhood. To prove this fact, we first consider the free operator $H_0(t_{\|})$ ($V_+ = 0$). Obviously, in this case the operator T is given by the formula:

$$T_0 = A_+^0 - A_-^0 = 2A_+^0.$$

Using formula (5.2.11) for A_+^0, it is not hard to show that

$$\|A_\pm^0(k^2 \pm i0, t_{\|})^{-1}\| = \rho_0^{-1/2}, \tag{5.3.2}$$

where

$$\rho_0(k, t_{\|}) = \min_{m_{\|} \in Z} \left| |t_{\|} + p_{m_{\|}}(0)|^2 - k^2 \right|. \tag{5.3.3}$$

Therefore, $\|T_0^{-1}\| < 1$, when $\rho_0 > 1$. We prove in Lemmas 5.27, 5.28 that this estimate holds when $t_{\|}$ is the projection of t of the nonsingular set $\chi_1(k, \delta, \delta)$. Moreover, for such $t_{\|}$:

$$\rho_0(k, t_{\|}) > k^{1/4}. \tag{5.3.4}$$

[7] $\chi_1(k, \delta, \delta)$ is the nonsingular set for the whole crystal (see Section 3.4). Its definition does not depend on V.

Thus, the operator $H_0(t_{\|})$ has no surface and quasisurface states with energy k^2, when $t_{\|}$ belongs to the projection of the nonsingular set $\chi_1(k,\delta,\delta)$ on $K_{\|}$. The main result of this section is that this assertion holds for the case of a nonzero potential, too.

Now we represent a condensed version of proving (5.3.1) for $t_{\|}$, which satisfy estimate (5.3.4). The main steps are the following:

1) We consider a solution of the equation:

$$(-\Delta + V_+ - k^2 - i0)u =_{x_1 \neq 0} 0,$$

which is a refracted wave in Q_+ and a reflected wave in Q_-. We look for a refracted wave as a linear combination of quasiperiodic functions whose quasimomenta t_1 are in the upper halfplane, or at real points coming from the upper plane as $k^2 + i\varepsilon \to k^2 + i0$. Similarly, a reflected wave is a linear combination of quasiperiodic functions whose quasimomenta t_1 are in the lower halfplane, or at real points coming from the lower plane as $k^2 + i\varepsilon \to k^2 + i0$.

2) We show that the linear span of the functions $\psi_-^0(k^2, t_{\|} + p_{m_{\|}}(0), x)$ is the complete set of reflected waves.

3) To describe the complete set of refracted waves, we construct a set of basic refracted waves $U_{m_{\|}}(x)$. We define these function as follows. If $m_{\|} : |t_{\|} + p_{m_{\|}}(0)|^2 < k^2$, then $U_{m_{\|}}$ is determined by the relations

$$(-\Delta + V - k^2 - i0)U_{m_{\|}}(x) =_{x_1 > 0} 0,$$

$$(U_{m_{\|}}|_{x_1=+0})_{n_{\|}} = \delta_{m_{\|} n_{\|}}. \tag{5.3.5}$$

If $m_{\|} : |t_{\|} + p_{m_{\|}}(0)|^2 > k^2$, then the function $U_{m_{\|}}$ by definition is the integral (5.3.11). Note that $|t_{\|} + p_{m_{\|}}(0)|^2 \neq k^2$, because we assume (5.3.4) holds. The main properties of the integral are that $U_{m_{\|}}(x) = \psi_+^0(k^2, t_{\|} + p_{m_{\|}}(0), x)$, when $V_+ = 0$, and that $U_{m_{\|}}(x)$ is asymptotically close to $\psi_+^0(k^2, t_{\|} + p_{m_{\|}}(0), x)$ at the boundary under condition (5.3.4) in the case of a nonzero potential (see (5.3.24), (5.3.25)). We show that any refracted wave can be represented as a linear combination of the functions $U_{m_{\|}}$:

$$\Psi_{refr} = \sum_{m_{\|}} d_{m_{\|}} U_{m_{\|}}(x). \tag{5.3.6}$$

4) Suppose $H(t_{\|})$ has a surface or quasisurface state. We show that for such a state, the nondecaying component of the reflected wave (a linear combination of the quasiperiodic solutions with real quasimomenta) is small, i.e.,

$$\Psi_{refl} = \Psi_{refl}^0 + o(1),$$

$$\Psi_{refl}^0 = \sum_{m_{\|} : |t_{\|} + p_{m_{\|}}(0)|^2 > k^2} c_{m_{\|}} \psi_-^0(k^2, t_{\|} + p_{m_{\|}}(0), x), \tag{5.3.7}$$

where the functions $\psi_-(k^2, t_{\|} + p_{m_{\|}}(0), x)$, $m_{\|} : |t_{\|} + p_{m_{\|}}(0)|^2 > k^2$, decay exponentially when $x_1 \to \infty$.

5) We use the continuity conditions on the boundary. We represent the refracted wave in the form (5.3.6). Taking into account that $U_{m_\|}(x)|_{x_1=+0} = \psi_+(k^2, t_\| + p_{m_\|}(0), x)|_{x_1=+0}$, when $m_\| : |t_\| + p_{m_\|}(0)|^2 < k^2$ (see (5.3.5)), and that $U_{m_\|}(x)|_{x_1=0}$ is asymptotically close to $\psi_+(k^2, t_\| + p_{m_\|}(0), x)|_{x_1=0}$ when $m_\| : |t_\| + p_{m_\|}(0)|^2 > k^2$, we obtain from the continuity condition on the boundary:

$$d_{m_\|} \approx \begin{cases} c_{m_\|} & \text{if } m_\| : |t_\| + p_{m_\|}(0)|^2 > k^2, \\ 0 & \text{otherwise.} \end{cases}$$

Thus,

$$\Psi_{refr} = \Psi^0_{refr} + o(1),$$

$$\Psi^0_{refr} = \sum_{m_\| : |t_\| + p_{m_\|}(0)|^2 > k^2} c_{m_\|} \psi^0_+(k^2, t_\| + p_{m_\|}(0), x). \tag{5.3.8}$$

It is clear that in $t_\|$-basis:

$$\left.\frac{\partial \Psi^0_{refr}}{\partial x_1}\right|_{x_1=+0} - \left.\frac{\partial \Psi^0_{refl}}{\partial x_1}\right|_{x_1=-0} = T_0 \mathbf{c}. \tag{5.3.9}$$

$$\mathbf{c}_{m_\|} = \begin{cases} c_{m_\|} & \text{if } m_\| : |t_\| + p_{m_\|}(0)|^2 > k^2, \\ 0 & \text{otherwise.} \end{cases}$$

6) We prove that operator $H(t_\|)$ has no surface or quasisurface state , when (5.3.4) holds. By definition, in a case of a surface or a quasisurface state, the functions Ψ_{refl}, Ψ_{refr} satisfy the continuity condition with high accuracy. This means that $\Psi^0_{refl}, \Psi^0_{refr}$ also satisfy the continuity condition with good accuracy. However, taking into account (5.3.9), we get that the continuity conditions for the derivatives of Ψ^0_{refl}, Ψ^0_{refr} can not be satisfied with good accuracy when (5.3.4) holds, and, therefore,

$$\|T_0^{-1}\| < k^{1/4}. \tag{5.3.10}$$

Thus, $H(t_\|)$ has no surface and quasisurface states.

Now we prove this result rigorously. We define the function $U_{m_\|}(k^2, x)$, $m_\| \in Z$, as follows:

$$U_{m_\|}(k^2, t_\|, x) = \tag{5.3.11}$$

$$a_{m_\|} \oint_{C_2} \sum_{n \in Z^2} (H(\tau) - k^2)^{-1}_{nm} \exp i(\mathbf{p}_n(\tau), x) d\tau_1,$$

where C_2 is the contour in the complex plane represented by Fig. 4, and $m = (0, m_\|)$, $\tau = (\tau_1, t_\|)$, the coefficients $a_{m_\|}$ are given by

$$a_{m_\|} = -\pi^{-1} A^0_+(k^2, t_\|)_{m_\| m_\|}. \tag{5.3.12}$$

We will show further (Lemma 5.4) that the integrand has no poles on C_2. The integral (5.3.11) is defined also for $k^2 \pm i\varepsilon$, ε being small enough, and

$$U_{m_\parallel}(k^2 + i0, t_\parallel, x) = U_{m_\parallel}(k^2 - i0, t_\parallel, x) = U_{m_\parallel}(k^2, t_\parallel, x). \tag{5.3.13}$$

Here we note only, that reasoning as in Proposition 5.6, it is not hard to show that the series on the right converges in the class $W_2^2(Q_+)$ uniformly in τ_1. Since the factor $\exp i(\tau, x)$ exponentially decays as $\Im\tau \to \infty$ when $x_1 > 0$, we can exchange summation and integration. It is easy to see that $U_{m_\parallel}(k^2, t_\parallel, x)$ satisfies quasiperiodic conditions (5.1.4), (5.1.7) and the equation:

$$(-\Delta + V - k^2)U_{m_\parallel}(k^2, t_\parallel, x) =$$

$$\left(\oint_{C_2} \exp i\tau_1 x_1 d\tau_1\right) \exp i(t_\parallel + p_{m_\parallel}(0), x) = 0. \tag{5.3.14}$$

Further, $U_{m_\parallel}(k^2, t_\parallel, x)$ exponentially decays as $x_1 \to \infty$, because of the factor $\exp i(\tau, x)$, $\Im\tau > 0$.

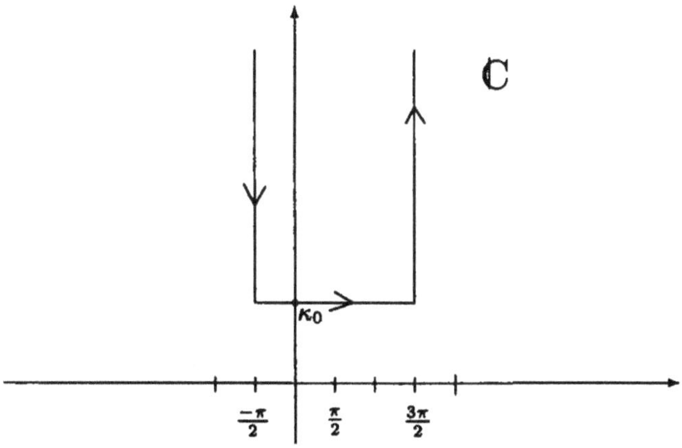

Fig.4. The contour $C_3, \kappa_0 = \kappa^{\frac{1}{4}}$

Lemma 5.2 . *Suppose $V = 0$, and $t_\parallel$ is such that $\rho_0(k, t_\parallel) > k^{1/4}$. Then, for ε small enough in absolute value, in particular, $\varepsilon = 0$,*

$$U_{m_\parallel}(k^2 + i\varepsilon, t_\parallel, x) = \psi_+^0(k^2 + i\varepsilon, t_\parallel + p_{m_\parallel}(0), x), \tag{5.3.15}$$

when $m_{\|} : |t_{\|} + p_{m_{\|}}(0)|^2 > k^2$. *Otherwise,*

$$U_{m_{\|}}(k^2 + i\varepsilon, t_{\|}) = 0. \tag{5.3.16}$$

Proof. Let $m_{\|} : |t_{\|} + p_{m_{\|}}(0)|^2 > k^2$. The operator $(H(\tau) - k^2 - i\varepsilon)^{-1}$ has a diagonal form and its matrix elements are: $((\tau_1 + 2\pi m_1 a_1^{-1})^2 + |t_{\|} + p_{m_{\|}}(0)|^2 - k^2 - i\varepsilon)^{-1}\delta_{mn}$. If $m_1 \neq 0$, then the matrix element is a holomorphic function inside C_2. If $m_1 = 0$, then the integrand has a simple pole at the point $\tau_1 = i\sqrt{(t_{\|} + p_{m_{\|}}(0))^2 - k^2 - i\varepsilon}$, $\Re\sqrt{\ } \geq 0$. From inequality (5.3.4), we obtain that this pole is inside C_2. We can evaluate the integral by calculating the residue, because the integrand contains the exponentially decreasing factor $\exp i\tau_1 x_1$. This calculation gives formula (5.3.15). In the case $|t_{\|} + p_{m_{\|}}(0)|^2 > k^2$, all the matrix elements are holomorphic inside the contour. Therefore, (5.3.16) holds. The lemma is proved.

We introduce the diagonal operators $L(t_1)$, $M(k^2 + i\varepsilon, t_{\|})$ in l_2^2 by the matrices:

$$L_{mm} = 2\pi m_1 a_1^{-1} + t_1, \tag{5.3.17}$$

$$M_{mm} = A^0_-(k^2 + i\varepsilon, t_{\|})_{m_{\|}m_{\|}}, \quad m \in Z^2. \tag{5.3.18}$$

It is easy to verify that for $t_1 \in C_2$

$$\|L^{-1}\| \leq (\Im t_1)^{-1}, \tag{5.3.19}$$

$$\|(L + iM)^{-1}\| \leq (\Im t_1)^{-1}. \tag{5.3.20}$$

The last inequality is based on the estimate $\Re A^0_-(k^2 + i\varepsilon, t_{\|})_{m_{\|}m_{\|}} \geq 0$ (see (5.2.11)).

Lemma 5.3 . *Suppose τ_1 lies on C_2, and $t_{\|}$ is such that $\rho_0(k, t_{\|}) > k^{1/4}$. Then,*

$$\|L(H_0(\tau) - k^2 - i\varepsilon)^{-1}\| \leq 1, \tag{5.3.21}$$

$$\|M(H_0(\tau) - k^2 - i\varepsilon)^{-1}\| \leq 1, \tag{5.3.22}$$

$$\tau = (\tau_1, t_{\|}),$$

$$L = L(\tau_1) \quad , M = M(k^2 + i\varepsilon, t_{\|}).$$

Corollary 5.3 .

$$\|(H_0(\tau) - k^2 - i\varepsilon)^{-1}\| \leq (\Im t_1)^{-1}. \tag{5.3.23}$$

Proof. We represent $\tau_1 \in C_2$ in the form $\tau_1 = \gamma_1 + i\gamma_2$, $\gamma_1, \gamma_2 \in R$, $\gamma_2 \geq b_0$, $b_0 = k^{1/8}$. It is clear that $|p_m^2(\tau) - k^2 - i\varepsilon|^2 = (p_m^2(\Re\tau) - k^2 - \gamma_2^2)^2 + (2(\gamma_1 + 2\pi m_1 a_1^{-1})\gamma_2 - \varepsilon)^2 > (2(\gamma_1 + 2\pi m_1 a_1^{-1})\gamma_2 - \varepsilon)^2$. If $(\gamma_1 + 2\pi m_1 a_1^{-1})^2 > 1$, then it is easy to verify the inequality $|(\tau_1 + 2\pi m_1 a_1^{-1})(p_m^2(t) - k^2 - i\varepsilon)^{-1})| < 1$. The estimate $(\gamma_1 + 2\pi m_1 a_1^{-1})^2 \leq 1$ can hold only on the horizontal part of the contour, that is when $\gamma_2 = b_0$, and $m_1 = 0$. In this case, we use the estimate $|p_m^2(t) - k^2 \pm i\varepsilon| > |p_m^2(\Re t) - k^2 - \gamma_2^2| = |\, |t_{\|} + p_{m_{\|}}(0)|^2 + \gamma_1^2 - k^2 - b_0^2|$. Taking into account the relation $\rho_0 > 2k^{1/4} = 2b_0^2$, we get $|p_m^2(t) - k^2| > b_0^2$. Noting that $|L_{mm}(\tau_1)| < 2b_0$, when $m_1 = 0$, we verify that $|(p_m^2(t) - k^2)^{-1} L_{mm}| < 1$ on the horizontal part of the contour. Thus, we have proved the last estimate for all m. Relation (5.3.21) is proved. Finally, let us prove (5.3.22). It is clear that

$$(L - iM)(H_0(\tau) - k^2 - i\varepsilon)^{-1} = (L + iM)^{-1}.$$

Using inequalities (5.3.20) and (5.3.21), we obtain (5.3.22).
The lemma is proved.

Next, we prove that the values of the functions $U_{m_\|}$ and $\frac{\partial}{\partial x_1} U_{m_\|}$ on the elementary cell of the boundary $x_1 = 0$ are asymptotically close to those of the unperturbed function. This means that

$$U_{m_\|}\Big|_{x_1=+0} = U^0_{m_\|}\Big|_{x_1=+0} + o(1), \tag{5.3.24}$$

$$\frac{\partial}{\partial x_1} U_{m_\|}\Big|_{x_1=+0} = \frac{\partial}{\partial x_1} U^0_{m_\|}\Big|_{x_1=+0} + o(1). \tag{5.3.25}$$

Firstly, we consider the operators J and D determined in $L_2(Q_\|)$ by the matrices (we use $t_\|$-basis):

$$J_{n_\| m_\|} = \left(U_{m_\|}\Big|_{x_1=+0} \right)_{n_\|}, \tag{5.3.26}$$

$$D_{n_\| m_\|} = \left(\frac{\partial}{\partial x_1} U_{m_\|}\Big|_{x_1=+0} \right)_{n_\|}, \tag{5.3.27}$$

where $U_{m_\|}$ is the integral (5.3.11). Taking into account the definition of the operator A_+^V (see (5.2.3)) and equality (5.3.14)), we obtain:

$$D = A_+^V(k^2 + i\varepsilon, t_\|) J. \tag{5.3.28}$$

From formulae (5.3.15) and (5.3.16) it follows that the operator J is a diagonal projection for $V = 0$:

$$J^0_{m_\| m_\|} = \begin{cases} 1, & \text{when } |t_\| + p_{m_\|}(0)|^2 \geq k^2; \\ 0, & \text{otherwise.} \end{cases} \tag{5.3.29}$$

It is clear that

$$D^0 = A_+^0(k^2 + i\varepsilon, t_\|) J^0. \tag{5.3.30}$$

From (5.2.11) it easily follows:

$$\lim_{\varepsilon \to 0} (A_-^0(k^2 + i\varepsilon, t_\|) - A_-^0(k^2 - i\varepsilon, t_\|))/2i = -iA_-^0(k^2 + i0, t_\|)(I - J^0). \tag{5.3.31}$$

The operator $-iA_-^0(k^2 + i0, t_\|)(I - J^0)$ is diagonal. Obviously, this operator is negatively determined in the subspace $(I - J^0)L_2(K_\|)$. Suppose $\rho_0(k^2, t_\|) > k^{1/4}$, then using (5.2.11), we obtain the stronger estimate:

$$(-iA_-^0(k^2 + i0, t_\|)(I - J^0)y, y)_{l_2^1} \leq -k^{1/8} \|(I - J^0)y\|^2_{l_2^{\frac{1}{2}}}. \tag{5.3.32}$$

Lemma 5.4 . *Suppose $t_\parallel$ is such that $\rho_0(k,t_\parallel) > k^{1/4}$. Then the operators $J(k^2+i\varepsilon,t_\parallel)$ and $D(k^2+i\varepsilon,t_\parallel)$ satisfy the estimates:*

$$\|J - J^0\| < ck^{-1/8}, \tag{5.3.33}$$

$$\|D - A^0_+ J^0\| < c, \tag{5.3.34}$$

$$\|A^0_-(J - J^0)\| < c, \qquad c \neq c(k). \tag{5.3.35}$$

Proof. First, we introduce the "strong" norm $\|F\|^*$ of an operator F in $L_2(Q)$:

$$4\|F\|^* = \|F\|^*_> + \|F\|^*_<,$$

$$\|F\|^*_> = \max_{n_\parallel} \sum_{m_\parallel} \left(\max_{n_1} \sum_{m_1} |F_{mn}| + \max_{m_1} \sum_{n_1} |F_{mn}| \right),$$

$$\|F\|^*_< = \max_{m_\parallel} \sum_{n_\parallel} \left(\max_{n_1} \sum_{m_1} |F_{mn}| + \max_{m_1} \sum_{n_1} |F_{mn}| \right).$$

It is obvious that

$$\|F\| \le \|F\|^* \tag{5.3.36}$$

and for diagonal operators $\|F\| = \|F\|^*$. Note that $\|V\|^* < \infty$ when V is a trigonometric polynomial. It is not hard to show that

$$\|F_1 + F_2\|^* \le \|F_1\|^* + \|F_2\|^*, \tag{5.3.37}$$

$$\|F_1 F_2\|^* \le \|F_1\|^* \|F_2\|^*. \tag{5.3.38}$$

Suppose $\|F\|^* < \infty$. Note that the operator $\hat{F}$, defined in $L_2(Q_\parallel)$ by the matrix

$$(\hat{F})_{n_\parallel m_\parallel} = \sum_{n_1} F_{nm}, \qquad m = (0, m_\parallel) \tag{5.3.39}$$

is bounded and

$$\|\hat{F}\| \le \|F\|^*. \tag{5.3.40}$$

Now, we define F by the formula:

$$F = \pi^{-1} \oint_{C_1} \left((H(\tau) - k^2 - i\varepsilon)^{-1} - (H_0(\tau) - k^2 - i\varepsilon)^{-1} \right) M(k^2 + i\varepsilon, t_\parallel) d\tau_1. \tag{5.3.41}$$

It follows from formulae (5.3.11), (5.3.18) and (5.3.26), that

$$J - J^0 = \hat{F}. \tag{5.3.42}$$

Relation (5.3.40) yields:

$$\|J - J^0\| \le \|F\|^*. \tag{5.3.43}$$

To estimate $\|F\|^*$ we represent $(H(t) - z)^{-1}$, $z = k^2 + i\varepsilon$, in the form:

$$(H(t)-z)^{-1}-(H_0(t)-z)^{-1}=\left(\sum_{r=1}^{\infty}B^r\right)(H_0(t)-z)^{-1}, \tag{5.3.44}$$

$$B=-(H_0(t)-z)^{-1}V.$$

Taking into account that $(H_0(t)-z)^{-1}$ is a diagonal operator, we obtain from estimates (5.3.23) and (5.3.38) that

$$\|B\|^* \le (\Im\tau_1)^{-1}\|V\|^*, \quad \tau_1\in C_2. \tag{5.3.45}$$

Using (5.3.44), we represent F in the form $F=F'+F''$, where

$$F'=\pi^{-1}\oint_{C_2}(H_0(\tau)-k^2-i\varepsilon)^{-1}V(H_0(\tau)-k^2-i\varepsilon)^{-1}M(k^2+i\varepsilon,t_{\|})d\tau_1,$$

$$F''=\pi^{-1}\oint_{C_2}B^2(I-B)^{-1}(H_0(\tau)-k^2-i\varepsilon)^{-1}M(k^2+i\varepsilon,t_{\|})d\tau_1. \tag{5.3.46}$$

Considering that V is a trigonometric polynomial, we obtain that $F'_{nm}=0$, when $|n-m|>R_0$. Therefore,

$$\|F'\|^* \le c(R_0)\max_{mn}|F'_{mn}|. \tag{5.3.47}$$

In the formula for F' the integrand has poles at the points:

$$\tau_1=\sqrt{k^2+i\varepsilon-|t_{\|}+p_{m_{\|}}(0)|^2},\quad \tau_1=\sqrt{k^2+i\varepsilon-|t_{\|}+p_{n_{\|}}(0)|^2}.$$

So, the integral can be replaced by an integral over a finite contour C_3 around these two points. We can choose this contour to be in a distance of order 1 from the points (possibly it contains two components). It is not hard to show that for $t=(\tau_1,t_{\|})$, $\tau_1\in C_3$, there is the estimate:

$$\|M(H_0(t)-k^2-i\varepsilon)^{-1}\|<c, \quad c\ne c(k). \tag{5.3.48}$$

It is not hard to show that (5.3.45) holds also for any $\tau_1\in C_3$. Using the definition of C_3, we get $\Im\tau>k^{1/8}$. Therefore,

$$\|V(H_0(t)-k^2+i\varepsilon)^{-1}\|<ck^{-1/8}, \quad c\ne c(k). \tag{5.3.49}$$

Considering inequalities (5.3.48) and (5.3.49), we obtain:

$$|F'_{mn}|<ck^{-1/8}. \tag{5.3.50}$$

From relations (5.3.47) and (5.3.50) it follows the estimate:

$$\|F'\|^*<ck^{-1/8}. \tag{5.3.51}$$

Next, we estimate F''. Using inequalities (5.3.22) and (5.3.45), we obtain:

$$\|B^2(I-B)(H_0(\tau)-k^2-i\varepsilon)^{-1}M(k^2+i\varepsilon,t_{\|})\|^*<c(\Im\tau_1)^{-2}.$$

Integrating both sides of the last estimate over the contour, we get:

$$\|F''\|^* < ck^{-1/8}. \tag{5.3.52}$$

Adding inequalities (5.3.51) and (5.3.52) yields:

$$\|F\|^* < ck^{-1/8}. \tag{5.3.53}$$

Considering relation (5.3.43), we get (5.3.33).

Let us prove (5.3.34) and (5.3.35). Using (5.3.11) and (5.3.27), it is not hard to show that

$$D - A_+^0 J^0 = i\pi^{-1}\hat{F}_1,$$

$$\hat{F}_1 = \oint_{C_1} L(\tau_1)\left((H(t) - k^2 - i\varepsilon)^{-1} - (H_0(t) - k^2 - i\varepsilon)^{-1}\right) M d\tau_1, \tag{5.3.54}$$

$$A_-^0(J - J^0) = \pi^{-1}\hat{F}_2,$$

$$\hat{F}_2 = \oint_{C_1} M\left((H(t) - k^2 - i\varepsilon)^{-1} - (H_0(t) - k^2 - i\varepsilon)^{-1}\right) M d\tau_1. \tag{5.3.55}$$

Arguing as in the proof of estimate (5.3.33), and using inequalities (5.3.21) and (5.3.22), we obtain (5.3.34) and (5.3.35). The lemma is proved.

From here on we define the function $U_{m_\|}$ by formula (5.3.11), when $|p_{m_\|}(0) + t_\||^2 > k^2$. In the case $|p_{m_\|}(0) + t_\||^2 \le k^2$, we define the function $U_{m_\|}$, $U_{m_\|} \in L_2(Q_+)$ by the equation:

$$(-\Delta + V - k^2 - i\varepsilon)U_{m_\|} = 0, \qquad x_1 > 0, \tag{5.3.56}$$

with the boundary condition

$$(U_{m_\|}|_{x_1=+0})_{n_\|} = \delta_{m_\| n_\|} \tag{5.3.57}$$

and with quasiperiodic conditions (5.1.4), (5.1.7). Now we consider the set of all the functions $U_{m_\|}$, $m_\| \in Z$. Note that the matrix $\hat{J}$, $\hat{J}_{n_\| m_\|} = (U_{m_\|}|_{x_1=0})_{n_\|}$, $m_\|, n_\| \in Z$, connecting the system of the functions $U_{m_\|}$ with $t_\|$-basis, can be represented in the form:

$$J = I + \Delta J, \qquad \Delta J = J - J_0. \tag{5.3.58}$$

According to Lemma 5.4, the estimate $\|\Delta J\| < k^{-1/8}$ holds. From this it easily follows:

Lemma 5.5 . *If $\rho_0(k, t_\| 0 > k^{1/4}$, then the set of functions $U_{m_\|}|_{x_1=+0}$ is complete in $L_2(Q_\|)$.*

Lemma 5.6 . *Suppose $t_\|$ is such that $\rho_0(k, t_\|) > k^{1/4}$. Then a solution y of the equation*

$$(A_+^V - A_-^0)(k^2 + i\varepsilon, t_\|)y = f, \tag{5.3.59}$$

$f \in L_2(Q_\|)$ satisfies the estimate:

$$\|J^0 A_-^0 y\| + k^{1/16}\|(I - J^0)y\| \le 2\|f\|, \tag{5.3.60}$$

which is uniform in ε, $0 < \varepsilon < \varepsilon_0$.

Corollary 5.4 . *The operator* $(A_+^V - A_-^0)^{-1}(k^2 + i\varepsilon, t_{\|})$ *is bounded, and*

$$\|(A_+^V - A_-^0)^{-1}(k^2 + i\varepsilon, t_{\|})\| < 2k^{-1/16}. \tag{5.3.61}$$

$$\|A_-^0(A_+^V - A_-^0)^{-1}(k^2 + i\varepsilon, t_{\|})\| < 2k^{15/16}. \tag{5.3.62}$$

The corollary easily follows from inequalities (5.3.60), (5.3.2) – (5.3.4) and the definition of J_0 (see (5.3.29)).

Proof. It is clear that

$$\Im\left((A_+^V - A_-^0)y, y\right) = \Im(f, y).$$

Using Proposition 5.9 and formula (5.3.31), we obtain that for sufficiently small positive ε the following estimate holds:

$$|(A_-^0(k^2 + i\varepsilon, t_{\|})(I - J^0)y, y)| \leq \|f\|\|y\|.$$

Taking into account the relation (5.3.32), we get

$$k^{1/8}\|(I - J^0)y\|^2 \leq \|f\|\|(I - J^0)y\| + \|f\|\|J^0 y\|. \tag{5.3.63}$$

Using that $\|(A_+^0)^{-1}\| < \rho^{-1/2} < k^{-1/16}$, we rewrite inequality (5.3.63) in the form:

$$k^{1/16}\|(I - J^0)y\| \leq \|f\|^{1/2}\|(I - J^0)y\|^{1/2} + k^{-1/16}\|f\|^{1/2}\|J_0 A_+^0 y\|^{1/2}.$$

Suppose we have proved the estimate:

$$\|J^0 A_+^0 y\| \leq \|f\| + c\|(I - J^0)y\|. \tag{5.3.64}$$

Adding the last two relations and introducing the notation $z = \|J^0 A_+^0 y\| + k^{1/16}\|(I - J_0)y\|$, we obtain

$$z \leq 2\|f\|^{1/2}k^{-1/32}z^{1/2} + ck^{-1/16}z + \|f\|. \tag{5.3.65}$$

Considering estimate (5.3.65) as an inequality for a quadratic polynomial of $z^{1/2}$, we get $z < 2\|f\|$, i.e., (5.3.61).

Thus, it remains to prove (5.3.64). Indeed, taking the scalar product of both parts of (5.3.59) with $U_{m_{\|}}|_{x_1=+0}(k^2 - i\varepsilon, t_{\|})$, $|t_{\|} + p_{m_{\|}}(0)|^2 > k^2$, we get:

$$(A_+^V(k^2 - i\varepsilon, t_{\|})U_{m_{\|}}|_{x_1=+0}, y) - (U_{m_{\|}}|_{x_1=+0}, A_-^0(k^2 + i\varepsilon, t_{\|})y) = (U_{m_{\|}}|_{x_1=+0}, f). \tag{5.3.66}$$

Using the definitions of J and D (see (5.3.26), (5.3.27)) we obtain:

$$(D(k^2 - i\varepsilon, t_{\|})\delta_{m_{\|}}, y) - (J(k^2 - i\varepsilon, t_{\|})\delta_{m_{\|}}, A_-^0(k^2 + i\varepsilon, t_{\|})y) = (J(k^2 - i\varepsilon, t_{\|})\delta_{m_{\|}}, f). \tag{5.3.67}$$

We rewrite (5.3.67) in the form:

$$\left((A_+^0(k^2 - i\varepsilon, t_{\|})J^0 - A_-^0(k^2 - i\varepsilon, t_{\|})J^0)\delta_{m_{\|}}, y\right) + (K_1\delta_{m_{\|}}, y) =$$

$$(J_0\delta_{m_\parallel}, f) + (K_2\delta_{m_\parallel}, f), \tag{5.3.68}$$

where

$$K_1 = D - A_+^0 J_0 - A_-^0(J - J_0), \quad K_2 = J - J_0.$$

Considering (5.3.33) – (5.3.35), we get

$$\|K_1\| < c, \tag{5.3.69}$$

$$\|K_2\| < ck^{-1/8}. \tag{5.3.70}$$

Considering that $A_+^0 = -A_-^0$ (see (5.2.11)) and using the definition of J^0 (5.3.29), we get

$$2A_+^0((k^2 - i\varepsilon, t_\parallel)J^0 y + J^0 K_1^* y = J^0 f + J^0 K_2^* f. \tag{5.3.71}$$

We rewrite the last relation in the form:

$$2A_+^0((k^2 - i\varepsilon, t_\parallel)J^0 y + 2K_3 A_+^0((k^2 - i\varepsilon, t_\parallel)J^0 y + K_4(I - J^0)y =$$

$$J^0 f + J^0 K_2^* f, \tag{5.3.72}$$

where $K_3 = J^0 K_1^*(A_+^0)^{-1} J_0$, $K_4 = J_0 K_1^*(I - J_0)$. Taking into account inequalities (5.3.2) and (5.3.69), we verify that $\|K_3\| < ck^{-1/8}$, $\|K_4\| < c$, $c \neq c(k)$. From this, using (5.3.70), we obtain (5.3.64). The lemma is proved.

Lemma 5.7 . *Suppose $t_\parallel$ is such that $\rho_0(k, t_\parallel) > k^{1/4}$. Then there exists a limit of the operator $T^{-1}(k^2 + i\varepsilon, t_\parallel)$ in the class S_2, when ε goes to zero and the following uniform in ε estimate holds:*

$$\|T^{-1}(k^2 + i\varepsilon, t_\parallel)\| < ck^{-1/16}. \tag{5.3.73}$$

Proof. From estimate (5.3.62) we get

$$\|A_-^0(k^2 + i0, t_\parallel)(A_+^V - A_-^0)^{-1}(k^2 + i0, t_\parallel)\| < ck^{15/16}. \tag{5.3.74}$$

It is clear that the bilinear form of the operator $A_-^0(k^2+i0, t_\parallel)(A_+^V - A_-^0)^{-1}(k^2 + i\varepsilon, t_\parallel)$ has the analytic properties similar to those of $(A_+^V - A_-^0)^{-1}(k^2 + i\varepsilon, t_\parallel)$. According to Lemma 5.1, there exists a limit ($\varepsilon \to 0$) of the operator $A_-^0(k^2 + i0)(A_+^V - A_-^0)^{-1}(k^2 + i\varepsilon)f$ in the sense of bilinear forms.

We prove that the operator $B_\delta(\varepsilon) = A_-^0(k^2 + i0, t_\parallel)^{1-\delta}(A_+^V - A_-^0)^{-1}(k^2 + i\varepsilon, t_\parallel)$ has a limit in the class of bounded operators when $\varepsilon \to 0$. It is clear that $B_\delta(\varepsilon)$ has a limit $B_\delta(0)$ in the sense of bilinear forms. We consider the diagonal projection P_N:

$$(P_N)_{m_\parallel m_\parallel} = \begin{cases} 1, & \text{when } (A_-^0)_{m_\parallel m_\parallel} < N; \\ 0, & \text{otherwise.} \end{cases}$$

It is obvious that $P_n B_\delta(\varepsilon)$ is a finite dimensional operator. Therefore, it is easy to see that $P_N B_\delta(\varepsilon) \to P_N B_\delta(0)$ in the class of bounded operators, i.e.,

$$\|P_N(B_\delta(\varepsilon) - B_\delta(0))\| < \varepsilon_1$$

when $\varepsilon < \varepsilon_0(\varepsilon_1, N)$ for an arbitrary small positive ε_1. From the definition of P_N and estimate (5.4.74), it follows:

$$\|(I - P_N)(B_\delta(\varepsilon) - B_\delta(0))\| < ck^{15/16}N^{-\delta}.$$

Therefore,

$$\|(B_\delta(\varepsilon) - B_\delta(0))\| < \varepsilon_1 + cN^{-\delta}k^{15/16}.$$

Choosing N and then ε, we make $\|(B_\delta(\varepsilon) - B_\delta(0))\|$ arbitrarily small. It is clear that $(A_+^V - A_-^0)^{-1} = (A_-^0)^{-1+\delta}B_\delta(\varepsilon)$. Note that $(A_-^0)^{-1+\delta}(k^2 + i0, t_{||}) \in S_2$ for δ small enough and does not depend on ε. Since $B_\delta(\varepsilon) \to B_\delta(0)$ in the class of bounded operators , we get $(A_+^V - A_-^0)^{-1}(k^2 + i\varepsilon, t_{||}) \to (A_+^V - A_-^0)^{-1}(k^2 + i0, t_{||})$ in S_2. The estimate (5.4.73) follows from (5.4.61). The lemma is proved.

In Section 5.5 we shall show that there exists a set $\Omega_0(k) \subset K_{||}$, such that $\rho_0(k^2, t_{||}) > k^{1/4}$, when $t_{||} \in \Omega_0(k)$. The set $\Omega_0(k)$ has an asymptotically full measure in $K_{||}$:

$$\frac{s(\Omega_0(k))}{s(K_{||})} \to_{k\to\infty} 1.$$

5.4 Elimination of Surface and Quasisurface States in the Three-Dimensional Case.

In the two-dimensional situation, the operator $H(t_{||})$ has no surface and quasisurface states with energy k^2, if $\rho_0(k, t_{||}) > k^{1/4}$ (see (5.4.3), (5.4.4)). To find out this condition, we considered $H(t_{||})$ as a perturbation of the free operator $H_0(t_{||})$. Namely, we constructed the solutions $U_{m_{||}}$ (see (5.4.11)) of the Schrödinger equation with potential $V(x)$, which are close at the boundary to those of the equation with the zero potential (see (5.4.24), (5.4.25)).

The situation becomes more complicated in the three-dimensional case. The perturbation formulae (5.4.24), (5.4.25) for $U_{m_{||}}$ do not hold any longer. However, it will be possible to prove that $U_{m_{||}} \approx U_{m_{||}}^1$ [8] for $|m_{||}| < k^\delta$, where $U_{m_{||}}^1$ are refracted waves constructed for an auxiliary operator H_{1+}. The Schrödinger operator H_{1+} corresponds to a potential $V_1(x_1)$ in R^3 depending only on one variable x_1. It is given by the formula: $H_{1+} = H_0 + V_{1+}$, where V_{1+} is the component of $V_+(x)$ depending only on x_1:

$$V_{1+}(x) = \begin{cases} V_1(x), & \text{when } x_1 \geq 0; \\ 0, & \text{when } x_1 < 0; \end{cases} \tag{5.4.1}$$

$$V_1(x) = \sum_{m_1 \in Z} v_{m_1 q_0} \exp(2\pi i m_1 a_1^{-1} x_1), \quad q_0 = (1, 0, 0). \tag{5.4.2}$$

The study of H_{1+} is obviously much simpler than that of H, because it can be reduced to the study of the Schrödinger operator in one-dimensional space.

[8] In the sense (5.4.24), (5.4.25)

Our plan is as follows. First, we eliminate the surface and quasisurface states for $H_{1+}(t_{||})$. This means that we formulate the corresponding conditions for $t_{||}$. We prove that the surface and quasisurface states are absent under conditions (5.4.133), (5.4.153). Then, we construct asymptotic formulae for $H(t_{||})$, considering $H_{1+}(t_{||})$ as the initial operator. We prove that $U_{m_{||}} \approx U^1_{m_{||}}$. To justify the new asymptotic formula, one has to put new conditions on $t_{||}$. Under these conditions, the description of surface and quasisurface states for operators $H(t_{||})$ and $H_{1+}(t_{||})$ are close: the operators have or have no such states together. So, eliminating the surface and quasisurface states for $H_{1+}(t_{||})$, we do this for $H(t_{||})$. Thus, we prove that the operator $H(t_{||})$ has no surface and quasisurface states under conditions of Lemma 5.13 (page 294).

5.4.1 One dimensional semicrystal.

Since $V_{1+}(x)$ depends only on x_1, the study of $H_{1+}(t_{||})$ in $L_2(R^3)$ can be reduced to that of the operator [9]

$$\tilde{H}_{1+}u = -\frac{\partial^2}{\partial x_1^2}u + V_{1+}(x_1)u \tag{5.4.3}$$

in $L_2(R)$. In this subsection we consider this operator. (Then we shall study H^1_+ in $L_2(R^3)$, and finally consider the general case of H_+.)

Let $\tilde{\psi}_+(z,x_1)$, $\Im z \neq 0$, be a square integrable over the positive axis solution of the equation

$$(-\frac{\partial^2}{\partial x_1^2} + V_1(x_1) - z)\tilde{\psi}_+(z,x_1) = 0, \; x_1 \geq 0, \tag{5.4.4}$$

(the Weyl solution). Further, let $\tilde{\psi}_-(z,x_1)$, $\Im z \neq 0$, be a square integrable over the negative axis solution of the equation

$$(-\frac{\partial^2}{\partial x_1^2} - z)\tilde{\psi}_-(z,x_1) = 0, \; x_1 \leq 0. \tag{5.4.5}$$

Both of these solutions exist for any nonreal z [Ti] and satisfy the quasiperiodic conditions

$$\begin{aligned} \tilde{\psi}_\pm(z,a_1) &= \tilde{\psi}_\pm(z,0)\exp(it_{1\pm}a_1), \\ \tilde{\psi}'_\pm(z,a_1) &= \tilde{\psi}'_\pm(z,0)\exp(it_{1\pm}a_1) \end{aligned} \tag{5.4.6}$$

with a quasimomentum t_{1+}, $\Im t_{1+} > 0$, for $\tilde{\psi}_+(z,x_1)$ and a quasimomentum t_{1-}, $\Im t_{1-} < 0$, for $\tilde{\psi}_-(z,x_1)$. We normalize $\tilde{\psi}_\pm(z,x_1)$ so that

$$\frac{1}{a_1}\int_0^{a_1}\left|\tilde{\psi}_\pm(z,x_1)\exp(-it_{1\pm}x_1)\right|^2 dx_1 = 1. \tag{5.4.7}$$

[9] here and below the sign $\tilde{}$ over an operator or a function means that this operator or function corresponds to the one-dimensional semicrystal with the potential $V_{1+}(x_1)$. In the three-dimensional case with the same potential, the sign is omitted.

The functions $\tilde{\psi}_\pm(z,x_1)$ and the quasimomenta $t_{1\pm}$ are uniquely determined (we suppose $\tilde{\psi}_\pm(z,x_1)$ to be positive at $x_1=0$). In the case $V_{1+}=0$, the functions $\tilde{\psi}_\pm(z,x_1)$ are equal to $\exp(\pm i\sqrt{z}x_1)$, $\Im\sqrt{z}>0$. It is proved in [Ti] that the quasimomenta $t_{1\pm}$ analytically depend on z in the complex plane with cuts on the real axis. The cuts form the spectrum Λ of the periodic operator

$$\tilde{H}_1 = -\frac{\partial^2}{\partial x_1^2} + V_1(x_1). \tag{5.4.8}$$

If $z=\bar{z}$, $z\notin\Lambda$, then $\sin(t_{1+}a_1)$ is purely imaginary. This means that $t_{1+}a_1$, $t_{1+}a_1-\pi$ or $t_{1+}a_1+\pi$ is purely imaginary. If $z\in\Lambda$, then the quasimomentum is real and has the opposite signs on both sides of a cut. Moreover, there exists a set of the inverse functions $z=\lambda_n(t_1)$, $n\in N$, being analytic in the complex plane with cuts. The cuts do not intersect the real axis. The spectrum Λ coincides with the ranges of the functions $\lambda_n(t_1)$, $n\in N$ for $t_1\in[-\pi a_1^{-1},\pi a_1^{-1}]$:

$$\Lambda = \cup_{n\in N, t_1\in[-\pi a_1^{-1},\pi a_1^{-1}]}\lambda_n(t_1).$$

Since the functions $\lambda_n(t_1)$ depend analytically on t_1, the relation $\lambda_n(t_1)=\lambda_m(t_1)$, $n\neq m$ can hold only on a denumerable set of t_1.

We denote by A, the matrix reducing $\tilde{H}_1(t_1)$ to diagonal form $\Lambda(t_1)$ (for ungenerated eigenvalues): $\tilde{H}_1(t_1)=A^{-1}\Lambda A$. Let L_1 be a diagonal matrix:

$$L_{nn}=t_1+\tilde{n}, \quad \tilde{n}=2\pi a_1^{-1}n. \tag{5.4.9}$$

Further, we denote by $\|A\|^*$, the "strong" norm of the operator A in l_2^1:

$$\|A\|^* = \frac{1}{2}\max_{n\in Z}\sum_{m\in Z}(|A_{nm}|+|A_{mn}|). \tag{5.4.10}$$

Now, we prove a few simple propositions concerning $\tilde{H}_1$, which will be needed in the further considerations.

Proposition 5.10 . *The matrix $A(t_1)$, $|\Re t_1|\le\pi a_1^{-1}$, $\Im t_1\neq 0$, reducing $\tilde{H}_1(t_1)$ to a diagonal form (for ungenerated eigenvalues), satisfies the following estimates:*

$$\|A\|^* < c(V_1), \ c\neq c(t_1), \tag{5.4.11}$$

$$\|A^{-1}\|^* < c(V_1). \tag{5.4.12}$$

Proof. It is well-known that the matrices A, A^{-1} are given by the formulae: $A_{nm}=(\tilde{\psi}_n^*)_m$, $A^{-1}_{nm}=(\tilde{\psi}_m)_n$, where $\tilde{\psi}_m$, $\tilde{\psi}_n^*$ are normalized eigenfunctions of the operators $\tilde{H}_1^*(t_1)$, $\tilde{H}_1(t_1)$, respectively. Let us estimate $\|A\|_1^*$. We represent the series $I\equiv\sum_{m\in Z}|(\tilde{\psi}_m)_n|$ in the form $I=I_1+I_2$,

$$I_1=\sum_{m\in M(n)}|(\tilde{\psi}_m)_n|, \ I_2=\sum_{m\notin M(n)}|(\tilde{\psi}_m)_n|, \tag{5.4.13}$$

$$M(n) = \{m : |(t_1 + \tilde{m})^2 - (t_1 + \tilde{n})^2| > 2\|V\|\}.$$

According to the general perturbation theory (see f.e. [Kato])

$$|(\tilde{\psi}_m)_n| < \frac{|(V_1\tilde{\psi}_m)_n|}{|\lambda_m(t_1) - (t_1 + \tilde{n})^2|}. \tag{5.4.14}$$

and for any fixed t_1, the eigenvalues $\lambda_n(t_1)$ can be labeled so that the inequality

$$|\lambda_n(t_1) - (t_1 + \tilde{n})^2| \le \|V_1\| \tag{5.4.15}$$

holds. Taking into account (5.4.15) and the definition of $M(n)$, we get

$$|(\tilde{\psi}_m)_n| < \left| \frac{2\|V_1\|}{|(t_1 + \tilde{m})^2 - (t_1 + \tilde{n})^2|} \right|. \tag{5.4.16}$$

Using (5.4.16), it is not hard to show that $|I_1| < c(V_1)$. Next, we consider I_2. It is obvious that the relation $|(t_1 + \tilde{m})^2 - (t_1 + \tilde{n})^2| \le 2\|V_1\|$ can be satisfied only if $\min\{|m|, |n|\} < c_1(V_1)$ or $m = \pm n$. Therefore, I_2 contains less than $c_2(V_1)$ terms. Taking into account that $|(\tilde{\psi}_m)_n| < 1$, we get $|I_2| < c_2(V_1)$. Thus, $|I| < c(V_1)$. The sum $\sum_{n\in Z} |(\tilde{\psi}_m)_n|$ is estimated similarly. Therefore, estimate (5.4.12) holds. Similar arguments give (5.4.11). The proposition is proved.

Proposition 5.11 . *Suppose $\lambda, t_1 \in C$, $\Im t_1 \ne 0$. Then the following estimate holds:*

$$\|(\tilde{H}_1(t_1) - \lambda)^{-1}\|^* \le c(V_1)\|(\tilde{H}_1(t_1) - \lambda)^{-1}\|. \tag{5.4.17}$$

Proof. Suppose the operator $\tilde{H}_1(t_1)$ can be reduced to diagonal form. Using the inequality

$$\|(\tilde{H}_1(t_1) - \lambda)^{-1}\|^* \le \|A^{-1}\|^*\|\Lambda\|^*\|A\|^*,$$

the relation $\|\Lambda\|_1^* = \|\Lambda\|$, and estimates (5.4.11), (5.4.12), we obtain (5.4.17). The proposition is proved.

Proposition 5.12 . *Suppose that $\lambda_0 \ge 0$, $\lambda_* \in C$, $|\lambda_* - \lambda_0| < 1$. Let t_0, t_* be the quasimomenta corresponding to λ_0, λ_*, such that $\Im t_* \ge 0$, $\Im t_0 \ge 0$. Then the following estimates are satisfied uniformly in λ_*, λ_0:*

$$|\sin^2(a_1 t_*) - \sin^2(a_1 t_0)| < c|\sqrt{\lambda_*} - \sqrt{\lambda_0}|, \tag{5.4.18}$$

$$c \ne c(\lambda_*, \lambda_0), \ Re\sqrt{\lambda_*} \ge 0, \ \sqrt{\lambda_0} \ge 0.$$

Proof. According to [Ti] the next relation holds:

$$\sin^2(a_1 t_*) = \tilde{F}(\kappa), \quad \kappa = \sqrt{\lambda_*}, \tag{5.4.19}$$

where $\tilde{F}(\kappa)$ is an analytic function of κ, satisfying the estimate $|\tilde{F}_\kappa(\kappa)| < c$ in the strip $|\Im\kappa| \le 1$. From this estimate (5.4.18) follows. The proposition is proved.

Proposition 5.13 . *Suppose t_0, $\Im t_0 \geq 0$, is a quasimomentum, corresponding to a positive λ_0, and t_* is such that $|t_* - t_0| < 1$, $\Im t_* \geq 0$. Then there exists λ_* such that (5.4.19), and the following inequality hold:*

$$|\sqrt{\lambda_*} - \sqrt{\lambda_0}| < c(V_1)|\sin^2(a_1 t_*) - \sin^2(a_1 t_0)|^{1/2}, \qquad (5.4.20)$$

$$c(V_1) \neq c(\lambda_*, \lambda_0),\ \Re\sqrt{\lambda_*} \geq 0,\ \Re\sqrt{\lambda_0} \geq 0.$$

Proof. It is known [Ti] that $\tilde{F}_\kappa(\kappa)$ has a denumerable number of zeros and they are simple. From the differentiable asymptotics of the function $\tilde{F}(\kappa)$, $\tilde{F}(\kappa) =_{|\kappa|\to\infty} \sin^2\kappa + o(\kappa)$, in the semistrip $|\Im\kappa| < 1, \Re\kappa \geq 0$, we see that

$$\min_{|\Im\kappa|<1} (|\tilde{F}_\kappa(\kappa)| + |\tilde{F}_{\kappa\kappa}(\kappa)|) > c_1(V_1) > 0.$$

Let us consider the function

$$G(\kappa_0, \varepsilon) = \tilde{F}_\kappa(\kappa_0)\varepsilon + \tilde{F}_{\kappa\kappa}(\kappa_0)\varepsilon^2/2.$$

From the relation

$$\max_{|\Im\kappa|<1} |\tilde{F}_{\kappa\kappa\kappa}(\kappa)| < c_2(V_1),$$

it follows that

$$\tilde{F}(\kappa) = \tilde{F}(\kappa_0) + G(\kappa_0, \kappa - \kappa_0) + \varphi(\kappa - \kappa_0),$$

where $|\varphi(\kappa - \kappa_0)| < c_3(V_1)|\kappa - \kappa_0|^3$ when $|\kappa^2 - \kappa_0^2| < 1$.

Let us consider solutions of the equation

$$G(\varepsilon) + \varphi(\varepsilon) = \delta, \quad \delta \equiv \sin^2(a_1 t_*) - \sin^2(a_1 t_0),$$

which is equivalent to the equation $\tilde{F}(\kappa) = \sin^2(a_1 t_*)$. There are two possibilities:
1) $\tilde{F}_\kappa(\kappa_0) \geq \mu_1|\delta|^{1/2}$,
2) $\tilde{F}_\kappa(\kappa_0) < \mu_1|\delta|^{1/2}$, $\mu_1 \equiv 8(c_2 + c_3 + c_4 + 1)$, where $c_4 = \max_{|\Im\kappa|<1} |\tilde{F}_{\kappa\kappa}(\kappa)|$.

In the first case, taking into account that $|\tilde{F}_{\kappa\kappa}(\kappa)| < \mu_1/2$, we obtain for $\varepsilon = |\delta|^{1/2} < 1$:

$$|G(k_0, \varepsilon)| > \mu_1|\delta|^{1/2}\varepsilon - (\mu_1/4)\varepsilon^2 > 3\mu_1|\delta|/4 > (\mu_1/4 + 1)|\delta| > |\varphi(\varepsilon) - \delta|.$$

Note that $G(0) = 0$. According to the Rouche theorem, the function $G(\varepsilon)+\varphi(\varepsilon)-\delta$ also has at least one zero inside the circle $|\varepsilon| < |\delta|^{1/2}$. Thus, if inequality 1) holds, then there exists $\kappa = \kappa_0 + \varepsilon$, $|\varepsilon| < |\delta|^{1/2}$, such that $\tilde{F}(\kappa) = \sin^2(a_1 t_*)$. Taking into account the definition of δ, we obtain (5.4.20).

Let us consider the second case. If $2|\delta|^{1/2} < c_1\mu_1^{-1}$, then $2|\tilde{F}_\kappa(\kappa_0)| < \mu_1|\delta|^{1/2} < c_1$. Therefore, $|\tilde{F}_{\kappa\kappa}(\kappa_0)| > c_1/2$. Setting $\varepsilon = b|\delta|^{1/2}$, $b = 1 + 8\mu_1 c_1^{-1}$, we get

$$|G(\varepsilon)| > c_1\varepsilon^2/4 - \mu_1|\delta|^{1/2}\varepsilon > b\delta(c_1 b/4 - \mu_1) > b\mu_1\delta.$$

For sufficiently small positive δ, $\delta < b^{-1}$, the following estimate holds:

$$b\mu_1\delta > \mu_1\delta^3 b^3/4 + \delta > |\varphi_1(\varepsilon) - \delta|.$$

Since $G(0) = 0$, applying the Rouche theorem, we get that $G(\varepsilon) + \varphi(\varepsilon) - \delta$ has at least one root in the circle $|\varepsilon| \leq b\delta^{1/2}$. Thus, if $|\delta| < b^{-1}$, $2|\delta|^{1/2} < c_1\mu_1^{-1}$, then there exists $\kappa = \kappa_0 + \varepsilon$, $|\varepsilon| < b|\delta|^{1/2}$, such that $\tilde{F}(\kappa) = \sin^2(a_1 t_*)$. From the last inequality, estimate (5.4.20) follows immediately. It remains to prove the assertion in the case $|\delta|^{1/2} > c_4(V)$, $c_4(V) = \min\{b^{-1}, c_1\mu_1\}$. It is easy to show that for any λ_0, there exists $\lambda_m(t)$, $m \in Z$, such that $|\sqrt{\lambda_m(t)} - \sqrt{\lambda_0}| < c_5(V_1)$, where $c_5 = 2(\|V_1\| + 2\pi a_1^{-1})$. It is clear that $c_5 < c_5 c_4^{-2}\delta$. Thus, we have obtained inequality (5.4.20) for all positive δ. The proposition is proved.

Remark 5.1. When $\lambda_0 > -2\|V_1\|$, $\Re\lambda_* > -2\|V_1\|$, estimates (5.4.18) and (5.4.20) can be rewritten in the form:

$$\left|\sin^2(a_1 t_*) - \sin^2(a_1 t_0)\right|^2 < c|\lambda_* - \lambda_0|^2(\lambda_* + \lambda_0 + 4\|V_1\|)^{-1}, \tag{5.4.21}$$

$$\frac{|\lambda_* - \lambda_0|^2}{\lambda_* + \lambda_0 + 4\|V_1\|} < c\left|\sin^2(a_1 t_*) - \sin^2(a_1 t_0)\right|, \tag{5.4.22}$$

$$c = c(V_1) < \infty.$$

Indeed, according to Propositions 5.12 and 5.13, estimates (5.4.18) and (5.4.20) hold for any positive $\tilde{\lambda}_*, \tilde{\lambda}_0$ and for the positive potential $\tilde{V}_1 = V_1 + 2\|V_1\|$. Substituting $\tilde{\lambda}_* = \lambda_* + 2\|V_1\|$ and $\tilde{\lambda}_0 = \lambda_0 + 2\|V_1\|$, and squaring both sides of the inequalities, we obtain estimates (5.4.21) and (5.4.22).

Now we consider the operators $\tilde{A}_{\pm}^{V_1}(z)$. They are one-dimensional operators. The single matrix element is $\tilde{A}_{\pm}^{V_1}(z) = \tilde{\psi}'_{\pm}(z,0)/\tilde{\psi}_{\pm}(z,0)$. Note that $\tilde{\psi}_{\pm}(z,0) \neq 0$, and $\tilde{\psi}'_{\pm}(z,0) \neq 0$ when $\Im z \neq 0$, because otherwise the Dirichlet or Neumann boundary problem has a nonreal eigenvalue. The functions $\tilde{A}_{\pm}^{V_1}(z)$ are analytic in the complex plane with the cuts on Λ. The poles of $\tilde{A}_{\pm}^{V_1}(z)$ coincide with the zeros of the function $\tilde{\psi}_{\pm}(z,0)$. Note that

$$\tilde{A}_{\pm}^{V_1}(\lambda + i0) = \tilde{A}_{\pm}^{V_1}(\lambda - i0) = \tilde{A}_{\pm}^{V_1}(\lambda + i0)^*, \text{ if } \lambda \notin \Lambda,\ \ \lambda = \bar{\lambda}. \tag{5.4.23}$$

It is clear that

$$\tilde{A}_{\pm}^{0}(z) = \mp\sqrt{-z}, \tag{5.4.24}$$

$\Re\sqrt{-z} > 0$. Thus, the boundary operator $\tilde{T}_1(z)$ is a one-dimensional operator with the single matrix element:

$$\tilde{T}_1(z) = \tilde{\psi}'_{+}(z,0)/\tilde{\psi}_{+}(z,0) + \sqrt{-z}. \tag{5.4.25}$$

The function $\tilde{T}_1(z)$ is analytic in the complex plane with the cuts on $\Lambda \cup [0,\infty)$ and the poles at the zeros of $\tilde{\psi}_{+}(z,0)$.

We denote by $g(\lambda)$ the distance from a point λ to the nearest of the points $\lambda_n(0), \lambda_n(\pi a_1^{-1}), n \in N$, which are the points of Λ corresponding to $t_1 = 0$ and $t_1 = \pi a_1^{-1}$. Thus,

$$g(\lambda) = \min\{g_0(\lambda), g_1(\lambda)\},$$

$$g_0(\lambda) = \min_{n\in N}|\lambda - \lambda_n(0)|, \quad g_1(\lambda) = \min_{n\in N}|\lambda - \lambda_n(\pi a_1^{-1})|.$$

Proposition 5.14 . *If $\lambda \in \Lambda$, then*

$$\Im \tilde{A}_{+}^{V_1}(\lambda + i0) > \frac{cg^3(\lambda)}{(|\lambda| + ||V_1||)^{5/2}}, \tag{5.4.26}$$

$$|\tilde{\psi}(\lambda + i0, 0)| > \frac{cg^2(\lambda)}{(|\lambda| + ||V_1||)^{3/2}}, \tag{5.4.27}$$

$$c = c(V_1) > 0.$$

Proof. It is clear that

$$\Im \tilde{A}_{+}^{V_1}(\lambda + i\varepsilon) = \Im \left(\tilde{\psi}'_{+}/\tilde{\psi}_{+}(\lambda + i\varepsilon, 0)\right) =$$

$$(2i)^{-1}\left(\tilde{\psi}'_{+}(\lambda + i\varepsilon, 0)\tilde{\psi}_{+}(\lambda - i\varepsilon, 0) - \tilde{\psi}_{+}(\lambda + i\varepsilon, 0)\tilde{\psi}'_{+}(\lambda - i\varepsilon, 0)\right) |\tilde{\psi}_{+}(\lambda + i\varepsilon, 0)|^{-2}.$$

Integrating the Weyl solutions $\tilde{\psi}_{+}$, $\tilde{\psi}_{-}$ by parts gives

$$\Im(\tilde{\psi}'_{+}\overline{\tilde{\psi}_{+}})(\lambda + i\varepsilon, 0) = \varepsilon \int_{R_+} |\tilde{\psi}_{+}(\lambda + i\varepsilon, x)|^2 dx. \tag{5.4.28}$$

Using relations (5.4.6), (5.4.7) and the notation

$$\Delta\lambda_{\varepsilon} = \frac{\varepsilon a_1}{1 - \exp(-2\Im t_1 a_1)}, \quad t_1 = t_1(\lambda + i\varepsilon), \tag{5.4.29}$$

we obtain that

$$\Im(\tilde{\psi}'_{+}\overline{\tilde{\psi}}_{+})(\lambda + i\varepsilon, 0) = \Delta\lambda_{\varepsilon}. \tag{5.4.30}$$

Hence,

$$\Im(\tilde{\psi}'_{+}/\tilde{\psi}_{+})(\lambda + i\varepsilon, 0) = \Delta\lambda_{\varepsilon}|\tilde{\psi}_{+}(\lambda + i\varepsilon, 0)|^{-2}. \tag{5.4.31}$$

It is obvious that

$$|\tilde{\psi}_{+}(\lambda + i\varepsilon, 0)| = \left|\int_0^{\infty} \tilde{\psi}'_{+}(\lambda + i\varepsilon, x_1)dx_1\right| \leq \tag{5.4.32}$$

$$\frac{a_1^{1/2}}{|1 - \exp it_1 a_1|}\left(\int_0^{a_1} |\tilde{\psi}'_{+}(\lambda + i\varepsilon, x_1)|^2 dx_1\right)^{1/2}.$$

From (5.4.4) it follows that for $\lambda \in \Lambda$ and $\varepsilon \ll 1$

$$\int_0^{a_1} |\tilde{\psi}'_{+}(\lambda + i\varepsilon, x)|^2 dx < (c + \sqrt{|\lambda|})^2 \int_0^{a_1} |\tilde{\psi}_{+}(\lambda + i\varepsilon, x)|^2 dx. \tag{5.4.33}$$

Using (5.4.7), (5.4.32) and (5.4.33), we obtain

$$|\tilde{\psi}_{+}(\lambda + i\varepsilon, 0)| < \frac{(c + \sqrt{|\lambda|})a_1}{|1 - \exp it_1 a_1|}, \quad t_1 = t_1(\lambda + i\varepsilon).$$

Using this inequality in (5.4.31) we get

$$|\Im(\tilde{\psi}'_+/\tilde{\psi}_+)(\lambda + i\varepsilon, 0)| > \Delta\lambda_\varepsilon \frac{|1 - \exp it_1 a_1|^2}{(c + \sqrt{|\lambda|})^2 a_1^2}, \quad t_1 = t_1(\lambda + i\varepsilon).$$

Taking into account that

$$2 \lim_{\varepsilon \to 0} \Delta_\varepsilon \lambda = \dot{\lambda}(t_1) \tag{5.4.34}$$

and the relation

$$|\dot{\lambda}(t_1)| > (c + \sqrt{|\lambda|})|\sin(a_1 t_1)|, \quad (\lambda \in \Lambda), \tag{5.4.35}$$

[Fir1] – [Fir4], we get

$$|\Im(\tilde{\psi}'_+/\tilde{\psi}_+)(\lambda + i\varepsilon, 0)| > c_1(c + \sqrt{|\lambda|})^{-1}|\sin^3(a_1 t_1)|. \tag{5.4.36}$$

Let us prove that

$$|\sin(a_1 t_1)| > c\frac{g(\lambda)}{\sqrt{\lambda + 2\|V\|}}. \tag{5.4.37}$$

Suppose this is not so, i.e., that for an arbitrary small positive $c = c(V_1)$, we can find λ, t_1 such that

$$|\sin(a_1 t_1)| < c\frac{g(\lambda)}{\sqrt{\lambda + 2\|V\|}}.$$

Let us set in Proposition 5.13 (page 262): $t_0 = t_1$, $\lambda_0 = \lambda$ and t_* is such that $\sin(t_* a_1) = 0$. Then, according to this proposition there is λ_* such that

$$F(\kappa) = 0, \quad \kappa = \sqrt{\lambda_*} \tag{5.4.38}$$

and

$$\frac{|\lambda_* - \lambda|^2}{\lambda_* + \lambda + 4\|V\|} < \frac{c^2(V_1) g^2(\lambda)}{(\lambda + 4\|V\|)^2}. \tag{5.4.39}$$

(see (5.4.22)). Note that λ_* is equal to $\lambda_n(0)$ or $\lambda_n(\pi a_1^{-1})$, because $\sin(t_* a_1) = 0$. Therefore, by the definition of $g(\lambda)$, we have $|\lambda - \lambda_*| \geq g(\lambda)$. It is not hard to show that the last inequality contradicts (5.4.39) when c is small enough. Thus, we have proved (5.4.37).

Using (5.4.36), we obtain (5.4.26) at the point $\lambda + i0$.

Let us prove (5.4.27). From relation (5.4.30), it is easy to get that

$$|\tilde{\psi}(\lambda + i\varepsilon, 0)| > \Delta_\varepsilon \lambda |\tilde{\psi}'(\lambda + i\varepsilon, 0)|^{-1}. \tag{5.4.40}$$

We estimate $|\tilde{\psi}'(\lambda + i\varepsilon)|$ from above. It is clear that

$$|\tilde{\psi}'(\lambda + i\varepsilon, 0)| = \frac{1}{|1 - \exp it_1 a_1|} \int_0^{a_1} \psi''(\lambda + i\varepsilon, x_1) dx_1 \leq$$

$$\frac{a_1^{1/2}}{|1 - \exp it_1 a_1|} \sqrt{\int_0^{a_1} |\psi''(\lambda + i\varepsilon, x_1)|^2 dx_1} < \frac{2a_1(|\lambda| + 2\|V_1\|)}{|1 - \exp it_1 a_1|}, \tag{5.4.41}$$

$$t_1 = t_1(\lambda + i\varepsilon).$$

Substituting (5.4.41) into (5.4.40), yields:

$$|\tilde{\psi}(\lambda + i\varepsilon, 0)| > \frac{\Delta_\varepsilon \lambda |1 - \exp it_1 a_1|}{2a_1(|\lambda| + \|V_1\|)}.$$

Using (5.4.34), (5.4.35) and (5.4.37) we obtain (5.4.27). The proposition is proved.

From (5.4.24) it follows that for positive λ

$$\Im \tilde{A}^0_-(\lambda \pm i0) = \mp\sqrt{\lambda}. \tag{5.4.42}$$

Proposition 5.15 . *The function $\tilde{T}_1(\lambda + i0)$ satisfies the inequalities:*

$$|\Im \tilde{T}_1(\lambda + i0)| \geq c_1 g^3(\lambda), \ \ \lambda \in \Lambda \setminus [0, \infty), \tag{5.4.43}$$

$$|\Im \tilde{T}_1(\lambda + i0)| \geq \sqrt{\lambda}, \ \ \lambda \in [0, \infty). \tag{5.4.44}$$

Proof. We obtain estimate (5.4.43) from (5.4.26) taking into account that $\Im A^0_- = 0$ and $|\lambda| \leq \|V\|$. We get estimate (5.4.44) noting that $\Im \tilde{A}^{V_1}_+ \geq 0$.

The proposition is proved.

Proposition 5.16 . *The function $\tilde{T}_1(z)$ can have only a finite number of zeros. They belong to the closure of the set $[-\|V_1\|, 0] \setminus \Lambda$.*

Proof. Since the operator $\tilde{H}_{1+}$ is selfadjoint and semibounded below, its surface states can lie only on the semiaxis $[-\|V_1\|, \infty)$. It follows from relation (5.4.44), that they cannot be on the positive axis. Inequality (5.4.43) means that surface states cannot be strictly inside a band of the spectrum of $\tilde{H}_{1+}$. Therefore, the zeros belong to the closure of the set $[-\|V_1\|, 0]\setminus\Lambda$. Since $\tilde{T}_1(z)$ is an analytic function, there is only a finite number of zeros in this set. The proposition is proved.

We denote by $\tilde{\lambda}_1, ..., \tilde{\lambda}_N$ the zeros of the function $\tilde{T}_1(z)$. They correspond to surface states. Next, we consider the quasisurface states: $|\tilde{T}_1(z)| < \delta$. It is obvious that for sufficiently small positive δ, $\delta < \delta_0(V)$, the set $\{z : |\tilde{T}_1(z)| < \delta\}$ is a neighborhood of the points $\tilde{\lambda}_1, ..., \tilde{\lambda}_N$; that is, $|\tilde{T}_1(z)| \geq \delta$ when $z \notin U(\delta)$, where

$$U(\delta) = \cup_{k=1}^{k=N} U(\delta_k, \tilde{\lambda}_k), \tag{5.4.45}$$

and $U(\delta_k, \tilde{\lambda}_k)$ are the δ_k-neighborhoods of $\tilde{\lambda}_k$. The size δ_k of the neighborhood is δ^{1/p_k}, p_k being the order of the zero $\tilde{\lambda}_k$ (it can be fractional if $\tilde{\lambda}_k$ coincides with an edge of a band). It is clear that

$$\delta_k < c\delta^{1/p}, \tag{5.4.46}$$

where $p = \max_{k=1,...,N} p_k$.

Further, for the construction of the perturbation theory it will be convenient to use a function $W(\lambda, x_1)$ instead of $\tilde{\psi}_+(\lambda, x_1)$:

$$W(\lambda, x_1) = \sum_{n \in Z} \frac{\sqrt{-\lambda}}{\pi} \int_{C_4} (\tilde{H}_1(t_1) - \lambda)^{-1}_{n0} \exp i(t_1 + 2\pi n a_1^{-1}) x_1 dt_1, \ \ x_1 > 0, \tag{5.4.47}$$

$\Re\sqrt{-\lambda} \geq 0$, the contour C_4 is shown in Fig.5.

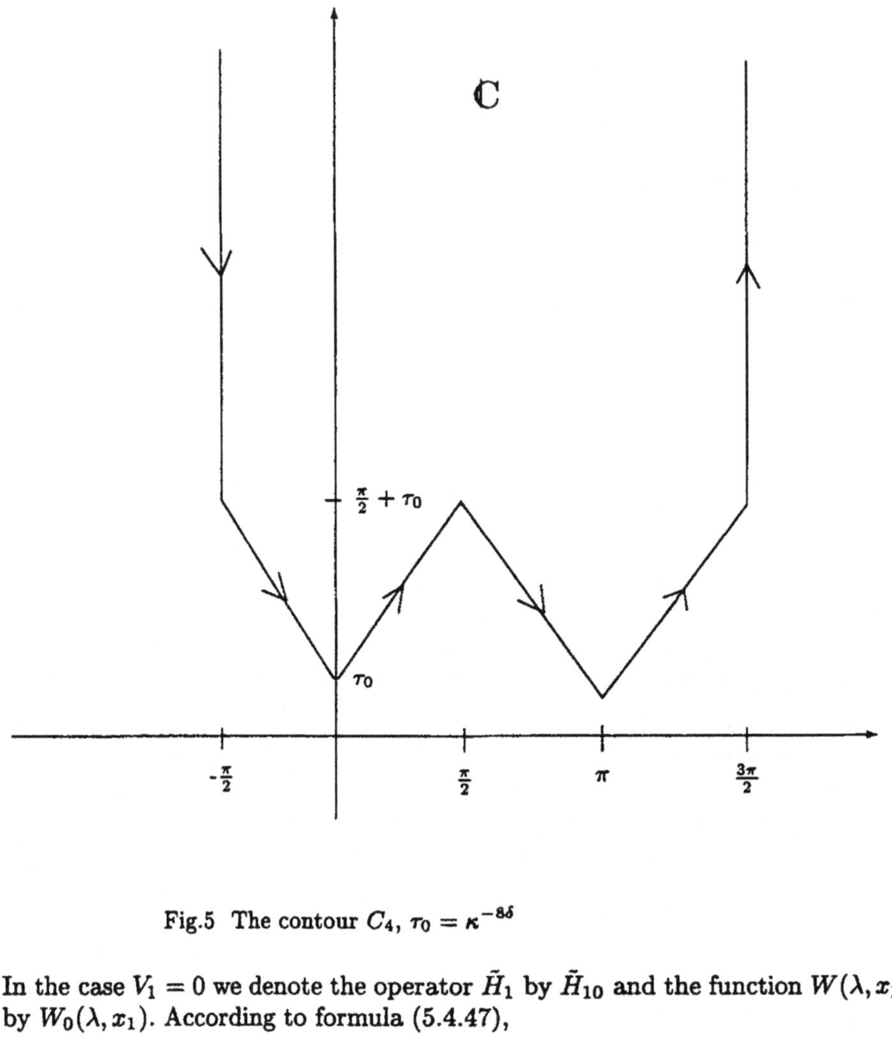

Fig.5 The contour C_4, $\tau_0 = \kappa^{-8\delta}$

In the case $V_1 = 0$ we denote the operator $\tilde{H}_1$ by $\tilde{H}_{10}$ and the function $W(\lambda, x_1)$ by $W_0(\lambda, x_1)$. According to formula (5.4.47),

$$W_0(\lambda, x_1) = \frac{\sqrt{-\lambda}}{\pi} \int_{C_4} (t_1^2 - \lambda)^{-1} \exp it_1 x_1 dt_1. \tag{5.4.48}$$

Suppose that $i\sqrt{-\lambda}$ is inside C_4. The integral converges because the integrand exponentially decays at infinity. Calculating the integral by residue, we obtain $W_0(\lambda, x_1) = \exp(-\sqrt{-\lambda}x_1)$. If $i\sqrt{-\lambda}$ is not inside C_4, then $W_0(\lambda, x) = 0$. In the case of a nonzero potential, there exists a unique $t_1(\lambda)$ in the upper half-plane for any λ in the plane with the cuts on Λ. The Riemann surfaces for the functions $t_1(\lambda)$ and $\lambda(t_1)$ are constructed by Firsova, N.E. [Fir1]. We denote by γ the set of λ corresponding to t_1 inside C_4. If $\lambda \in \gamma$, then the integral on the right of (5.4.47) converges. Arguing as in proving Proposition 5.6 (page 243), we verify that the series converges in $W_2^1(R_+)$ uniformly in t_1. Therefore, one can exchange the summation and the integration. It is easy to see that $W(\lambda, x_1)$ satisfies quasiperiodic condition (5.4.6) and the equation $(-\Delta + V_1 - \lambda)W = 0$. Moreover, $W(\lambda, x_1)$ exponentially decays as $x_1 \to \infty$ because of the factor $\exp(it_1x_1)$. Therefore, $W(\lambda, x_1) \in L_2(0, \infty)$ and,

$$W(\lambda, x_1) = c(\lambda)\tilde{\psi}_+(\lambda, x_1), c(\lambda) \in C. \tag{5.4.49}$$

We denote by $\tilde{L}_1$ the diagonal operator:

$$(\tilde{L}_1)_{nn}(t_1) = t_1 + \tilde{n}, \quad \tilde{n} \equiv 2\pi a_1^{-1} n_1. \tag{5.4.50}$$

We denote by $\tilde{H}_{01}$ the operator $\tilde{H}_1$ corresponding to $V_1 = 0$.

Proposition 5.17 . *Suppose that* $\tau_0 < \pi a_1^{-1}/4$, $\lambda \in R$, $|\lambda| > 16\tau_0^2$, *and* $t_1 \in C_4(\tau_0)$. *Then the following estimates hold:*

$$\|(\tilde{H}_{01}(t_1) - \lambda)^{-1}\|^* \leq \left(\Im t_1 + \sqrt{|\lambda|}\right)^{-1} \tau_0^{-1}, \tag{5.4.51}$$

$$\|\tilde{L}_1(\tilde{H}_{01}(t_1) - \lambda)^{-1}\|^* \leq \left(\Im t_1 + \sqrt{|\lambda|}\right)^{-1} \left(\sqrt{|\lambda|}\tau_0^{-1} + 1\right). \tag{5.4.52}$$

If $\lambda < -\pi a_1^{-1}$, *then these estimates can be improved:*

$$\|(\tilde{H}_{01}(t_1) - \lambda)^{-1}\|^* \leq \left(\Im t_1 + \sqrt{|\lambda|}\right)^{-1} 4a_1/\pi, \tag{5.4.53}$$

$$\|\tilde{L}_1(\tilde{H}_{01}(t_1) - \lambda)^{-1}\|^* \leq \left(\Im t_1 + \sqrt{|\lambda|}\right)^{-1} \left(\sqrt{|\lambda|}4a_1/\pi + 1\right). \tag{5.4.54}$$

Proof. First, we prove estimate (5.4.51) in the case of a positive λ. We represent t_1 in the form $t_1 = \tau_1 + i\tau_2$, $\tau_1, \tau_2 \in R$. Then,

$$(\tilde{H}_{01}(t_1) - \lambda)_{nn} = (\tau_1 + \tilde{n} + i\tau_2 + \sqrt{\lambda})(\tau_1 + \tilde{n} + i\tau_2 - \sqrt{\lambda}). \tag{5.4.55}$$

Suppose $\tau_1 + \tilde{n} > 0$. Then

$$|\tau_1 + \tilde{n} + i\tau_2 + \sqrt{\lambda}| > \tau_2 + \sqrt{\lambda}. \tag{5.4.56}$$

Since $\tau_2 > \tau_0$, we have

$$|\tau_1 + \tilde{n} + i\tau_2 - \sqrt{\lambda}| > \tau_0.$$

In the case $\tau_1 + \tilde{n} < 0$, we argue similarly, replacing the first factor by the second and conversely. Thus, we obtain

$$\left|(\tilde{H}_{01}(t_1) - \lambda)_{nn}\right| > |\Im t + \sqrt{\lambda}|\tau_0,$$

i.e., (5.4.51) holds for positive λ.

Let us prove (5.4.51) for negative λ. It is clear that

$$(\tilde{H}_{01}(t_1)-\lambda)_{nn}=(\tau_1+\tilde{n}+i\tau_2+i\sqrt{|\lambda|})(\tau_1+\tilde{n}+i\tau_2-i\sqrt{|\lambda|}) \quad (5.4.57)$$

and

$$|\tau_1+\tilde{n}+i\tau_2+i\sqrt{|\lambda|}|>\tau_2+\sqrt{|\lambda|}. \quad (5.4.58)$$

We estimate the second factor on the right of (5.4.57). Indeed, if $\tau_2>2\tau_0$, then $3\pi/(2a_1)>|\tau_1|>\tau_0$ or $|\tau_1-\pi a_1^{-1}|>\tau_0$. Two last inequalities give: $|\tau_1+\tilde{n}|>\tau_0$. Hence,

$$|\tau_1+\tilde{n}+i\tau_2-i\sqrt{|\lambda|}|>\tau_0. \quad (5.4.59)$$

If $\tau_2<2\tau_0$, then, using the condition on λ gives $|\tau_2-\sqrt{|\lambda|}|>2\tau_0$. Thus, inequality (5.4.59) holds in this case, too. Using estimates (5.4.58) and (5.4.59) we verify that

$$\left|(\tilde{H}_{01}(t_1)-\lambda)^{-1}_{nn}\right|\le(\Im t_1+\sqrt{|\lambda|})^{-1}\tau_0^{-1}. \quad (5.4.60)$$

Taking into account that for a diagonal operator, the "strong" norm is equal to the norm in the class of bounded operators, we obtain estimate (5.4.51) for negative λ, too. Therefore, (5.4.51) holds for $\lambda\in R$, $|\lambda|>16\tau_0^2$.

Estimate (5.4.53) can be proved in a similar fashion. Indeed, if $\tau_2>\pi/2a_1$, then $|\tau_1+\tilde{n}|>\pi/4a_1$ for any n. Suppose $\tau_2<\pi/2a_1$. Note that $|\sqrt{|\lambda|}-\tau_2|>\pi/2a_1$. Therefore, for any t_1 on C_4 we have:

$$|\tau_1+\tilde{n}+i\tau_2-i\sqrt{|\lambda|}|>\pi/4a_1. \quad (5.4.61)$$

From relations (5.4.58) and (5.4.61) it follows (5.4.53).

To prove inequality (5.4.52) for positive λ, we represent the operator $\tilde{L}_1$ in the form:

$$\tilde{L}_1=\tilde{L}_1-\sqrt{\lambda}I+\sqrt{\lambda}I. \quad (5.4.62)$$

It is clear that

$$\left(\tilde{L}_1(\tilde{H}_{01}(t_1)-\lambda)^{-1}\right)_{nn}=$$

$$\left((\tilde{L}_1-\sqrt{\lambda}I)(\tilde{H}_{01}(t_1)-\lambda)^{-1}\right)_{nn}\mp\sqrt{\lambda}\left(\tilde{H}_{01}(t_1)-\lambda)^{-1}\right)_{nn}. \quad (5.4.63)$$

If $\lambda>0$, $\tau_1+\tilde{n}>0$, then

$$|(t_1+\tilde{n}-\sqrt{\lambda})(\tilde{H}_{01}(t_1)-\lambda)^{-1}_{nn}|=|\tau_1+\tilde{n}+i\tau_2+\sqrt{\lambda}|^{-1}\le(\tau_2+\sqrt{|\lambda|})^{-1}. \quad (5.4.64)$$

If $\tau_1+\tilde{n}\le 0$, then

$$|(\tau_1+\tilde{n}+\sqrt{\lambda})(\tilde{H}_{01}(t_1)-\lambda)^{-1}_{nn}|=|\tau_1+\tilde{n}-i\tau_2-\sqrt{\lambda}|^{-1}\le(\tau_2+\sqrt{|\lambda|})^{-1}. \quad (5.4.65)$$

Now, combining (5.4.63), (5.4.64), (5.4.65) and (5.4.60) we get

$$|(\tilde{L}_1)(\tilde{H}_{01}(t_1)-\lambda)^{-1}_{nn}|=(\sqrt{\lambda}\tau_0^{-1}+1)(\tau_2+\sqrt{\lambda})^{-1}. \quad (5.4.66)$$

Hence, (5.4.52) holds for any positive λ.

Suppose λ is negative. We represent the operator $\tilde{L}_1$ in the form:

$$\tilde{L}_1 = \tilde{L}_1 - i\sqrt{|\lambda|}I + i\sqrt{|\lambda|}I. \tag{5.4.67}$$

It is clear that

$$\|\tilde{L}_1(\tilde{H}_{01}(t_1) - \lambda)^{-1}\|^* \leq$$
$$\|(\tilde{L}_1 - i\sqrt{|\lambda|}I)(\tilde{H}_{01}(t_1) - \lambda)^{-1}\|^* + \sqrt{\lambda}\|(\tilde{H}_{01}(t_1) - \lambda)^{-1}\|^*. \tag{5.4.68}$$

It is not hard to verify:

$$|(\tilde{L}_1 - i\sqrt{|\lambda|}I)(\tilde{H}_{01}(t_1) - \lambda)^{-1}_{nn}| = |\tau_1 + \tilde{n} + i\tau_2 + i\sqrt{|\lambda|}|^{-1} \leq (\tau_2 + \sqrt{|\lambda|})^{-1}. \tag{5.4.69}$$

From this it follows that

$$\|(\tilde{L}_1 - i\sqrt{|\lambda|}I)(\tilde{H}_{01}(t_1) - \lambda)^{-1}\|^* \leq (\Im t_1 + \sqrt{|\lambda|})^{-1}. \tag{5.4.70}$$

Considering estimates (5.4.68), (5.4.51) and (5.4.70), we obtain (5.4.52) for non-positive λ, too. Thus, (5.4.52) holds for $\lambda \in R$, $|\lambda| > 16\tau_0^2$. Using inequalities (5.4.70) and (5.4.53) gives (5.4.54). The proposition is proved.

Proposition 5.18 . *For $\lambda < \lambda_0$, $\lambda_0 = -(16a_1^2\|V_1\|^{*2}\pi^{-2} + \pi a_1^{-1} + 4\|V_1\|)$ and $t \in C_4$, the following estimates hold:*

$$\|(\tilde{H}_1(t_1) - \lambda)^{-1} - (\tilde{H}_{01}(t_1) - \lambda)^{-1}\|^* \leq 32a_1^2\pi^{-2}(\Im t_1 + \sqrt{|\lambda|})^{-2}\|V_1\|^*, \tag{5.4.71}$$

$$\|\tilde{L}_1\left((\tilde{H}_1(t_1) - \lambda)^{-1} - (\tilde{H}_{01}(t_1) - \lambda)^{-1}\right)\|^* \leq$$
$$32a_1^2\pi^{-2}(\Im t_1 + \sqrt{|\lambda|})^{-2}(\sqrt{|\lambda|} + \pi(4a_1)^{-1})\|V_1\|^*. \tag{5.4.72}$$

$$\|(\tilde{H}_1(t_1) - \lambda)^{-1}\|^* \leq 8a_1\pi^{-1}(\Im t_1 + \sqrt{|\lambda|})^{-1}, \tag{5.4.73}$$

$$\|\tilde{L}_1(\tilde{H}_1(t_1) - \lambda)^{-1}\|^* \leq 2(\Im t_1 + \sqrt{|\lambda|})^{-1}(\sqrt{|\lambda|} + 1)(8a_1\pi^{-1} + 1). \tag{5.4.74}$$

Proof. Using the Hilbert identity

$$(\tilde{H}_1(t_1) - \lambda)^{-1} = (\tilde{H}_{01}(t_1) - \lambda)^{-1} + (\tilde{H}_{01}(t_1) - \lambda)^{-1}V(\tilde{H}_1(t_1) - \lambda)^{-1}$$

and taking into account that $\|V(\tilde{H}_1(t_1) - \lambda)^{-1}\|^* < 1/2$ when $\lambda < \lambda_0$ (see Proposition 5.11, page 262), we get

$$\|(\tilde{H}_1(t_1) - \lambda)^{-1}\|^* < 2\|(\tilde{H}_{01}(t_1) - \lambda)^{-1}\|^*.$$

Using again the Hilbert identity and estimates (5.4.53), (5.4.54), we get (5.4.71) and (5.4.72). Estimates (5.4.73) and (5.4.74) we easily obtain from (5.4.53), (5.4.54) and (5.4.71), (5.4.72). The proposition is proved.

Proposition 5.19 . *A point λ of the set $(-\infty, 0]\backslash A$ belongs to $\gamma(\tau_0)$ if λ satisfies the inequalities:*

$$\min_{n:|\lambda_n(0)| \leq 2\|V_1\|} |\lambda - \lambda_n(0)| > \tau_0^{7/8}, \tag{5.4.75}$$

$$\min_{n:|\lambda_n(\pi a_1^{-1})| \leq 2\|V_1\|} |\lambda - \lambda_n(\pi a_1^{-1})| > \tau_0^{7/8}, \tag{5.4.76}$$

and τ_0 is sufficiently small, $\tau_0 < T(V_1)$.

Proof. Since $\lambda \in (\infty, 0]\backslash \Lambda$, the corresponding quasimomentum is imaginary or its real part is πa_1^{-1}, while the imaginary part is nonzero. Let t_1 be purely imaginary. We prove that $|t_1| > \tau_0$. Suppose this is not so. Then, according to (5.4.22), we have $\min_n |\lambda - \lambda_n(0)| < \tau_0\sqrt{c(V_1)}$. Since λ is negative, then $\min_n |\lambda - \lambda_n(0)| = \min_{n:|\lambda_n(0)|<2\|V_1\|} |\lambda - \lambda_n(0)|$, because there exists at least one point $\lambda_n(0)$ in the interval $[-\|V_1\|, \|V_1\|]$. Thus, $\min_{n:|\lambda_n(0)|<2\|V_1\|} |\lambda - \lambda_n(0)| < \tau_0\sqrt{c(V_1)}$. However, the last relation contradicts inequality (5.4.75), if $\tau_0 < T(V_1)$. Therefore, $|t_1| > \tau_0$. This means that t_1 lies inside C_4. In the case $\Re t_1 = \pi$ we argue similarly, using condition (5.4.76). The proposition is proved.

Proposition 5.20 . *Suppose a real λ satisfies the inequalities*

$$\min_{n:|\lambda_n(0)|<\tau_0^{-1/8}} |\lambda - \lambda_n(0)| > \tau_0^{1/4}, \tag{5.4.77}$$

$$\min_{n:|\lambda_n(0)|<\tau_0^{-1/8}} |\lambda - \lambda_n(\pi)| > \tau_0^{1/4}, \tag{5.4.78}$$

for any $|\lambda_n(\pi)| < \tau_0^{-1/8}$ and for sufficiently small τ_0, $\tau_0 < T(V_1)$. Then the following estimate holds:

$$\max_{t_1 \in C_4} \|(\tilde{H}_1(t_1) - \lambda)^{-1}\|^* \leq c_1(V)\tau_0^{-3}, \tag{5.4.79}$$

Proof. According to Proposition 5.11 (page 262), it suffices to verify the relation $\|(\tilde{H}_1(t_1) - \lambda)^{-1}\| \leq \tau_0^{-3}$. Suppose the last estimate does not hold. Then, there exists $n \in N$, $t \in C_4$, such that

$$|\lambda_n(t_1) - \lambda| < \tau_0^3. \tag{5.4.80}$$

Let t_2 be the quasimomenta corresponding to $\lambda : \lambda_m(t_2) = \lambda$, $\Im t_2 \geq 0$. Since λ is real, the quasimomenta t_2 is purely real, imaginary or its real part is πa_1^{-1}, while the imaginary part is nonzero [10] .

First, suppose that $t_2 = \bar{t}_2$, i.e., $\lambda \in \Lambda$. Since $|\lambda_n(t_1) - \lambda_m(t_2)| < c\tau_0^3$, according to (5.4.21) we have

$$|\sin^2(a_1t_1) - \sin^2(a_1t_2)| < c^2\tau_0^3(\lambda + \lambda_n(t_1) + 4\|V_1\|)^{-1/2} < c_1^2\tau_0^3. \tag{5.4.81}$$

Hence,

$$|t_1 - t_2| < c_1^{1/2}\tau_0^{3/2}. \tag{5.4.82}$$

Since $\Im t_1 > \tau_0$, because $t_1 \in C_4$, and $\Im t_2 = 0$, we have

$$|t_1 - t_2| > \tau_0. \tag{5.4.83}$$

For sufficiently small τ_0, $\tau_0 < c_1^{-1}$, inequalities (5.4.82) are (5.4.83) are contradictory. Therefore, (5.4.79) holds when $\lambda \in \Lambda$.

[10] We exclude a denumerable set of λ where it is not so.

Next, suppose $\lambda \notin \Lambda$ and $\lambda < \lambda_0$, $\lambda_0 = -(16a_1^2\pi^{-2}\|V_1\|^2 + \pi a_1^{-1} + 4\|V_1\|$. According to Proposition 5.18, estimate (5.4.73) holds. For τ_0 small enough inequalities (5.4.73) and (5.4.80) contradict each other.

Next, suppose $\lambda \notin \Lambda$ and $\lambda_0 \leq \lambda < \tau_0^{-1/8}$. Then the inequality similar to (5.4.21) holds:

$$|\sin^2(a_1t_1) - \sin^2(a_1t_2)| < c^2\tau_0^3(\lambda + \lambda_n(t_1) + 4\|V_1\| + 4|\lambda_0|)^{-1/2}. \tag{5.4.84}$$

From this, (5.4.82) follows. Since the real part of t_2 is equal to 0 or π, we have

$$|\Re t_1| < c_1^{1/2}\tau_0^{3/2} \quad \text{or} \quad |\Re t_1 - \pi| < c_1^{1/2}\tau_0^{3/2}. \tag{5.4.85}$$

From this, using that $t_1 \in C_4$, we get $\tau_0 < \Im t_1 < \tau_0 + c^{1/2}\tau_0^{3/2}$. From the last inequality, using (5.4.82), we obtain the estimate

$$\tau_0/2 < \Im t_2 < 3/2\tau_0, \tag{5.4.86}$$

while the real part of t_2 is equal to 0 or π. According to (5.4.22) and the inequality $\lambda < \tau_0^{-1/8}$, there exists l such that

$$|\lambda - \lambda_l(0)|^2 \leq c\tau_0^{15/8} < \tau_0^{1/2},$$

or

$$|\lambda - \lambda_l(\pi)|^2 \leq c\tau_0^{15/8} < \tau_0^{1/2}.$$

But the last inequalities contradict to condition (5.4.77) or (5.4.78). Therefore, (5.4.80) does not hold for $\lambda \notin \Lambda$ and $\lambda_0 \leq \lambda < \tau_0^{-1/8}$.

It remains to consider the case $\lambda \notin \Lambda$ and $\lambda \geq \tau_0^{-1/8}$. First, we check (5.4.86). We can do this in a similar way, as we did this for the previous case. Second, note that λ belongs to a gap of the spectrum. Since $\tau_0^{-1/8}$ is large enough and a length of a gap for C^∞-potential decreases faster than any power of λ (when $\lambda \to \infty$), in particular faster than λ^{-16} [LaPan] there exists l such that $|\lambda_l(0) - \lambda| < c\lambda^{-16}$ or $|\lambda_l(\pi) - \lambda| < c\lambda^{-16}$. Now, using relation (5.4.21), we obtain $|t_2| < c\lambda^{-33/4} < \tau_0^{33/32}$, or, $|t_2 - \pi| < c\lambda^{-33/4} < \tau_0^{33/32}$. This contradicts inequality (5.4.86). We obtain that (5.4.80) can not be satisfied for any real λ and natural n. Thus, (5.4.79) holds. The proposition is proved.

Proposition 5.21 . *For the function W, given by (5.4.47), the following asymptotic formulae hold:*

$$W(\lambda, 0)_{\lambda \to -\infty} = 1 + O(|\lambda|^{-1/2}), \tag{5.4.87}$$

$$W'(\lambda, 0) =_{\lambda \to -\infty} -\sqrt{\lambda}\left(1 + O(|\lambda|^{-1/2})\right). \tag{5.4.88}$$

Proof. The function W_0 is given by (5.4.48). It is clear that for a negative λ

$$W_0(\lambda, 0) = 1, \quad W_0'(\lambda, 0) = -\sqrt{|\lambda|}.$$

We represent W in the form:

$$W = W_0 + I_1 + I_2,$$

$$I_1 = \frac{\sqrt{-\lambda}}{\pi} \sum_{n \in Z} \int_{C_4} \left((\tilde{H}_{01}(t_1) - \lambda)^{-1} V_1 (\tilde{H}_{01}(t_1) - \lambda)^{-1} \right)_{n0} e^{i(t_1+\tilde{n})x_1} dt_1,$$

$$I_2 = \frac{\sqrt{-\lambda}}{\pi} \times$$

$$\sum_{n \in Z} \int_{C_4} \left((\tilde{H}_{01}(t_1) - \lambda)^{-1} V_1 ((\tilde{H}_1(t_1) - \lambda)^{-1} - (\tilde{H}_{01}(t_1) - \lambda)^{-1}) \right)_{n0} e^{i(t_1+\tilde{n})x_1} dt_1,$$

$$\tilde{n} = 2\pi n a_1^{-1}.$$

Calculating the first integral by the residue at the pole $t_1 = i\sqrt{|\lambda|}$, we get $I_1 < |\lambda|^{-1/2}$. Using estimates (5.4.71) and (5.4.73) and integrating them over C_4, we get $I_2 = O(|\lambda|^{-1/2})$. Adding the estimates for I_1 and I_2, we obtain (5.4.87). We prove estimate (5.4.88) similarly, using (5.4.72) and (5.4.74). The proposition is proved.

Proposition 5.21 shows that a sufficiently large in absolute value negative λ belongs to γ, because otherwise $W(\lambda, 0) = 0$. Since $W(\lambda, 0)$ depends analytically on λ in γ, the relation $W(\lambda, 0) = 0$ can be satisfied at a denumerable set of points. Moreover, only a finite number of points belongs to the set $(-\infty, 0] \cap \gamma$, because of the asymptotic (5.4.87). We denote them by $\alpha_1, ..., \alpha_M$, $M < \infty$. If $\lambda \in (-\infty, 0] \cap \gamma$, $\lambda \neq \alpha_k$, $k = 1, ..., M$, then $W(\lambda, x) = c(\lambda)\tilde{\psi}_+(\lambda, x)$, $c(\lambda) \neq 0$. Note that the points $\alpha_1, ..., \alpha_M$ can be of two types. In the first case, $W(\alpha_k, 0) = W'(\alpha_k, 0) = c(\lambda) = 0$. These points we could remove by a modification of the definition of $W(\lambda, x)$. This would be followed by essential, but not really needed overloading formulae. For the sake of simplicity, we use the definition of $W(\lambda, x)$ given above. In the second case,, $W(\alpha_k, 0) = 0$, $W'(\alpha_k, 0) \neq 0$, $c(\lambda) \neq 0$. In this case α_k is an eigenvalue of the Dirichlet boundary problem on the positive axis. It cannot be removed by a permissible modification of the definition of $W(\lambda, x)$.

Now we consider the function $p(\lambda)$:

$$p(\lambda) = W'(\lambda, 0) + \sqrt{-\lambda} W(\lambda, 0), \qquad \Re\sqrt{-\lambda} > 0, \ \lambda \in \gamma. \tag{5.4.89}$$

We recall that $\tilde{\lambda}_k$, $k = 1, ..., N$ are the zeros of the function $\tilde{T}_1(\lambda)$ (see (5.4.25)).

Proposition 5.22 . *Suppose that $\lambda \in (-\infty, 0) \setminus \Lambda$, $\lambda \neq \tilde{\lambda}_k$, $k = 1, ..., N$, $\lambda \neq \alpha_m$, $m = 1, ..., M$, and relations (5.4.75), (5.4.76) are satisfied. Then,*

$$p(\lambda) \neq 0. \tag{5.4.90}$$

Proof. According to Proposition 5.19, $\lambda \in \gamma(\tau_0)$. Therefore, $W(\lambda, x) = = c(\lambda)\tilde{\psi}_+(\lambda, x)$. The relation $c(\lambda) = 0$ or $\tilde{\psi}_+(\lambda, x) = 0$ can be satisfied only at the points $\alpha_1, ..., \alpha_M$. Since $\lambda \neq \alpha_k$, $k = 1, ..., M$, we have $c(\lambda) \neq 0$. Hence, λ is a root of the function $\tilde{T}_1(\lambda)$, i.e., $\lambda = \lambda_k$, $k = 1, ..., N$. But this contradicts the hypothesis of the proposition. The proposition is proved.

Next, we consider the points $t_1(\lambda_k)$, $t_1(\alpha_k)$, $\Im t_1 \geq 0$.

Proposition 5.23 . *If τ_0 is small enough, $\tau_0 < T(V_1)$, then the points $t_1(\lambda_k)$, $t_1(\alpha_k)$ are separated from the contour C_4. Moreover,*

$$\min_{t_1 \in C_4, k=1,\dots,N} |t_1(\tilde{\lambda}_k) - t_1| > \tau_0, \tag{5.4.91}$$

$$\min_{t_1 \in C_4, k=1,\dots,M} |t_1(\alpha_k) - t_1| > \tau_0. \tag{5.4.92}$$

Proof. Since $\tilde{\lambda}_k$, α_k are real, the quasimomenta $t_1(\tilde{\lambda}_k)$, $t_1(\alpha_k)$ are real, imaginary, or their real parts can be equal to $\pm\pi$. In the case where $t_1(\tilde{\lambda}_k) = \bar{t}_1(\tilde{\lambda}_k)$, the inequality

$$\min_{t_1 \in C_4} |t_1(\tilde{\lambda}_k) - t_1| > \tau_0 \tag{5.4.93}$$

follows from the definition of the contour C_4. If $t_1(\tilde{\lambda}_k)$ is not real, then $\Im t_1(\tilde{\lambda}_k) > c_0 > 0$, because we have only a finite number of the points $\tilde{\lambda}_k$. Therefore, for τ_0 small enough, inequality (5.4.93) holds, too. The same can be said in the case when $\Re t_1(\tilde{\lambda}_k) = \pm\pi$. Thus, (5.4.91) is proved. Estimate (5.4.92) is proved in a similar way. The proposition is proved.

Let $s_0 = \max\{q, p\} + 1$, q being the maximal multiplicity of zeros of the function $W(\lambda, 0)$, and p being the maximal multiplicity of the zeros of the function $\tilde{T}_1(\lambda)$.

Proposition 5.24 . *Suppose that $\lambda \in \gamma$, $\varepsilon > 0$, and*

$$\min_{k=1,\dots,N} |\lambda - \tilde{\lambda}_k| > \varepsilon^{1/s_0}, \tag{5.4.94}$$

$$\min_{k=1,\dots,M} |\lambda - \alpha_k| > \varepsilon^{1/s_0}. \tag{5.4.95}$$

Then, for a sufficiently small ε, $\varepsilon < \varepsilon_0(V)$

$$|p(\lambda)| > \varepsilon, \tag{5.4.96}$$

$$|W(\lambda, 0)| > \varepsilon. \tag{5.4.97}$$

Proof. The function $p(\lambda)$ depends analytically on λ in γ, and has roots only at the points λ_k and α_k inside γ. Using relation (5.4.87), it is easy to show that

$$p(\lambda)_{\lambda \to -\infty} = |\lambda|^{1/2} \left(1 + O(|\lambda|^{-1/2})\right).$$

Therefore, for sufficiently small positive ε, $\varepsilon < \varepsilon_0(V)$, the set $U_0(\delta) = \{\lambda : |p(\lambda)| \le \varepsilon\}$ lies in a small neighborhood of the points $\tilde{\lambda}_k$, α_i:

$$U_0(\delta) \subset \cup_{k=1}^{N} U(\tilde{\lambda}_k, \tilde{\varepsilon}_k) \cup_{i=1}^{M} U(\alpha_i, \varepsilon_i),$$

where $\tilde{\varepsilon}_k \le c_1(V_1)\varepsilon^{1/p} \le \varepsilon^{1/s_0}$, $\varepsilon_i \le c_1 \varepsilon^{1/q} < \varepsilon^{1/s_0}$. Thus, $|p(\lambda)| > \varepsilon$ when (5.4.94) and (5.4.95) are satisfied. Similarly, $W(\lambda, 0)$ has roots at the points α_i. Moreover, $W(\lambda, 0) =_{\lambda \to -\infty} 1 + O(|\lambda|^{-1/2})$. Therefore, under condition (5.4.95) inequality (5.4.97) holds. The proposition is proved.

As stated above, quasisurface states are absent, when $|\tilde{T}_1(\lambda)| \geq \delta$, i.e., when λ does not belong to a small neighborhood $U(\delta)$ (see (5.4.45)) of the points $\tilde{\lambda}_k$, $k = 1, \ldots, N$. Thus, quasisurface states are absent when (5.4.94) is satisfied. Now we introduce a stronger definition instead of the previous one.

DEFINITION. Surface and quasisurface states are absent when (5.4.94) and (5.4.95) are satisfied.

This definition means that surface and quasisurface states are absent when λ does not belong to the neighborhood of the points $\tilde{\lambda}_k$, $k = 1, \ldots, N$, and α_i, $i = 1, \ldots, M$.

The latter definition is more convenient, because after a natural generalization for the three-dimensional operator $H_1 = -\Delta + V_{1+}(x_1)$, this definition turns out to be stable with respect to a perturbation $V - V_1$, V being a trigonometric polynomial of a general form. This means that under some additional conditions on $t_{||}$, the operators $H_1(t_{||})$ or $H(t_{||})$ have or have no quasisurface states together.

5.4.2 The case of a potential depending only on x_1.

Next, we consider the operators $H_1 = -\Delta + V_1(x_1)$ and $H_{1+} = -\Delta + V_{1+}(x_1)$ in $L_2(R^3)$. The variables are separable in this case. The matrix of $H_1(t)$ has the form

$$H_1(t)_{mn} = \tilde{H}_1(t_1)_{m_1 n_1} \delta_{m_{||} n_{||}} + |t_{||} + p_{m_{||}}(0)|^2 \delta_{mn}. \tag{5.4.98}$$

We assign to any function $\varphi(x)$ of $L_2(Q_+)$ its components $\varphi_{m_{||}}(x_1)$ in $t_{||}$-basis. This means that

$$\varphi(x) = \sum_{m_{||} \in Z^2} \varphi_{m_{||}}(x_1) \exp i(t_{||} + p_{m_{||}}(0), x_{||}), \quad x_1 \geq 0. \tag{5.4.99}$$

The matrix of the operator $H_{1+}(t_{||})$ defined by the differential expression $H_{1+}(t_{||}) = -\Delta + V_{1+}$ and the quasiperiodic conditions (5.4.4), has the diagonal form in $t_{||}$-basis:

$$H_{1+}(t_{||})_{m_{||} n_{||}} = \left(-\frac{d^2}{dx_1^2} + V_{1+}(x_1) + |t_{||} + p_{m_{||}}(0)|^2 \right) \delta_{m_{||} n_{||}} =$$

$$\left(\tilde{H}_{1+} + |t_{||} + p_{m_{||}}(0)|^2 \right) \delta_{m_{||} n_{||}}. \tag{5.4.100}$$

Thus, we have reduced the study of the operator $H_{1+}(t_{||})$ in $L_2(Q_+ \cup Q_-)$ to that of the family of the operators $\tilde{H}_{1+} + |t_{||} + p_{m_{||}}(0)|^2 I$, $m_{||} \in Z^2$ in $L_2(R)$.

Let us consider the functions in $L_2(Q_\pm)$:

$$\Psi_\pm(k^2 + i\varepsilon, t_{||} + p_{m_{||}}(0), x) =$$

$$\tilde{\psi}_\pm(k^2 + i\varepsilon - |t_{||} + p_{m_{||}}(0)|^2, x_1) \exp i(t_{||} + p_{m_{||}}(0), x_{||}), \tag{5.4.101}$$

$\tilde{\psi}_\pm$ being the Weyl solutions in the one-dimensional situation (see page 260). It is clear that the functions $\Psi_\pm$ satisfy quasiperiodic conditions (5.4.4) and the equation $(-\Delta + V_1 - k^2 - i\varepsilon)\Psi_\pm = 0$ in $L_2(Q_\pm)$. Recalling the definition of the operators $A_\pm^{V_1}$ in $L_2(Q_{||})$, we obtain

$$A_\pm^{V_1}(k^2 + i\varepsilon, t_{||})u_\pm = v_\pm, \tag{5.4.102}$$

where

$$u_\pm(x_{||}) = \Psi_\pm(k^2 + i\varepsilon, t_{||} + p_{m_{||}}(0), x)\big|_{x_1=0} =$$

$$\tilde{\psi}_\pm(k^2 + i\varepsilon - |t_{||} + p_{m_{||}}(0)|^2, 0) \exp i(t_{||} + p_{m_{||}}(0), x_{||}), \tag{5.4.103}$$

$$v_\pm(x_{||}) = \frac{\partial}{\partial x_1}\Psi_\pm(k^2 + i\varepsilon, t_{||} + p_{m_{||}}(0), x)\Big|_{x_1=0} =$$

$$\tilde{\psi}'_\pm(k^2 + i\varepsilon - |t_{||} + p_{m_{||}}(0)|^2, 0) \exp i(t_{||} + p_{m_{||}}(0), x_{||}).$$

Using $t_{||}$-basis we obtain

$$(u_\pm)_{m_{||}} = \tilde{\psi}_\pm(k^2 + i\varepsilon - |t_{||} + p_{m_{||}}(0)|^2, 0),$$

$$(v_\pm)_{m_{||}} = \tilde{\psi}'_\pm(k^2 + i\varepsilon - |t_{||} + p_{m_{||}}(0)|^2, 0). \tag{5.4.104}$$

Thus, the matrices $A_\pm^{V_1}(k^2 + i\varepsilon, t_{||})$ are diagonal in this basis. Their elements are

$$A_\pm^{V_1}(k^2 + i\varepsilon, t_{||})_{m_{||}m_{||}} = \tilde{A}_\pm^{V_1}(k^2 + i\varepsilon - |t_{||} + p_{m_{||}}(0)|^2) =$$

$$(\tilde{\psi}'/\tilde{\psi})(k^2 + i\varepsilon - |t_{||} + p_{m_{||}}(0)|^2, 0). \tag{5.4.105}$$

Note that $\tilde{\psi}(k^2 + i\varepsilon - |t_{||} + m_{||}|^2, 0) \neq 0$ when $\varepsilon \neq 0$, because otherwise the Dirichlet operator has a nonreal eigenvalue. The matrix A_-^0 is given by (5.4.11).

We define the diagonal operator J_0 in the three-dimensional situation also by (5.4.29). It is clear that relation (5.4.31) holds.

Next, we consider $\rho_1(k^2, t_{||})$:

$$\rho_1(k^2, t_{||}) = \min_{m_{||} \in Z^2} \left|k^2 - |t_{||} + p_{m_{||}}(0)|^2\right|. \tag{5.4.106}$$

In Section 5.5 we prove that there exists a set $\Omega_1(k, \delta) \subset K_{||}$, such that

$$\rho_1(k^2, t_{||}) > k^{-2\delta}, \tag{5.4.107}$$

when $t_{||} \in \Omega_1(k, \delta)$. The set $\Omega_1(k, \delta)$ has an asymptotically full measure in $K_{||}$ [11]:

$$\frac{s(\Omega_1(k, \delta))}{s(K_{||})} \to_{k \to \infty} 1. \tag{5.4.108}$$

It follows from this result (see Section 5.5) that there exists a subset $\tilde{\Omega}_1(k, \delta)$ of the sphre S_k, such that

$$\rho_1(k^2, k_{||}) > k^{-2\delta}, \tag{5.4.109}$$

[11] In fact, we prove a stronger result.

when $\mathbf{k} \in \tilde{\Omega}_1(k,\delta)$. The set $\tilde{\Omega}_1(k,\delta)$ has an asymptotically full measure on S_k:

$$\frac{s\left(\tilde{\Omega}_1(k,\delta)\right)}{s(S_k)} \to_{k\to\infty} 1. \tag{5.4.110}$$

Proposition 5.25 . *If $\rho_1(k,t_{||}) > k^{-2\delta}$, then*

$$-\Im(A^0_-(k^2+i0,t_{||})x,x) \geq k^{-\delta}\|(I-J_0)x\|^2, \tag{5.4.111}$$

$$\|A^0_-(k^2+i0)^{-1}\| < k^{\delta}. \tag{5.4.112}$$

Proof. Estimates (5.4.111) and (5.4.112) immediately follow from relations (5.4.11), (5.4.31) and the estimate for $\rho_1(k,\delta)$. The proposition is proved.

Now, we introduce a diagonal projection $J'(k^2,t_{||})$ in $L_2(K_{||})$:

$$J'_{m_{||}m_{||}} = \begin{cases} 1, & \text{if } k^2 - |t_{||}+p_{m_{||}}(0)|^2 \notin \Lambda; \\ 0, & \text{otherwise.} \end{cases} \tag{5.4.113}$$

It is clear that J' is equal to J_0, given by (5.4.29), when $V_1 = 0$.

Proposition 5.26 . *For the operators $A^{V_1}_{\pm}$ the following relation holds:*

$$\lim_{\varepsilon\to 0}\left(A^{V_1}_{\pm}(k^2+i\varepsilon,t_{||}) - A^{V_1}_{\pm}(k^2-i\varepsilon,t_{||})\right)J' = 0. \tag{5.4.114}$$

Proof. This assertion immediately follows from a diagonal form of the operators $A^{V_1}_{\pm}$, and formula (5.4.23). The proposition is proved.

Next, we introduce a diagonal projection J_1 in $L_2(Q_{||})$: $J_1 = (I-J')J_0$, i.e.,

$$(J_1)_{m_{||}m_{||}} = \begin{cases} 1, & \text{if } k^2 - |t_{||}+p_{m_{||}}(0)|^2 \in (-\infty,0]\cap\Lambda; \\ 0, & \text{otherwise.} \end{cases} \tag{5.4.115}$$

We consider $\rho_2'(k^2,t_{||})$, $\rho_2''(k^2,t_{||})$, $\rho_2(k^2,t_{||})$:

$$\rho_2'(k^2,t_{||}) = \min_{n:|\lambda_n(0)|<k^{\delta},\ m_{||}\in Z^2} \left|k^2 - |t_{||}+p_{m_{||}}(0)|^2 - \lambda_n(0)\right|, \tag{5.4.116}$$

$$\rho_2''(k^2,t_{||}) = \min_{n:|\lambda_n(\pi a_1^{-1})|<k^{\delta},\ m_{||}\in Z^2} \left|k^2 - |t_{||}+p_{m_{||}}(0)|^2 - \lambda_n(\pi)\right|, \tag{5.4.117}$$

$$\rho_2 = \min\{\rho_2', \rho_2''\}. \tag{5.4.118}$$

In Section 5.5 we prove that there exists a set $\Omega_2(k,\delta) \subset K_{||}$, such that

$$\rho_2(k^2,t_{||}) > k^{-2\delta}, \tag{5.4.119}$$

when $t_{||} \in \Omega_2(k,\delta)$. The set $\Omega_2(k,\delta)$ has an asymptotically full measure in $K_{||}$:

$$\frac{s\left(\Omega_2(k,\delta)\right)}{s(K_{||})} \to_{k\to\infty} 1. \tag{5.4.120}$$

It follows from this result (see Section 5.5) that there exists a subset $\tilde{\Omega}_2(k,\delta)$ of the sphere S_k, such that

$$\rho_2(k^2, k_{||}) > k^{-2\delta}, \qquad (5.4.121)$$

when $\mathbf{k} \in \tilde{\Omega}_2(k,\delta)$. The set $\tilde{\Omega}_2(k,\delta)$ has an asymptotically full measure on $S_0(k)$:

$$\frac{s\left(\tilde{\Omega}_2(k,\delta)\right)}{s(S_0(k))} \to_{k\to\infty} 1. \qquad (5.4.122)$$

Proposition 5.27 . *If* $\rho_2(k,t_{||}) > k^{-2\delta}$, *then*

$$\Im(A_+^{V_1}(k^2+i0,t_{||})x,x) \geq k^{-7\delta}||J_1x||^2, \qquad (5.4.123)$$

and

$$\left|\Psi_\pm(k^2+i0,t_{||}+p_{m_{||}}(0),x)\Big|_{x_1=0}\right| > k^{-4\delta}, \qquad (5.4.124)$$

$$\left|\frac{d}{dx_1}\Psi_\pm(k^2+i0,t_{||}+p_{m_{||}}(0),x)\Big|_{x_1=0}\right| < k^{3\delta}, \qquad (5.4.125)$$

when

$$k^2 - |t_{||}+m_{||}|^2 \in (-\infty,0]\cap\Lambda.$$

Proof. Let us prove (5.4.123). Since the operators $A_+^{V_1}$ and J_1 are diagonal, and $(2i)^{-1}(A_+^{V_1}(k^2+i0) - A_+^{V_1*}(k^2+i0))$ is non-negatively determined, it suffices to show that

$$(2i)^{-1}(A_+^{V_1} - A_+^{V_1*})(k^2+i0,t_{||})_{m_{||}m_{||}} > k^{-12\delta}, \qquad (5.4.126)$$

when $(J_1)_{m_{||}m_{||}} = 1$. Using Proposition 5.14 (formula (5.4.26)), we obtain:

$$(2i)^{-1}(A_+^{V_1} - A_+^{V_1*})(k^2+i0,t_{||})_{m_{||}m_{||}} >$$

$$g^3(k^2-|t_{||}+p_{m_{||}}(0)|^2)\left(\left||t_{||}+p_{m_{||}}(0)|^2-k^2\right|+||V||\right)^{-5/2}. \qquad (5.4.127)$$

From the definition of J_1 we have $-||V|| \leq k^2 - |t_{||}+p_{m_{||}}(0)|^2 \leq 0$, when $(J_1)_{m_{||}m_{||}} = 1$. Noting also that

$$g(k^2-|t_{||}+p_{m_{||}}(0)|^2) \geq \rho_2(k,t_{||}) \geq k^{-2\delta}, \qquad (5.4.128)$$

we obtain (5.4.123). Relation (5.4.124) follows from (5.4.27). Relation (5.4.125) follows from (5.4.37) and (5.4.41).
The proposition is proved.

Proposition 5.28 . *If* $\rho_1(k^2,t_{||}) > k^{-2\delta}$, $\rho_2(k^2,t_{||}) > k^{-2\delta}$, *then*

$$\Im(T_1(k^2\pm i0)x,x) \geq k^{-7\delta}||(I-J_0+J_1)x||^2. \qquad (5.4.129)$$

Proof. Considering the relation

$$\Im(T_1(k^2+i0)x,x) = \Im(A_+^{V_1}(k^2+i0)x,x) - \Im(A_-^0(k^2+i0)x,x),$$

using estimates (5.4.111), (5.4.123) and, taking into account that the projections J_0, J_1 are mutually orthogonal, we obtain (5.4.129). The proposition is proved.

The next step is to consider surface and quasisurface states of the operator H_{1+}. Since $T_1(\lambda)$ is diagonal, a surface state exists if and only if $T_1(\lambda)_{m_\| m_\|} = \tilde{T}_1(k^2 - |t_\| + p_{m_\|}(0)|^2) = 0$ for some $m_\|$, i.e., when

$$k^2 - |t_\| + p_{m_\|}(0)|^2 = \lambda_k \tag{5.4.130}$$

for some $m_\| \in Z^2$ and $k = 1, ..., N$. There exists a quasisurface state if

$$|T_1(k^2)_{m_\| m_\|}| = |\tilde{T}_1(k^2 - |t_\| + m_\||^2)| \le k^{-\delta} \tag{5.4.131}$$

for some $m_\|$. This relation can be satisfied only if

$$\left|k^2 - |t_\| + p_{m_\|}|^2 - \lambda_k\right| < k^{-\delta/p} \tag{5.4.132}$$

for some $m_\| \in Z^2$ and $k = 1, ..., N$ (see (5.4.45), (5.4.46)). Naturally, surface and quasisurface states are absent ($\|T_1^{-1}(k^2)\| < k^{\delta}$) when

$$\rho_3(k^2, t_\|) > k^{-\delta/p}, \tag{5.4.133}$$

where

$$\rho_3(k^2, t_\|) = \min_{m_\| \in Z^2,\ k=1,...,N} \left|k^2 - |t_\| + p_{m_\|}(0)|^2 - \tilde{\lambda}_k\right|, \tag{5.4.134}$$

$\tilde{\lambda}_k$, $k = 1, ..., N$, being the zeros of the function $\tilde{T}(\lambda)$, p – their maximal multiplicity. In Section 5.5 we prove that there exists a set $\Omega_3(k,\delta) \subset K_\|$, such that relation (5.4.133) holds when $t_\| \in \Omega_3(k,\delta)$. The set $\Omega_3(k,\delta)$ has an asymptotically full measure in $K_\|$:

$$\frac{s\left(\Omega_3(k,\delta)\right)}{s(K_\|)} \to_{k\to\infty} 1. \tag{5.4.135}$$

It follows from this result (see Section 5.5) that there exists a subset $\tilde{\Omega}_3(k,\delta)$ of the sphere S_k, such that

$$\rho_3(k^2, k_\|) > k^{-\delta/p}, \tag{5.4.136}$$

when $\mathbf{k} \in \tilde{\Omega}_3(k,\delta)$. The set $\tilde{\Omega}_3(k,\delta)$ has an asymptotically full measure on S_k:

$$\frac{s\left(\tilde{\Omega}_3(k,\delta)\right)}{s(S_k)} \to_{k\to\infty} 1. \tag{5.4.137}$$

Now we introduce the function

$$U_{m_{\|}}^{(1)}(k^2+i\varepsilon,t_{\|},x)=a_{m_{\|}}\sum_{n\in Z^3}\int_{C_4}(H_1(t)-k^2-i\varepsilon)^{-1}_{nm}\exp i(t+p_{m_{\|}}(0),x)dt_1, \tag{5.4.138}$$

where ε is small enough in absolute value,

$$m_{\|}:k^2-|t_{\|}+p_{m_{\|}}(0)|^2\in(-\infty,0)\setminus\Lambda,$$

$$t=(t_1,t_{\|}),\qquad m=(0,m_{\|}),$$

and the coefficients $a_{m_{\|}}$ are given by $a_{m_{\|}}=\pi^{-1}A_+^0(k^2+i\varepsilon,t_{\|})_{m_{\|}m_{\|}}$. Since the matrix $H(t)_{mn}$ is diagonal with respect to $\|$-component of indeces (see (5.4.100)), it is easy to see that

$$U_{m_{\|}}^{(1)}(k^2-i\varepsilon,t_{\|},x)=W(k^2-i\varepsilon-|t_{\|}+p_{m_{\|}}(0)|^2,x_1)\exp i(t_{\|}+p_{m_{\|}}(0),x_{\|}), \tag{5.4.139}$$

W being given by (5.4.47). It is obvious that $U_{m_{\|}}^{(1)}(k^2+i\varepsilon,t_{\|},x)$ belongs to $L_2(Q_+)$ (because $W\in L_2(0,\infty)$) and satisfies quasiperiodic condition (5.4.4) and the equation $H_{1+}U_{m_{\|}}^{(1)}=(k^2+i\varepsilon)U_{m_{\|}}^{(1)}$. The function $U_{m_{\|}}^{(1)}(k^2+i\varepsilon,t_{\|},x)$ does not vanish when $k^2+i\varepsilon-|t_{\|}+p_{m_{\|}}(0)|^2\in\gamma$ and $k^2+i\varepsilon-|t_{\|}+p_{m_{\|}}(0)|^2\neq\alpha_i$, $i=1,\ldots,N$.

Let L, M, P' be the diagonal matrices:

$$L_{mm}=t_1+\tilde m_1,\quad m_1\in Z^3;\quad \tilde m_1=2\pi a_1^{-1}m_1, \tag{5.4.140}$$

$$M_{mm}=\left(A_+^0(k^2+i\varepsilon,t_{\|})\right)-m_{\|}m_{\|}=-\sqrt{-k^2-i\varepsilon+|t_{\|}+p_{m_{\|}}(0)|^2},\quad m\in Z^3. \tag{5.4.141}$$

Proposition 5.29 . *Suppose that $\rho_1(k^2,t_{\|})>k^{-2\delta}$, $t_1\in C_4(\tau_0)$, $\tau_0=k^{-8\delta}$. Then the following estimates hold:*

$$\|(H_0(t)-k^2)^{-1}\|^*\leq(\Im t_1+k^{-\delta})^{-1}k^{8\delta}, \tag{5.4.142}$$

$$\|(H_0(t)-k^2)^{-1}\|^*\leq k^{9\delta}, \tag{5.4.143}$$

$$\|M(H_0(t)-k^2)^{-1}\|^*\leq k^{8\delta}, \tag{5.4.144}$$

$$\|L(H_0(t)-k^2)^{-1}\|^*\leq k^{9\delta}. \tag{5.4.145}$$

Proof. Taking into account that H_0 is a diagonal operator and using the relation $H_0(t)_{mm}=H_{01}(t_1)_{m_1m_1}+|t_{\|}+p_{m_{\|}}(0)|^2$, we obtain

$$\|(H_0(t)-k^2)^{-1}\|^*=\max_{m_{\|}}\left\|\left(\tilde H_{01}(t_1)-k^2+|t_{\|}+p_{m_{\|}}(0)|^2\right)^{-1}\right\|. \tag{5.4.146}$$

Note that for any $m_{\|}$

$$\left|k^2-|t_{\|}+p_{m_{\|}}(0)|^2\right|>\rho_1(k^2,t_{\|})>k^{-2\delta}>16\tau_0^2 \tag{5.4.147}$$

Moreover, for k large enough we have the inequality: $\tau_0=k^{-8\delta}<\pi(4a_1)^{-1}$. Thus, the hypothesis of Proposition 5.17 (page 269) holds. According to (5.4.51),

$$\left\|(\tilde{H}_{01}(t_1) - k^2 + |t_{||} + m_{||}|^2)^{-1}\right\| \leq (\Im t_1 + \left|k^2 - |t_{||} + p_{m_{||}}(0)|^2\right|^{1/2})^{-1} k^{8\delta} \leq \tag{5.4.148}$$
$$(\Im t_1 + \rho_1^{1/2})^{-1} k^{8\delta} \leq (\Im t_1 + k^{-\delta})^{-1} k^{8\delta}.$$

Relation (5.4.142) follows from relations (5.4.146) and (5.4.148).

Since $\Im t_1 \geq k^{-8\delta}$, relation (5.4.143) holds.

From the definition of M, and estimate (5.4.51), it follows that

$$\|M(H_0(t) - k^2)^{-1}\|^* \leq$$
$$\max_{m_{||}} \left(|\sqrt{k^2 - |t_{||} + p_{m_{||}}(0)|^2} \left\|(\tilde{H}_{01}(t_1) - k^2 + |t_{||} + p_{m_{||}}(0)|^2)^{-1}\right\|\right) \tag{5.4.149}$$
$$\leq \max_{m_{||}} \left(|\sqrt{k^2 - |t_{||} + p_{m_{||}}(0)|^2} (\Im t_1 + \sqrt{k^2 - |t_{||} + p_{m_{||}}(0)|^2}\right)^{-1} \tau_0^{-1}.$$

Estimate (5.4.144) easily follows from the last relation.

Similarly, using estimate (5.4.52) we obtain (5.4.145).

The proposition is proved.

Proposition 5.30 . *If $\rho_2(k^2, t_{||}) > k^{-2\delta}$, then for sufficiently large k, $k > k_0(V, \delta)$, the following estimate holds:*

$$\max_{t_1 \in C_4} \|(H_1(t) - k^2)^{-1}\|^* \leq k^{25\delta}, \tag{5.4.150}$$

Proof. Taking into account that $H_1(t)$ is diagonal with respect to $||$-indeces, we obtain

$$\|(H_1(t) - k^2)^{-1}\|^* \leq \max_{m_{||}} \|(\tilde{H}_1(t) - k^2 + |t_{||} + p_{m_{||}}(0)|^2)^{-1}\|^* \tag{5.4.151}$$

By hypothesis, $\rho_2(k^2, t_{||}) > k^{-2\delta}$, i.e.,

$$\left|k^2 - |t_{||} + p_{m_{||}}(0)|^2 - \lambda_n(0)\right| > k^{-2\delta},$$
$$\left|k^2 - |t_{||} + p_{m_{||}}(0)|^2 - \lambda_n(\pi a_1^{-1})\right| > k^{-2\delta}$$

for all $m_{||} \in Z^2$ and $n : |\lambda_n(0)| < k^{\delta}$. Applying Proposition 5.20 (estimate (5.4.79)), we get (5.4.150). The proposition is proved.

Proposition 5.31 . *Suppose that $\rho_2(k^2, t_{||}) > k^{-3\delta}$, and the point $\lambda = k^2 - |t_{||} + p_{m_{||}}(0)|^2$ belongs to $(-\infty, 0] \setminus \Lambda$. Then $\lambda \in \gamma$.*

Proof. It is clear that

$$\min_{n:|\lambda_n(0)| \leq 2\|V\|} |\lambda - \lambda_n(0)| > \rho_2(k^2, t_{||}) > k^{-3\delta} = \tau_0^{3/8}.$$

Thus, λ satisfies (5.4.75). Similarly, we prove (5.4.76). By Proposition 5.19 $\lambda \in \gamma$. The proposition is proved.

Next, we consider $\rho_4(k^2, t_{||})$:

$$\rho_4(k^2, t_{||}) = \min_{i=1,\dots,M;\ m_{||}\in Z^2} \left| k^2 - |t_{||} + p_{m_{||}}(0)|^2 - \alpha_i \right|. \quad (5.4.152)$$

In Section 5.5 we prove that there exists a set $\Omega_4(k,\delta) \subset K_{||}$, such that

$$\rho_4(k^2, t_{||}) > k^{-\delta/s_0}, \quad (5.4.153)$$

when $t_{||} \in \Omega_4(k,\delta)$, and α_i, s_0 being defined on pages 274, 275. The set $\Omega_4(k,\delta)$ has an asymptotically full measure in $K_{||}$:

$$\frac{s\left(\Omega_4(k,\delta)\right)}{s(K_{||})} \rightarrow_{k\rightarrow\infty} 1. \quad (5.4.154)$$

It follows from this result (see Section 5.5) that there exists a subset $\tilde{\Omega}_4(k,\delta)$ of the sphere S_k, such that

$$\rho_4(k^2, k_{||}) > k^{-2\delta}, \quad (5.4.155)$$

when $\mathbf{k} \in \tilde{\Omega}_4(k,\delta)$. The set $\tilde{\Omega}_4(k,\delta)$ has an asymptotically full measure on S_k:

$$\frac{s\left(\tilde{\Omega}_4(k,\delta)\right)}{s(S_k)} \rightarrow_{k\rightarrow\infty} 1. \quad (5.4.156)$$

Proposition 5.32 . *Suppose that* $\rho_2(k^2, t_{||}) > k^{-3\delta}$, $\rho_3(k^2, t_{||}) > k^{-3\delta/s_0}$, $\rho_4(k^2, t_{||}) > k^{-3\delta/s_0}$ *and the point* $\lambda = k^2 - |t_{||} + p_{m_{||}}(0)|^2$ *belongs to* $(-\infty, 0] \setminus \Lambda$. *Then,*

$$U^{(1)}_{m_{||}}(0, x_{||}) = W(\lambda, 0) \exp i(t_{||} + p_{m_{||}}(0), x_{||}), \quad (5.4.157)$$

with $W(\lambda, 0)$ *satisfying the estimate:*

$$|W(\lambda, 0)| > k^{-3\delta}. \quad (5.4.158)$$

Proof. Taking into account that $\rho_2(k^2, t_{||}) > k^{-3\delta}$, and the point $\lambda = k^2 - |t_{||} + p_{m_{||}}(0)|^2$ belongs to $(-\infty, 0] \setminus \Lambda$ and using Proposition 5.31, we obtain that $\lambda \in \gamma$. Using (5.4.100), the definition (5.4.138) of $U^{(1)}_{m_{||}}$ and the definition (5.4.47) of $W(\lambda, x)$, we obtain (5.4.157). The hypothesis of Proposition 5.24 (page 275) is satisfied, since $\rho_3(k^2, t_{||}) > k^{-3\delta/s_0}$ and $\rho_4(k^2, t_{||}) > k^{-3\delta/s_0}$. Therefore, (5.4.158) holds. The proposition is proved.

Now we consider the sets M_0, M_1, M_2:

$$M_0 = \{m_{||} :\ k^2 - |t_{||} + p_{m_{||}}(0)|^2 \in (-\infty, 0]\},$$

$$M_1 = \{m_{||} :\ k^2 - |t_{||} + p_{m_{||}}(0)|^2 \in (-\infty, 0] \cap \Lambda\}, \quad (5.4.159)$$

$$M_2 = \{m_{||} :\ k^2 - |t_{||} + p_{m_{||}}(0)|^2 \in (-\infty, 0] \setminus \Lambda\}.$$

It is clear that $M_0 = M_1 \cup M_2$, $M_1 \cap M_2 = \emptyset$, $M_0 = M_1 \cup M_2$.

We define the diagonal projection J_2 in $L_2(Q_{||})$ as follows:

$$(J_2)_{m_{||}m_{||}} = \begin{cases} 1 & \text{if } m_{||} \in M_2, \\ 0 & \text{otherwise.} \end{cases} \quad (5.4.160)$$

From the definitions of J^0, J_1 (see (5.4.29), (5.4.115)), we see that J_0, J_1 are determined similarly with respect to M_0, M_1. Hence, $J_0 = J_1 + J_2$. From now on we define the functions $U^{(1)}_{m_\parallel}(k^2+i\varepsilon, x)$ $(0<|\varepsilon|<k^{-4})$ by formula (5.4.138), when $m_\parallel \in M_2$. If $m_\parallel \notin M_2$, then we define $U^{(1)}_{m_\parallel}(k^2-i\varepsilon, x)$ as follows:

$$U^1_{m_\parallel}(k^2+i\varepsilon, x) = \varphi(k^2+i\varepsilon-|t_\parallel + p_{m_\parallel}(0)|^2, x_1)\exp i(t_\parallel + p_{m_\parallel}(0), x), \quad (5.4.161)$$

$$\varphi(z, x_1) \equiv \tilde{\psi}_+(z, x_1)/\tilde{\psi}_+(z, 0), \quad m_\parallel \notin M_2,$$

where $\tilde{\psi}_+(z, x_1)$ is defined on page 260.

We introduce the operators $\hat{J}_1(k^2+i\varepsilon, t_\parallel)$ and $D_1(k^2+i\varepsilon, t_\parallel)$ in $L_2(Q_\parallel)$ by the matrices:

$$(\hat{J}_1)_{n_\parallel m_\parallel} = \left(U^{(1)}_{m_\parallel}(k^2+i\varepsilon, x)|_{x_1=0}\right)_{n_\parallel}, \quad (5.4.162)$$

$$(D_1)_{n_\parallel m_\parallel} = \left(\frac{\partial U^{(1)}_{m_\parallel}(k^2+i\varepsilon, x)}{\partial x_1}|_{x_1=0}\right)_{n_\parallel}. \quad (5.4.163)$$

If $m_\parallel \in M_2$, then

$$\hat{J}_1(k^2+i0)_{n_\parallel m_\parallel} = \hat{J}_1(k^2-i0)_{n_\parallel m_\parallel} \equiv \hat{J}_1(k^2)_{n_\parallel m_\parallel}, \quad (5.4.164)$$

$$D_1(k^2+i0)_{n_\parallel m_\parallel} = D_1(k^2-i0)_{n_\parallel m_\parallel} = D_1(k^2)_{n_\parallel m_\parallel}. \quad (5.4.165)$$

From the new definition of $U^{(1)}_{m_\parallel}$, and the definition of $\hat{J}_1$ we see that

$$\hat{J}_1(I-J_2) = \hat{J}_1^{-1}(I-J_2) = I - J_2, \quad (5.4.166)$$

i.e., the operator $\hat{J}_1$ acts in the subspace $(I-J_2)l_2^2$ as the identity. It is easy to see that matrices $\hat{J}_1$, D_1 are diagonal. Using (5.4.105), (5.4.49), (5.4.139) and (5.4.161), we obtain

$$A^{V_1}_+ = D_1\hat{J}_1^{-1}. \quad (5.4.167)$$

Proposition 5.33 . *Suppose $\rho_2(k^2, t_\parallel) > k^{-2\delta}$, $\rho_3(k^2, t_\parallel) > k^{-8\delta/s_0}$, $\rho_4(k^2, t_\parallel) > k^{-8\delta/s_0}$. Then the following estimates hold:*

$$\left\|(\hat{J}_1)_{m_\parallel m_\parallel}(k^2, t_\parallel)\right\| > k^{-8\delta}, \qquad m_\parallel \in M_2, \quad (5.4.168)$$

$$\Im(D_1)(k^2+i0, t_\parallel)_{m_\parallel m_\parallel} > k^{-7\delta}, \qquad m_\parallel \in M_1. \quad (5.4.169)$$

Proof. From formulae (5.4.157) and (5.4.162), it follows

$$(\hat{J}_1)_{m_\parallel m_\parallel}(k^2, t_\parallel) =$$

$$W(k^2 - |t_\parallel + p_{m_\parallel}(0)|^2, 0), \quad (5.4.170)$$

when $m_\parallel \in M_2$. According to Proposition 5.31, $\lambda \in \gamma$. The hypothesis of Proposition 5.32 holds. Using (5.4.158) we get (5.4.168).

Next, we consider $D_1 J_1$. From relations (5.4.166) and (5.4.167) it follows that

$$D_1 J_1 = D_1 \hat{J}_1^{-1} J_1 = A_+^{V_1} J_1.$$

According to Proposition 5.27 (see (5.4.124))

$$\Im(A_+^{V_1})_{m_{||}m_{||}} > k^{-7\delta}, \qquad m_{||} \in M_1.$$

Hence we immediately obtain (5.4.169). The proposition is proved.

Using the definition of the operator T_1, and relation (5.4.167), we get:

$$T_1 = D_1 \hat{J}_1^{-1} - A_-^0. \tag{5.4.171}$$

We consider that surface and quasisurface states are absent when $\|T_1(\lambda)^{-1}\| \leq k^{-\delta}$, i.e., when the points $k^2 - |t_{||} + p_{m_{||}}(0)|^2$ lie sufficiently far from the points $\tilde{\lambda}_k$, $k = 1, ..., N$, precisely , if $\rho_3(k^2, t_{||}) > k^{-3\delta/p}$ (see (5.4.133)). By analogy to the one-dimensional case, we introduce a stronger definition of the absence of surface and quasisurface states. We consider the operator

$$P = D_1 - A_-^0 \hat{J}_1. \tag{5.4.172}$$

Note that its matrix is diagonal in $t_{||}$-basis and

$$P(k^2 + i\varepsilon, t_{||})_{m_{||}m_{||}} = p(k^2 + i\varepsilon - |t_{||} + p_{m_{||}}(0)|^2), \tag{5.4.173}$$

where p is the function defined by formula (5.4.89).

DEFINITION. The operator $H_{1+}(t_{||})$ has no surface and quasisurface states with energy k^2 if the inequality $\|P^{-1}(k^2 + i\varepsilon, t_{||})\| < k^{7\delta}$ holds.

Note that the last inequality is equivalent to the next one:

$$\min_{m_{||}} |p(k^2 - |t_{||} + p_{m_{||}}(0)|^2)| > k^{-7\delta}. \tag{5.4.174}$$

It turns out that for this equality to hold an additional condition similar to (5.4.95) must be satisfied, i.e., the points $k^2 - |t_{||} + p_{m_{||}}(0)|^2$ must lie sufficiently far from the points α_i, $i = 1, ..., M$.

Lemma 5.8 . *Suppose $\varepsilon > 0$, $\rho_1(k^2 + i\varepsilon, t_{||}) > k^{-3\delta}$, $\rho_2(k^2 + i\varepsilon, t_{||}) > k^{-3\delta}$, $\rho_3(k^2 + i\varepsilon, t_{||}) > k^{-3\delta/s_0}$, $\rho_4(k^2 + i\varepsilon, t_{||}) > k^{-3\delta/s_0}$. Then the following estimates hold:*

$$\|P^{-1}(k^2 + i\varepsilon, t_{||})\| < k^{8\delta}, \tag{5.4.175}$$

$$|P(k^2 + i\varepsilon, t_{||})_{m_{||}m_{||}}| > k^{-4\delta} |(A_-^0)_{m_{||}m_{||}}|, \quad \text{when } m_{||} \in M_2. \tag{5.4.176}$$

Proof. We prove inequality (5.4.174), which gives estimate (5.4.175). Indeed, let $m_{||} \in M_2$. From the inequality for ρ_2 it follows that the point $\lambda = k^2 - |t_{||} + p_{m_{||}}(0)|^2$ belongs to γ (Proposition 5.31, page 282)). From the inequalities for ρ_3, ρ_4 it follows that the hypothesis of Proposition 5.24 holds with the same λ. Therefore,

$$|p(k^2 - |t_{||} + p_{m_{||}}(0)|^2)| > k^{-3\delta}, \qquad m_{||} \in M_2. \tag{5.4.177}$$

Let $m_{||} \in Z^2 \setminus M_2$. Then taking into account (5.4.166) we obtain $D_1(I - J_2) = A_+^{V_1}(I - J_2)$, $P(I - J_2) = (A_+^{V_1} - A_-^0)(I - J_2)$. Suppose $m_{||} \in M_1$. Considering that $\Im(A_-^0)_{m_{||},m_{||}} = 0$ for such $m_{||}$, we obtain $\Im P(k^2 + i\varepsilon, t_{||})_{m_{||},m_{||}} = \Im(D_1)_{m_{||}m_{||}}$. Using (5.4.169) we arrive at

$$|P(k^2 + i\varepsilon, t_{||})_{m_{||}m_{||}}| > k^{-7\delta}, \quad m_{||} \in M_1. \tag{5.4.178}$$

If $m_{||} \in Z^2 \setminus M_0$, then using that $\Im A_+^{V_1} > 0$, we obtain

$$|P(k^2 + i\varepsilon, t_{||})_{m_{||}m_{||}}| > |\Im(A_-^0)_{m_{||}m_{||}}| > \rho_1 > k^{-3\delta}, m_{||} \in Z^2 \setminus M_0. \tag{5.4.179}$$

From relations (5.4.177) – (5.4.179) we get

$$|P(k^2 + i\varepsilon, t_{||})_{m_{||}m_{||}}| > k^{-7\delta}, m_{||} \in Z^2. \tag{5.4.180}$$

Thus, (5.4.175) holds.

Next, we prove (5.4.176). If $m_{||} \in M_2$, then, according to Proposition 5.32, formula (5.4.157) holds. Therefore,

$$P(k^2 \pm i0, t_{||})_{m_{||}m_{||}} =$$

$$W'(k^2 - |t_{||} + p_{m_{||}}(0)|^2) - \sqrt{|t_{||} + p_{m_{||}}(0)|^2 - k^2} W(k^2 - |t_{||} + p_{m_{||}}(0)|^2).$$

Using (5.4.87) and (5.4.88) we show that

$$3|P(k^2, t_{||})_{m_{||}m_{||}}| > \sqrt{|t_{||} + p_{m_{||}}(0)|^2 - k^2} \tag{5.4.181}$$

for sufficiently large values of $|t_{||} + p_{m_{||}}(0)|^2 - k^2$:

$$|t_{||} + p_{m_{||}}(0)|^2 - k^2 > c(V).$$

If the last relation does not hold, then we use estimate (5.4.177). We rewrite it in the form:

$$2|P(k^2, t_{||})_{m_{||}m_{||}}| > k^{-4\delta}\sqrt{|t_{||} + p_{m_{||}}(0)|^2 - k^2}. \tag{5.4.182}$$

Estimates (5.4.181) and (5.4.182) together give (5.4.176).

The proposition is proved.

5.4.3 The general case $H_+ = -\Delta + V_+$.

Let us define the functions $U_{m_{||}}$, $m_{||} \in Z^2$. Considering that $Z^2 = (Z^2 \setminus M_0) \cup M_1 \cup M_2$, we define $U_{m_{||}}$ for each of these sets separately.

If $m_{||} \notin M_0$, then $U_{m_{||}}(k^2 + i\varepsilon, t_{||}, x)$, $\varepsilon \neq 0$, is uniquely defined by the equation

$$H_+U_{m_{||}} = (k^2 + i\varepsilon)U_{m_{||}}, \quad x_1 > 0, \tag{5.4.183}$$

by the quasiperiodic conditions (5.4.4) and the conditions

$$U_{m_{||}} \in L_2(Q_+), \qquad \left(U_{m_{||}}|_{x_1=+0}\right)_{j_{||}} = \delta_{m_{||}j_{||}}. \tag{5.4.184}$$

The following lemma is to define $U_{m_{||}}$ when $m_{||} \in M_1$.

In section 5.5 we show that there is a set $\Omega(k,\delta)$ of asymptotically full measure on $K_{||}$, such that the following lemma holds.

Lemma 5.9 . *Suppose $t_{||}$ belongs to $(k^{-2-2\delta})$-neighborhood of the set $\Omega(k,\delta)$, and $\rho_2(k^2,t_{||}) > k^{-8\delta/s_0}$, $\varepsilon \neq 0$. Then for any $m_{||} \in M_1$, there exists a function $U_{m_{||}}(k^2+i\varepsilon,t_{||},x)$ satisfying equation (5.4.183), quasiperiodic conditions (5.4.4), and the following asymptotic formula:*

$$U_{m_{||}}(k^2 + i\varepsilon, t_{||}, x) =_{k\to\infty} U^{(1)}_{m_{||}}(k^2 + i\varepsilon, t_{||}, x) + o(k^{-4/5+10\delta}), \tag{5.4.185}$$

here $o(k^{-4/5+10\delta})$ is infinitely small in the class $L_2(K_{||})$.

This asymptotic formula is differentiable with respect to x_1:

$$\frac{\partial U_{m_{||}}(k^2 + i\varepsilon, t_{||}, x)}{\partial x_1} =_{k\to\infty} \frac{\partial U^{(1)}_{m_{||}}(k^2 + i\varepsilon, t_{||}, x)}{\partial x_1} + o(k^{-4/5+20\delta}). \tag{5.4.186}$$

The asymptotic formulae (5.4.185) and (5.4.186) are uniform in x_1 on any finite interval of x_1. The function $U_{m_{||}}(k^2+i\varepsilon,t_{||},x)$ analytically depends on $k^2+i\varepsilon$ in both upper and lower semineighborhoods of the point k^2, and decays exponentially as $x_1 \to \infty$ for $\varepsilon \neq 0$.

The proof of the lemma is given in Section 5.5.

In the case $m_{||} \in M_2$, we define $U_{m_{||}}$ by the following formula:

$$U_{m_{||}}(k^2+i\varepsilon,t_{||},x) = a_{m_{||}} \sum_{n\in Z^3} \oint_{C_4} (H(t)-k^2-i\varepsilon)^{-1}_{nm} \exp i(t+\tilde{n}_1,x)dt_1, \tag{5.4.187}$$

where $m = (0, m_{||})$. We introduce the operators J, D in $L_2(Q_{||})$ by the formulae:

$$J_{n_{||}m_{||}} = (U_{m_{||}}(0,x_{||}))_{n_{||}}, \tag{5.4.188}$$

$$D_{n_{||}m_{||}} = \left(\frac{\partial U_{m_{||}}}{\partial x_1}(0,x_{||})\right)_{n_{||}}. \tag{5.4.189}$$

By the definitions of the function $U_{m_{||}}$ and operator J_0 we have

$$(I - J_0)J = (I - J_0)J^{-1} = I - J_0. \tag{5.4.190}$$

Proposition 5.34 . *If $t_{||} \in \Omega(k,\delta)$, then the number of points in M_1 does not exceed $4ck^{1/5+10\delta}$.*

The proof is given in Section 5.5.

Lemma 5.10 . *Suppose* $\rho_2(k,\delta) > k^{-3\delta}$ *and* w *is a linear combination of the functions* $U_{m_\|}(k^2+i\varepsilon, t_\|, x)|_{x_1=0}$, $m_\| \in M_1$. *Then,*

$$\Im(A_+^V w, w) \geq k^{-8\delta}\|J_1 w\|^2. \qquad (5.4.191)$$

Proof. We consider the function $U_{m_\|}(k^2-i\varepsilon, t_\|, x)|_{x_1=0}$, $m_\| \in M_1$. Using the definitions of A_+^V, $A_+^{V_1}$ we rewrite (5.4.186) in the form :

$$A_+^V U_{m_\|}(k^2+i\varepsilon, t_\|, x)\Big|_{x_1=0} = A_+^{V_1} U_{m_\|}^{(1)}(k^2+i\varepsilon, t_\|, x)\Big|_{x_1=0} + O(k^{-4/5+20\delta}). \qquad (5.4.192)$$

Hence, using formula (5.4.161) and estimates (5.4.124) and (5.4.125), it is not hard to show that $|A_+^{V_1}U_{m_\|}^1| < k^{9\delta}$. Using the last inequality and (5.4.185), we obtain

$$\Im(A_+^V U_{m_\|} U_{\hat{m}_\|}) = \Im(A_+^{V_1}U_{m_\|}^{(1)}, U_{\hat{m}_\|}^{(1)}) + O(k^{-4/5+20\delta}), \quad m_\|, \hat{m}_\| \in M_1. \quad (5.4.193)$$

Let w and w_1 be linear combinations of $U_{m_\|}$ and $U_{m_\|}^{(1)}$ with the same coefficients:

$$w = \sum_{m_\| \in M_1} c_{m_\|} U_{m_\|}\Bigg|_{x_1=+0},$$

$$w_1 = \sum_{m_\| \in M_1} c_{m_\|} U_{m_\|}^{(1)}\Bigg|_{x_1=+0}.$$

From relation (5.4.193) it follows

$$\Im(A_+^V w, w) = \Im(A_+^{V_1} w_1, w_1) + \varphi(k),$$

where

$$|\varphi(k)| < C(V)k^{-4/5+20\delta}\left(\sum_{m_\| \in M_1} |c_{m_\|}|\right)^2.$$

Applying Proposition 5.34, we get

$$|\varphi(k)| < C(V)k^{-3/5+30\delta} \sum_{m_\| \in M_1} |c_{m_\|}|^2. \qquad (5.4.194)$$

Using relation (5.4.123), we obtain

$$\Im(A_+^V w, w) \geq k^{-7\delta}\|J_1 w_1\|^2 - |\varphi(k)|.$$

From the definition of $U_{m_\|}^{(1)}$ it immediately follows that

$$\|w_1\|^2 = \|J_1 w_1\|^2 = \sum_{m_\| \in M_1} |c_{m_\|}|^2. \qquad (5.4.195)$$

Therefore, considering (5.4.194), we get:

$$2\Im(A_+^V w, w) \geq k^{-7\delta}\|J_1 w_1\|^2. \tag{5.4.196}$$

Using (5.4.185), Proposition 5.34 and (5.4.195), it is easy to show that

$$\|w - w_1\| \leq c(V)k^{-3/5+20\delta}\|w_1\|.$$

Taking into account that $J_1 w_1 = w_1$, we get

$$\|J_1 w - J_1 w_1\| \leq c(V)k^{-3/5+20\delta}\|J_1 w_1\|.$$

Hence, $2\|J_1 w_1\| \geq \|J_1 w\|$. Substituting the last estimate in the right-hand side of (5.4.196), we obtain (5.4.191). The lemma is proved.

Next, we consider $\rho_5(k^2, t_{||})$:

$$\rho_5(k^2, t_{||}) = \min_{n_{||}, m_{||} \in Z^2; n_{||} \neq m_{||}, |n_{||} - m_{||}| < k^{\delta}} \left|k^2 - |t_{||} + p_{m_{||}}(0)|^2\right| \left|k^2 - |t_{||} + p_{n_{||}}(0)|^2\right|. \tag{5.4.197}$$

In Section 5.5 we prove that there exists a set $\Omega_5(k, \delta) \subset K_{||}$, such that

$$\rho_5(k^2, t_{||}) > k^{1-9\delta}, \tag{5.4.198}$$

when $t_{||} \in \Omega_5(k, \delta)$. The set $\Omega_5(k, \delta)$ has an asymptotically full measure in $K_{||}$:

$$\frac{s\left(\Omega_5(k, \delta)\right)}{s(K_{||})} \to_{k \to \infty} 1. \tag{5.4.199}$$

It follows from this result (see Section 5.5) that there exists a subset $\tilde{\Omega}_5(k, \delta)$ of S_k, such that

$$\rho_5(k^2, k_{||}) > k^{1-9\delta} \tag{5.4.200}$$

when $\mathbf{k} \in \tilde{\Omega}_5(k, \delta)$. The set $\tilde{\Omega}_5(k, \delta)$ has an asymptotically full measure on S_k:

$$\frac{s\left(\tilde{\Omega}_5(k, \delta)\right)}{s(S_k)} \to_{k \to \infty} 1. \tag{5.4.201}$$

We introduce the notations:

$$\Delta J = (J - \hat{J}_1)J_2, \qquad \Delta D = (D - D_1)J_2, \tag{5.4.202}$$

J, D, $\hat{J}_1$,D_1 and J_2 being given by (5.4.188), (5.4.189), (5.4.162), (5.4.163) and (5.4.160).

Lemma 5.11 . *Suppose $t_1 \in C_4$, and $t_{||}$ satisfies the estimates: $\rho_1(k^2, t_{||}) > k^{-2\delta}$, $\rho_2(k^2, t_{||}) > k^{-2\delta}$, $\rho_5(k^2, t_{||}) > k^{1-9\delta}$. Then*

$$\Delta J(k^2 + i0) = \Delta J(k^2 - i0) \equiv \Delta J(k^2), \tag{5.4.203}$$

$$\Delta D(k^2+i0)=\Delta D(k^2-i0)\equiv \Delta D(k^2), \tag{5.4.204}$$

and the following estimates hold:

$$\|\Delta J\|^* < k^{-1/4+120\delta}, \tag{5.4.205}$$

$$\|\Delta D\|^* < k^{70\delta}, \tag{5.4.206}$$

$$\|A_+^0\Delta J\|^* < k^{60\delta}, \tag{5.4.207}$$

$$\|(A_+^0)^{-1}\Delta D\|^* < k^{-1/4+120\delta}. \tag{5.4.208}$$

Proof. We represent $(H(t)-k^2)^{-1}$ in the form:

$$(H(t)-k^2)^{-1}=$$

$$(H_1(t)-k^2)^{-1}+B+\sum_{k=1}^{\infty}(BV_2)^k\left(I+(H_1(t)-k^2)^{-1}V_2\right)(H_1(t)-k^2)^{-1}, \tag{5.4.209}$$

$$B=(H_1(t)-k^2)^{-1}V_2(H_1(t)-k^2)^{-1}, \qquad V_2=V-V_1.$$

From the definition of V_1, it easily follows that $(V_2)_{nm}=0$, if $n_{||}-m_{||}=0$. We prove that

$$\|B\|^* < \left(k^{1/2-20\delta}+(\Im t_1)^2\right)^{-1}k^{69\delta}, \tag{5.4.210}$$

when $t_1\in C_4$. From relation (5.4.150) and the obvious relation $\|H_1-H_0\|\le \|V\|^*$, we get

$$\|(H_1(t)-k^2)^{-1}(H_0(t)-k^2)\|^* < ck^{25\delta}. \tag{5.4.211}$$

Therefore, it suffices to check the estimate:

$$\|B_0\|^* < \left(k^{1/2-20\delta}+(\Im t_1)^2\right)^{-1}k^{19\delta}, \tag{5.4.212}$$

where

$$B_0=(H_0(t)-k^2)^{-1}V_2(H_0(t)-k^2)^{-1}.$$

Suppose $|\Im t_1|\ge k^{1/4-10\delta}$. Using estimate (5.4.142) we obtain

$$\|B_0\|^* < ck^{16\delta}(\Im t_1+k^{-\delta})^{-2} < c(V)k^{16\delta}(k^{1/2-20\delta}+(\Im t_1)^2)^{-1}.$$

Next, we consider the case $|\Im t_1|\le k^{1/4-10\delta}$. Since $(B_0)_{nm}=0$, when $|n-m|>R_0$, we have

$$\|B_0\|^* < c(R_0)\max_{n,m}|(B_0)_{nm}|. \tag{5.4.213}$$

We estimate $|(B_0)_{nm}|$. Suppose $|n_1|\ge k^{1/4-10\delta}$, $n_1\in Z$. Since $|n_1-m_1|<R_0$, we have $2|m_1|\ge k^{1/4-10\delta}$. Taking into account that $\Im t_1\ge \tau_0=k^{-8\delta}$, we get:

$$|(\tilde n_1+t_1)^2+|t_{||}+p_{n_{||}}(0)|^2-k^2|\ge|\Im(t_1+\tilde n_1)^2|=2|n_1|\Im t_1\ge k^{1/4-18\delta},$$

where $\tilde n_1=2\pi n_1/a_1$, $\tilde m_1=2\pi m_1/a_1$. A similar inequality holds for m_1. Therefore,

$$|(B_0)_{nm}| < ck^{-1/2+36\delta} < 2(k^{1/2-20\delta} + (\Im t)^2)^{-1}k^{16\delta}.$$

It remains to consider the case $|\Im t_1| \leq k^{1/4-10\delta}$, $|n_1| < k^{1/4-10\delta}$. Since $|n_1 - m_1| < R_0$, we have $|m_1| < 2k^{1/4-10\delta}$. We assume for the definiteness that

$$\left|(\tilde{n}_1 + t_1)^2 + |t_{||} + p_{n_{||}}(0)|^2 - k^2\right| \leq \left|(\tilde{m}_1 + t_1)^2 + |t_{||} + p_{m_{||}}(0)|^2 - k^2\right|. \tag{5.4.214}$$

First, suppose that

$$|(\tilde{n}_1 + t_1)^2 + |t_{||} + p_{n_{||}}(0)|^2 - k^2| \geq k^{1/2-\delta}.$$

From (5.4.214), it follows that a similar inequality holds for m_1. Therefore,

$$|(B_0)_{nm}| < ck^{-1+2\delta} < (k^{1/2-20\delta} + (\Im t_1)^2)^{-1}.$$

If, at last,

$$|(\tilde{n}_1 + t_1)^2 + |t_{||} + p_{n_{||}}(0)|^2 - k^2| < k^{1/2-\delta},$$

then

$$||t_{||} + p_{n_{||}}(0)|^2 - k^2| \leq k^{1/2-\delta}, \tag{5.4.215}$$

because $|(\tilde{n}_1 + t_1)|^2 < k^{1/2-20\delta}$. Taking into account that $\rho_5(k, t_{||}) > k^{1-9\delta}$ (see (5.4.197)), we obtain

$$||t_{||} + p_{m_{||}}(0)|^2 - k^2| > k^{1/2-8\delta}.$$

Since $|(\tilde{m}_1 + t_1)|^2 < k^{1/2-20\delta}$, we have

$$2\left|(m_1 + t_1)^2 + |t_{||} + p_{m_{||}}(0)|^2 - k^2\right| > k^{1/2-8\delta}.$$

From relation (5.4.143), it follows

$$|(\tilde{n}_1 + t_1)^2 + |t_{||} + p_{n_{||}}(0)|^2 - k^2| > k^{-9\delta}.$$

The last two inequalities together give:

$$|(B_0)_{nm}| < ck^{-1/2+17\delta} < k^{9\delta}(k^{-1/2-20\delta} + (\Im t_1)^2)^{-1}. \tag{5.4.216}$$

Relation (5.4.216) and, therefore (5.4.212) are proved in the case $|\Im t_1| \leq k^{1/4-10\delta}$, too. From the definitions of the operators ΔJ, J, $\hat{J}_1$ (see formulae (5.4.202)), (5.4.160), (5.4.188) and (5.4.162)) we get that

$$(\Delta J)_{n_{||}m_{||}} = \left((U_{m_{||}} - U^{(1)}_{m_{||}})|_{x_1=0}\right)_{n_{||}}, \tag{5.4.217}$$

when $m_{||} \in M_2$, and

$$(\Delta J)_{n_{||}m_{||}} = 0, \tag{5.4.218}$$

when $m_{||} \notin M_2$.

Using formulae (5.4.138) and (5.4.187) for $U^{(1)}_{m_{||}}$ and $U_{m_{||}}$, we obtain

$$\Delta J = \oint_{C_4} \left((H(t)-k^2)^{-1} - (H_1(t)-k^2)^{-1}\right) M\,dt_1. \tag{5.4.219}$$

Using (5.4.209), (5.4.211) and (5.4.143) it is not hard to show

$$\|\Delta J\| < I_1 + I_2,$$

$$I_1 = \left\| \oint_{C_4} (H_1(t)-k^2)^{-1} V_2 (H_1(t)-k^2)^{-1} M\,dt_1 \right\|^*,$$

$$I_2 = k^{35\delta} \sum_{r=1}^{\infty} \|V_2\|^r \oint_{C_4} \|B\|^{*r} \|(H_0(t)-k^2)^{-1} M\|^* dt_1. \tag{5.4.220}$$

Let us estimate I_1. It is easy to see that $I_1 = I_1' + I_1'' + I_1'''$, where

$$I_1' = \left\| \oint_{C_4} \left((H_1(t)-k^2)^{-1} - (H_0(t)-k^2)^{-1}\right) V_2 (H_0(t)-k^2)^{-1} M\,dt_1 \right\|,$$

$$I_1'' = \left\| \oint_{C_4} (H_0(t)-k^2)^{-1} V_2 \left((H_1(t)-k^2)^{-1} - (H_0(t)-k^2)^{-1}\right) M\,dt_1 \right\|,$$

$$I_1''' = \left\| \oint_{C_4} (H_0(t)-k^2)^{-1} V_2 (H_0(t)-k^2)^{-1} M\,dt_1 \right\|^*.$$

It is easy to show, using (5.4.211), that

$$I_1' < k^{25\delta} \left\| \oint_{C_4} \|B\| \|V_2\| \|(H_0(t)-k^2)^{-1} M\| dt_1 \right\|^*.$$

Taking into account (5.4.144), (5.4.143) and (5.4.210), we get $I_1' < k^{-1/4+105\delta}$. Similarly, $I_1'' < k^{-1/4+105\delta}$.

Let us estimate I_1'''. Considering that V is a trigonometric polynomial, we obtain that $F'_{nm} = 0$, when $|n-m| > R_0$. Therefore,

$$\|I_1'''\|^* \le c(R_0) \max_{mn} |(I_1''')_{mn}|. \tag{5.4.221}$$

In the formula for I_1''', the integrand has poles at the points:

$$\tau_1 = a_{m_\parallel} = \sqrt{k^2 - |t_\parallel + p_{m_\parallel}(0)|^2}, \quad \tau_1 = a_{n_\parallel} = \sqrt{k^2 - |t_\parallel + p_{n_\parallel}(0)|^2}.$$

So, the integral can be replaced by the integral over a finite contour C_5 around these two points. We can choose this contour to be in a distance of order 1 from the points. This means that,

$$\min_{t_1 \in C_3} \left\{ \left|t_1 - a_{m_\parallel}\right|, \left|t_1 - a_{n_\parallel}\right| \right\} \approx 1.$$

If $|a_{m_\parallel} - a_{n_\parallel}| \gg 1$, then the contour contains two components. It is not hard to show that

$$(H_0(t)-k^2)_{m_\parallel m_\parallel} > a_{m_\parallel}, \tag{5.4.222}$$

$$(H_0(t)-k^2)_{n_\| n_\|} > a_{n_\|}, \tag{5.4.223}$$

when $t=(\tau_1,t_\|)$, $\tau_1 \in C_5$. From these inequalities it follows:

$$\left((H_0(t)-k^2)^{-1}M\right)_{m_\| m_\|} < 1. \tag{5.4.224}$$

Suppose $a_{n_\|} > k^{1/2-18\delta}$. Then, using (5.4.224), we get

$$|(I_1''')_{nm}| < ck^{-1/2+18\delta}. \tag{5.4.225}$$

If $a_{n_\|} < k^{1/2-18\delta}$, then $a_{m_\|} > k^{1/2}$, because otherwise $\rho_5(k,\delta) < k^{1-9\delta}$. In this case, the contour C_5 consists of two parts separated by the distance r of order $k^{1/2}$. Let us consider the part where $|t_1 - a_{m_\|}| \approx 1$. Then, $|(H_0(t)-k^2)_{n_\| n_\|}| > k$. Taking into account (5.4.224), we get

$$\left|\left((H_0(t)-k^2)^{-1}V_2(H_0(t)-k^2)^{-1}M\right)_{nm}\right| < ck^{-1}. \tag{5.4.226}$$

Let us consider the part where $|t_1 - a_{n_\|}| \approx 1$. In this case

$$\left|(H_0(t)-k^2)^{-1}_{n_\| n_\|}\right| < 1,$$

$$\left|\left((H_0(t)-k^2)^{-1}M\right)_{m_\| m_\|}\right| < \frac{a_{m_\|}}{a_{m_\|}^2 - a_{n_\|}^2} < \frac{2}{a_{m_\|}} < 2k^{1/2}.$$

We have obtained (5.4.225) for all n and m. Thus, $I_1 < k^{-1/4+110\delta}$. Using (5.4.210) and (5.4.144), we prove the same estimate for I_2. Thus, (5.4.205) is proved.

The proof of (5.4.208) is similar to that of (5.4.205), but in the proof of (5.4.208) we use the formula

$$(A_+^0)^{-1}\Delta D = \oint_{C_4} \left((H(t)-k^2)^{-1} - (H_1(t)-k^2)^{-1}\right) L\,dt_1$$

and the estimate (5.4.145) instead of (5.4.144). We obtain estimate (5.4.206) by similar considerations using the formula

$$\Delta D = \oint_{C_4} L\left((H(t)-k^2)^{-1} - (H_1(t)-k^2)^{-1}\right) M\,dt_1.$$

To prove (5.4.207) we use the formula

$$A_+^0 \Delta J = \oint_{C_4} M\left((H(t)-k^2)^{-1} - (H_1(t)-k^2)^{-1}\right) M\,dt_1,$$

(5.4.144) and (5.4.143). We see that all formulae are stable with respect to a small perturbation of k^2. Taking into account formulae (5.4.164), (5.4.165), we get (5.4.203) and (5.4.204). The lemma is proved.

Lemma 5.12 . *Under conditions of the previous lemma, the following inequality holds:*

$$\|J - \hat{J}_1\| < k^{-1/4+120\delta}. \tag{5.4.227}$$

Proof. From the definitions of J, $\hat{J}_1$, J_0, $U^{(1)}_{m_{\|}}$ and $U_{m_{\|}}$ (see (5.4.188), (5.4.162), (5.4.29), (5.4.161) and (5.4.183)), it follows that

$$(J - \hat{J}_1)(I - J_0) = 0.$$

The definitions of ΔJ (5.4.202) and the previous lemma give us the inequality

$$\|(J - \hat{J}_1)J_2\| < k^{-1/4+120\delta}.$$

Using (5.4.185) and Proposition 5.34, it is not hard to show that

$$\|(J - \hat{J}_1)J_1\| < ck^{-3/5+20\delta}.$$

The last three inequalities give us (5.4.227). The lemma is proved.

Lemma 5.13 . *Suppose $t_{\|}$ is such that $\rho_1(k^2, t_{\|}) > k^{-2\delta}$, $\rho_2(k^2, t_{\|}) > k^{-2\delta}$, $\rho_3(k^2, t_{\|}) > k^{-3\delta/s_0}$, $\rho_4(k^2, t_{\|}) > k^{-3\delta/s_0}$, $\rho_5(k^2, t_{\|}) > k^{1-9\delta}$, and $t_{\|} \in \Omega(k, \delta)$. Then a solution x of the equation*

$$(A^V_+ - A^0_-)(k^2 + i\varepsilon, t_{\|})x = f, \quad f \in l^2_2 \tag{5.4.228}$$

satisfies the uniform in ε estimate:

$$\|(I - J_2)x\| + \|A^0_- J_2 x\| \leq k^{200\delta}\|f\|. \tag{5.4.229}$$

The proof is similar to that of Lemma 5.6 (the two-dimensional case). Indeed, for any solution of the equation (5.4.228), we have

$$\Im\left((A^V_+ - A^0_-)x, x\right) = \Im(f, x).$$

Since the operator $(A^V_+(k^2 + i\varepsilon) - A^V_+(k^2 - i\varepsilon))/2i$ is positively determined, and $(A^0_-(k^2 + i\varepsilon) - A^0_-(k^2 - i\varepsilon))/2i$ is negatively determined (Proposition 5.9, 247), we obtain that for sufficiently small ε the following inequality holds:

$$|\Im(A^0_-(k^2 + i\varepsilon)x, x)| \leq \|f\|\|x\|.$$

Now, considering inequality (5.4.111), we see that

$$k^{-\delta}\|(I - J_0)x\|^2 \leq \|f\|\|x\|. \tag{5.4.230}$$

Similarly,

$$|\Im(A^V_+(k^2 + i\varepsilon)x, x)| \leq \|f\|\|x\|.$$

Using relation (5.4.123), we obtain

$$k^{-7\delta}\|J_1 x\|^2 \leq \|f\|\|x\|. \tag{5.4.231}$$

We add inequalities (5.4.230) and (5.4.231). Considering that $I - J_0$ and J_1 are orthogonal projections, we get:

$$k^{-7\delta}\|(I - J_0 - J_1)x\|^2 \le 2\|f\|\|x\|. \tag{5.4.232}$$

Furthermore, multiplying both sides of the equation (5.4.228) by $U_{m_\parallel}(k^2 - i\varepsilon, t_\parallel)\big|_{x_1=0}$, $m_\parallel \in M_2$, we obtain

$$(x, A_+^V(k^2 - i\varepsilon)U_{m_\parallel}) - (A_-^0(k^2 + i\varepsilon)x, U_{m_\parallel}) = (f, U_{m_\parallel}). \tag{5.4.233}$$

Taking into account that

$$(A_+^V(k^2 - i\varepsilon)U_{m_\parallel})_{n_\parallel} = D_{n_\parallel m_\parallel}(k^2 - i\varepsilon), \quad m_\parallel \in M_2, \tag{5.4.234}$$

we obtain

$$(x, A_+^V(k^2 - i\varepsilon)U_{m_\parallel}) = \sum_{n_\parallel} x_{n_\parallel}\overline{D_{n_\parallel m_\parallel}} = (D^*(k^2 - i\varepsilon)x)_{m_\parallel}. \tag{5.4.235}$$

Similarly, using the definition of J, we get

$$(A_-^0(k^2 + i\varepsilon)x, U_{m_\parallel}) = (J^*(k^2 - i\varepsilon)A_-^0 x)_{m_\parallel}. \tag{5.4.236}$$

Considering relations (5.4.235) and (5.4.236), we rewrite (5.4.233) in the form:

$$J_2(D^* - J^*A_-^0)x = J_2J^*f.$$

Taking into account notations (5.4.202), we get

$$J_2(D_1^* - \hat{J}_1^*A_-^0) + K_1J_2A_-^0x + K_2(I - J_2)x = J_2f + \Delta J^*f, \tag{5.4.237}$$

where $K_1 = \Delta\hat{J}^* + \Delta D^*(A_-^0)^{-1}$, $K_2 = \Delta J^*A_-^0 + \Delta D^*$. From estimates (5.4.205) - (5.4.208), it follows

$$\|K_1\| < 2k^{-1/4+120\delta}, \qquad \|K_2\| < 2k^{70\delta}. \tag{5.4.238}$$

Using (5.4.176), we get

$$\|J_2(D_1^* - J_1^*A_-^0)x\| \ge k^{-4\delta}\|J_2A_-^0x\|. \tag{5.4.239}$$

From (5.4.237), using inequalities (5.4.238) and (5.4.239), we obtain

$$k^{-3\delta}\|J_2A_-^0x\| < 2\|f\| + 2k^{70\delta}\|(I - J_2)x\|.$$

From the last inequality it follows

$$\|J_2A_-^0x\|^{1/2} < k^{2\delta}\|f\|^{1/2} + k^{40\delta}\|(I - J_2)x\|^{1/2}. \tag{5.4.240}$$

Using that $J_0 - J_1 = J_2$, we transform inequality (5.4.232) to the form

$$k^{-4\delta}\|(I - J_2)x\| \le \|f\|^{1/2}\|(I - J_2)x\|^{1/2} + \|f\|^{1/2}\|J_2x\|^{1/2}. \tag{5.4.241}$$

Taking into account that $\|(A^0_-)^{-1}\| \le \rho_1^{-1/2} < k^{\delta}$, we obtain

$$k^{-4\delta}\|(I-J_2)x\| \le \|f\|^{1/2}\|(I-J_2)x\|^{1/2} + k^{\delta}\|f\|^{1/2}\|J_2A^0_- x\|^{1/2}. \quad (5.4.242)$$

Using estimate (5.4.240) in the last inequality, we obtain

$$k^{-4\delta}\|(I-J_2)x\| \le \|f\|^{1/2}\|(I-J_2)x\|^{1/2} + k^{3\delta}\|f\|^{1/2}(\|f\|^{1/2} + k^{40\delta}\|(I-J_2)x\|^{1/2}). \quad (5.4.243)$$

Considering the last inequality as quadratic with respect to $\|(I-J_2)x\|^{1/2}$, we easily show that

$$\|(I-J_2)x\| < k^{50\delta}\|f\|. \quad (5.4.244)$$

From relation (5.4.240) it follows

$$\|A^0_- J_2 x\| < k^{190\delta}\|f\|. \quad (5.4.245)$$

Relations (5.4.244) and (5.4.245) together give (5.4.229). The lemma is proved.

Lemma 5.14 . *Suppose $t_{\|}$ is such that $\rho_1(k^2, t_{\|}) > k^{-2\delta}$, $\rho_2(k^2, t_{\|}) > k^{-2\delta}$, $\rho_3(k^2, t_{\|}) > k^{-3\delta/s_0}$, $\rho_4(k^2, t_{\|}) > k^{-3\delta/s_0}$, $\rho_5(k^2, t_{\|}) > k^{1-9\delta}$, and $t_{\|} \in \Omega(k,\delta)$. Then there exists the limit of the operator $(A^V_+ - A^0_-)(k^2 + i\varepsilon, t_{\|})$ as $\varepsilon \to 0$ in the class $\mathbf{S}_2$. Moreover, the estimates*

$$\|T^{-1}(k^2 + i\varepsilon, t_{\|})\| < k^{200\delta} \quad (5.4.246)$$

$$\|A^0_- T^{-1}(k^2 + i\varepsilon, t_{\|})\| < k^{1+200\delta}. \quad (5.4.247)$$

hold.

Proof of the lemma is quite similar to that of Lemma 5.7 (page 258) for the two-dimensional case and also the obvious inequality $\|A^0_-(I-J_2)\| < k$. Here we use inequality (5.4.229) instead of (5.3.60) for the two-dimensional case. The lemma is proved.

To construct reflected and refracted waves we shall use the following modification of the previous lemma.

Lemma 5.15 . *Suppose $\mathbf{k} \in S_k$ is such that $\rho_1(k^2, k_{\|}) > k^{-2\delta}$, $\rho_2(k^2, k_{\|}) > k^{-2\delta}$, $\rho_3(k^2, k_{\|}) > k^{-3\delta/s_0}$, $\rho_4(k^2, k_{\|}) > k^{-3\delta/s_0}$, $\rho_5(k^2, k_{\|}) > k^{1-9\delta}$, and $\mathcal{K}_{\|}\mathbf{k} \in \Omega(k,\delta)$. Then there exists the limit of the operator $(A^V_+ - A^0_-)(k^2 + i\varepsilon, k_{\|})$ as $\varepsilon \to 0$ in the class $\mathbf{S}_2$. Moreover, the estimate*

$$\|T^{-1}(k^2 + i\varepsilon, k_{\|})\| < k^{200\delta} \quad (5.4.248)$$

holds.

Proof. We consider $t_{\|} = \mathcal{K}_{\|}\mathbf{k}$. Note that $H_+(k_{\|}) = H_+(t_{\|})$ and $\rho_i(k^2, k_{\|}) = \rho_i(k^2, t_{\|})$, $i = 1,2,3,4,5$. Therefore, we are in the conditions of the previous lemma for such $t_{\|}$. It is not hard to show that the coordinates of a function f in $k_{\|}$-basis and $t_{\|}$-basis coincide up to the shift of indices by $m_{\|} = k_{\|} - t_{\|}$. Therefore, the matrices of the operators $A^V_+(k^2 + i\varepsilon, k_{\|})$ and $A^V_+(k^2 + i\varepsilon, t_{\|})$ also coincide up to the same shift of indices. Operators $T^{-1}(k^2 + i\varepsilon, k_{\|})$ and $T^{-1}(k^2 + i\varepsilon, t_{\|})$ are also so simply linked. Therefore, from estimate (5.4.246) it follows (5.4.248). The lemma is proved.

5.5 Geometric Constructions.

5.5.1 Geometric Construction for the Three-Dimensional Case.

We found in Section 5.4 that the operator $H(t_{\|})$ has no surface and quasisurface states corresponding to eigenvalue k^2, if $\rho_1(k^2,t_{\|}) > k^{-2\delta}$, $\rho_2(k^2,t_{\|}) > k^{-2\delta}$, $\rho_3(k^2,t_{\|}) > k^{-3\delta/s_0}$, $\rho_4(k^2,t_{\|}) > k^{-3\delta/s_0}$, $\rho_5(k^2,t_{\|}) > k^{1-9\delta}$ and $t_{\|} \in \Omega(k,\delta)$. [12]

Lemma 5.16 . *For any positive δ and sufficiently large k, $k > k_0(\delta, a_2, a_3)$, there exists a set $\Omega_1 \subset K_{\|}$, such that $\rho_1(k,t_{\|}) > k^{-\delta}$, when $t_{\|}$ belongs to the $(k^{-1-2\delta})$-neighborhood of $\Omega_1(k,\delta)$. The set $\Omega_1(k,\delta)$ has an asymptotically full measure on $K_{\|}$. Moreover,*

$$s(K_{\|} \setminus \Omega_1) = O(k^{-\delta}). \tag{5.5.1}$$

Proof. We consider $x \in R^2$ such that

$$|k^2 - |x|^2| < 2k^{-\delta}. \tag{5.5.2}$$

This inequality is satisfied if x belongs to Q_2:

$$Q_2 = \{x : |k - |x|| < k^{-\delta-1}\}.$$

We denote by $\mathcal{K}_{\|}$ the mapping $\mathcal{K}_{\|} : R^n \to K_{\|}$

$$\mathcal{K}_{\|}\mathbf{p}_m(t) = t_{\|} \tag{5.5.3}$$

(in the present case n=3). We define Ω_1 by the formula:

$$\Omega_1 = K_{\|} \setminus \mathcal{K}_{\|}Q_2. \tag{5.5.4}$$

Now we prove that $\rho_1(k,t_{\|}) > 2k^{-\delta}$, when $t_{\|}$ belongs to $\Omega_1(k,\delta)$. Suppose this is not so. Then, there exists $m_{\|} \in Z^2$ such that

$$|k^2 - |t_{\|} + p_{m_{\|}}(0)|^2| < 2k^{-\delta}, \quad p_{m_{\|}}(0) = \left(\frac{2\pi m_2}{a_2}, \frac{2\pi m_3}{a_3}\right). \tag{5.5.5}$$

This means that $t_{\|} + p_{m_{\|}}(0) \in Q_2$, i.e., $t_{\|} \in \mathcal{K}_{\|}Q_2$. The last relation contradicts the hypothesis $t_{\|} \in \Omega_1$. Therefore, $\rho_1(k,t_{\|}) > 2k^{-\delta}$. It is easy to see this estimate is stable with respect to a perturbation of order $k^{-1-2\delta}$. Hence, $\rho_1(k,t_{\|}) > k^{-\delta}$, when $t_{\|}$ belongs to the $(k^{-1-2\delta})$-neighborhood of $\Omega_1(k,\delta)$.

Next, we prove that Ω_1 has an asymptotically full measure on $K_{\|}$. It is clear that $s(\mathcal{K}_{\|}Q_2) \leq s(Q_2)$. Hence,

$$s(\Omega_1) \geq s(K_{\|}) - s(Q_2). \tag{5.5.6}$$

Since $s(Q_2) \leq 2\pi k^{-\delta}$, relation (5.5.1) holds. The lemma is proved.

---

[12] The functions ρ_i are defined by formulae (5.4.106), (5.4.116) – (5.4.118), (5.4.134), (5.4.152), (5.4.197), the set $\Omega(k,\delta)$ is such that Lemma 5.9 holds. It will be defined by formula (5.5.18).

Lemma 5.17 . *For any positive δ and sufficiently large k, $k > k_0(\delta, a_2, a_3)$, there exists a set $\Omega_2 \subset K_{\|}$, such that $\rho_2(k, t_{\|}) > k^{-2\delta}$, when $t_{\|}$ belongs to the $(k^{-1-2\delta})$-neighborhood of $\Omega_2(k,\delta)$. The set $\Omega_2(k,\delta)$ has an asymptotically full measure on $K_{\|}$. Moreover,*

$$s(K_{\|} \setminus \Omega_2) = O(k^{-\delta}). \tag{5.5.7}$$

Proof. Let $\tilde{\Omega}_2^{(i)}$ be the subset of $K_{\|}$ defined by the inequality

$$\min_{m_{\|}} |k^2 - \lambda_i(0) - |t_{\|} + p_{m_{\|}}(0)|^2| > 4k^{-2\delta}.$$

Arguing as in the proof of the previous lemma (up to replacement of k^2 with $k^2 + \lambda_i(0)$), we show that $\tilde{\Omega}_2^{(i)}$ is given by the formula:

$$\tilde{\Omega}_2^{(i)} = K_{\|} \setminus \mathcal{K}_{\|}\tilde{Q}_2^{(i)},$$

$$\tilde{Q}_2^{(i)} = \{x : \left||x| - \sqrt{k^2 + \lambda_i(0)}\right| < 4k^{-1-2\delta}\},$$

and

$$s(K_{\|} \setminus \tilde{\Omega}_2^{(i)}) = O(k^{-2\delta}). \tag{5.5.8}$$

We denote by $\tilde{\Omega}_2$ the set

$$\tilde{\Omega}_2 = \cap_{i:|\lambda_i(0)|<k^\delta} \tilde{\Omega}_2^{(i)}.$$

It is clear that $\rho_2'(k, t_{\|}) > 2k^{-2\delta}$ when $t_{\|} \in \tilde{\Omega}_2$. It is easy to see that this estimate is stable with respect to a perturbation of order $k^{-1-2\delta}$. Hence, $\rho_2'(k, t_{\|}) > k^{-2\delta}$, when $t_{\|}$ belongs to the $(k^{-1-2\delta})$-neighborhood of $\tilde{\Omega}_2(k,\delta)$. Now we estimate $s(K_{\|} \setminus \tilde{\Omega}_2)$. It is obvious that $K_{\|} \setminus \tilde{\Omega}_2 \subset \cup_i (K_{\|} \setminus \tilde{\Omega}_2^{(i)})$. Therefore

$$s(K_{\|} \setminus \tilde{\Omega}_2) \leq \sum_{i:|\lambda_i(0)|<k^\delta} S(K_{\|} \setminus \tilde{\Omega}_2^{(i)}).$$

Taking into account that the number of indices i, satisfying the inequality $|\lambda_i(0)| < k^\delta$, does not exceed $c(V)k^{\delta/2}$, and using estimate (5.5.8), we get:

$$s(K_{\|} \setminus \tilde{\Omega}_2) < c(V)k^{-3\delta/2}.$$

Similarly, we construct the set $\hat{\Omega}_2$, such that $\rho_2''(k, t_{\|}) > k^{-2\delta}$, when $t_{\|}$ belongs to the $(k^{-1-2\delta})$-neighborhood of $\hat{\Omega}_2(k,\delta)$ and

$$s(K_{\|} \setminus \hat{\Omega}_2) < c(V)k^{-3\delta/2}.$$

Let $\Omega_2 = \tilde{\Omega}_2 \cap \hat{\Omega}_2$. It is clear that $\rho_2(k, t_{\|}) > k^{-2\delta}$, when $t_{\|}$ belongs to the $(k^{-1-2\delta})$-neighborhood of $\Omega_2(k,\delta)$ and relation (5.5.7) holds. The lemma is proved.

Lemma 5.18 . *For any positive δ and sufficiently large k, $k > k_0(\delta, a_2, a_3)$, there exists a set $\Omega_3 \subset K_{\|}$, such that $\rho_3(k, t_{\|}) > k^{-\delta}$, when $t_{\|}$ belongs to the $(k^{-1-2\delta})$-neighborhood of $\Omega_3(k, \delta)$. The set $\Omega_3(k, \delta)$ has an asymptotically full measure on $K_{\|}$. Moreover,*

$$s(K_{\|} \setminus \Omega_3) = O(k^{-\delta}). \tag{5.5.9}$$

Lemma 5.19 . *For any positive δ and sufficiently large k, $k > k_0(\delta, a_2, a_3)$, there exists a set $\Omega_4 \subset K_{\|}$, such that $\rho_4(k, t_{\|}) > k^{-\delta}$, when $t_{\|}$ belongs to the $(k^{-1-2\delta})$-neighborhood of $\Omega_4(k, \delta)$. The set $\Omega_4(k, \delta)$ has an asymptotically full measure on $K_{\|}$. Moreover,*

$$s(K_{\|} \setminus \Omega_4) = O(k^{-\delta}). \tag{5.5.10}$$

Proofs of Lemmas 5.18 and 5.19 are similar to that of Lemma 5.17 up to replacement of $\lambda_i(0)$ with $\tilde{\lambda}_k$ or α_i.

Lemma 5.20 . *For any positive δ and sufficiently large k, $k > k_0(\delta, a_2, a_3)$, there exists a set $\Omega_5 \subset K_{\|}$, such that $\rho_5(k, t_{\|}) > k^{1-9\delta}$, when $t_{\|}$ belongs to the $(k^{-1-2\delta})$-neighborhood of $\Omega_5(k, \delta)$. The set $\Omega_5(k, \delta)$ has an asymptotically full measure on $K_{\|}$. Moreover,*

$$s(K_{\|} \setminus \Omega_5) = O(k^{-\delta}). \tag{5.5.11}$$

Proof. The proof is very similar to the proof of Lemma 3.2 (page 74) in the case of $n = 2$. Indeed, we define Ω_5 as follows:

$$\Omega_5 = (K_{\|} \setminus Q_5) \cap \Omega_1,$$

$$Q_5 = \cup_{m=0}^{M} T_m, \tag{5.5.12}$$

T_m being defined by formula (3.4.2) with $\beta = \delta, n = 2$. We prove that $\rho_5(k, t_{\|}) > k^{1-9\delta}$, when $t_{\|}$ belongs to $\Omega_5(k, \delta)$, i.e., for any $i \in Z^2 \setminus \{0\}$ and $q \in Z^2, |q| < k^{\delta}$, the following relation holds:

$$|(p_i^2(t_{\|}) - k^2)(p_{i+q}^2(t_{\|}) - k^2)| > k^{1-3\delta}. \tag{5.5.13}$$

Suppose,

$$|p_i^2(t_{\|}) - k^2| < |p_{i+q}^2(t_{\|}) - k^2)|. \tag{5.5.14}$$

If $|p_i^2(t_{\|}) - k^2| \geq k^{1/2-4\delta}$, then the same inequality holds for $|p_{i+q}^2(t_{\|}) - k^2|$. Therefore, (5.5.13) is satisfied. In the case of the opposite inequality $|p_i^2(t_{\|}) - k^2| < k^{1/2-3/2\delta}$, we note that $|p_i^2(t_{\|}) - k^2| > k^{-\delta}$, because $t_{\|} \in \Omega_1$. Therefore,

$$k^{-1-\delta} \leq 2|p_i(t_{\|}) - k| \leq 4k^{-1/2-4\delta}.$$

This means that $\mathbf{p}_i(t_{\|}) \in \Upsilon_1 \cup ... \cup \Upsilon_M$, here Υ_1, ... Υ_M are the rings given by (3.4.8). Since the rings Υ_m, $m = 1, ..., M$ do not intersect with each other, we have $\mathbf{p}_i(t_{\|}) \in \Upsilon_m$ for some m. We have proved in Lemma 3.2, that for such $\mathbf{p}_i(t_{\|})$, the inequality

$$|p_i^2(t_{\|}) - p_{i+q}^2(t_{\|})| > 4k^{1-\delta(m+1)-2\delta} \tag{5.5.15}$$

holds (see (3.4.9) for $\beta = \delta$, $n = 2$), otherwise $t_{\|} \in Q_5$. Since $\mathbf{p}_i(t_{\|}) \in \Upsilon_m$ (see (3.4.10)), we have

$$|p_i^2(t_{\|}) - k^2| > k^{(m-1)\delta}. \tag{5.5.16}$$

Multiplying inequalities (5.5.15) and (5.5.16), we obtain

$$|p_i^2(t_{\|}) - p_{i+q}^2(t_{\|})||p_i^2(t_{\|}) - k^2| > 4k^{1-5\delta}. \tag{5.5.17}$$

Note that for any $m \leq M$ (M is given by formula (3.4.2) for $n = 2$): $1-\delta(m+1)-\delta \geq (m-1)\delta$. Therefore, taking into account inequalities (5.5.15) and (5.5.16), we get

$$|p_i^2(t_{\|}) - p_{i+q}^2(t_{\|})| > 4|p_i^2(t_{\|}) - k^2|.$$

Now it is easy to see that

$$2|p_{i+q}^2(t_{\|}) - k^2| > |p_i^2(t_{\|}) - p_{i+q}^2(t_{\|})|.$$

Applying the last inequality to the left side of (5.5.17), we obtain (5.5.13). If (5.5.14) does not hold, then by making the transformation $i' = i+q$, $i'+q' = i$, we arrive at the previous case. Thus, $\rho_5(k, t_{\|}) > 2k^{1-9\delta}$, when $t_{\|}$ belongs to $\Omega_5(k, \delta)$. It is easy to see this estimate to be stable with respect to a perturbation of order $k^{-1-2\delta}$. Hence, $\rho_5(k, t_{\|}) > k^{1-9\delta}$, when $t_{\|}$ belongs to the $(k^{-1-2\delta})$-neighborhood of $\tilde{\Omega}_5(k, \delta)$.

Note that the estimate $s(Q_5) < c(V)k^{-\delta}$ easily follows fromthe definitions of Q_5 and T_m (see (5.5.12), (3.4.2) and (3.4.12) for $\beta = \delta$, $n = 2$. Taking into account that $s(K_{\|} \setminus \Omega_1) \leq k^{-\delta}$, we get (5.5.11).

The lemma is proved.

Let us consider the set $\Omega(k, \delta) \subset K_{\|}$ given by the formula:

$$K_{\|} \setminus \Omega(k, \delta) = K_{\|} \left(\hat{\mu}_{q_0}(k, \delta) \setminus \hat{\chi}_{q_0}^2(k, V, \delta)\right), \tag{5.5.18}$$

where $q_0 = (1, 0, 0)$ and $\hat{\mu}_{q_0}(k, \delta)$ is given by (4.6.5), $\hat{\chi}_{q_0}^2$ is described by (4.13.28).

Lemma 5.21 . *The set $\Omega(k, \delta)$ has an asymptotically full measure in $K_{\|}$. Moreover,*

$$s(K_{\|} \setminus \Omega) = O(k^{-3\delta}). \tag{5.5.19}$$

Proof. Using formula (4.6.5) for $\hat{\mu}_q(k, \delta)$, we easily get the normal $\mathbf{n}$ to $\hat{\mu}_q(k, \delta)$:

$$\mathbf{n} = \frac{\mathbf{g}}{|\mathbf{g}|}, \quad \mathbf{g} = \mathbf{p}_j(t) + \mathbf{g}_0,$$

$$\mathbf{g_0} = \left(\frac{\partial \Delta \hat{\Lambda}_{jj}(t)}{\partial t_1}, 0, 0\right).$$

The projection of the surface element $d\hat{S}$ of $\hat{\mu}_q(k,\delta)$ onto $K_\parallel$ is given by the formula

$$\mathcal{K}_\parallel d\hat{S} = |\mathbf{g_0} + \mathbf{p}_{j_1}(t_1)||\mathbf{g}|^{-1} d\hat{S} = O(k^{\delta-1}), \quad \mathbf{p}_{j_1}(t_1) = \left(2\pi j_1 a_1^{-1}, 0, 0\right).$$

From this relation and formula (5.5.18) it follows:

$$S(K_\parallel \setminus \Omega) = s\left(\mathcal{K}_\parallel\left(\hat{\mu}_q(k,\delta) \setminus \hat{\chi}^2_{q_0}(k,V,\delta)\right)\right) =$$

$$O(k^{\delta-1}) s\left(\hat{\mu}_q(k,\delta) \setminus \hat{\chi}^2_{q_0}(k,V,\delta)\right).$$

Using formula (4.6.15), we obtain (5.5.19). The lemma is proved.

The proof of Lemma 5.9. If $m_\parallel \in M_1$, then $k^2 - |t_\parallel + p_{m_\parallel}(0)|^2 \in (-\infty, 0] \cap \Lambda$, i.e., there exist t_{01} and m_1, such that

$$k^2 - |t_\parallel + p_{m_\parallel}(0)|^2 = \lambda_{m_1}(t_{01}), \tag{5.5.20}$$

where

$$|\lambda_{m_1}(t_{01})| \leq \|V\|.$$

Let $t_0 = (t_{01}, t_\parallel)$. From the definition of $\hat{\mu}_{q_0}(k,\delta)$ (see (4.6.5)), we have $t_0 \in \hat{\mu}_{q_0}(k,\delta)$. Now we prove that $t_0 \in \hat{\chi}^2_{q_0}(k,V,\delta)$ when $t_\parallel \in \Omega$. Suppose this is not so. Then, $t_0 \in \hat{\mu} \setminus \hat{\chi}^2_{q_0}(k,V,\delta)$, and, therefore, $t_\parallel \in \mathcal{K}_\parallel\left(\hat{\mu}(k,\delta) \setminus \hat{\chi}^2_{q_0}(k,V,\delta)\right)$. This means that $t_\parallel \notin \Omega$. This contradiction proves that $t_0 \in \hat{\chi}^2_{q_0}(k,V,\delta)$.

Let us prove that

$$|\dot{\lambda}_{m_1}(t_{01})| > k^{-1/5-\delta}. \tag{5.5.21}$$

Indeed, from the definitions of $\hat{\chi}^2_{q_0}(k,V,\delta)$ (see (4.13.28)) and $S_H^{q_0}$, we obtain that there is $t \in \hat{\chi}^0_{q_0}(k,V,\delta)$ such that $|t - t_0| < k^{-8/5+12\delta}$. According to (4.13.22) $|sin(t_{01}a_1)| > k^{-3\delta}$. From this it follows that (5.5.21) holds. This means that t_{01} and, therefore t_1, can be chosen in two different ways, corresponding to different signs of t_{01}. We fix our choice assuming that

$$\dot{\lambda}_{m_1}(t_{01}) > k^{-3\delta}. \tag{5.5.22}$$

According to the definition of $\hat{\chi}^2_{q_0}(k,V,\delta)$ given by (4.13.28), there exists t, belonging to $S_H^{q_0}$
such that $t = (t_1, t_\parallel)$, $|t_1 - t_{10}| < k^{-8/5+12\delta}$. The hypothesis of Theorem 4.20 (page 229) holds for $j = m$, $m = (m_1, m_\parallel)$, $m_\parallel \in M_1$. This means that there is an eigenfunction of the operator $H(t_\parallel)$ satisfying the asymptotic formulae

$$\psi(t,x)|_{x_1=0} = \psi_1(t_0,x)|_{x_1=0} + O(k^{-1}), \tag{5.5.23}$$

$$\left.\frac{\partial \psi(t,x)}{\partial x_1}\right|_{x_1=0} = \left.\frac{\partial \psi_1(k^2,x)}{\partial x_1}\right|_{x_1=0} + O(k^{-1}), \tag{5.5.24}$$

where $\psi_1(k^2,x)$ is the solution of the equation $H_1\psi_1 = k^2\psi_1$, corresponding to the quasimomentum t_0. Let us prove that $\psi_1(k^2,x)$ is a refracted wave. Indeed, from (5.5.22) it follows that the equation $\lambda_{m_1}(t_1) = k^2 + i\varepsilon$ has a solution $t_1(k^2 + i\varepsilon)$ for small nonnegative ε, such that $t_1(k^2+i0) = t_{01}$ and $\Im t_1(k^2+i\varepsilon) > 0$ for a positive ε. This means that the $\psi_1(k^2,x) = \lim_{\varepsilon\to 0}\psi_1(k^2+i\varepsilon,)$, where $\psi_1(k^2+i\varepsilon)$ is the solution of the equation $H_1\psi_1 = (k^2+i\varepsilon)\psi_1$ with the quasimomentum $(t_1(k^2+i\varepsilon), t_{\|})$. The solution $\psi_1(k^2+i\varepsilon,)$ exponentially decays when $x_1\to\infty$, because $\Im t_1(k^2+i\varepsilon) > 0$ for a positive ε. Thus, $\psi_1(k^2,x)$ is a refracted wave. In the notations of the present chapter (see (5.4.101) and page 258), this means that

$$\psi_1(k^2,x) = \Psi_+(k^2+i0, t_{\|} + p_{m_{\|}}(0), x). \tag{5.5.25}$$

As in Chapter 4, we denote by $\hat{\lambda(t)}$ the eigenvalue corresponding to $\psi(t,x)$. From Theorem 4.19 and notations (4.3.29) – (4.3.31), it follows that

$$\frac{\partial\hat{\lambda}(t)}{\partial t_1} = \dot{\lambda}_{m_1}(t_{01})(1+o(1)).$$

This means that $\partial\hat{\lambda}(t)/\partial t_1$ is also positive. Hence, the solution $\psi(k^2+i\varepsilon,x)$ exponentially decays when $x_1\to\infty$, because $\Im t_1(k^2+i\varepsilon) > 0$ for a positive ε.

According to Proposition 5.27, $\Psi_+(k^2+i0, t_{\|}+p_{m_{\|}}(0), x) > ck^{-8\delta}$, when $k^2 - |t_{\|}+p_{m_{\|}}(0)|^2 \in (-\infty,0]\cap\Lambda$ and $\rho_2(k^2,t_{\|}) > k^{-2\delta}$. Dividing both sides of (5.5.23) into $\Psi_+(k^2+i0, t_{\|}+p_{m_{\|}}(0), x)$, and using (5.5.25) and the definition of $U^1_{m_{\|}}$ (see (5.4.161)), we obtain (5.4.185) and (5.4.186). Note, that all the estimates are stable with respect the perturbation of order $k^{-2-2\delta}$. The lemma is proved.

Lemma 5.22 . *For any positive δ and sufficiently large k, $k > k_0(\delta,a_2,a_3)$, there exists a set $\Omega_6\subset K_{\|}$, such that, $\rho_1(k^2,t_{\|}) > k^{-2\delta}$, $\rho_2(k^2,t_{\|}) > k^{-2\delta}$, $\rho_3(k^2,t_{\|}) > k^{-3\delta/s_0}$, $\rho_4(k^2,t_{\|}) > k^{-3\delta/s_0}$, $\rho_5(k^2,t_{\|}) > k^{1-9\delta}$ and $t_{\|}\in\Omega(k,\delta)$ when $t_{\|}$ belongs to the $(k^{-1-2\delta})$-neighborhood of $\Omega_6(k,\delta)$. The set $\Omega_6(k,\delta)$ has an asymptotically full measure on $K_{\|}$. Moreover,*

$$s(K_{\|}\setminus\Omega_6) = O(k^{-s_1\delta}),\quad s_1 = \min\{\delta, 3\delta/s_0\}. \tag{5.5.26}$$

Proof. We define Ω_6 as follows:

$$\Omega_6(k,\delta) = (\cap_{i=1,2,5}\Omega_i(k,\delta))\cap(\cap_{i=3,4}\Omega_i(k,3\delta/s_0))\cap\Omega. \tag{5.5.27}$$

Using Lemmas 5.16 – 5.21, we get all the inequalities for ρ_i, $i = 1,2,3,4,5$. We obtain (5.5.26), using similar inequalities for Ω_i, $i = 1,...,5$ and Ω (see (5.5.1), (5.5.7), (5.5.9) – (5.5.11) and (5.5.19)). The lemma is proved.

Theorem 5.1 . *If $\delta > 0$ and $t_{\|}$ is in the $(k^{-2-2\delta})$-neighborhood of $\Omega_6(k,\delta)$, then for sufficiently large k, $k > k_0(\delta,a_2,a_3)$, there exists $\lim_{\varepsilon\to 0}T^{-1}(k^2+i\varepsilon,t_{\|})$ in the class of compact operators; and the following estimate holds:*

$$\|T^{-1}(k^2+i0,t_{\|})\| < k^{200\delta}. \tag{5.5.28}$$

Corollary 5.5 . *Operator $H(t_{\|})$ has no surface or quasisurface states.*

The theorem immediately follows from Lemmas 5.15 (page 296 and 5.22. The corollary follows from the definitions of surface and quasisurface states and estimate (5.5.28).

We introduce the set $S_6(k,\delta) \subset S_k$:

$$S_6 = \{\mathbf{k} : \mathcal{K}_{\|}\mathbf{k} \in \Omega_6(k,\delta)\}. \tag{5.5.29}$$

Lemma 5.23 . *If $\delta > 0$ and $\mathbf{k}$ is in the $(k^{-2-2\delta})$-neighborhood of $S_6(k,\delta)$, then for sufficiently large k, $k > k_0(\delta, a_2, a_3)$, the following relations hold: $\rho_1(k^2, k_{\|}) > k^{-2\delta}$, $\rho_2(k^2, k_{\|}) > k^{-2\delta}$, $\rho_3(k^2, k_{\|}) > k^{-3\delta/s_0}$, $\rho_4(k^2, k_{\|}) > k^{-3\delta/s_0}$, $\rho_5(k^2, k_{\|}) > k^{1-9\delta}$, and $\mathcal{K}_{\|}\mathbf{k} \in \Omega(k,\delta)$. The set $S_6(k,\delta)$ has an asymptotically full measure on $S_{\mathbf{k}}$. Moreover,*

$$\frac{s(S_k \setminus S_6)}{s(S_k)} = O(k^{-\delta s_1/2}). \tag{5.5.30}$$

Proof. If $\mathbf{k}$ belongs to the $(k^{-2-2\delta})$- neighborhood of $S_6(k,\delta)$, then $t_{\|} \equiv \mathcal{K}_{\|}\mathbf{k}$ lies in the $(k^{-2-2\delta})$-neighborhood of $\Omega_6(k,\delta)$. Therefore, taking into account that $\rho_i(k^2, k_{\|}) = \rho_i(k^2, t_{\|})$, we obtain that the estimates $\rho_1(k^2, k_{\|}) > k^{-2\delta}$, $\rho_2(k^2, k_{\|}) > k^{-2\delta}$, $\rho_3(k^2, k_{\|}) > k^{-3\delta/s_0}$, $\rho_4(k^2, k_{\|}) > k^{-3\delta/s_0}$, $\rho_5(k^2, k_{\|}) > k^{1-9\delta}$ hold in the $(k^{-2-2\delta})$- neighborhood of $S_6(k,\delta)$.

To prove estimate (5.5.30), we break R^2 into elementary cells of the dual lattice and construct Ω_6 in each cell. Thus, we obtain a "parquet". We denote by $\tilde{\Omega}_6$ the intersection of this "parquet" with the ring $|k_{\|}| \le k$:

$$\tilde{\Omega}_6 = \{k_{\|} : \mathcal{K}_{\|}k_{\|} \in \Omega_6, |k_{\|}| \le k\}. \tag{5.5.31}$$

It is easy to show that $\mathbf{k} \in S_6(k,\delta)$ if and only if $k_{\|} \in \tilde{\Omega}_6$. Indeed, if $\mathbf{k} \in S_6(k,\delta)$, then $|k_{\|}| \le k$ and $\mathcal{K}_{\|}k_{\|} = \mathcal{K}_{\|}\mathbf{k} \in \Omega_6$; therefore $k_{\|} \in \tilde{\Omega}_6$. Inversely, if $\mathbf{k} \in S_k$ and $\mathbf{k} : k_{\|} \in \tilde{\Omega}_6$, then $\mathcal{K}_{\|}\mathbf{k} = \mathcal{K}_{\|}k_{\|} \in \Omega_6$; therefore $\mathbf{k} \in S_6(k,\delta)$.

Let S_6' be the complement of S_6 in S_k and $\tilde{\Omega}_6'$ be the complement of $\tilde{\Omega}_6$ in the disk $|x_{\|}| \le k$. It follows from the definitions of the sets that

$$\chi(S_6, \mathbf{k}) = \chi(\tilde{\Omega}_6, k_{\|}), \quad \chi(S_6', \mathbf{k}) = \chi(\tilde{\Omega}_6', k_{\|}), \tag{5.5.32}$$

χ being the characteristic function of the sets. Since Ω_6 has an asymptotically full measure on $K_{\|}$ (see (5.5.26)), we can conclude that $\tilde{\Omega}_6$ is a subset of an asymptotically full measure in the disk $|x_{\|}| \le k$. Moreover, the following estimate holds:

$$\frac{s(\tilde{\Omega}_6')}{\pi k^2} < 2ck^{-\delta s_1}. \tag{5.5.33}$$

It is clear that

$$s(S_6') = \int_{S_6'} \chi(S_6', \mathbf{k}) dS_k(\mathbf{k}).$$

Taking into account that $dS_k(\mathbf{k}) = k(k^2 - |k_{\|}|^2)^{-1/2}dk_{\|}$, we obtain

$$s(S_6') = k\int_{|k_{\|}|\le k} \chi(\tilde{\Omega}_6', k_{\|})(k^2 - |k_{\|}|^2)^{-1/2}dk_{\|}.$$

We break the region of integration into two parts: the smaller disk $|k_{\|}| < k - k^{1-\delta_{s1}}$ and the ring $k - k^{1-\delta_{s1}} \le |k_{\|}| \le k$. It is not hard to show that the integral over the ring is less than $2\pi k^{2-\delta_{s1}/2}$. To estimate another integral, we take into account the inequality $\sqrt{k^2 - |k_{\|}|^2} > k^{1-\delta_{s1}/2}$. Then,

$$s(S_6') \le 2\pi k^{2-\delta_{s1}/2} + s(\tilde{\Omega}_6')k^{\delta_{s1}/2}.$$

Now, using inequality (5.5.33), we arrive at relation (5.5.26).

The lemma is proved.

DEFINITION. Points $\mathbf{k}$ *and* $\mathbf{k}_0$ *belonging to sphere* S_k*, are said to be* N*-neighbors if* $k_{\|} - k_{0\|} = \mathbf{p}_{m_{\|}}(0)$*, where* $|m_{\|}| \le N$.

It is clear that each point can have only a finite number of N-neighbors.

DEFINITION. The set $\tilde{S} \subset S_k$ *is called* N*-intensification of* $S \subset S_k$ *if it consists of the points belonging together with their* N*-neighbors to the set* S.

It is clear that 0-intensification of S coincides with S, and $(N+1)$-intensification belongs to N-intensification.

Lemma 5.24 . *Suppose* $\delta > 0$ *and* S *is a subset of* S_k *such that*

$$\frac{s(S_k \setminus S)}{s(S_k)} < k^{-4\delta}. \tag{5.5.34}$$

Then, $\tilde{S}$*, being the* k^{δ}*-intensification of* S*, satisfies the inequality:*

$$\frac{s(S_k \setminus \tilde{S})}{s(S_k)} < ck^{-\delta/2}, \quad c \ne c(k). \tag{5.5.35}$$

Proof. We define Q, $\tilde{Q}$ as follows: $Q = S_k \setminus S$, $\tilde{Q} = S_k \setminus \tilde{S}$. Let $Q_{\|}$, $\tilde{Q}_{\|}$ be the projections of Q and $\tilde{Q}$, respectively, on the plane $x_1 = 0$. We prove first that

$$\chi(\tilde{Q}_{\|}, k_{\|}) \le \sum_{|m_{\|}|<k^{\delta}} \chi(Q_{\|}, k_{\|} + \mathbf{p}_{m_{\|}}(0)). \tag{5.5.36}$$

Indeed, suppose $k_{\|} \in \tilde{Q}_{\|}$. Then, there exists $\mathbf{k} = (k_1, k_{\|})$ such that $\mathbf{k} \in S_k \setminus \tilde{S}$. Since $\mathbf{k} \notin \tilde{S}$, there exists a k^{δ}- neighbor $\mathbf{k}_0$ of $\mathbf{k}$ which does not belong to S, i.e., $\mathbf{k}_0 \in S_k \setminus S = Q$. This means that $k_{0\|} = k_{\|} + p_{m_{\|}}(0)$ for some $m_{\|} : \ |m_{\|}| < k^{\delta}$ (by the definition of a k^{δ}-neighbor) and $k_{0\|} \in Q_{\|}$. Thus, if $k_{\|} \in \tilde{Q}_{\|}$, then there exists $m_{\|} : \ |m_{\|}| < k^{\delta}$ such that $k_{\|} + \mathbf{p}_{m_{\|}}(0) \in Q_{\|}$. Therefore, relation (5.5.36) holds.

Considering the spherical layer

$$S_k' = \{x : |x| = k, |x_1| < k^{1-\delta}\},$$

we easily show that

$$s(\tilde{Q}) \leq \int_{S_k \setminus S'_k} \chi(\tilde{Q}, \mathbf{k}) dS_k + ck^{2-\delta}. \tag{5.5.37}$$

Now, we estimate the integral on the right. Taking into account that $dS_k(\mathbf{k}) = k(k^2 - |k_{\|}|^2)^{-1/2} dk_{\|}$, we show::

$$s(\tilde{Q}) \leq k^{\delta} \int_{|k_{\|}|<k-k^{\delta}} \chi(\tilde{Q}_{\|}, k_{\|}) dk_{\|} + ck^{2-\delta}. \tag{5.5.38}$$

Using relation (5.5.36), and considering that the sum contains about $ck^{2\delta}$ terms, we obtain

$$\int_{|k_{\|}|<k-k^{\delta}} \chi(\tilde{Q}_{\|}, k_{\|}) dk_{\|} \leq ck^{2\delta} \max_{m_{\|}:|m_{\|}|<k^{\delta}} \int_{|k_{\|}|<k} \chi(Q_{\|}, k_{\|} + p_{m_{\|}}(0)) dk_{\|}. \tag{5.5.39}$$

Introducing $\tilde{k}_{\|} = k_{\|} + p_{m_{\|}}(0)$ as the new variable of integration leads to the following estimate:

$$\int_{|k_{\|}|<k-k^{\delta}} \chi(\tilde{Q}_{\|}, k_{\|}) dk_{\|} \leq ck^{2\delta} \int_{|k_{\|}|<k} \chi(Q_{\|}, \tilde{k}_{\|}) d\tilde{k}_{\|}. \tag{5.5.40}$$

Taking into account (5.5.34), it is not hard to verify the relation

$$\int \chi(Q_{\|}, k_{\|}) dk_{\|} < ck^{2-4\delta}. \tag{5.5.41}$$

Using inequality (5.5.41) in estimate (5.5.40), and taking into account (5.5.38), we obtain (5.5.35). The lemma is proved.

Next, we consider the nonsingular set $\chi_3(k, V, 5\delta)$ of the whole crystal, which by definition belongs to $\mathcal{K}S_k$. Since inequality (4.3.12) holds, the set $\chi_3(k, V, 15\delta)$ can be uniquely mapped, by parallel shifting pieces, on the sphere S_k. We denote the image of $\chi_3(k, V, 10\delta)$ on S_k by $\chi'_3(k, V, 10\delta)$. Let $\tilde{\chi}'_3(k, V, 10\delta)$ be the $k^{3\delta}$-intensification of $\chi'_3(k, V, 10\delta)$. We define $S^{(3)}$ as follows:

$$S^{(3)}(k, V, \delta) = \tilde{\chi}'_3(k, V, 10\delta) \cap S_6(k, \delta). \tag{5.5.42}$$

Let $B_r(k, k_{\|} + p_{m_{\|}}(0), x)$ be the functions $B_r(k, t_{0m_{\|}}, x)$ introduced in Chapter 4 (see (4.14.30)), where

$$t_{0m_{\|}} = \mathcal{K}\left(k_1(m_{\|}), k_{\|} + p_{m_{\|}}(0)\right), \quad k_1(m_{\|}) = \sqrt{k^2 - \left|k_{\|} + p_{m_{\|}}(0)\right|^2}, \ \Re\sqrt{\ } > 0. \tag{5.5.43}$$

The following theorem is the combination of Theorems 4.18 (page 227)and Lemmas 4.43, 4.44 (page 226) and 5.15 (page 296).

Theorem 5.2 . *If $\delta > 0$ and $\mathbf{k}$ is in the $(k^{-2-\delta})$-neighborhood of $S^{(3)}(k, V, \delta)$, then for sufficiently large k, $k > k_0(\delta, V)$, there exists $\lim_{\varepsilon \to 0} T^{-1}(k^2 + i\varepsilon, k_{\|})$ in the class of compact operators; and the following estimate uniform in ε holds:*

$$\|T^{-1}(k^2+i\varepsilon,k_{\|})\| < k^{200\delta}. \tag{5.5.44}$$

There exists a set of eigenfunctions $\{\Psi(k,k_{\|}+p_{m_{\|}}(0),x)\}_{m_{\|}\in Z^2,|m_{\|}|<k^{3\delta}}$ *of the operator* H*, which satisfy quasiperiodic conditions (5.1.4), (5.1.7) and admit the following asymptotic expansions:*

$$\Psi(k,k_{\|}+p_{m_{\|}}(0),x) =_{k\to\infty} \frac{1}{2k_1(m_{\|})}\Psi_0^{+}(k,k_{\|}+p_{m_{\|}}(0),x)+$$

$$\sum_{r=1}^{R_2} B_r(k,k_{\|}+p_{m_{\|}}(0),x) + C(k^2,k_{\|}+p_{m_{\|}}(0),x), \tag{5.5.45}$$

$$R_2=[k^{3\delta}/R_0], \quad |m_{\|}|<k^{3\delta}.$$

The functions B_r, C *satisfy the estimates* [13]

$$\|A_+^0 B_r\|_{2,M} < c(M)k^{-(1-8\delta)r}, \tag{5.5.46}$$

$$\left\|\frac{\partial B_r}{\partial x_1}\right\|_{2,M} < c(M)k^{-(1-8\delta)r}, \tag{5.5.47}$$

$$\|A_+^0 C\|_{2,M} < c(M)k^{-(1-8\delta)(R_2+1)}, \tag{5.5.48}$$

$$\left\|\frac{\partial C}{\partial x_1}\right\|_{2,M} < c(M)k^{-(1-8\delta)(R_2+1)}, \quad c(M)<\infty. \tag{5.5.49}$$

For any x_1 *and* r

$$B_r(k^2,k_{\|}+p_{m_{\|}}(0),x_1)_{q_{\|}}=0, \tag{5.5.50}$$

when $|m_{\|}-q_{\|}|>rR_0$. *For any* r *the function* B_r *can be represented in the form:*

$$B_r(k^2,k_{\|}+p_{m_{\|}}(0),x)=$$

$$\sum_{q\in Z^3,|q|<rR_0}\sum_{l=0}^{r} a_{ql}^{m_{\|}r}\exp\left(i(\mathbf{p}_q(0),x)+i(k_{\|}+p_{m_{\|}}(0),x_{\|})+ik_1(m_{\|})x_1\right)\frac{x_1^l}{l!}, \tag{5.5.51}$$

$$m=(0,m_{\|}), \quad k_1(m_{\|})=\sqrt{k^2-|k_{\|}+p_{m_{\|}}(0)|^2},$$

where the coefficients $a_{ql}^{m_{\|}r}$ *satisfy the estimates:*

$$|a_{ql}^{m_{\|}r}| < k^{-1-(1-8\delta)r+(2+\delta)l}. \tag{5.5.52}$$

The set $S^{(3)}(k,V,\delta)$ *has an asymptotically full measure on* $\mathcal{K}S_k$*:*

$$\frac{s(\mathcal{K}S_k\setminus S^{(3)})}{s(\mathcal{K}S_k)} < k^{-\delta}. \tag{5.5.53}$$

[13] $\|\cdot\|_{2,M}$ is given by (4.14.29).

Proof. Suppose $\mathbf{k} \in S^{(3)}(k,V,\delta)$. Then $\mathbf{k} \in S_6$ (see (5.5.42)). Therefore, $t_{\|} \in \Omega_6$ (see (5.5.29)). According to Lemma 5.14 (page 296) there exists $\lim_{\varepsilon\to 0} T^{-1}(k^2+i\varepsilon, t_{\|})$ in the class of compact operators; and estimate (5.5.44) uniform in ε holds. Taking into account that the matrices $T^{-1}(k^2+i\varepsilon, t_{\|})$ and $T^{-1}(k^2+i\varepsilon, k_{\|})$ coincide up to the shift of indices on $m_{\|}$, we get the same result for the latter one.

Since $\mathbf{k} \in \chi_3'(k,V,10\delta)$, we have $\mathcal{K}\mathbf{k} \in \chi_3(k,V,10\delta)$. According to Theorem 4.18 (page 227), there exists eigenfunction $\Psi(k,k_{\|},x)$ of H, which satisfies quasiperiodic condition (5.1.4), (5.1.7) and admits asymptotic expansion (5.5.45) with $m_{\|}=0$ (we consider the new notations for B_r on page 305) and (4.14.40). From Lemma 4.43 (page226) we get (5.5.50) with $m_{\|}=0$. Estimates (5.5.47) and (5.5.49) for B_r and C immediately follows from (4.14.50) and (4.14.51).

Now we prove inequalities (5.5.46) and (5.5.48), which are a somewhat stronger than (4.14.48) and (4.14.49). Indeed, it is obvious that

$$\|A_+^0 B_r\|_{2,M} = \max_{x_1} \sum_{n_{\|}} (A_+^0)^2_{n_{\|}n_{\|}} \Big|\sum_{n_{\|}} B_r(x_1)_{n_{\|}}\Big|^2.$$

Taking into account that $|n_{\|}| < rR_0$, we get

$$|(A_+^0)_{n_{\|}n_{\|}}| < 2k.$$

Hence we obtain (5.5.46). We prove estimate (5.5.49) similarly, taking into account that $\hat{B}_r(x_1)_{n_{\|}} = 0$, when $|n_{\|}| > k^{1/5}r$ (see (4.14.32).

Using Lemma 4.44 (page 227)and introducing the new notations $q=n-j$, $a_{ql}^{0r} = a_{nl}^{(r)}$, we obtain (5.5.51) and (5.5.52) for $m_{\|}=0$.

Since $\mathbf{k} \in \tilde{\chi}_3'(k,V,\delta)$, we have $\mathcal{K}\mathbf{k}_0 \in \chi_3(k,V,\delta)$, $\mathbf{k}_0$ being any k^δ-neighbor of $\mathbf{k}$. Considering, that $k_{0\|} = k_{\|} + p_{m_{\|}}(0)$, we obtain formula (5.5.45) and all estimates for $k_{0\|}$. The theorem is proved.

Let us introduce a scalar factor α in front of the potential V_+, i.e., we consider the operator:

$$H_{+\alpha} = -\Delta + \alpha V_+, \qquad 0 \le \alpha \le 1.$$

Lemma 5.25 . *Suppose* $0 \le |\alpha| \le k^{-1-2\delta}$, $0<\delta<1/300$. *Then* $S^{(3)}(k,\alpha V,\delta) \subset S^{(3)}(k,V,\delta)$.

Proof. Using the definition of $S^{(3)}(k,\alpha V,\delta)$ (see (5.5.42), (5.5.29) and (5.5.27)), we obtain that it suffices to prove the relations:

$$\chi_3(k,\alpha V,\delta) \subset \chi_3(k,V,\delta), \tag{5.5.54}$$

$$\Omega_1(k,\alpha V,\delta) = \Omega_1(k,V,\delta), \tag{5.5.55}$$

$$\Omega_2(k,\alpha V,\delta) = \Omega_1(k,V,\delta), \tag{5.5.56}$$

$$\Omega_3(k,\alpha V,\delta) \subset \Omega_1(k,V,\delta), \tag{5.5.57}$$

$$\Omega_4(k,\alpha V,\delta)\subset\Omega_1(k,V,\delta), \tag{5.5.58}$$

$$\Omega_5(k,\alpha V,\delta)=\Omega_5(k,V,\delta), \tag{5.5.59}$$

$$\Omega(k,\alpha V,\delta)=K_{\|}. \tag{5.5.60}$$

To prove the convergence of the series for $|\alpha|<k^{-1-\delta}$, it suffices to satisfy only condition 1^0 in the definition of $\chi_3(k,\alpha V,\delta)$, because in this case the potential is smaller than the distance to the nearest eigenvalue, and the regular perturbation theory works. This condition does not depend on the potential and is satisfied when $t\in\chi_3(k,V,\delta)$. Therefore, (5.5.54) holds. The definitions of Ω_1 and Ω_5 (see (5.4.106), (5.4.107), (5.4.197), (5.4.198)) do not depend on the potential. Therefore, (5.5.55) and (5.5.59) hold. For α small enough, $\rho_2\approx\rho_1$ (see (5.4.116) – (5.4.118)), because both $\lambda_n(0)$ and $\lambda_n(\pi a_1^{-1})$ are of order $O(k^{-1-\delta})$. Therefore, (5.5.56) holds. It is easy to show that $W(\lambda,x)=W_0(\lambda,x)+O(k^{-1+16\delta})$, when $|\alpha|<k^{-1-\delta}$ and $t_{\|}\in\Omega_1(k,V,\delta)$. From this it easily follows that $W(\lambda,x)$ and $T_1(\lambda)$ do not vanish inside the contour C_4. This means that (5.5.57) and (5.5.58) hold.

Note that the set M_2, given by (5.4.159), is empty for such a small potential, because $\rho_1>k^{-\delta}$. This mean that (5.5.60) holds. The lemma is proved.

From the last lemma and Theorem 5.2, the following theorem immediately follows.

Theorem 5.2A *If $\delta>0$, $\alpha=1$ or $0\le|\alpha|<k^{-1-2\delta}$, and $\mathbf{k}$ is in the $(k^{-2-\delta})$-neighborhood of $S^{(3)}(k,V,\delta)$, then for sufficiently large k, $k>k_0(\delta,V)$, there exists $\lim_{\varepsilon\to0}T_\alpha^{-1}(k^2+i\varepsilon,k_{\|})$ in the class of compact operators; and the following estimate uniform in ε holds:*

$$\|T_\alpha^{-1}(k^2+i\varepsilon,k_{\|})\|<k^{200\delta}. \tag{5.5.61}$$

There exists a set of eigenfunctions $\{\Psi_\alpha(k,k_{\|}+p_{m_{\|}}(0),x)\}_{m_{\|}\in Z^2,|m_{\|}|<k^{3\delta}}$ of the operator H_α, $\alpha\in[0,1]$, which satisfy quasiperiodic conditions (5.1.4), (5.1.7) and admit the following asymptotic expansions:

$$\Psi_\alpha(k,k_{\|}+p_{m_{\|}}(0),x)=_{k\to\infty}\frac{1}{2k_1(m_{\|})}\Psi_0^+(k,k_{\|}+p_{m_{\|}}(0),x)+$$

$$\sum_{r=1}^{R_2}\alpha^rB_r(k,k_{\|}+p_{m_{\|}}(0),x)\;+\;\alpha^{R_2+1}C(\alpha,k^2,k_{\|}+p_{m_{\|}}(0),x), \tag{5.5.62}$$

$$R_2=[k^{3\delta}/R_0],\quad |m_{\|}|<k^{3\delta}.$$

The functions B_r, C satisfy the estimates

$$\|A_+^0B_r\|_{2,M}<c(M)k^{-(1-8\delta)r}, \tag{5.5.63}$$

$$\left\|\frac{\partial B_r}{\partial x_1}\right\|_{2,M}<c(M)k^{-(1-4\delta)r}, \tag{5.5.64}$$

$$\|A_+^0C(1,k^2,k_{\|}+p_{m_{\|}}(0),x)\|_{2,M}<c(M)k^{-(1-8\delta)(R_2+1)}, \tag{5.5.65}$$

$$\left\| \frac{\partial C}{\partial x_1}(1, k^2, k_{\|} + p_{m_{\|}}(0), x) \right\|_{2,M} < c(M) k^{-(1-8\delta)(R_2+1)}, \quad c(M) < \infty. \quad (5.5.66)$$

The function $C(\alpha, k^2, k_{\|} + p_{m_{\|}}(0), x)$ is a holomorphic function of α in some neighborhood of zero.

For any x_1 and r

$$B_r(k^2, k_{\|} + p_{m_{\|}}(0), x_1)|_{q_{\|}} = 0, \quad (5.5.67)$$

when $|m_{\|} - q_{\|}| > rR_0$. The function B_r can be represented in the form:

$$B_r(k^2, k_{\|} + p_{m_{\|}}(0), x) =$$

$$\sum_{q \in Z^3, |q| < rR_0} \sum_{l=0}^{r} a_{ql}^{m_{\|} r} \exp\left(i(\mathbf{p}_q(0), x) + i(k_{\|} + p_{m_{\|}}(0), x_{\|}) + ik_1(m_{\|})x_1\right) \frac{x_1^l}{l!},$$

$$(5.5.68)$$

$$m = (0, m_{\|}), \quad k_1(m_{\|}) = \sqrt{k^2 - |k_{\|} + p_{m_{\|}}(0)|^2},$$

where the coefficients $a_{ql}^{m_{\|} r}$ satisfy the estimates:

$$|a_{ql}^{m_{\|} r}| < k^{-1-(1-8\delta)r+(2+\delta)l}. \quad (5.5.69)$$

The set $S^{(3)}(k, V, \delta)$ has an asymptotically full measure on $\mathcal{K}S_k$:

$$\frac{s(\mathcal{K}S_k \setminus S^{(3)})}{s(\mathcal{K}S_k)} < k^{-\delta}. \quad (5.5.70)$$

Proof. For $\alpha = 1$, the assertion of the theorem coincides with that of Theorem 5.2. Thus, we obtain (5.5.63), (5.5.64), (5.5.67) – (5.5.69), which do not depend on α and (5.5.65), (5.5.66) where $\alpha = 1$. Suppose $0 \leq |\alpha| < k^{-1-2\delta}$. From consideration of the last Lemma 5.25, it follows that the perturbation series for such a small potential αV converges with respect to the free operator, i.e.,

$$\Psi_\alpha(k, k_{\|} + p_{m_{\|}}(0), x) =_{k \to \infty} \Psi_0^+(k, k_{\|} + p_{m_{\|}}(0), x) +$$

$$\sum_{r=1}^{\infty} \alpha^r B_r(k, k_{\|} + p_{m_{\|}}(0), x), \quad (5.5.71)$$

$$\|B_r\|_{2,M} < k^{-1+(1+\delta)r}.$$

from this formulae we get (5.5.62), where

$$C(\alpha, k^2, k_{\|} + p_{m_{\|}}(0), x) = \sum_{r=R_2+1}^{\infty} \alpha^{r-R_2-1} B_r(k, k_{\|} + p_{m_{\|}}(0), x), \quad (5.5.72)$$

$$|\alpha| < k^{-1-2\delta}.$$

It is clear that the function $C(k^2, t_0, x)$ depends analytically on α in the $(k^{-1-2\delta})$ neighborhood of zero for a fixed t_0. The theorem is proved.

5.5.2 Geometric Constructions in the Two-Dimensional Case.

We proved in Section 5.3 that surface and quasisurface states are absent for given quasimomentum $t_{\|}$ and energy k^2 if the inequality

$$\rho_0(k, t_{\|}) > k^{1/4}, \tag{5.5.73}$$

holds, where

$$\rho_0(k, t_{\|}) = \min_{m_{\|} \in Z} \left| |t_{\|} + p_{m_{\|}}(0)|^2 - k^2 \right|.$$

Lemma 5.26 . *For any positive δ and sufficiently large k, $k > k_0(\delta, a_2)$, there exists a set $\Omega_0 \subset [0, 2\pi a_1^{-1})$, such that $\rho_0(k, t_{\|}) > k^{1/4}$, when $t_{\|}$ belongs to the $(k^{-3/4-2\delta})$-neighborhood of $\Omega_0(k, \delta)$. The set $\Omega_0(k, \delta)$ has an asymptotically full measure on $[0, 2\pi a_1^{-1})$. Moreover,*

$$s([0, 2\pi a_1^{-1}) \setminus \Omega_0) = O(k^{-3/4}). \tag{5.5.74}$$

Proof. The proof is similar to that of Lemma 5.16 (page 297). Indeed, let us consider $x \in R$ such that

$$|k^2 - |x|^2| < 2k^{1/4}. \tag{5.5.75}$$

This inequality is satisfied if $x \in Q_0$,

$$Q_0 = [k - k^{-3/4},\ k + k^{-3/4}].$$

We define Ω_0 as follows

$$\Omega_0 = K_{\|} \setminus \mathcal{K}_{\|} Q_0. \tag{5.5.76}$$

Now we prove that $\rho_0(k, t_{\|}) > 2k^{1/4}$ when $t_{\|}$ belongs to $\Omega_0(k, \delta)$. Suppose this is not so. Then, there exists $m_{\|}$ such that

$$|k^2 - |t_{\|} + p_{m_{\|}}(0)|^2| \le 2k^{-1/4}. \tag{5.5.77}$$

This means that $t_{\|} + m_{\|} \in Q_0$, i.e., $t_{\|} \in \mathcal{K}_{\|} Q_0$. This contradicts to the hypothesis $t_{\|} \in \Omega_0$. Therefore, $\rho_0(k, t_{\|}) > 2k^{1/4}$. It is easy to see that this estimate is stable with respect to a perturbation of order $k^{-3/4-2\delta}$. Hence, $\rho_0(k, t_{\|}) > k^{1/4}$, when $t_{\|}$ belongs to the $(k^{-1-2\delta})$-neighborhood of $\Omega_0(k, \delta)$.

Let us prove that Ω_0 has an asymptotically full measure on $[0, 2\pi a_1^{-1})$. It is clear that $s(\mathcal{K}_{\|} Q_0) \le s(Q_0)$. Therefore,

$$s(\Omega_0) \ge s([0, 2\pi a_1^{-1})) - s(Q_0) \ge 2\pi a_2^{-1} - 2k^{-3/4}. \tag{5.5.78}$$

Thus, relation (5.5.74) holds. The lemma is proved.

Let us consider $\tilde{\Omega}_0(k) \in S_k$:

$$\tilde{\Omega}_0(k) = \{\mathbf{k} \in S_k,\ \ \mathcal{K}_{\|} \mathbf{k} \in \Omega_0, \}.$$

Considering as in the proving Lemma 5.23 (page 303), we obtain the following lemma.

Lemma 5.27 . *If* $\mathbf{k} \in \tilde{\Omega}_0(k)$, *then*

$$\rho(k^2, k_{\|}) > k^{1/4}.$$

The set $\tilde{\Omega}_0(k)$ *has an asymptotically full measure on* S_k. *Moreover,*

$$\frac{s(S_k \setminus \tilde{\Omega}_0)}{s(S_k)} < ck^{-\delta}, c \neq c(k). \tag{5.5.79}$$

Let us recall that the nonsingular set $\chi_1(k, \delta, \delta) \subset S_k$ can be uniquely (by the parallel shift) mapped onto the circle S_k, because (3.4.3) holds on $\chi_1(k, \delta, \delta)$, n=2. We denote by $\chi_1'(k, \delta, \delta)$ the analog of $\chi_1(k, \delta, \delta)$ on S_k. If $\mathbf{k} \in \chi_1'(k, \delta, \delta)$, then $t = \mathcal{K}\mathbf{k} \in \chi_1(k, \delta, \delta)$ and $\mathbf{k} - t = \mathbf{p}_m(0)$.

Lemma 5.28 . *For sufficiently large* k, $k > k_0(V)$:

$$\mathcal{K}_{\|}\chi_1(k, \delta, \delta) \subset \Omega_0, \tag{5.5.80}$$

$$\chi_1'(k, \delta, \delta) \subset \tilde{\Omega}_0. \tag{5.5.81}$$

<u>Proof.</u> Let us prove that for any $\mathbf{k}$ belonging to the nonsingular set $\chi_1(k, \delta, \delta)$ the following relation holds:

$$\rho_0(k, k_{\|}) > k^{1/4}. \tag{5.5.82}$$

Suppose this is not the case. Then, there exists $m_{\|}$ such that:

$$|k^2 - |k_{\|} + p_{m_{\|}}(0)|^2| < k^{1/4}. \tag{5.5.83}$$

Let us consider $t = K\mathbf{k}$, and $\mathbf{p}_m(t) = (t_1, k_{\|} + p_{m_{\|}}(0))$. Obviously, that

$$|p_m^2(t) - k^2| < 2k^{1/4}.$$

Let $i = (1, 0, 0)$. We easily see

$$|p_{m+i}^2(t) - p_m^2(t)| < c, \qquad c \neq c(k).$$

On the other hand, since $K\mathbf{k}$ is in the $(k^{-1-2\delta})$-neighborhood of $\chi_1(k, \delta, \delta)$, we have

$$|p_{m+i}^2(t) - p_m^2(t)| > k^{3/4-9\delta}$$

(see inequality (3.2.3) for $\beta = \delta$). The last two inequalities contradict to each other. Thus, (5.5.82) is proved. This means that $k_{\|} \in \Omega_0$ by the definition of Ω_0. Thus, (5.5.80) is proved. Suppose $t_{\|} = \mathcal{K}_{\|} k_{\|}$. It is clear that

$$\rho_0(k, t_{\|}) = \rho_0(k, k_{\|}). \tag{5.5.84}$$

Thus, if $\mathcal{K}\mathbf{k} \in \chi_1(k, \delta, \delta)$, then $t_{\|} = \mathcal{K}_{\|} k_{\|} \in \Omega_0(k, \delta)$. Therefore, (5.5.81) holds. <u>The lemma is proved.</u>

Theorem 5.3 . *If $t_{\|}$ is in the $(k^{-3/4-\delta})$-neighborhood of $\Omega_0(k,\delta)$, then for sufficiently large k, $k > k_0(\delta, a_2)$, there exists $\lim_{\varepsilon\to 0} T^{-1}(k^2+i\varepsilon, t_{\|})$ in the class of compact operators and the following estimate holds:*

$$\|T^{-1}(k^2+i0, t_{\|})\| < ck^{-1/16}. \tag{5.5.85}$$

Corollary 5.6 . *Operator $H(t_{\|})$ has no surface or quasisurface states.*

The theorem follows from Lemmas 5.7 and 5.26 (pages 258 and 310). Corollary follows from the definitions of surface and quasisurface states and estimate (5.5.85).

Let $S_2(k,\delta)$ be the (k^{δ})-intensification of $\chi_1'(k,\delta,10\delta)$. Let us consider the operator $H_\alpha = H_0 + \alpha V$, $|\alpha| \le 1$. Note that the definition of $S_2(k,\delta)$ does not depend on α in the two-dimensional case.

Thus, in the two-dimensional case the following theorem holds:

Theorem 5.4 . *If $\delta > 0$, $\alpha \in [0,1]$ and $\mathbf{k}$ is in the $(k^{-1-2\delta})$-neighborhood of $S^{(2)}(k,\delta)$ then for sufficiently large k, $k > k_0(\delta, V)$, there exists $\lim_{\varepsilon\to 0} T_\alpha^{-1}(k^2+i\varepsilon, k_{\|})$ in the class of compact operators and the following estimate uniform in ε is valid:*

$$\|T_\alpha^{-1}(k^2+i\varepsilon, k_{\|})\| < ck^{-1/16} \tag{5.5.86}$$

There exists a set of eigenfunctions $\{\Psi_\alpha(k, k_{\|}+p_{m_{\|}}(0), x)\}_{m_{\|}\in Z, |m_{\|}|<k^{3\delta}}$ of the operator H_α, which satisfy quasiperiodic condition (5.1.4), (5.1.7) and admit the following asymptotic expansions

$$\Psi_\alpha(k, k_{\|}+p_{m_{\|}}(0), x) =_{k\to\infty} \frac{1}{2k_1(m_{\|})}\Psi_0^+(k, k_{\|}+p_{m_{\|}}(0), x)+$$

$$\sum_{r=1}^{R_2} \alpha^r B_r(k, k_{\|}+p_{m_{\|}}(0), x) + \alpha^{R_2+1}C(\alpha, k^2, k_{\|}+p_{m_{\|}}(0), x), \tag{5.5.87}$$

$$R_2 = [k^{3\delta}R_0^{-1}].$$

The functions B_r, C satisfy the estimates

$$\|A_+^0 B_r\|_{2,M} < c(M)k^{-(1-8\delta)r}. \tag{5.5.88}$$

$$\left\|\frac{\partial B_r}{\partial x_1}\right\|_{2,M} < c(M)k^{-(1-8\delta)r}. \tag{5.5.89}$$

$$\|A_+^0 C\|_{2,M} < c(M)k^{-(1-8\delta)(R_2+1)}. \tag{5.5.90}$$

$$\left\|\frac{\partial C}{\partial x_1}\right\|_{2,M} < c(M)k^{-(1-8\delta)(R_1+1)}, \quad c(M) < \infty. \tag{5.5.91}$$

The function $C(\alpha, k^2, k_{\|}+p_{m_{\|}}(0), x)$ is holomorphic function of α when $|\alpha| < 1$.

For any x_1

$$B_r(k^2, k_{\|}+p_{m_{\|}}(0), x_1)|_{q_{\|}} = 0 \tag{5.5.92}$$

when $|m_{\|} - q_{\|}| > rR_0$. *The function* B_r *can be represented in the form:*

$$B_r(k^2, k_{\|} + p_{m_{\|}}(0), x) =$$

$$\sum_{q \in Z^2, |q| < rR_0} \sum_{l=0}^{r} a_{ql}^{m_{\|} r} \exp i \left((\mathbf{p}_q(0), x) + (k_{\|} + p_{m_{\|}}(0), x_{\|}) + k_1(m_{\|}) x_1 \right) \frac{x_1^l}{l!}, \tag{5.5.93}$$

$$m = (0, m_{\|}), \quad k_1(m_{\|}) = \sqrt{k^2 - |k_{\|} + p_{m_{\|}}(0)|^2},$$

where $a_{ql}^{m_{\|} r}$ *satisfy the estimates:*

$$|a_{ql}^{m_{\|} r}| < k^{-1-(1-8\delta)r+(1+\delta)l}. \tag{5.5.94}$$

The set $S^{(2)}(k, \delta)$ *has an asymptotically full measure on* $\mathcal{K}S_k$*:*

$$\frac{s(\mathcal{K}S_k \setminus S^{(2)})}{s(\mathcal{K}S_k)} < k^{-\delta}. \tag{5.5.95}$$

This theorem is the combination of Theorems 3.10, 3.11 and Lemma 5.7 (pages 94 and 258).

We shall use the notation:

$$S^{(n)}(k, \alpha V, \delta) = \begin{cases} S^{(3)}(k, V, \delta) & \text{if } n = 3; \\ S_1^{(2)}(k, \delta), & \text{if } n = 2. \end{cases}$$

As we already noted before, the definition of the set $S_1^{(n)}(k, \alpha V, \delta)$, in the case $n = 2$, does not depend on α. This fact together with Lemma 5.25 (page 307) gives us the following lemma:

Lemma 5.29 . *Suppose* $|\alpha| < k^{-1-\delta}$, $0 < \delta < 1/300$. *Then* $S^{(n)}(k, \alpha V, \delta) \subset S^{(n)}(k, V, \delta)$.

5.6 Asymptotic Formulae for the Reflected and Refracted Waves.

The aim of this section is to study the reflection and refraction of the plane wave $\exp i(\mathbf{k}, x) \equiv \Psi_+^0(k^2, k_{\|}, x)$, $k_1 > 0$, by the semicrystal l $H_+ = -\Delta + \alpha V$, in the cases $n = 2, 3$. In fact we are interested in the situation where $\alpha = 1$. However, in the proofs we will need to consider also the case $|\alpha| < k^{-1-2\delta}$. So, we write the coefficient α in front of the potential, keeping in mind that α is equal to 1 or $|\alpha| < k^{-1-2\delta}$. Let $\mathbf{k}$ belong to $S^{(n)}(k, V, \delta)$. According to Lemma 5.29, this means that $\mathbf{k}$ also belongs to $S^{(n)}(k, \alpha V, \delta)$ when $|\alpha| < k^{-1-2\delta}$. Firstly, we determine Ψ_{refl} and Ψ_{refr} approximately, taking the approximate solution Ψ_{refl}^0 as a linear combination of the functions $\Psi_-^0(k^2, k_{\|} + p_{m_{\|}}(0), x)$, $|m_{\|}| < k^{3\delta}$:

$$\Psi^0_{refl} = \sum_{|m_\|| < k^{3\delta}} \beta^0_{m_\|} \Psi^0_-(k^2, k_\| + p_{m_\|}(0), x), \tag{5.6.1}$$

and the approximate refracted wave ψ^0_{refr} as a linear combination of the functions $\psi_\alpha(k^2, k_\| + p_{m_\|}(0), x)$, $|m_\|| < k^{3\delta}$:

$$\Psi^0_{refr} = \sum_{|m_\|| < k^{3\delta}} \zeta^0_{m_\|} a_{m_\|} \Psi_\alpha(k^2, k_\| + p_{m_\|}(0), x), \tag{5.6.2}$$

Ψ_α being given by (5.5.62) for $\alpha = 1$, $a_{m_\|}$ being the coefficients equal to $a_{m_\|} = -\pi^{-1} A^0_+(k^2 + i0, t_\|)_{m_\|, m_\|}$. As we show below, the coefficients $\beta^0_{m_\|}$, $\zeta^0_{m_\|}$ can be chosen so that the function (5.6.1), (5.6.2) satisfy the boundary conditions with accuracy to $k^{-k^{2\delta} R_0^{-1}}$. Since $\mathbf{k} \in S^{(n)}(k, V, \delta)$, the operator $H(k_\|)$ has no surface and quasisurface states (see Theorems 5.2 and 5.4on pages 305 and 312). Using this fact, we shall show that Ψ^0_{refl} approximates the reflected wave with accuracy to $O(k^{-k^{2\delta} R_0^{-1}})$ for any fixed x_1. Similarly, Ψ^0_{refr} approximates the refracted wave with accuracy to $(k^{-k^{2\delta} R_0^{-1}})$ for any fixed x_1.

Thus, we begin with the construction of the approximate solutions Ψ^0_{refl}, Ψ^0_{refr}. According to Theorems 5.2A and 5.4, the functions $\Psi_\alpha(k, k_\| + p_{m_\|}(0), x)$, $|m_\|| < k^{3\delta}$ admit the asymptotic expansion (5.5.62) – (5.5.69) as $k \to \infty$, when $\mathcal{K}\mathbf{k}$ is in the $(k^{-n+1-2\delta})$-neighborhood of $S^{(n)}(k, \alpha V, \delta)$. In particular, one can write the corresponding asymptotic formulae for the boundary values ($x_1 = 0$) of functions $\Psi_\alpha(k, k_\| + p_{m_\|}(0), x)$ and their derivatives with respect to x_1. The components of these boundary values with respect to $t_\|$-basis are:

$$\left(\Psi_\alpha(k, k_\| + p_{m_\|}(0), x)|_{x_1=0}\right)_{q_\|} =$$

$$\left(2A^0_+(k^2 + i0, k_\|)\right)^{-1}_{m_\| m_\|} \delta_{m_\| q_\|} + \sum_{r=1}^{R_{m_\|}} \alpha^r (F^{(0)}_r)_{q_\| m_\|} + \alpha^{R_{m_\|}+1} Q^{(0)}_{q_\| m_\|}, \tag{5.6.3}$$

$$\left(\frac{\partial \Psi_\alpha(k, k_\| + p_{m_\|}(0), x)}{\partial x_1}|_{x_1=0}\right)_{q_\|} =$$

$$(1/2)\delta_{m_\| q_\|} + \sum_{r=1}^{R_{m_\|}} \alpha^r (F^{(1)}_r)_{q_\| m_\|} + \alpha^{R_{m_\|}+1} Q^{(1)}_{q_\| m_\|}, \tag{5.6.4}$$

$$R_{m_\|} = [(k^{3\delta} - |m_\||) R_0^{-1}].$$

The matrix-valued functions $F^{(0)}_r$, $F^{(1)}_r$, $Q^{(0)}$, $Q^{(1)}$ are defined by the formulae:

$$(F^{(0)}_r)_{j_\| m_\|} = \left(B_r(k^2, k_\| + p_{m_\|}(0), x)|_{x_1=0}\right)_{j_\|}, \tag{5.6.5}$$

$$(F^{(1)}_r)_{j_\| m_\|} = \left(\frac{\partial B_r(k^2, k_\| + p_{m_\|}(0), x)}{\partial x_1}|_{x_1=0}\right)_{j_\|}, \tag{5.6.6}$$

$$(Q^{(0)})_{j_{\|}m_{\|}} = \left(\tilde{C}(\alpha, k^2, k_{\|} + p_{m_{\|}}(0), x)\Big|_{x_1=0} \right)_{j_{\|}}, \tag{5.6.7}$$

$$(Q^{(1)})_{j_{\|}m_{\|}} = \left(\frac{\partial \tilde{C}(\alpha, k^2, k_{\|} + p_{m_{\|}}(0), x)}{\partial x_1}\Bigg|_{x_1=0} \right)_{j_{\|}}, \tag{5.6.8}$$

$$\tilde{C}(\alpha, k^2, k_{\|} + p_{m_{\|}}(0), x) =$$

$$\sum_{r=R_{m_{\|}}+1}^{R_2} \alpha^{r-R_{m_{\|}}-1} B_r(k, k_{\|}+p_{m_{\|}}(0), x) + \alpha^{R_2+1-R_{m_{\|}}-1} C(\alpha, k^2, k_{\|}+p_{m_{\|}}(0), x).$$

The functions $Q^{(0)}_{j_{\|}m_{\|}}$, $Q^{(1)}_{j_{\|}m_{\|}}$ depend analytically on α in a small neighborhood of zero.

We supplement the finite matrices $(F^{(0)}_r)_{j_{\|}m_{\|}}$, $(F^{(1)}_r)_{j_{\|}m_{\|}}$, $(Q^{(0)})_{j_{\|}m_{\|}}$, $(Q^{(0)})_{j_{\|}m_{\|}}$ ($|j_{\|}| < k^{3\delta}$, $|m_{\|}| < k^{3\delta}$) with zeros to make them infinite. Thus, we have $(F^{(0)}_r)_{j_{\|}m_{\|}} = 0$ if $|j_{\|}| \geq k^{3\delta}$ or $|m_{\|}| \geq k^{3\delta}$. There correspond operators in $L_2(K_{\|})$ to these matrices.

Let P be a diagonal projection:

$$P_{m_{\|}m_{\|}} = \begin{cases} 1, \text{if } |m_{\|}| < k^{3\delta}; \\ 0, \text{if } |m_{\|}| \geq k^{3\delta}. \end{cases} \tag{5.6.9}$$

Lemma 5.30 . *If* $\mathbf{k} \in S^{(n)}(k, V, \delta)$, *then for the operators* $A^+_0(k^2 + i0)$, $F^{(0)}_r(k^2, k_{\|})$, $F^{(1)}_r(k^2, k_{\|})$ *the following estimates hold:*

$$\|(A^0_+)^{-1}\|_2 < k^{-1+8\delta}, \tag{5.6.10}$$

$$\|PA^0_+\|_2 < 2k, \tag{5.6.11}$$

$$\|A^0_+ F^{(0)}_r(k^2, k_{\|})\|_2 < k^{-(1-8\delta)r}, \tag{5.6.12}$$

$$\|F^{(1)}_r(k^2, k_{\|})\|_2 < k^{-(1-8\delta)r}, \tag{5.6.13}$$

The matrix elements $(F^{(0)}_r)_{q_{\|}m_{\|}}$, $(F^{(1)}_r)_{q_{\|}m_{\|}}$ *satisfy the relations*

$$(F^{(0)}_r)_{q_{\|}m_{\|}} = (F^{(1)}_r)_{q_{\|}m_{\|}} = 0, \tag{5.6.14}$$

when $|q_{\|} - m_{\|}| > rR_0$.

Corollary 5.7 . *There are estimates:*

$$\left|(A^0_+ F^{(0)}_r)_{j_{\|}m_{\|}}\right| < k^{-(1-8\delta)|j_{\|}-m_{\|}|R_0^{-1}}, \tag{5.6.15}$$

$$\left|(F^{(1)}_r)_{j_{\|}m_{\|}}\right| < k^{-(1-8\delta)|j_{\|}-m_{\|}|R_0^{-1}}. \tag{5.6.16}$$

Proof of the corollary. Indeed, if $|j_{\|}-m_{\|}|R_0^{-1}<r$, then inequality (5.6.15) immediately follows from (5.6.12). Otherwise, (5.6.14) holds, therefore (5.6.15) is satisfied. Inequality (5.6.16) is proved similarly.

Proof of the lemma. Let us prove that

$$\left||\mathbf{k}+p_{m_{\|}}(0)|^2-k^2\right|>k^{1-8\delta}, \tag{5.6.17}$$

when $|m_{\|}|<k^{3\delta}$. We represent $\mathbf{k}$ in the form $\mathbf{k}=\mathbf{p}_j(t)$. From the definition of $S^{(n)}(k,V,\delta)$, it follows that $t\in\chi_3(k,V,\delta)$, when $n=3$, and $t\in\chi_1(k,\delta,\delta)$, when $n=2$. This means that the inequality

$$\left||\mathbf{p}_j(t)+p_{m_{\|}}(0)|^2-p_j^2(t)\right|>k^{1-8\delta} \tag{5.6.18}$$

holds (see (3.2.2) for $\beta=\delta$ and (4.3.39)). Taking into account that $\mathbf{p}_j(t)=\mathbf{k}$, we get (5.6.17). The estimate (5.6.10) easily follows from (5.6.17). We get estimate (5.6.11) from the definitions of P and A_+^0 (see (5.2.11) and (5.6.9)).

Using estimates (5.5.63), (5.5.64) for $n=3$, and (5.5.88), (5.5.89) for $n=2$, and $|m_{\|}|<k^{3\delta}$, and considering that the number of indices to satisfy the last inequality does not exceed $ck^{3\delta n}$, we obtain (5.6.12), (5.6.13).

According to Theorems 5.2 and 5.4:

$$\left(B_r(k^2,k_{\|}+p_{m_{\|}}(0))|_{x_1=0}\right)_{q_{\|}}=0,$$

when $|q_{\|}-m_{\|}|>rR_0$. The same relations are valid for the derivatives of B_r. Taking into account the definitions of $F_r^{(0)}$, $F_r^{(1)}$, we get (5.6.14).

The lemma is proved.

Lemma 5.31 . *Under the conditions of the previous lemma, the following estimates for $Q^{(0)}$, $Q^{(1)}$ hold:*

$$\left(\sum_{j_{\|}\in Z^{n-1}}|(A_+^0Q^{(0)})_{j_{\|}m_{\|}}|^2\right)^{1/2}<k^{-(R_{m_{\|}}+1)(1-8\delta)}, \tag{5.6.19}$$

$$\left(\sum_{j_{\|}\in Z^{n-1}}(Q^{(1)})_{j_{\|}m_{\|}}|^2\right)^{1/2}<k^{-(R_{m_{\|}}+1)(1-8\delta)}. \tag{5.6.20}$$

Proof. Estimates (5.6.19), (5.6.20) easily follow from the definitions of $Q^{(0)}$, $Q^{(1)}$ (see (5.6.7), (5.6.8)) and (5.5.63) - (5.5.66) and (5.5.88) - (5.5.91). The lemma is proved.

We introduce the matrices $F^{(0)}$, $F^{(1)}$, $\mathcal{E}_r$ and $\mathcal{E}$:

$$F_{j_{\|}m_{\|}}^{(0)}=\sum_{r=1}^{R_{m_{\|}}}\alpha^r(F_r^{(0)})_{j_{\|}m_{\|}},\quad F_{j_{\|}m_{\|}}^{(1)}=\sum_{r=1}^{R_{m_{\|}}}\alpha^r(F_r^{(1)})_{j_{\|}m_{\|}}, \tag{5.6.21}$$

$$\mathcal{E}_r = A_+^0 F_r^{(0)} + F_r^{(1)}, \tag{5.6.22}$$

$$\mathcal{E} = A_+^0 F^{(0)} + F^{(1)}. \tag{5.6.23}$$

Using (5.6.14) and (5.6.21) we get:

$$\mathcal{E}_{j_\| m_\|} = \sum_{r \geq |j_\| - m_\| | R_0^{-1}}^{R_{m_\|}} \alpha^r (\mathcal{E}_r)_{j_\| m_\|} = \left(A_+^0 F^{(0)} + F^{(1)} \right)_{j_\| m_\|} \tag{5.6.24}$$

It is not hard to show that

$$\mathcal{E} = \sum_{r=1}^{M_1} \alpha^r \tilde{\mathcal{E}}_r, \tag{5.6.25}$$

where $M_1 = [k^{3\delta}]$,

$$\left(\tilde{\mathcal{E}}_r\right)_{m_\| j_\|} = \begin{cases} (\mathcal{E}_r)_{j_\| m_\|}, & \text{if } r < R_{m_\|}, \\ 0, & \text{if } r \geq R_{m_\|}. \end{cases}$$

Let coefficients $\beta^0 \equiv \{\beta^0_{m_\|}\}_{|m_\||<k^{3\delta}}$ and $\zeta^0 \equiv \{\zeta^0_{m_\|}\}_{|m_\||<k^{3\delta}}$ be given by the formulae:

$$\beta^0_{m_\|} = \tau_{m_\| 0}, \tag{5.6.26}$$

$$\zeta^0_{m_\|} = \tilde{\tau}_{m_\| 0}, \tag{5.6.27}$$

$$\tau = -P(A_+^0)^{-1}(\mathcal{E} - 2A_+^0 F^{(0)})(I + \mathcal{E})^{-1} A_+^0, \tag{5.6.28}$$

$$\tilde{\tau} = 2(A_+^0)^{-1}(I + \mathcal{E})^{-1}(A_+^0). \tag{5.6.29}$$

Now, we prove that the continuity conditions on the surface for the functions Ψ^0_{refl}, Ψ^0_{refr}, given by (5.6.1), (5.6.2), are satisfied with an accuracy to $O(k^{-k^{2\delta} R_0^{-1}})$.

We introduce the vectors Δ_0, Δ_1 as follows:

$$\Delta_0(k^2, k_\|, x_\|)_{j_\|} = \left((\Psi^0_{refr} - \Psi^0_{refl} - \exp i(\mathbf{k}, x))\big|_{x_1=0} \right)_{j_\|}, \tag{5.6.30}$$

$$\Delta_1(k^2, k_\|, x_\|)_{j_\|} = \left(\frac{\partial(\Psi^0_{refr} - \Psi^0_{refl} - \exp i(\mathbf{k}, x))}{\partial x_1}\bigg|_{x_1=0} \right)_{j_\|}, \tag{5.6.31}$$

where functions Ψ^0_{refl}, Ψ^0_{refr} are given by (5.6.1), (5.6.2), (5.6.26) – (5.6.29).

Lemma 5.32 . *If* $\mathbf{k}$ *is in the* $(k^{-n+1-6\delta})$*-neighborhood of* $S^{(n)}(k, \alpha V, \delta)$, $\alpha \in [0,1]$, *then for sufficiently large* k, $k > k_0(V, \delta)$, *the vectors* $\Delta_0(k^2, k_\|, x_\|)$, $\Delta_1(k^2, k_\|, x_\|)$ *satisfy the estimates:*

$$|A_+^0 \Delta_0| < (\alpha k^{-1+31\delta})^{R_2}, \tag{5.6.32}$$

$$|\Delta_1| < (\alpha k^{-1+21\delta})^{R_2}, \tag{5.6.33}$$

$$R_2 = k^{3\delta} R_0^{-1}.$$

Proof. First, we prove the estimate:

$$|(I+\mathcal{E})^{-1}_{m_\| j_\|}| < \alpha^{|m_\| - j_\||R_0^{-1}} k^{-(1-28\delta)|m_\| - j_\||R_0^{-1}}. \tag{5.6.34}$$

Indeed,

$$(I+\mathcal{E})^{-1}_{m_\| j_\|} = \sum_{l=0}^{\infty} (-\mathcal{E}^l)_{m_\| j_\|}. \tag{5.6.35}$$

Taking into account estimates (5.6.14) – (5.6.16) and formulae (5.6.22) – (5.6.24), we get the following estimate for $\mathcal{E}$:

$$|\mathcal{E}_{m_\| j_\|}| \le \left(\alpha k^{-(1-15\delta)}\right)^{|j_\| - m_\||R_0^{-1}}. \tag{5.6.36}$$

Next, we prove that the operator $\mathcal{E}^l$ satisfies a similar inequality:

$$|(\mathcal{E}^l)_{m_\| j_\|}| \le \left(\alpha k^{-(1-27\delta)}\right)^{|j_\| - m_\||R_0^{-1}}. \tag{5.6.37}$$

Indeed, from (5.6.12) and (5.6.13) it follows:

$$\|\mathcal{E}\| < k^{-(1-8\delta)},$$

and, therefore,

$$|(\mathcal{E}^l)_{m_\| j_\|}| \le \alpha^l k^{-(1-8\delta)l}. \tag{5.6.38}$$

Suppose $l > |j_\| - m_\||R_0^{-1}$. Then, from (5.6.38) we easily get (5.6.37). In the case of the opposite inequality $l \le |j_\| - m_\||R_0^{-1}$, we represent $\mathcal{E}^l_{m_\| j_\|}$ in the form:

$$\mathcal{E}^l_{m_\| j_\|} = \sum_{i_1,\dots,i_{l-1} \in Z^{n-1}} \mathcal{E}_{m_\| i_1} \dots \mathcal{E}_{i_{l-1} j_\|}. \tag{5.6.39}$$

Since $\mathcal{E}_{m_\| j_\|} = 0$, when $|j_\|| > k^{3\delta}$ or $|m_\|| > k^{3\delta}$, the number of nonzero elements of the matric $\mathcal{E}$ does not exceed $k^{6\delta(n-1)}$. Therefore, the sum on the right of (5.6.39) contains less than $k^{6\delta(n-1)l}$ terms. Hence,

$$|\mathcal{E}_{m_\| j_\|}| \le k^{6\delta(n-1)l} \max_{i_1,\dots,i_{l-1} \in Z^{n-1}} |\mathcal{E}_{m_\| i_1} \dots \mathcal{E}_{i_{l-1} j_\|}|. \tag{5.6.40}$$

Applying estimate (5.6.36) to each factor on the right and taking into account that

$$|m_\| - i_1| + \sum_{k=1}^{l-2} |i_k - i_{k+1}| + |i_{l-1} - j_\|| \ge |j_\| - m_\||,$$

we get the inequality

$$|\mathcal{E}^l_{m_\| j_\|}| \le k^{6\delta(n-1)l} \left(\alpha k^{-(1-15\delta)}\right)^{|j_\| - m_\||R_0^{-1}}. \tag{5.6.41}$$

Noting that $l \le |j_{\|}-m_{\|}|R_0^{-1}$ and $n \le 3$, we obtain (5.6.37). Considering (5.6.35), and applying (5.6.37) when $l \le |j_{\|}-m_{\|}|R_0^{-1}$ and (5.6.38) when $l > |j_{\|}-m_{\|}|R_0^{-1}$, and also the obvious inequality $|j_{\|}-m_{\|}| < 2k^{3\delta}$, we get (5.6.34).

Now, taking into account that $|(A_+)_{j_{\|}j_{\|}}| < 2|(A_+)_{m_{\|}m_{\|}}|$ for any $j_{\|}$, $m_{\|}$, and formulae (5.6.27), (5.6.29) for ζ^0, we obtain the estimate:

$$|\zeta^0_{m_{\|}}| < \alpha^{|m_{\|}|R_0^{-1}} k^{-(1-28\delta)|m_{\|}|R_0^{-1}}. \qquad (5.6.42)$$

Next, we consider the functions:

$$\varphi(k^2, k_{\|} + p_{m_{\|}}(0), x) = \sum_{r=0}^{R_{m_{\|}}} \alpha^r B_r(k^2, k_{\|} + p_{m_{\|}}(0), x), \quad |m_{\|}| < k^{3\delta}, \qquad (5.6.43)$$

$$B_0(k^2, k_{\|} + p_{m_{\|}}(0), x) = \left(2A_+^0(k^2 + i0, k_{\|})\right)^{-1}_{m_{\|}m_{\|}} \Psi_+^0(k^2, k_{\|} + p_{m_{\|}}(0), x).$$

From the definitions of $F^{(0)}$, $F^{(1)}$, $F_r^{(0)}$, $F_r^{(1)}$ (see (5.6.5), (5.6.6), (5.6.21)) it is easy to see that

$$\left(\varphi(k^2, k_{\|} + p_{m_{\|}}(0), x)|_{x_1=0}\right)_{j_{\|}} = (F^{(0)} + (2A_+^0)^{-1})_{j_{\|}m_{\|}},$$

$$\left(\frac{\partial \varphi}{\partial x_1}(k^2, k_{\|} + p_{m_{\|}}(0), x)|_{x_1=0}\right)_{j_{\|}} = (F^{(1)} + P/2)_{j_{\|}m_{\|}}. \qquad (5.6.44)$$

Let

$$\tilde\varphi(k^2, k_{\|} + p_{m_{\|}}(0), x) = \sum_{|m_{\|}|<k^{3\delta}} \zeta^0_{m_{\|}} a_{m_{\|}} \varphi(k^2, k_{\|} + p_{m_{\|}}(0), x), \qquad (5.6.45)$$

where ζ^0, $a_{m_{\|}}$ are given by formulae (5.6.27), (5.3.12). By inspection, using relations (5.6.30) and (5.6.29), we verify that the function $\tilde\varphi$ and Ψ^0_{refl} satisfy the continuity condition on the boundary:

$$\tilde\varphi\,|_{x_1=0} = (\Psi^0_{refl} + \exp i(\mathbf{k}, x))\big|_{x_1=0},$$

$$\frac{\partial\tilde\varphi}{\partial x_1}|_{x_1=0} = \left.\frac{\partial(\Psi^0_{refl} + \exp i(\mathbf{k}, x))}{\partial x_1}\right|_{x_1=0}. \qquad (5.6.46)$$

From relations (5.6.30) and (5.6.31), it immediately follows:

$$\Delta_0 = (\Psi^0_{refr} - \tilde\varphi)|_{x_1=0}, \quad \Delta_1 = \left.\frac{\partial(\Psi^0_{refr} - \tilde\varphi)}{\partial x_1}\right|_{x_1=0}. \qquad (5.6.47)$$

We estimate Δ_0. From the definitions of Ψ^0_{refr} and $\tilde\varphi$ (see (5.6.2) and (5.6.45)), we get

$$(\Psi^0_{refr} - \tilde\varphi)(k^2, k_{\|}, x) = \sum_{m_{\|}} \zeta^0_{m_{\|}} a_{m_{\|}} (\psi_\alpha - \varphi)(k^2, k_{\|} + p_{m_{\|}}(0), x). \qquad (5.6.48)$$

Relations (5.6.3) and (5.6.43) give:

$$\left((\psi_\alpha - \varphi)(k^2, k_{\|} + p_{m_{\|}}(0), x)|_{x_1=0}\right)_{j_{\|}} = \alpha^{R_{m_{\|}}+1} Q^{(0)}_{j_{\|} m_{\|}}. \tag{5.6.49}$$

Hence it follows that

$$(\Delta_0)_{j_{\|}} = \sum_{m_{\|}} \zeta^0_{m_{\|}} a_{m_{\|}} \alpha^{R_{m_{\|}}+1} Q^{(0)}_{j_{\|} m_{\|}}. \tag{5.6.50}$$

From this formula, using (5.6.19) and (5.6.42), we obtain the estimate (5.6.32).
Similarly, one can show that

$$(\Delta_1)_{j_{\|}} = \sum_{m_{\|}} \zeta^0_{m_{\|}} a_{m_{\|}} \alpha^{R_{m_{\|}}+1} (Q^{(1)})_{j_{\|} m_{\|}}. \tag{5.6.51}$$

Using estimates (5.6.20), (5.6.34), we get inequality (5.6.33).
The lemma is proved. Naturally, we look for $\Psi_r efl$ in the form:

$$\Psi_{refl} = \sum_{m_{\|} \in Z^{n-1}} \beta_{m_{\|}} \psi_0^+(k^2, k_{\|} + p_{m_{\|}}(0), x). \tag{5.6.52}$$

We introduce the notations:

$$\delta\Psi_{refl} = \Psi_{refl} - \Psi^0_{refl},$$

$$\delta\Psi_{refr} = \Psi_{refr} - \Psi^0_{refr}. \tag{5.6.53}$$

From formulae (5.6.52), (5.6.1) it follows that

$$\delta\Psi_{refl} = \sum_{m_{\|} \in Z^{n-1}} (\beta_{m_{\|}} - \beta^0_{m_{\|}}) \psi_0^+(k^2, k_{\|} + p_{m_{\|}}(0), x), \tag{5.6.54}$$

where $\beta^0_{m_{\|}} = 0$ when $|m_{\|}| > k^{3\delta}$.

Theorem 5.5 . *If* $\mathbf{k}$ *is in the* $(k^{-n+1-2\delta})$*-neighborhood of* $S^{(n)}(k, \alpha V, \delta)$*, then for sufficiently large* k*,* $k > k_0(k, V, \delta)$*, the functions* $\delta\Psi_{refl}$*,* $\delta\Psi_{refr}$ *satisfy the following estimates:*

$$\left\| \delta\Psi_{refl}|_{x_1=0} \right\|_{L_2(K_{\|})} < (\alpha/k)^{k^\delta}, \tag{5.6.55}$$

$$\left\| \left. \frac{\partial \delta\Psi_{refl}}{\partial x_1} \right|_{x_1=0} \right\|_{L_2(K_{\|})} < (\alpha/k)^{k^\delta}, \tag{5.6.56}$$

$$\left\| \delta\Psi_{refr}|_{x_1=0} \right\|_{L_2(K_{\|})} < (\alpha/k)^{k^\delta}, \tag{5.6.57}$$

$$\left\| \left. \frac{\partial \delta\Psi_{refr}}{\partial x_1} \right|_{x_1=0} \right\|_{L_2(K_{\|})} < (\alpha/k)^{k^\delta}. \tag{5.6.58}$$

Proof. We prove that $\delta\Psi_{refl}$ and $\delta\Psi_{refr}$ satisfy the boundary conditions:

$$(\delta\Psi_{refl} - \delta\Psi_{refr})|_{x_1=0} = \Delta_0; \quad \frac{\partial}{\partial x_1}(\delta\Psi_{refl} - \delta\Psi_{refr})|_{x_1=0} = \Delta_1. \tag{5.6.59}$$

Indeed, relation (5.6.59) easily follows from (5.6.53), (5.6.46), (5.6.47) and the continuity conditions on the boundary:

$$\Psi_{refr}|_{x_1=0} = (\Psi_{refl} + \exp i(\mathbf{k}, x))|_{x_1=0},$$

$$\frac{\partial}{\partial x_1}\Psi_{refr}\Big|_{x_1=0} = \frac{\partial}{\partial x_1}(\Psi_{refl} + \exp i(\mathbf{k}, x))\Big|_{x_1=0}. \tag{5.6.60}$$

We introduce $z \in l_2^{(n-1)}$ by the formula

$$z_{j_\|} = (\delta\Psi_{refr}|_{x_1=0})_{j_\|}. \tag{5.6.61}$$

From equations (5.6.59), taking into account the definitions of the operators A_+^V, A_-^0, we obtain

$$A_-^0 z - A_+^V z = \Delta_2, \tag{5.6.62}$$

$$\Delta_2 = \Delta_1 - A_-^0 \Delta_0. \tag{5.6.63}$$

Hence,

$$z = T^{-1}\Delta_2. \tag{5.6.64}$$

Using Theorem 5.2A in the case $n = 3$, and Theorem 5.4 in the case $n = 2$, we get

$$\|z\| < k^{200\delta}(\|\Delta_1\| + \|A_-^0\Delta_0\|), \tag{5.6.65}$$

i.e.,

$$\|\delta\Psi_{refr}|_{x_1=0}\|_{L_2(K_\|)} < k^{200\delta}(\|\Delta_1\| + \|A_-^0\Delta_0\|). \tag{5.6.66}$$

We introduce the notation:

$$y_{j_\|} = (\delta\Psi_{refl}|_{x_1=0})_{j_\|}. \tag{5.6.67}$$

From relations (5.6.59), (5.6.93) it follows that

$$\|y\| < 2k^{200\delta}(\|\Delta_1\| + \|A_-^0\Delta_0\|), \tag{5.6.68}$$

i.e.,

$$\|\delta\Psi_{refl}|_{x_1=0}\|_{L_2(K_\|)} < 2k^{200\delta}(\|\Delta_1\| + \|A_-^0\Delta_0\|). \tag{5.6.69}$$

Taking into account estimates (5.6.32) and (5.6.33), and the obvious estimate $R_0 < k^\delta$, we get (5.6.55) and (5.6.57).

To estimate the derivative of $\delta\Psi_{refr}$, we remark that

$$\left(\frac{\partial}{\partial x_1}\delta\Psi_{refr}\right)_{j_\|} = (A_+^V z)_{j_\|} = -\Delta_2 + A_-^0 z. \tag{5.6.70}$$

From inequalities (5.6.65) and (5.6.11), it follows that

$$\|A_+^V z\| < k^{1+200\delta}\|\Delta_2\|.$$

Taking into account the estimate for Δ_2 (see (5.6.32), (5.6.33) and (5.6.63)), we get inequality (5.6.58). Inequality (5.6.56) follows from estimate (5.6.58) and formula (5.6.59). The theorem is proved.

Theorem 5.6 . *If* **k** *is in the* $(k^{-n+1-2\delta})$*-neighborhood of* $S^{(n)}(k,\alpha V,\delta)$, $0<\alpha<1$, *then for sufficiently large* k, $k>k_0(k,V,\delta)$, *the vector of reflection coefficients* β *satisfies the following estimate:*

$$\|\beta-\beta_0\|_{l_2^{(n-1)}} < (\alpha/k)^{-k^\delta}. \tag{5.6.71}$$

Proof. Inequality (5.6.71) follows immediately from formula (5.6.54) and inequality (5.6.55). The theorem is proved.

Next, we look for the refracted wave in the form:

$$\Psi_{refr} = \sum_{|m_\parallel|<k^{3\delta}} \zeta_{m_\parallel} a_{m_\parallel} \psi_\alpha(k^2, k_\parallel + p_{m_\parallel}(0), x) + \sum_{|m_\parallel|\geq k^{3\delta}} \zeta_{m_\parallel} a_{m_\parallel} U_{m_\parallel}(x), \tag{5.6.72}$$

the functions $U_{m_\parallel}$ being defined in Sections 5.3, 5.4 (see (5.4.183), (5.4.185), (5.4.187)) for $n=3$ and (5.3.11), (5.3.56), (5.3.57) for $n=2$) $\psi_\alpha(k^2, k_\parallel + p_{m_\parallel}(0), x)$ being given by (5.5.62). Using formulae (5.6.2) and (5.6.53), we get:

$$\delta\Psi_{refr} = \sum_{|m_\parallel|<k^{3\delta}} (\zeta_{m_\parallel} - \zeta^0_{m_\parallel}) a_{m_\parallel} \psi_\alpha(k^2, k_\parallel + p_{m_\parallel}(\dot{0}), x) + \sum_{|m_\parallel|\geq k^{3\delta}} \zeta_{m_\parallel} a_{m_\parallel} U_{m_\parallel}(x), \tag{5.6.73}$$

where $\zeta^0_{m_\parallel}$ is given by (5.6.27).

Theorem 5.7 . *If* **k** *is in the* $(k^{-n+1-2\delta})$*-neighborhood of* $S^{(n)}(k,\alpha V,\delta)$, *then for sufficiently large* k, $k>k_0(k,V,\delta)$, *the vector of refraction coefficients* ζ *satisfies the following estimate:*

$$\|\zeta-\zeta_0\|_{l_2^{(n-1)}} < (\alpha/k)^{k^\delta}, \tag{5.6.74}$$

$(\zeta_0)_{m_\parallel} = \tilde{\tau}_{m_\parallel 0}$, *when* $|m_\parallel| < k^{3\delta}$ *and* $\zeta^0_{m_\parallel} = 0$ *otherwise;* $\tilde{\tau}$ *is given by (5.6.29).*

Proof. We use notation (5.6.61) and estimate (5.6.65). Taking into account (5.6.32), (5.6.33), we get:

$$\|z\| < (\alpha/k)^{-k^\delta}. \tag{5.6.75}$$

From the definitions of the functions $U_{m_\parallel}$ and ψ_α, it easily follows that

$$z = \hat{I}\zeta, \tag{5.6.76}$$

$$\hat{I} = J(I-P) + \left(I + (F^{(0)} + Q^{(0)})A_+\right)P.$$

Suppose $n = 3$. Taking into account the inequalities (5.4.227), (5.4.168) and relation(5.4.166), it is easy to see that

$$\|J(I-P)\zeta_0\| > k^{-8\delta}\|(I-P)\zeta_0\|. \tag{5.6.77}$$

Using (5.6.12), (5.6.21) and (5.6.19), we obtain

$$2\|\left(I + (F^{(0)} + Q^{(0)})A_+\right)P\zeta_0\| > \|P\zeta_0\|. \tag{5.6.78}$$

From (5.6.77) and (5.6.78) it follows:

$$\|\hat{I}^{-1}\| < k^{8\delta} \tag{5.6.79}$$

From this inequality, (5.6.74) follows. Similar considerations give the proof for $n = 2$. The theorem is proved.

We note that all obtained estimates hold not only for the incident wave $\psi_0^+(k^2, k_{\|}, x)$, but also for the incident waves: $\psi_0^+(k^2, k_{\|} + p_{q_{\|}}(0), x)$, $|q_{\|}| < k^{\delta}$. We denote by fi $_{q_{\|}}$ and ı $_{q_{\|}}$ the vectors of reflection and refraction coefficients, respectively, corresponding to $\psi_0^+(k^2, k_{\|} + p_{q_{\|}}(0), x)$. Thus, the following generalization of Theorems 5.6, 5.7 holds:

Theorem 5.8 . *If* k *is in the* $(k^{-n+1-2\delta})$*-neighborhood of* $S^{(n)}(k, \alpha V, \delta)$*, then for sufficiently large* k*,* $k > k_0(k, V, \delta)$*, the vectors of reflection coefficients* $\beta_{q_{\|}}$*,* $|q_{\|}| < k^{\delta}$*, satisfy the following estimate:*

$$\|\beta_{q_{\|}} - \beta^0_{q_{\|}}\|_{l_2^{(n-1)}} < (\alpha/k)^{k^{\delta}}, \tag{5.6.80}$$

$$\|\zeta_{q_{\|}} - \zeta^0_{q_{\|}}\|_{l_2^{(n-1)}} < (\alpha/k)^{k^{\delta}}, \tag{5.6.81}$$

$$(\beta^0_{q_{\|}})_{m_{\|}} = \tau_{m_{\|} q_{\|}}, \tag{5.6.82}$$

$$(\zeta^0_{q_{\|}})_{m_{\|}} = \tilde{\tau}_{m_{\|} q_{\|}}. \tag{5.6.83}$$

The vectors τ, $\tilde{\tau}$ are given by (5.6.28), (5.6.29).

Next, we consider the kernel $R_z(x, y)$ of the resolvent of the operator $H_+(k_{\|})$ at a complex point $z = k^2 + i\varepsilon$. Suppose $y = (0, y_{\|})$, i.e., y lies on the boundary of the crystal. Considering $R_z(x, y)$ as a function of $y_{\|}$, we denote by $r_z(x)_{q_{\|}}$ its components with respect to the $k_{\|}$-basis. It is clear that $r_z(x)_{q_{\|}}$ satisfies the equation:

$$(-\Delta + V_+ - z)r_z(x)_{q_{\|}} = \delta(x_1)\exp i(k_{\|} + p_{q_{\|}}(0), x_{\|}).$$

Hence,

$$r_z(x)_{q_{\|}}|_{x_1=+0} - r_z(x)_{q_{\|}}|_{x_1=-0} = 0,$$

$$\frac{\partial}{\partial x_1} r_z(x)_{q_{\|}}\Big|_{x_1=+0} - \frac{\partial}{\partial x_1} r_z(x)_{q_{\|}}\Big|_{x_1=-0} = \exp i(k_{\|} + \mathbf{p}_{q_{\|}}(0), x_{\|}). \tag{5.6.84}$$

We consider the function:

$$\psi = (\psi_+^0 - \psi_-^0)(z, k_\parallel, x)\chi_-(x_1) + 2ik_1 r_z(x)_0, \tag{5.6.85}$$

where $\chi_-(\cdot)$ is the indicator of the negative axis. It is clear that ψ satisfies the continuity conditions on the boundary:

$$\psi|_{x_1=+0} = \psi|_{x_1=-0},$$
$$\frac{\partial}{\partial x_1}\psi|_{x_1=+0} = \frac{\partial}{\partial x_1}\psi|_{x_1=-0}, \tag{5.6.86}$$

and the equation $(-\Delta + V_+ - z)\psi = 0$ in the strip $Q_+ \cup Q_-$. Moreover, since $z = k^2 + i\varepsilon$, $\varepsilon > 0$, and ψ_-^0 exponentially decays when $x_1 \to -\infty$, we have $\psi - \psi_+^0 \in L_2(R^n)$. Thus, $\psi - \psi_+^0$ satisfies the definitions of reflected and refracted waves. Therefore

$$\Psi_{refl} = (\psi - \psi_+^0)\chi_-(x_1) = -\psi_-^0(k^2, k_\parallel, x) + 2ik_1 r_z(x)_0, \quad x_1 \le 0,$$
$$\Psi_{refr} = 2ik_1 r_z(x)_0, \quad x_1 \ge 0. \tag{5.6.87}$$

It is clear that the reflection coefficients are given by the formula

$$\beta_{m_\parallel} = -\delta_{m_\parallel 0} + 2ik_1 r_z(0)_{0m_\parallel}, \tag{5.6.88}$$

where $r_z(0)_{0m_\parallel}$ are the components of the function $r_z(x)_0|_{x_1=0}$ of $x_\parallel$ with respect to the $k_\parallel$-basis.

We denote by $f(x_1, y_1)_{n_\parallel m_\parallel}$ the components of a function $f(x, y)$, $x, y \in Q_- \cup Q_+$ with respect to the $k_\parallel$-basis, for fixed x_1, y_1.

We have calculated β^0, ζ^0 (see (5.6.26) – (5.6.29)). Let us find another form of an asymptotic expansion for the reflection coefficients.

Lemma 5.33 . *Suppose* **k** *is in the* $(k^{-n+1-6\delta})$*-neighborhood of* $S^{(n)}(k, \alpha_0 V, \delta)$ *for some* $\alpha_0 \in [0, 1]$. *Then for sufficiently large* k, $k > k_0(k, V, \delta)$, *the vector of the reflection coefficients* fi *can be represented in the form:*

$$\beta_{m_\parallel}(k^2, \alpha_0 V, k_\parallel) =$$

$$2i\sqrt{k^2 - |k_\parallel|^2}\sum_{r=1}^{M}\alpha_0^r \left(R_{k^2+i0}^0(V_+ R_{k^2+i0}^0)^r\right)_{m_\parallel 0}(0,0) + O(\alpha_0^M k^{-M}), \tag{5.6.89}$$

where $M = [k^\delta]$.

The following estimates hold:

$$\left|\left(R_{k^2}^0(V R_{k^2}^0)^r\right)_{m_\parallel 0}(0,0)\right| < k^{-(1-12\delta)r-1}. \tag{5.6.90}$$

Proof. According to Theorem 5.6, $\beta_{m_\parallel}(k^2, \alpha_0 V, k_\parallel)$ can be represented in the form

$$\beta_{m_\parallel}(k^2, \alpha_0 V, k_\parallel) = \beta_{m_\parallel}^0(k^2, \alpha_0 V, k_\parallel) + O(\alpha_0^M k^{-M}). \tag{5.6.91}$$

Now we check that β^0 can be expanded in the series:

$$\beta^0_{m_\|}(\alpha_0 V, k^2+i\varepsilon) = \sum_{r=1}^{\infty} \gamma_{m_\| r}\alpha_0^r, \tag{5.6.92}$$

where

$$|\gamma_{m_\| r}| < k^{-(1-12\delta)r}. \tag{5.6.93}$$

We use formulae (5.6.28) and (5.6.26). In fact, expanding $(I+\mathcal{E})^{-1}$ in a series in powers of $\mathcal{E}$ and using formulae (5.6.22) and (5.6.25), we obtain:

$$(I+\mathcal{E})^{-1} = \sum_{p=0}^{\infty}(-1)^p \sum_{r_1 \dots r_p=1}^{M_1} \tilde{\mathcal{E}}_{r_1}\dots\tilde{\mathcal{E}}_{r_p}\alpha_0^{r_1+\dots+r_p}.$$

Introducing a new index of the summation $r = r_1+\dots+r_p$, we get:

$$(I+\mathcal{E})^{-1} = I + \sum_{r=1}^{\infty} b_r \alpha^r,$$

$$b_r = \sum_{p=0}^{r}(-1)^p \sum_{r_1 \dots r_{p-1}=1}^{M_1} \tilde{\mathcal{E}}_{r_1}\dots\tilde{\mathcal{E}}_{r-r_1-\dots-r_{p-1}}, \quad M_1 = [k^{3\delta}].$$

From the definition of $\tilde{\mathcal{E}}_r$ and estimates (5.6.12), (5.6.13), it follows $\|\tilde{\mathcal{E}}_{r_i}\| < k^{-(1-8\delta)r_i}$. Since $M_1 = k^{3\delta}$, it is easy to verify that

$$\|b_r\| < k^{-(1-12\delta)r}.$$

Similarly, we obtain:

$$\left(\mathcal{E} - A_+^0 F^{(0)}\right)(I+\mathcal{E})^{-1} = \sum_{r=1}^{\infty} \tilde{b}_r \alpha_0^r,$$

$$\|\tilde{b}_r\| < k^{-(1-11\delta)r}.$$

Taking into account that $\|A_+^0 P\| < k$, $\|(A_+^0)^{-1}P\| < k^{-1+\delta}$ (P is a diagonal projector given by (5.6.9)) and using (5.6.28), we arrive at formula (5.6.92) and estimate (5.6.93). Using relation (5.6.91), we verify that

$$\beta_{m_\|}(k^2+i\varepsilon, \alpha_0 V, k_\|) = \sum_{r=1}^{M} \gamma_{m_\| r}\alpha_0^r + O(\alpha_0^M k^{-M}). \tag{5.6.94}$$

Note that these formulae and estimates are valid in some upper neighborhood of k^2. Now we show that

$$\gamma_{m_\| r} = 2i\sqrt{k^2-|k_\||^2}\left(R^0_{k^2+i0}(V_+ R^0_{k^2+i0})^r\right)_{m_\| 0}(0,0). \tag{5.6.95}$$

According to Lemmas 5.25 and 5.29 $S^{(n)}(k,\alpha_0 V,\delta) \supset S^{(n)}(k,\alpha V,\delta)$ for any $0<\alpha<\min\{\alpha_0, k^{-1-\delta}\}$. Considering as in proving (5.6.94), we obtain a similar formula for $\beta_{m_\|}(k^2+i\varepsilon,\alpha V,k_\|)$:

$$\beta_{m_\|}(k^2+i\varepsilon,\alpha V,k_\|)=\sum_{r=1}^{M}\gamma_{m_\| r}\alpha^r+O(\alpha^M k^{-M}) \tag{5.6.96}$$

with the same coefficients $\gamma_{m_\| r}$. For α small enough, the resolvent R_z, $z=k^2+i\varepsilon$, can be expanded in a series in powers of α. From (5.6.88) it follows:

$$\beta_{m_\|}(z,\alpha V,k_\|)=-\delta_{m_\| 0}+2i\sqrt{k^2-k_\|^2}\sum_{r=0}^{\infty}\alpha^r\left(R_z^0(V_+R_z^0)^r\right)_{m_\| 0}(0,0).$$

Taking into account that $2i\sqrt{k^2-k_\|^2}\left(R_z^0\right)_{m_\| 0}(0,0)=\delta_{m_\| 0}$, we get

$$\beta_{m_\|}(\alpha V,z)=2i\sqrt{k^2-k_\|^2}\sum_{r=1}^{\infty}\alpha^r\left(R_z^0(V_+R_z^0)^r\right)_{m_\| 0}(0,0). \tag{5.6.97}$$

Comparing formula (5.6.94) for $|\alpha|<\varepsilon\|V\|^{-1}$ with relation (5.6.97) gives (5.6.95). The lemma is proved.

Thus, we have obtained k^δ leading terms of the expansion of the vector of the reflection coefficients in a series in powers of α and the estimate for the remainder. It is in order k^{-k^δ}. The vector coefficients at powers of α satisfy estimates (5.6.90). However, these estimates are rather rough. The following two lemmas provide us more precise asymptotic formulae.

Lemma 5.34 . *If* k *is in the* $(k^{-n+1-2\delta})$*-neighborhood of* $S^{(n)}(k,V,\delta)$ *and* $|m_\||<k^{3\delta}$, $r\le k^\delta$, *then the following formula holds:*

$$2ik_1R^0_{k^2+i0}\left(V_+R^0_{k^2+i0}\right)^r(x_1,0)_{m_\| 0}=$$

$$\sum_{j_1:j_1\in Z,|j_1|<rR_0}\;\sum_{q_\|:|q_\||<k^{3\delta},|m_\|-q_\||<rR_0}\;\sum_{p_1=0}^{r}b^{(r,m_\|)}_{p_1,q_\|,j_1}\frac{x_1^{p_1}}{p_1!}\exp i(k_1(q_\|)+2\pi a_1^{-1}j_1)x_1, \tag{5.6.98}$$

where

$$k_1(q_\|)\equiv\sqrt{k^2-|k_\|+p_{q_\|}(0)|^2} \tag{5.6.99}$$

and the complex coefficients $b^{(r,m_\|)}_{p_1,q_\|,j_1}$ *satisfy the estimates*

$$\left|b^{(r,m_\|)}_{p_1,q_\|,j_1}\right|\le k^{-(1-13\delta)r}. \tag{5.6.100}$$

Proof. Suppose $z=k^2+i\varepsilon$, $\varepsilon>0$ and $\alpha\ll\varepsilon\|V\|^{-1}$. Then, using the expression of Ψ_{refr} by the resolvent (see (5.6.87)) and expanding the resolvent in a series in powers of α, we obtain a Taylor series for the refracted wave:

$$(\Psi_{\alpha\ refr})_{m_\|}(x_1) = 2ik_1 \sum_{r=0}^{\infty} \alpha^r \left(R_z^0(V_+R_z^0)^r\right)_{m_\| 0}(x_1, 0). \qquad (5.6.101)$$

The series converges in the disk $|\alpha| < \varepsilon\|V\|^{-1}$. On the other hand, for α small enough, $S^{(n)}(k,\delta,\alpha V) \subset S^{(n)}(k,\delta,V)$ and therefore the assertion of Theorem 5.5 holds for αV, i.e., $\Psi_{refr}(\alpha V) = \Psi^0_{refr}(\alpha V) + \delta\Psi_{refr}(\alpha V)$, the function $\Psi^0_{refr}(\alpha V)$ being given by formula (5.6.2) and $\delta\Psi_{refr}(\alpha V)$ satisfying estimates (5.6.57), (5.6.58).

Using relations (5.6.74) and (5.6.101), we show that

$$\left(\Psi^0_{\alpha\ refr}\right)_{m_\|}(x_1) = 2ik_1 \sum_{r=0}^{M} \alpha^r \left(R_z^0(V_+R_z^0)^r\right)_{m_\| 0}(x_1, 0) + O(\alpha^{M+1}) \qquad (5.6.102)$$

for any fixed x_1.

Next, we obtain the expansion of Ψ^0_{refr} in a series in powers of α. Substituting expansions (5.6.94) for $\beta_{q_\|}$ and (5.5.62) for $\Psi_\alpha(k^2, k_\| + p_{q_\|}(0), x)$ in formula

$$\Psi^0_{refr} = \sum_{|q_\||<k^{3\delta}} \beta_{q_\|} a_{q_\|} \Psi_\alpha(k^2, k_\| + p_{q_\|}(0), x), \qquad (5.6.103)$$

we get:

$$\left(\Psi^0_{refr}\right)_{m_\|} = \sum_{r=0}^{M} \alpha^r a_r(k^2, k_\|, x_1)_{m_\|} + O(\alpha^{M+1}), \qquad (5.6.104)$$

$$a_r(k^2, k_\|, x_1)_{m_\|} = \sum_{|q_\||<k^{3\delta}} \sum_{r_1=0}^{r} \gamma_{q_\| r_1} B_{r-r_1}(V, k^2, k_\| + p_{q_\|}(0), x_1)_{m_\|}. \qquad (5.6.105)$$

Comparing formulae (5.6.104) and (5.6.102), we obtain that the coefficients at the powers of α are equal to each other:

$$2ik_1 \left(R_z^0(V_+R_z^0)^r\right)_{m_\| 0} = a_r(k^2, k_\|, x_1)_{m_\|}. \qquad (5.6.106)$$

We consider $a_r(k^2, k_\|, x_1)_{m_\|}$. According to Theorems 5.2 and 5.4, B_{r-r_1} can be represented in the form:

$$B_{r-r_1}(V, k^2, k_\| + p_{q_\|}(0), x) =$$

$$\sum_{j\in Z^n, |j|<(r-r_1)R_0} \sum_{p_1=0}^{r-r_1} a_{jp_1}^{(q_\|, r-r_1)} \frac{x_1^{p_1}}{p_1!} \exp i(k_1(q_\|)x_1 + (k_\| + p_{q_\|}(0), x_\|) + (\mathbf{p}_j(0), x)), \qquad (5.6.107)$$

where the coefficients $a_{jp_1}^{(q_\|, r-r_1)}$ satisfy the estimates:

$$\left|a_{jp_1}^{(q_\|, r-r_1)}\right| < k^{-1-(r-r_1)(1-8\delta)}. \qquad (5.6.108)$$

Substituting the right side of formula (5.6.107) in (5.6.105), we arrive at the formula:

$$a_r =$$

$$\sum_{|q_\parallel|<k^{3\delta}} \sum_{j\in Z^n,\ |j|<rR_0} \sum_{p_1=0}^{r} \tilde{b}^{(r)}_{p_1,q_\parallel j} \frac{x_1^{p_1}}{p_1!} \exp i\left((k_1(q_\parallel)x_1 + (k_\parallel + p_{q_\parallel}(0), x_\parallel) + (\mathbf{p}_j(0), x)\right),$$

where

$$\tilde{b}^{(r)}_{p_1,q_\parallel j} = \sum_{r_1=0}^{r} a_{q_\parallel} \gamma_{q_\parallel r_1} a_{jp_1}^{(q_\parallel, r-r_1)};$$

here we define $a_{jp_1}^{(q_\parallel, r-r_1)} = 0$, when $|j| \geq (r-r_1)R_0$. From estimates (5.6.93) and (5.6.108), it follows:

$$\left|\tilde{b}^{(r)}_{p_1,q_\parallel j}\right| < rk^{-(1-12\delta)r} < k^{-(1-13\delta)r}. \tag{5.6.109}$$

Introducing a new index of summation, $m_\parallel = q_\parallel + j_\parallel$, yields:

$$a_r = \sum_{q_\parallel:|q_\parallel|<k^{3\delta}} \sum_{m_\parallel:|m_\parallel - q_\parallel|<rR_0} \sum_{|j_1|<rR_0} \sum_{p_1=0}^{r} b^{(r,m_\parallel)}_{p_1,q_\parallel,j_1} \frac{x_1^{p_1}}{p_1!} \times \tag{5.6.110}$$

$$\exp i\left(k_1(q_\parallel)x_1 + 2\pi a_1^{-1} j_1 x_1 + (k_\parallel + p_{m_\parallel}(0), x_\parallel)\right),$$

where

$$b^{(r,m_\parallel)}_{p_1,q_\parallel,j_1} = \tilde{b}^{(r)}_{p_1,q_\parallel,j}, \quad j = (j_1, m_\parallel - q_\parallel). \tag{5.6.111}$$

Hence

$$(a_r)_{m_\parallel} = \sum_{q_\parallel:|q_\parallel|<k^{3\delta}} \sum_{j_1:|j_1|<rR_0} \sum_{p_1=0}^{r}{}' \, b^{(r,m_\parallel)}_{p_1,q_\parallel,j_1} \frac{x_1^{p_1}}{p_1!} \exp i(k_1(q_\parallel)x_1 + 2\pi a_1^{-1} j_1 x_1). \tag{5.6.112}$$

Passing to limits ($\varepsilon \to 0$) in all the relations, and taking into account that all the estimates are uniform with respect to ε, we obtain (5.6.98) from (5.6.106), (5.6.112) and (5.6.100) from (5.6.109), (5.6.111). The lemma is proved.

We consider the vector-valued function $\mathbf{Y}_r(x_1)$:

$$\mathbf{Y}_r(x_1)_{m_\parallel} = 2ik_1 \exp(-ik_1(m_\parallel)x_1)(R^0_{k^2+i0}(V + R^0_{k^2+i0})^r)(x_1, 0)_{m_\parallel 0}, \quad x_1 > 0, \tag{5.6.113}$$

$k_1(m_1)$ being given by (5.6.99).

It is clear that

$$\mathbf{Y}_0(x_1) = \delta_{m_\parallel 0}. \tag{5.6.114}$$

Lemma 5.35 . *If* $\mathbf{k}$ *is in the* $(k^{-n+1-6\delta})$*-neighborhood of* $S^{(n)}(k, V, \delta)$ *then for sufficiently large* k, $k > k_0(k, V, \delta)$, *the* p*-derivative of the vector-valued function* $\mathbf{Y}_r(x_1)$ *can be represented in the form:*

$$\frac{d^p \mathbf{Y}_r(x_1)}{dx_1^p} = \sum_{p_1=0}^{r} \mathbf{g}_{p_1}^{(p,r)}(x_1)\frac{x_1^{p_1}}{p_1!}, \quad p \geq 0, \quad r \geq 1, \tag{5.6.115}$$

where l-derivative, $l \geq 0$ of the vector-valued functions $\mathbf{g}_{p_1}^{(p,r)}(x_1)$ satisfy the uniform in ε estimates:

$$\left\| \frac{d^l \mathbf{g}_{p_1}^{(p,r)}(x_1)}{dx_1^l} \right\|_{L_2(Q_\parallel)} < 2^p k^{-(1-13\delta)r+3\delta(l+p+8)} \tag{5.6.116}$$

for any fixed x_1.

Corollary 5.8 .

$$\left\| \left. \frac{d^p \mathbf{Y}_r(x_1)}{dx_1^p} \right|_{x_1=0} \right\|_{L_2(Q_\parallel)} \leq 2^p k^{-(1-13\delta)r+3\delta(p+8)}. \tag{5.6.117}$$

Proof. In accordance with formulae (5.6.98) and (5.6.113), relation (5.6.115) is valid for $p = 0$, where

$$\left(\mathbf{g}_{p_1}^{(0,r)}\right)_{m_\parallel} =$$

$$\sum_{q_\parallel : |q_\parallel| < k^{3\delta}, |m_\parallel - q_\parallel| < rR_0, j_1 \in Z : |j_1| < rR_0} b_{p_1 q_\parallel j_1}^{(r,m_\parallel)} \exp i(k_1(q_\parallel) - k_1(m_\parallel) + 2\pi a_1^{-1} j_1)x_1.$$

Estimate (5.6.116) holds for $\mathbf{g}_{p_1}^{(0,r)}$, since $|j_1| < rR_0 < MR_0 < k^{\delta/2}$; and $|k_1(q_\parallel) - k_1(m_\parallel)| < k^{3\delta}$, because $|q_\parallel| < k^{3\delta}$, $|m_\parallel| < k^{3\delta}$; and relations (5.6.100) are satisfied. We prove formulae (5.6.115) and (5.6.116) for $p > 0$ by induction. Indeed, suppose we have proved the formula for $p = p_0 - 1$, $p_0 \geq 1$. We verify that it holds for $p = p_0$. In fact,

$$\frac{d^{p_0} \mathbf{Y}_r}{dx_1^{p_0}} =$$

$$\sum_{p_1=0}^{r} \frac{d\mathbf{g}_{p_1}^{(p_0-1,r)}}{dx_1} \frac{x_1^{p_1}}{p_1!} + \sum_{p_1=1}^{r} \mathbf{g}_{p_1}^{(p_0-1,r)} \frac{x_1^{p_1-1}}{(p_1-1)!}.$$

Setting

$$\mathbf{g}_{p_1}^{(p_0,r)} = \frac{d\mathbf{g}_{p_1}^{(p_0-1,r)}}{dx_1} + \mathbf{g}_{p_1+1}^{(p_0-1,r)}, \tag{5.6.118}$$

we get (5.6.115). From estimates (5.6.116) for $p = p_0 - 1$ and formula (5.6.118), it follows (5.6.116) for $p = p_0$. The lemma is proved.

Next, we consider the functions $V_{m_\parallel}(x_1)$, which are the Fourier coefficients of $V(x)$ with respect to $x_\parallel$. It is convenient also to introduce the functions $\tilde{V}_{m_\parallel q_\parallel}(x_1)$:

$$\tilde{V}_{m_\parallel q_\parallel}(x_1) = V_{m_\parallel - q_\parallel}(x_1) \exp i\left((k_1(q_\parallel) - k_1(m_\parallel))x_1\right). \tag{5.6.119}$$

It is clear that l-th derivative of $\tilde{V}_{m_\parallel q_\parallel}$ at $x_1 = 0$ can be represented as a linear combination of i-th derivatives $i \leq l$ of the functions $V_{m_\parallel q_\parallel}$ at $x_1 = 0$.

Lemma 5.36 . *If* $\mathbf{k}$ *is in the* $(k^{-n+1-6\delta})$*-neighborhood of* $S^{(n)}(k,V,\delta)$ *then for sufficiently large* k, $k > k_0(k,V,\delta)$, *the vectors* $\mathbf{Y}_1^{(p)}(0)$, $= 0,1,...$ *can be represented in the form of the series:*

$$\mathbf{Y}_1(0)_{m_\|} = \frac{1}{2ik_1(m_\|)}\sum_{l=0}^{\infty}\frac{i^{l+1}V^{(l)}(0)_{m_\|}}{(k_1(m_\|)+k_1(0))^{l+1}}, \tag{5.6.120}$$

$$\mathbf{Y}_1(0) = \frac{1}{2ik_1(m_\|)}\sum_{l=0}^{\infty}\frac{i^{l+1}\tilde{V}^{(l)}(0)_{m_\|0}}{(2k_1(m_\|))^{l+1}}, \tag{5.6.121}$$

where [14]

$$k_1(m_\|) > k^{1-3\delta}, \quad k_1(0) > k^{1-3\delta}, \tag{5.6.122}$$

$$|V^{(l)}(0)| < (\|V\|R_0)^l, \quad \left|\tilde{V}^{(l)}(0)_{m_\|0}\right| < (\|V\|R_0k^{3\delta})^l. \tag{5.6.123}$$

Proof. From the definition of $\mathbf{Y}_1$ (see (5.6.113)), it follows that

$$\mathbf{Y}_1(0)_{m_\|} = \frac{1}{2ik_1(m_\|)}\int_0^\infty \exp(i(k_1(m_\|)+k_1(0))y)\; V_{m_\|}(y)dy.$$

Integrating by parts gives series (5.6.120). Since $\mathbf{k}$ is in the $(k^{-n+1-6\delta})$-neighborhood of $S^{(n)}(k,V,\delta)$, estimates (5.6.122) hold. Inequalities (5.6.123) are satisfied because V is a trigonometric polynomial. We obtain the series (5.6.121) using the formula:

$$\mathbf{Y}_1(0) = \frac{1}{2ik_1(m_\|)}\int_0^\infty \exp(2ik_1(m_\|)y)\; \tilde{V}_{m_\|0}(y)dy.$$

Note that $|k_1(m_\|) - k_1(0)| < k^{3\delta}$, because $|m_\|| < R_0$; otherwise $(\mathbf{Y}_1)_{m_\|} = 0$. Using this relation, we easily obtain (5.6.123). The lemma is proved.

Let

$$\mathbf{Y}_r^{(p)} = \left\{\frac{d^p}{dx_1^p}Y_r(x_1)_{m_\|}\bigg|_{x_1=0}\right\}_{m_\|\in Z^{n-1}}. \tag{5.6.124}$$

Theorem 5.9 . *If* $\mathbf{k}$ *is in the* $(k^{-n+1-6\delta})$*-neighborhood of* $S^{(n)}(k,V,\delta)$, *then for sufficiently large* k, $k > k_0(k,V,\delta)$, *the vector* $\mathbf{Y}_r^{(p)}(0)$ *can be represented in the form of the series:*

$$\left(\mathbf{Y}_r^{(p)}\right)_{m_\|} = -\sum_{l=0}^{\infty}\frac{1}{(-2ik_1(m_\|))^{l+1}}(\tilde{V}\mathbf{Y}_{r-1})_{m_\|}^{(p+l-1)}, \tag{5.6.125}$$

$(\tilde{V}Y_{r-1}^{(-1)})(0) \equiv 0$, *the following estimates being valid:*

$$\left|(\tilde{V}\mathbf{Y}_{r-1})_{m_\|}^{(p+l-1)}\right| \le k^{6\delta(p+l+8)}k^{-(1-3\delta)(r-1)}. \tag{5.6.126}$$

[14] Note that $k_1(0)$ corresponds to $m_\| = 0$ i.e., $k_1(0) \equiv k_1 = \sqrt{k^2 - k_\|^2}$.

Corollary 5.9 .

$$\left(\mathbf{Y}_r^{(p)}\right)_{m_\parallel} = -\sum_{l=0}^{\infty} \frac{1}{(-2ik_1(m_\parallel))^{l+1}} \sum_{p_1=0}^{p+l-1} C_{p+l-1}^{p_1} (\tilde{V}^{(p+l-1-p_1)}(0)\mathbf{Y}_{r-1}^{(p_1)})_{m_\parallel}. \tag{5.6.127}$$

Corollary 5.10 .

$$\left(\mathbf{Y}_1^{(p)}\right)_{m_\parallel} = -\sum_{l=0}^{\infty} \frac{1}{(-2ik_1(m_\parallel))^{l+1}} (\tilde{V}^{(p+l-1)})_{m_\parallel 0}. \tag{5.6.128}$$

Proofof the Corollaries. Formula (5.6.127) is obtained by differentiating a product. Using the fact $(\tilde{V}Y_{r-1}^{(-1)})(0) \equiv 0$, it is not hard to show that formula (5.6.128) coincides with (5.6.121) for $p = 0$. Formula (5.6.128) follows from (5.6.125) and (5.6.114). For $p = 0$ it coincides with (5.6.121) when taking into account $V(-1) \equiv 0$.

Proof. Estimate (5.6.126) easily follows from (5.6.117). Next, we prove formula (5.6.125). From the definition of $\mathbf{Y}_r$ (see (5.6.113)), we obtain:

$$\mathbf{Y}_r(x_1)_{m_\parallel} = \sum_{q_\parallel} \exp(-ik_1(m_\parallel)x_1) \times$$

$$\int_0^\infty \frac{\exp ik_1(m_\parallel)|x_1 - y_1|}{2ik_1(m_\parallel)} V_{m_\parallel - q_\parallel}(y_1) \left(\exp ik_1(q_\parallel)y_1\right) \ \mathbf{Y}_{r-1}(y_1)_{q_\parallel} dy_1, \tag{5.6.129}$$

$$= I_1 + I_2,$$

where

$$I_1 = \int_0^{x_1} \frac{(\tilde{V}\mathbf{Y}_{r-1})(y_1)_{m_\parallel}}{2ik_1(m_\parallel)} dy_1,$$

$$I_2 = \int_{x_1}^\infty \frac{\exp 2ik_1(m_\parallel)(y_1 - x_1)}{2ik_1(m_\parallel)} (\tilde{V}\mathbf{Y}_{r-1})_{m_\parallel}(y_1) dy_1 =$$

$$\int_0^\infty \frac{\exp 2ik_1(m_\parallel)z}{2ik_1(m_\parallel)} (\tilde{V}\mathbf{Y}_{r-1})_{m_\parallel}(z + x_1) dz.$$

It is clear that

$$\left.\frac{d^p}{dx_1^p}(I_1)_{m_\parallel}\right|_{x_1=0} = \frac{1}{2ik_1(m_\parallel)} \left(\tilde{V}\mathbf{Y}_{r-1}\right)_{m_\parallel}^{(p-1)}, \tag{5.6.130}$$

$$\left.\frac{d^p}{dx_1^p}(I_2)_{m_\parallel}\right|_{x_1=0} = \int_0^\infty \frac{\exp 2ik_1(m_\parallel)z}{2ik_1(m_\parallel)} (\tilde{V}\mathbf{Y}_{r-1})_{m_\parallel}^{(p)}(z) dz. \tag{5.6.131}$$

Integrating by parts yields the series

$$\frac{d^p}{dx_1^p}(I_2)_{m_\parallel}|_{x_1=0} = -\sum_{l=0}^{\infty} \frac{(\tilde{V}\mathbf{Y}_{r-1})_{m_\parallel}^{(p+l)}(0)}{(-2ik_1(m_\parallel))^{l+2}}. \tag{5.6.132}$$

This series converges because estimate (5.6.126) and the inequality $k_1(m_{\|}) > k^{1-3\delta}$ hold. The last inequality is satisfied because $\mathcal{K}\mathbf{k} \in S^n(k, V, \delta)$. Adding relations (5.6.130), (5.6.132), we get (5.6.125). The lemma is proved.

Lemma 5.37 . *If* $\mathbf{k}$ *is in the* $(k^{-n+1-6\delta})$*-neighborhood of* $S^{(n)}(k, V, \delta)$*, then for sufficiently large* k*,* $k > k_0(k, V, \delta)$*, the vector* $\mathbf{Y}_r^{p_0}(0)$*,* $r \geq 1$ *can be represented in the form:*

$$\left(\mathbf{Y}_r^{(p_0)}\right)_{q_0} = -\sum_{l_1,\dots,l_r=0}^{\infty} \sum_{p_1=0}^{l_1+p_0-1} \cdots \sum_{p_j=0}^{l_j+p_{j-1}-1} \cdots \sum_{p_r=0}^{l_r+p_{r-1}-1} \sum_{q_1,\dots,q_r \in Z^{n-1}} \delta_{p_r 0}\delta_{q_r 0} \times \tag{5.6.133}$$

$$\prod_{s=1}^{r} \frac{1}{(-2ik_1(q_{s-1}))^{l_s+1}} C^{p_s}_{p_{s-1}+l_s-1} \tilde{V}^{(p_{s-1}+l_s-1-p_s)}_{q_{s-1}q_s}(0),$$

where $\delta_{p_r 0}$ *and* δ_{q0} *are respectively one and three dimensional Kronecker symbols.*

Proof. Suppose $r = 1$. In this case formula (5.6.133) coincides with (5.6.127). For $r > 1$ we prove relation (5.6.133) by induction. Indeed, let (5.6.133) be valid for $r-1$. Using relation (5.6.127) and formula (5.6.133) for $\mathbf{Y}_{r-1}^{(p_1)}$, by direct calculation we verify that (5.6.133) holds for r. The lemma is proved.

Lemma 5.38 . *If* $\mathbf{k}$ *is in the* $(k^{-n+1-6\delta})$*-neighborhood of* $S^{(n)}(k, V, \delta)$ *and* $r < M$*,* $M = [k^{\delta}]$*, then*

$$(\mathbf{Y}_r^{(p_0)})_{q_0} = \sum_{l=0}^{\infty} \frac{\Phi_l^{(r,p_0)}(\tilde{V}^{(0)}, \dots, \tilde{V}^{(l+p_0-r)})_{q_0}}{(-2ik_1(q_0))^{l+r}}, \tag{5.6.134}$$

where $\Phi_l^{(r,p_0)}$ *depend on* $\tilde{V}^{(0)}, \dots, \tilde{V}^{(l+p_0-r)}$.

If $l + p_0 - r < 0$*, then* $\Phi_l^{(r,p_0)} = 0$*. In the case of the opposite inequality* $l + p_0 - r \geq 0$*, the following formula holds:*

$$\left(\Phi_l^{(r,p_0)}\right)_{q_0} =$$

$$-(-2ik_1(q_0))^{l+r} \sum_{l_1,\dots,l_r \in N;\ \sum l_j = l} \sum_{p_1=0}^{l_1+p_0-1} \cdots \sum_{p_j=0}^{l_j+p_{j-1}-1} \cdots \sum_{p_r=0}^{l_r+p_{r-1}-1} \sum_{q_j \in Z^n} \delta_{p_r 0}\delta_{q_r 0} \times$$

$$\prod_{s=1}^{r} \frac{1}{(-2ik_1(q_{s-1}))^{l_s+1}} C^{p_s}_{p_{s-1}+l_s-1} \tilde{V}^{(p_{s-1}+l_s-1-p_s)}_{q_{s-1}q_s}(0). \tag{5.6.135}$$

The coefficients $(\Phi_l^{(r,p_0)})_{q_0}$ *satisfy the estimates*

$$|(\Phi_l^{(r,p_0)})_{q_0}| < k^{9\delta(l+p_0)}(l)^r. \tag{5.6.136}$$

Proof. Formula (5.6.134) coincides with (5.6.133) up to the notation (5.6.135). Note that

$$p_{s-1}+l_s-p_s-1\geq 0,\quad s=1,...,r.$$

Summing over s the last inequalities, we get

$$p_0+l-r>0,$$

otherwise $\Phi_l^{(r,p_0)}=0$. Note that $|q_{s-1}-q_s|<R_0$. Therefore, $|q_0-q_s|<rR_0<MR_0<k^{\delta/2}$. From this it easily follows that $|k_1(q_0)/k_1(q_{s-1})|<2$. Moreover, since $R_0<k^{\delta/2}$, we have

$$\sum_{q_{s-1}}|\tilde{V}^{(r)}_{q_{s-1},q_s}|\leq k^{3\delta(r+1)}.$$

Therefore,

$$\left|\left(\Phi_l^{(r,p_0)}\right)_{q_0}\right|\leq 2^{l+r}k^{3\delta r}\times$$

$$\sum_{l_1,...,l_r\in N;\ \sum l_j=l}\ \sum_{p_1=0}^{l_1+p_0-1}\cdots\sum_{p_j=0}^{l_j+p_{j-1}-1}\cdots\sum_{p_r=0}^{l_r+p_{r-1}-1}\prod_{s=1}^{r}C^{p_s}_{p_{s-1}+l_s-1}k^{3\delta(p_{s-1}+l_s-1-p_s)}. \tag{5.6.137}$$

We prove that

$$\sum_{p_1=0}^{l_1+p_0-1}\cdots\sum_{p_j=0}^{l_j+p_{j-1}-1}\cdots\sum_{p_r=0}^{l_r+p_{r-1}-1}\prod_{s=1}^{r}C^{p_s}_{p_{s-1}+l_s-1}k^{3\delta(p_{s-1}+l_s-1-p_s)}\leq$$

$$((r+1)k^{3\delta})^{(l+p_0-r)},\quad l=l_1+...+l_r. \tag{5.6.138}$$

Indeed, suppose $r=1$. Then the left side of (5.6.138) has the form:

$$\sum_{p_1=0}^{l_1+p_0-1}C^{p_1}_{p_0+l_1-1}k^{3\delta(p_0+l_1-1-p_1)}=(k^{3\delta}+1)^{p_0+l_1-1}\leq(2k^{3\delta})^{(p_0+l_1-1)}.$$

Next, suppose relation (5.6.138) is valid for $r=r'-1$. Then, we rewrite the left side of (5.6.137) in the form:

$$\sum_{p_1=0}^{l_1+p_0-1}C^{p_1}_{p_0+l_1-1}k^{3\delta(p_0+l_1-1-p_1)}\left(\sum_{j=2}^{r}\sum_{p_j}^{l_j+p_{j-1}-1}\prod_{s=2}^{r}C^{p_s}_{p_{s-1}+l_s-1}k^{3\delta(p_{s-1}+l_s-1-p_s)}\right).$$

Estimating the expression in the parentheses by the value $(rk^{3\delta})^{(p_1+\sum_{j=2}^{r}l_j-r+1)}$, we obtain that the sum can be estimated by the value

$$(k^{3\delta}+rk^{3\delta})^{p_0+l_1-1}(rk^{3\delta})^{\left(\sum_{j=2}^{r}l_j-r+1\right)},$$

i.e., by $((r+1)k^{3\delta})^{l+p_0-r}$. From (5.6.137), (5.6.138) we obtain

$$\left|(\Phi_l^{(r,p)})_{q_0}\right| \le 2^{l+r} k^{3\delta(l+p_0)} \left((r+1)\right)^{l+p_0-r} \sum_{l_1,\dots,l_r;\ l_1+\dots l_r=l} 1 \le$$

$$2^{l+r} k^{3\delta(l+p_0)} \left((r+1)\right)^{l+p_0-r} l^r.$$

Taking into account that $r < l + p_0$ and $r < k^\delta$, we get

$$2^{l+r} k^{3\delta(l+p_0)} (2k^{5\delta})^{l+p_0-r} l^r < 2^{l+r} k^{8\delta(l+p_0)} 2^{l+p_0-r} l^r < k^{9\delta(l+p_0)} l^r.$$

Using estimate (5.6.136) and $k(q_0) > k^{1-3\delta}$, it is not hard to show that series (5.6.134) converge. The lemma is proved.

Theorem 5.10 . *If* $\mathbf{k}$ *is in the* $(k^{-n+1-6\delta})$*-neighborhood of* $S^{(n)}(k, V, \delta)$*, then the reflection coefficient* $\beta_{m_\parallel}$ *admits the following asymptotic expansion:*

$$\beta_{m_\parallel} = -\sum_{l=0}^{M} \frac{\tilde V^{(l)}_{m_\parallel 0}(0) + \tilde\Phi_l(\tilde V(0), \dots, \tilde V^{(l-2)}(0))_{m_\parallel}}{(-2ik_1(m_\parallel))^{l+2}} + O(k^{-k^\delta}), \tag{5.6.139}$$

where $M = [k^\delta]$,

$$\tilde\Phi_l(\tilde V(0), \dots, \tilde V^{(l-2)}(0)) = \sum_{r=2}^{[l/2]+1} \Phi^{(r,0)}_{l-r+2}(\tilde V(0), \dots, \tilde V^{(l-2r+2)}(0)), \tag{5.6.140}$$

$$\tilde\Phi_0 = 0.$$

There are estimates for $\tilde\Phi_l$*:*

$$\left|(\tilde\Phi_l)_{m_\parallel}\right| \le k^{12\delta l}. \tag{5.6.141}$$

Proof. From formula (5.6.89) and the definition (5.6.113) of $\mathbf{Y}_r$ we see that

$$\beta_{m_\parallel}(k^2) = \sum_{r=1}^{M} \alpha^r \mathbf{Y}_r(0)_{m_\parallel} + O(\alpha^{k^\delta} k^{-k^\delta}).$$

Substituting instead of $\mathbf{Y}_r$ its asymptotic expansion (see (5.6.134) for $p_0 = 0$) and (5.6.121), we get

$$\beta_{m_\parallel} = \sum_{r=2}^{M}\sum_{l=r}^{\infty} \frac{\Phi_l^{(r,0)}(\tilde V(0), \dots, \tilde V^{(l-r)}(0))_{m_\parallel}}{(-2ik_1(m_\parallel))^{l+r}} - \sum_{l=0}^{\infty} \frac{\tilde V^{(l-1)}_{m_\parallel 0}(0)}{(-2ik_1(m_\parallel))^{l+2}} + O(k^{-k^\delta}).$$

Taking into account the estimates for $\Phi_l^{(r,0)}$ and $\tilde V_{m_\parallel 0}$ it is not hard to show that the tails of the series corresponding to $l > M$ are of order k^{-k^δ}. Therefore we can cut the summation at $l = M$, i.e.,

$$\beta_{m_\parallel} = \sum_{r=2}^{M}\sum_{l=r}^{M} \frac{\Phi_l^{(r,0)}(\tilde V(0), \dots, \tilde V^{(l-r)}(0))_{m_\parallel}}{(-2ik_1(m_\parallel))^{l+r}} - \sum_{l=0}^{M} \frac{\tilde V^{(l-1)}_{m_\parallel 0}(0)}{(-2ik_1(m_\parallel))^{l+2}} + O(k^{-k^\delta}).$$

Let us consider the first sum. Making replacement of l by l', $l' = r + l - 2$ and changing the order of the summations, we get, that the sum is equal to

$$\sum_{l'=2}^{M}\sum_{r=2}^{L}\frac{\Phi_{l'-r+2}^{(r,0)}(\tilde{V}(0),...,\tilde{V}^{(l'-2r+2)}(0))_{m_{\|}}}{(-2ik_1(m_{\|}))^{l'+2}},$$

where

$$L = [l'/2] + 1.$$

Now it is easy to see that representation (5.6.139) is valid for $\beta_{m_{\|}}$, $\tilde{\Phi}_l$ being given by (5.6.140). Note that orders of the derivatives on the right of (5.6.140) do not exceed $l-2$, because $r \geq 2$. Using estimates (5.6.136) for $\Phi_l^{(r,0)}$ and taking into account that $r < l < k^{\delta}$ we easily prove

The theorem is proved.

Theorem 5.11 . *If* **k** *is in the* $(k^{-n+1-6\delta})$*-neighborhood of* $S^{(n)}(k,V,\delta)$*, then the reflection coefficient* $\beta_{m_{\|}}$ *admits the following asymptotic expansion:*

$$\beta_{m_{\|}} = -\sum_{l=0}^{M}\frac{V_{m_{\|}}^{(l)}(0) + \hat{\Phi}_l(V(0),...,V^{(l-1)}(0))_{m_{\|}}}{(-2ik_1(m_{\|}))^{l+2}} + O(k^{-k^{\delta}}), \qquad (5.6.142)$$

where

$$\hat{\Phi}_0 = 0,$$

$$\hat{\Phi}_l(V(0),...,V^{(l-1)}(0))_{m_{\|}} =$$

$$\sum_{q=1}^{l} C_q^l(i(k_1(0) - k_1(m_{\|})))^q V_{m_{\|}}^{(l-q)}(0) + \tilde{\Phi}_l(V(0),...,V^{(l-2)}(0))_{m_{\|}}. \qquad (5.6.143)$$

There are estimates for $\hat{\Phi}_l$*:*

$$\left|\hat{\Phi}_l\right| \leq 2k^{12\delta l}. \qquad (5.6.144)$$

Proof. Taking into account that $\tilde{V}^{(l)}$ can be expressed by $V^{(i)}$, $i = 1,...,l$, (see (5.6.119) we easily obtain the theorem from the previous one.

The theorem is proved.

5.7 Solution of the Inverse Problem.

The formulas obtained in the previous section enable us to determine the potential from the asymptotics of reflection coefficients in the high energy region, if it is known in advance that the potential is a trigonometric polynomial. We suppose also that one of the periods is orthogonal to the surface and other period(s).

To begin with, we determine the periods of the potential in the directions $x_{\|}$, i.e., a_2 in the two-dimensional case, and a_2, a_3 in the three-dimensional case. It is clear that Ψ_{refl} satisfies the quasiperiodic conditions (5.1.4), (5.1.7). This means that $\exp(-i(k_{\|},x_{\|}))\Psi_{refl}$ is a periodic function in the direction(s) $x_{\|}$ with a period(s) of the potential. Nevertheless, we, generally speaking, cannot determine the period(s) of the potential from the period(s) of $\exp(-i(k_{\|},x_{\|}))\Psi_{refl}$, because it can happen that a_2 (a_2, a_3) is (are) not the fundamental period(s) of $\exp(-i(k_{\|},x_{\|}))\Psi_{refl}$, i.e., the fundamental period(s) of the refracted wave can be several times less than a_2 (a_2, a_3). We prove now that there is a set $\mu_0 \subset S_k$ of the asymptotically full measure on S_k, such that the fundamental periods of $\exp(-i(k_{\|},x_{\|}))\Psi_{refl}$ and $V(x)$ coincide. This means that, to obtain the fundamental period(s) of the potential, one has to consider the fundamental period(s) of $\exp(-i(k_{\|},x_{\|}))\Psi_{refl}$ for $\mathbf{k} \in S_k$. For some poor set of $\mathbf{k}$, this period(s) probably can be several times less than the fundamental period(s) of the potential. However, for others it will be exactly the period(s) of the potential. In the three-dimensional case:

$$a_2 = \max_{\mathbf{k}\in S_k} a_2(\mathbf{k}), \tag{5.7.1}$$

$$a_3 = \max_{\mathbf{k}\in S_k} a_3(\mathbf{k}); \tag{5.7.2}$$

here a_2, a_3 are the fundamental periods of the potential, $a_2(\mathbf{k})$, $a_3(\mathbf{k})$ are the fundamental periods of the function $\exp(-i(k_{\|},x_{\|}))\Psi_{refl}$ for given $\mathbf{k}$. Note that the maximum is reached on the set $\mu_0 \subset S_k$ of the asymptotically full measure on S_k. In the two-dimensional situation, formula (5.7.1) works. To obtain (5.7.1), (5.7.2), it suffices to prove the following lemma:

Lemma 5.39 . *If* $\mathbf{k} \in S^{(n)}(k,V,\delta)$, *then* $\exp(-i(k_{\|},x_{\|}))\Psi_{refl}$ *is periodic in the direction(s)* $x_{\|}$, *and its fundamental period(s) is (are) equal to the fundamental period(s) of the potential* $V(x)$.

Proof. Let us represent $V(x)$ in the form:

$$V(x) = \sum_{m_{\|}} V_{m_{\|}}(x_1)\exp(i(p_{m_{\|}}(0),x_{\|})).$$

Suppose we are in the two-dimensional situation, and a_2 is the fundamental period of $V(x)$. This means that a_2 is the smallest mutual period of the functions $\exp(i(p_{m_{\|}}(0),x_{\|}))$, such that $V_{m_{\|}}(x_1) \not\equiv 0$. Let us prove that the reflection coefficient $\beta_{m_{\|}}$ is not equal to zero, when $V_{m_{\|}}(x_1) \not\equiv 0$ and, moreover, it satisfies the asymptotic:

$$\beta_{m_{\|}} = \frac{c_{m_{\|}}}{(-2ik_1(0))^{\tilde{l}+2}}(1+o(1)), \quad \tilde{l} = \tilde{l}(m_{\|}), \tag{5.7.3}$$

where $c_{m_{\|}} > 0$, $c_{m_{\|}} \neq c_{m_{\|}}(k)$. In fact, if $\mathbf{k} \in S^{(n)}(k,V,\delta)$, then the asymptotic expansion (5.6.142) holds. Suppose $V_{m_{\|}}(0) \neq 0$. Taking into account that $k_1(m_{\|}) \approx k_1(0)$, when $|(m_{\|}| < R_0 < k^{\delta}$, we get formula (5.7.3) for $\tilde{l} = 0$. Suppose $V_{m_{\|}}(0) = ... = V_{m_{\|}}^{(\tilde{l}-1)}(0) = 0$, $V_{m_{\|}}^{(\tilde{l})}(0) \neq 0$, $\tilde{l} \geq 1$. In this case, formula (5.7.3) also easily follows from (5.6.142). If $V_{m_{\|}}^{(\tilde{l})}(0) = 0$ for all natural $\tilde{l}$ (in fact, even only for $\tilde{l} < cR_0^{n-1}$), then $V_{m_\|}(x_1) \equiv 0$, because $V_{m_\|}(x_1)$ is a trigonometric polynomial. Formula (5.7.3) is proved. Note that a_2 is the smallest mutual period of the functions $\exp(-i(k_\|, x_\|))\Psi_-(k^2, k_\| + p_{m_\|}(0), x)$ when $m_\| \in \mathcal{M}$, $\mathcal{M} = \{m_\| : V_{m_\|}(x_1) \not\equiv 0\}$, because a_2 is the smallest mutual period of the functions $\exp(i(p_{m_\|}(0), x_\|))$ when $m_\| \in \mathcal{M}$. Using $\beta_{m_\|} \neq 0$, we obtain that a_2 is the fundamental period of the function:

$$\exp(-i(k_\|, x_\|)) \sum_{m_\| \in \mathcal{M}} \beta_{m_\|} \Psi_-(k^2, k_\| + p_{m_\|}(0), x). \tag{5.7.4}$$

Let us prove that a_2 is a fundamental period of $\exp(-i(k_\|, x_\|))\Psi_{refl}$. Suppose it is not so, i.e., that the period of the function is a_2/q, $q \in N$. This means that the function can be represented as a linear combination of $\Psi_-(k^2, k_\| + p_{qm_\|}(0), x)$. In this case, a_2/q is also a period for the function (5.7.4). However, this is not true, because a_2 is a fundamental period of this function. Thus, a_2 is a fundamental period of $\exp(-i(k_\|, x_\|))\Psi_{refl}$.

The three-dimensional situation is considered in a similar way.

Next, according to Theorem 5.11, there exists a set $S^{(n)}(k, V, \delta)$ of an asymptotically full measure on S_k, such that the reflection coefficients $\beta_{m_\|}$, $|m_\|| < k^\delta$ have the asymptotic:

$$\beta_{m_\|} = -\frac{V_{m_\|}(0)}{(-2ik_1(m_\|)^2} + O(k_1(0)^{-3}).$$

Considering that $k_1 \approx k(m_\|)$ for $|m_\|| < k^\delta$, we can represent the reflected wave in the form:

$$\Psi_{refl} = -(-2ik_1(0))^{-2} \sum_{|m_\|| < R_0} V_{m_\|}(0)\Psi_-^0(k^2, k_\| + p_{m_\|}(0), x) + O(k^{-3+20\delta}).$$

It is easy to find $V_{m_\|}(0)$ from the last relation. However, the problem is that the set $S^{(n)}(k, V, \delta)$ depends on the potential, which is not given. To solve this problem, first we consider the set μ_0 of $\mathbf{k}$, such that $\beta_{m_\|}$, $|m_\|| < k^{3\delta}$, have the asymptotic

$$\beta_{m_\|} = O(k_1(0)^{-2}),$$

when $\mathbf{k} \in \mu_0$. It is clear that $\mu_0 \supset S^{(n)}(k, V, \delta)$ and $\mu_0 \setminus S^{(n)}(k, V, \delta)$ is the set of an asymptotically small measure on μ_0. Therefore, averaging the value $\mathrm{fl}_0 = \{4k_1(0)^2\beta_{m_\|}\}_{|m_\||<k^{3\delta}}$ over μ_0, and passing to the limit, we get

$$\mathbf{V}(0) = \lim_{k \to \infty} \frac{\int_{\mu_0} \mathrm{fl}_0(\mathbf{k})d\mathbf{k}}{s(\mu_0)}.$$

Further, if $t \in S^{(n)}(k, V, \delta)$, then the reflection coefficients satisfy the asymptotic expansion:

$$\beta_{m_\|} - = \frac{V_{m_\|}(0)}{(-2ik_1(m_\|))^2} - \frac{V_{m_\|}^{(1)}(0) + \hat{\Phi}_1(\mathbf{V}(0))_{m_\|}}{(-2ik_1(m_\|))^3} + O(k_1(0)^{-4}).$$

Note that there exists a set μ_1, $\mu_1 \supset S^{(n)}(k, V, \delta)$ of an asymptotically full measure on S_k, such that $\beta_{m_\parallel}$ satisfies the asymptotic formula:

$$\beta_{m_\parallel} = -\frac{V_{m_\parallel}(0)}{(-2ik_1(m_\parallel))^2} + O(k_1(0)^{-3}).$$

Since $\mu_1 \supset S^{(n)}(k, V, \delta)$, the set $\mu_1 \setminus S^{(n)}(k, V, \delta)$ has an asymptotically small measure on μ_1. Therefore, averaging the value

$$\text{fl}_1 = \left\{ - \left(\beta_{m_\parallel} - \frac{V_{m_\parallel}(0)}{(-2ik_1(m_\parallel))^2} - \frac{\hat{\Phi}_1(\mathbf{V}(0))_{m_\parallel}}{(-2ik_1(m_\parallel))^3} \right) (-2ik_1(m_\parallel))^3 \right\}_{|m_\parallel| < k^{3\delta}}$$

over μ_1, and passing to the limit, we get

$$\mathbf{V}(0)^{(1)} = \lim_{k \to \infty} \frac{\int_{\mu_1} \text{fl}_1(\mathbf{k}) d\mathbf{k}}{s(\mu_1)}.$$

Repeating the similar procedure r times, we get

$$\mathbf{V}^{(r)}(0) = \lim_{k \to \infty} \frac{\int_{\mu_r} \text{fl}_r(\mathbf{k}) d\mathbf{k}}{s(\mu_r)},$$

where

$$\text{fl}_r = \left\{ - \left(\beta_{m_\parallel} + \frac{\hat{\Phi}_r(\mathbf{V}(0), ..., \mathbf{V}^{(r-1)}(0))_{m_\parallel}}{(-2ik_1(m_\parallel))^{r+2}} \right. \right.$$

$$\left. \left. + \sum_{l=0}^{r-1} \left(\frac{V_{m_\parallel}^{(l)}(0)}{(-2ik_1(m_\parallel))^{l+2}} + \frac{\hat{\Phi}_l(\mathbf{V}(0), ..., \mathbf{V}^{(l-1)}(0))_{m_\parallel}}{(-2ik_1(m_\parallel))^{l+2}} \right) (-2ik_1(m_\parallel))^{r+2} \right\}_{|m_\parallel| < k^{3\delta}},$$

μ_r being a set of an asymptotically full measure on S_k, such that $\beta_{m_\parallel}$, $|m_\parallel| < k^{3\delta}$, have the asymptotics:

$$\beta_{m_\parallel} = -\sum_{l=1}^{r-1} \frac{V_{m_\parallel}^{(l)}(0) + \hat{\Phi}_l(\mathbf{V}(0), ..., \mathbf{V}^{(l-1)})_{m_\parallel}}{(-2ik_1(m_\parallel))^{l+2}} + O(k_1(0)^{-(r+2)}).$$

It easily follows from Theorem 5.11 that μ_r has an asymptotically full measure on S_k; and $S_k \setminus \mu_r$ has an asymptotically small measure with respect to μ_r. Thus, we determine $\mathbf{V}(0)$, $\mathbf{V}^{(1)}(0)$,...,$\mathbf{V}^{(r)}(0)$, $r < k^\delta$. Note that $k^\delta \to \infty$ when $k \to \infty$. Therefore, we can determine any number of the derivatives. We suppose $V(x)$ to be a trigonometric potential, i.e., an analytical function. Hence,

$$\mathbf{V}(x_1)_{m_\parallel} = \sum_{l=0}^{\infty} V(0)_{m_\parallel}^{(l)} \frac{x_1^l}{l!}.$$

Thus, we have determined the potential.

All considerations of this chapter can be generalized to the case of an analytic potential. Using this method, it is also possible to construct asymptotic formulae for surface and quasisurface states. However, it is necessary to finish the considerations at a point. Let it be here. ♡

References

[A] Adachi, T. *On the Spectrum of Periodic Schrödinger Operators and Tower of Coverings.* Bull. London. Math. Soc., **27** (1995), 2, pp. 173 - 176.

[Ad]Adams, R.A. *Sobolev Spaces.* Academic Press, 1978.

[Ae]Aerts, E. *Surface States of One Dimensional Crystals.(I,II,III)* Physica, **26** (1960), pp. 1047 - 1072.

[Ag]Agmon, S. *On Positive Solutions of Elliptic Equations with Periodic Coefficients in R^n, Spectral Results and Extensions to Elliptic Operators on Riemannian Manifolds.* Proc. Internat. Conf. on Differential Equations, North-Holland Math. Studies, **92**, 1984, pp. 7 - 17.

[AgMi]Agranovich, V. and Mills, D. (eds) *Surface Polaritons.* North Holland, 1982.

[Ar]Arscott, F.M. *Periodic Differential Equations.* Macmillan, NY, 1964.

[Av1]Avron, J., *On the Spectrum of $p^2 + V(x) + \varepsilon x$ with V Periodic and ε Complex.* J. Phys. A., **12** (1979), p. 2393.

[Av2]Avron, J.E. *Bragg Scattering from Point Interactions. An Explicit Formula for the Reflection Coefficients.* Mathematical problems in Theoretical Physics, Proceedings, Berlin (West), 1981, Ed. by Schrader, Seiler, D.A., Uhlenbrock D.A. Lecture Notes in Physics, **153**, Springer Verlag, 1982, pp. 126 - 128.

[AvGrHø]Avron, J.E., Grossmann, A., Høegh-Krohn, R. *The Reflection from a Semi-infinite Crystal of Point Scatterers.* Phys. Lett., **94A** (1983), pp. 42 - 44.

[AvGrRo1]Avron, J., Grossmann, A., Rodriguez, R. *Hamiltonians in the One Electron Theory of Solids,I.* Preprint, Centre de Physique Theoriqué, C.N.R.S., Marseille, 1972

[AvGrRo2]Avron, J., Grossmann, A., Rodriguez, R. *Hamiltonians in the One Electron Theory of Solids, I.* Rep. Math. Phys., **5** (1974), 1, pp. 113 -120.

[AvSi1]Avron, J.E., Simon, B. *Analytic Properties of Band Functions.* Annals of Physics, **110** (1978), 1, pp. 85 -101.

[AvSi2]Avron, J.E., Simon, B. *Stability of Gaps for Periodic Potential under Variation of a Magnetic Field.* J. Phys. A, **18** (1985), 12, pp. 2199 - 2205.

[Bä1]Bättig, D. *A Toroidal Compactification of the Two-dimensional Bloch variety.* Thesis, ETH, Zürich, 1988.

[Bä2]Bättig, D. *A Toroidal Compactification of the Fermi Surface for the Discrete Schrödinger Operator.* Comment. Math. Helv., **67** (1996), 1, pp. 1 - 16.

[BäKnTr]Bättig, D., Knörrer, H., Trubowitz, E. *A Directional Compactification of the Complex Fermi Surface.* Composito Mathematica, **79** (1991), pp. 205 - 229.

[Bel1]Bellissard, J. *Gap Labelling Theorems for Schrödinger Operators.* in "From Number Theory to Physics." Waldschmidt, M., Moussa, P., Luck, J.-M., Itzykson, C. (eds), Springer- Verlag, 1992.

[BelBovChe]Bellissard, J., Bovier, A., Chez, J.M. *Spectral Properties of a Tight Binding Hamiltonian with a Period Doubling Potential.* Comm. Math. Phys., **135** (1991), 2, pp. 379 – 399.

[BenLiPa]Benssousan, A., Lions, J.L., Papanicolaou, G. *Asymptotic Analysis for Periodic Structures.* North Holland, Amsterdam, 1978.

[Bent1]Bentosela, F. *Scattering from Impurities in Crystal.* Comm. Math. Phys., **46** (1976), 2, pp. 153 - 166.

[Bent2]Bentosela, F. *Magnetic Bloch Functions.* Nuovo Cim., **16B** (1973), pp. 115 – 126.

[Ber]Berthier, A.M., *Spectral Theory and Wave Operators for the Schrödinger Equation (Chapter X)*, Research Notes in Mathematics, **71**, Pitman, 1982.

[BS]Bethe, G., Sommerfeld, A. *Elektronentheorie der Metalle.* Berlin, New-York, Springer-Verlag, 1967.

[Bi1]Birman M.Sh. *On Discrete Spectrum in Gaps of a Second Order Perturbed Periodic Operator.* Functional Anal. i Prilozhen., **25** (1991), 2, pp. 89 – 92; Engl. transl.: Functional Anal. Appl., **25** (1991), 2, pp. 158 – 161.

[Bi2]Birman, M.Sh. *Discrete Spectrum of the Periodic Schrödinger Operator for a Non-negative Perturbations.* Proceedings of the International Conference (Blossin 1993): Oper. Theory Adv. App, **70**, Bikhäuser, Basel, 1993.

[BiYa]Birman, M.Sh., Yafaev, D.R. *The Scattering Matrix for the Perturbation of a Periodic Schrödinger Operator by a Decaying Potential.* Algebra i Analiz, **6** (1994), 3, pp. 17 – 39; Engl. Transl.: St. Petersburg Math. J., **6**, 3, pp. 453 – 474.

[Bi3]Birman, M.Sh. *The Discrete Spectrum in Gaps of the Perturbed Periodic Schrödinger Operator. I.* Regular perturbations, Adv. Partial Differential Equations, **2**, Academic Verlag, Berlin, 1995, 334-352.

[Bi4]Birman, M. Sh. *The Discrete Spectrum of the Periodic Schrödinger Operator, Perturbed by a Decaying Potential.* Algebra i Analiz, **8** (1996), 1, pp. 3-20; Engl. Transl.: St. Petersburg Math. J., **8**, 1.

[BiSus]Birman, M.Sh., Suslina, T.A. *Two Dimensional Periodic Magnetic Hamiltonian is Absolutely Continuous.* Algebra i Analiz, **9** (1997), 1, 31-47; Engl. Transl.: St. Petersburg Math. J., **9**, 1.

[Bl]Bloch, F. *Über die Quantenmechanik der Elektronen in Kristallgittern.* Z. Phys., **52** (1928), pp. 555 - 600.

[Bri]Brillouin, L *Wave Propagation in Periodic Structures, Electric Filters and Crystal Lattices.* Dover, NY, 1953.

[BrSu1]Brüning, J., Sunada, T. *On the Spectrum of Periodic Elliptic Operators.* Nagoya Math. J., **126** (1992), pp. 159 – 171.

[BrSu2]Brüning, J., Sunada, T. *The Schrödinger Operator with a Periodic Magnetic Field.* Preprint 1992.

[Bu1]Buslaev, V.S. *Semiclassical Approximation for Equations with Periodic Coefficients.* Sov. Math. Uspekhi, **42**, (1987), 6, pp.77 – 98; Engl. transl.: Russ. Math. Surv., **42**, (1987), 6, pp. 97 – 125.

[Bu2]Buslaev, V.S. *Adiabatic Perturbation of a Periodic Potential.* Teor. i Mat. Fiz., **58** (1984), 2, pp. 233 – 243, Engl. transl.: Theor. and Math. Phys., **58** (1984), 2, pp. 153 – 159.

[BuDm1]Buslaev V.S., Dmitrieva L.A. *Adiabatic Perturbation of a Periodic Potential.* Teor. i Mat. Fiz., **73** (1987), 3. pp. 430 – 442, Engl. transl.: Theor. and Math. Phys., **73** (1988), 3, pp. 1320 – 1329.

[BuDm2]Buslaev V.S., Dmitrieva L.A. *Bloch Electrons in an External Electric Field.* Schrödinger Operator Standard a Non-standard, (Dubna 1988). World-Scientific, 1989, Ed. P. Exner and P. Seba, pp. 103 – 130.

[BuDm3]Buslaev V.S., Dmitrieva L.A. *Geometrical Aspects of the Bloch Electrons Theory in External Fields.* Topological Phases in Quantum theory (Dubna 1988). World Scientific, 1989, Ed. Markovski, S. Vinitsky, pp. 218 – 250.

[Ca]Callaway, J. *Energy Band Theory.* Academic Press, New-York/London, 1964.

[C1]Carvey, D. Mc. *Operations Commuting with Translation by One. Representation Theorems.* J. of Math. Anal. and Appl., **4** (1962), pp. 336-410.

[C2]Carvey, D. Mc. *Differential Operators with Periodical Coefficients in* $L_p(-\infty,\infty)$. J. of Math. Anal. and Appl., **11** (1965), pp. 564-596.

[C3]Carvey, D. Mc. *Perturbation Results for Periodic Differential Operators.* J. of Math. Anal and Appl., **12** (1965), pp. 187-234.

[Chu1]Chuburin, Yu.P. *On Solutions of the Schrödinger Equation in the Case of a Semibounded Crystal.* Preprint FTI, Sverdlovsk, 1986 (in Russian).

[Chu2]Chuburin, Yu.P. *Scattering of the Schrödinger Operator in the Case of Crystal Film.* Teor. i Mat. Fiz., **72** (1987), 1, pp. 120 – 131; Engl. Transl.: Theor. and Math. Phys., **72** (1988), 1, pp. 764 – 773.

[Chu3]Chuburin, Yu.P. *Floquet Asymptotic Solutions of the Schrödinger Equation in the Case of a Semibounded Crystal.* Teor. i Mat. Fiz., **77** (1988), 3, pp. 472 – 477; Engl. Transl.: Theor. and Math. Phys., **77** (1988), 3, pp. 1331 – 1336.

[Chu4]Chuburin, Yu.P. *On Scattering for the Schrödinger Operator in the Case of a Semi-Bounded Crystal.* In Problems of Theory of Periodic Motions, Izhevsk, 1988, **9**, pp. 63 – 72 (in Russian).

[Chu5]Chuburin, Yu.P. *On the Schrödinger Operator in the Case of two Joint Crystals.* In Problems of Theory of Periodic Motions, Izhevsk, 1990, **10**, pp. 49 – 55 (in Russian).

[Ch]Chulaevsky, V. *On Perturbations of a Schrödinger Operator with a Periodic Potential.* Usp. Mat. Nauk, **36**, 5, pp. 203 – 204; Engl. transl.: Russian Math. Surveys, **36** (1981), 5, p. 143.

[Cl1]Cloizeaux, J. *Energy Bands and Projection Operators in a Crystal. Analytic and Asymptotic Properties.* Phys. Rev. (2), **135** (1964), pp. A685 – A697.

[Cl2]Cloizeaux, J. *Analytical Properties of N-Dimensional Energy Bands and Wannier Functions.* Phys. Rev. (2), **135** (1964), pp. A698 – A707.

[CrWo]Cracknell, A.P., Wong, K.S. *The Fermi Surface.* Clarendon Press, Oxford, 1973.

[DahTr]Dahlberg, B.E.J., Trubowitz, E. *A Remark on Two Dimensional Periodic Potentials.* Comment. Math. Helvetici, **57** (1982), pp. 130 –134.

[Dau]Daumer, F. *Equation de Schrödinger Dans l'Approximation du Tight-binding.* Thése de Doctorat, Université de Nantes, 1990.

[DavSi]Davies, E.B., Simon, B. *Scattering Theory for Systems with Different Spatial Asymptotics on the Left and Right.* Commun. Math. Phys., **63** (1978), pp. 277-301.

[DaLe]Davison, S.G., Levine, J.D.: *Surface States.* Solid State Physics, **25** (1970), pp. 1 – 149.

[Di]Dimitrushenkov, V.A. *Spectrum of the Schrödinger Operator with a Periodic Potential in a Two-Dimensional Strip.* Functional. Anal. i Prilozhen., **16** (1982), 4, pp. 66-68; Engl. transl.: Funct. Anal. Appl., **16** (1983), pp. 297 – 301.

[DuNo]Dubrovin, B.A., Novikov, S.P. *Ground States in a Peridic Field. Magnetic Bloch Functions and Vector Bundles.* Soviet Math. Dokl., **22** (1980), pp. 240 – 244.

[DS]Dunford, N., Schwarz, J. *Linear Operators, Spectral Theory.* Interscience Publishers, New-York – London, 1963.

[DyPe]Dyakin, V.V., Petrukhnovskii, S.I. *On the Discreteness of the Spectrum of Some Operator Sheaves Associated with the Periodic Schrödinger Equation.* Teor. i Mat. Fiz., **74** (1988), 1, pp. 94 – 102; Engl. Transl.: Theor. and Math. Phys., **74** (1988), 1, pp. 66 – 72.

[Ea1]Eastam, M.S.P. *The Schrödinger Equation with a Periodic Potential.* Proc. Roy. Soc., **69a** (1971), pp. 125 – 131.

[Ea2]Eastam, M.S.P. *The Spectral Theory of Periodic Differential Equations.* Edinburg: Scottish Academic Press, 1973.

[EsRaTr1]Eskin, G., Ralston, J., Trubowitz E. *On Isospectral Periodic Potential in R^n, I.* Commun. Pure Appl. Math., **37** (1984), pp. 647 -676.

[EsRaTr2]Eskin, G., Ralston, J., Trubowitz, E. *On Isospectral Periodic Potential in R^n, II.* Commun. Pure Appl. Math., **37** (1984), pp. 715 – 753.

[Es1]Eskin, G. *Inverse Spectral Problem for the Schrödinger Operator with Periodic Magnetic and Electric Potentials.* Seminaire sur les Equations aux Derivees Partielles, 1988 – 1989, Exp. No. XII, 6, Ecole Polytechnic, Palaiseau, 1989.

[Es2]Eskin, G. *Inverse Spectral Problem for the Schrödinger Equation with Periodic Vector Potentials.* Comm. Math. Phys., **185** (1989), 2, pp. 263 – 300.

[EvPav]Evstratov V.V., Pavlov B.S. *Lattice Models of Solids.* Collection of papers devoted to the memory of R. Høegh-Krohn. Springer, 1990.

[FeKnTr1]Feldman, J., Knörrer, H., Trubowitz, E. *Perturbatively Stable Spectrum of a Periodic Schrödinger Operator.* Invent. Math., **100** (1990), pp. 259 – 300.

[FeKnTr2]Feldman, J., Knörrer, H., Trubowitz, E. *Perturbatively Unstable Eigenvalues of Periodic Schrödinger Operator.* Comment. Math. Helvetici, **66** (1991), pp. 557 – 579.

[FigKu1]Figotin, A., Kuchment, P. *Band-Gap Structure of Spectra of Periodic Dielectric and Acoustic media, I. Scalar Model.* SIAM J. Appl. Math., **56** (1996), 1, pp. 1561 – 1620.

[FigKu2]Figotin, A., Kuchment, P. *Band-Gap Structure of Spectra of Periodic Dielectric and Acoustic media, II. Two-Dimensional Photonic Crystals.* SIAM J. Appl. Math., **56** (1996), 6, pp. 1561 – 1620.

[Fir1]Firsova, N.E. *Riemann Surface of Quasimomentum and the Scattering Theory for a Perturbed Hill's Operator.* Notes of LOMI Sci. Seminars, **51** (1975), 7, pp. 183 – 196. (in Russian).

[Fir2]Firsova, N.E. *Some Spectral Identities for the One-Dimensional Hill Operator.* Teor. i. Mat. Fiz., **37** (1978), 2, pp. 281 – 288; Engl. transl.: Theor. and Math. Phys., **37** (1978), 2, pp. 1022 – 1027.

[Fir3]Firsova, N.E. *Direct and Inverse Problem of Scattering for One Dimensional Perturbed Hill Operator.* Mat. Sb., **130(172)**, 3(7) (1986), pp. 349 – 385; Engl. transl.: Mathematics of the USSR-Sbornik, **58** (1987), 2, pp. 351 – 388.

[Fir4]Firsova, N.E. *On the Effective Mass for One Dimensional Hill operator.* Vestnik Leningradskogo Universiteta (fizika), **10**, 2 (1979), pp. 13 – 18.

[FirKor]Firsova, N.E., Korotyaev, E.L., *Diffusion in Layered Media at Large Time.* Teor. i Mat. Fiz., **98** (1994), pp. 106 – 148, English transl.: Theor. and Math. Phys., **98** (1994), pp. 72 – 99.

[Fri]Friedlander, L. *On the Spectrum for the Periodic Problem for the Schrödinger Operator.* Communications in Partial Differential Equations, **15** (1990), pp. 1631 – 1647.

[Fl]Floquet, G. *Sur les Equations Differentielles Lineaires a Coefficients Periodiques.* Ann Ecole Norm., 2 (1883), 12, pp. 47 – 89.

[FroPav1]Frolov, S.V., Pavlov, B.S. *A Formula for the Sum of the Effective Masses of a Multidimensional Lattice.* Teor. i Mat. Fiz., **87** (1991), 3, pp. 456 – 472; Engl. transl.: Theor. and Math. Phys., **87** (1991), 3, pp. 657 – 668.

[FroPav2]Frolov, S.V., Pavlov, B.S. *Spectral Identities for the Zone Spectrum in the One Dimensional Case.* Teor. i Mat. Fiz., **89** (1991), 1, pp. 3 – 10; Engl. transl.: Theor. and Math. Phys., **89** (1991), 1, pp. 1013 – 1019.

[Fro]Frolov, S.V. *A Spectral Identity for a Multidimensional Schrödinger Equation with a Periodic Potential.* Teor. i Mat. Fiz., **91** (1992), 3, pp. 524 – 528; Engl. transl.: Theor. and Math. Phys., **91** (1992), 3, pp. 692 – 695.

[GaOlPav]Galunov, G.V., Oleinik V.L., Pavlov B.S. *Estimates for Negative Spectral Bands of Three Dimensional Periodic Schrödinger Operator.* J. Math. Phys., **34** (1993), 3, pp. 936 – 942.

[GaTr1]Garnett, J.,Trubowitz, E. *Gaps and Bands of One Dimensional Periodic Schrödinger Operator.I* Comment. Math. Helv., **59** (1984), pp. 258-312.

[GaTr2]Garnett, J.,Trubowitz, E. *Gaps and Bands of One Dimensional Periodic Schrödinger Operator II.* Comment. Math. Helv., **62** (1987), pp. 18 – 37.

[GeilMarg1]Geiler, V.A., Margulis, V.A. *Spectrum of a Bloch Electron in a Magnetic Field in a Two-Dimensional Lattice.* Teor. and Mat. Fiz., **58** (1984), 3, pp. 461 – 473; Engl. Transl.: Theor. and Math. Phys., **58** (1984), 3, pp. 302 – 310.

[GeilMarg2]Geiler, V.A., Margulis, V.A. *Structure of the Spectrum of a Bloch Electron in a Magnetic Field in a Two-Dimensional Lattice.* Teor. and Mat. Fiz., **61** (1984), pp. 140 – 150; ; Engl. Transl.: Theor. and Math. Phys., **61** (1985), 1, pp. 1049 – 1056.

[Gelf]Gel'fand, I.M. *Expansion in Eigenfunctions of an Equation with Periodic Coefficients.* Dokl. Akad. Nauk SSSR, **73** (1950), pp. 1117-1120. (in Russian)

[Gér]Gérard, C. *Resonance Theory for Periodic Schrödinger operator.* Bull. Soc. Math. France, **118** (1990), pp. 27 – 54.

[GéMaSj]Gérard, C., Martinez, A., Sjostrand, J. *A Mathematical Approach to the Effective Hamiltonian in Perturbed Periodic Problems.* Comm. Math. Phy., **142** (1991), 2, pp. 217 – 242.

[GiKnTr1]Gieseker, D., Knörrer, H., Trubowitz, E. *An Overview of the Geometry of Algebraic Fermi Curves.* Contemporary Mathematics, **116** (1991), pp. 19 – 46.

[GiKnTr2]Gieseker, D., Knörrer, H., Trubowitz, E. *The Geometry of Algebraic Fermi Curves.* Perspectives in Mathematics, **14**, Academic Press Inc., Boston 1993.

[GoKr]Gohberg, I.Ts., Krein, M.G., *Introduction to the Theory of Linear Nonselfadjoint Operators in Hilbert Space.* "Nauka", Moskow, 1965; English transl.: Amer. Math. Soc., Providence, R.I., 1969.

[GorKapp1]Gordon, C.S., Kappeler, T. *On Isospectral Potentials on Flat Tori. I*, Duke Math. J., **63** (1991), 1, pp. 217 – 233.

[GorKapp2]Gordon, C.S., Kappeler, T. *On Isospectral Potentials on Flat Tori. II*, Comm. PDE, **29** (1995), 3 – 4, pp. 709 – 728.

[GriMoh]Grigis, A., Mohamed, A. *Finitude des Lacunes Dans le Spectre de L'opérateur de Schrödinger et de Celui de Dirac Aves des Potentiels Electrique at Magnétique Périodiques.* J. Math. Kyoro Univ., **33**, 4, pp. 1073 – 1098.

[GroHøMe]Grossman, A., Høegh-Krohn, R., Mebkhout, M. *The One Particle Theory of Periodic Point Interactions.* Commun. Math. Phys., **77** (1980), pp. 87 – 110.

[GuRaTr1]Guillot, J.-C., Ralston, J, Trubowitz E. *Semi-Classical Approximations in Solid State Physics.* Comm. Math. Phys., **116** (1988), 3, pp. 401 – 415.

[GuRaTr2]Guillot, J.-C., Ralston, J, Trubowitz E. *Semi-Classical Approximations in Solid State Physics.* Lecture Notes in Mathematics, **1324**, pp. 263 – 269.

[Gun]Guillemin, V. *Inverse Spectral Results on Two-Dimensional Tori*, J. Amer. Math. Soc., **3** (1990), 2, pp. 375 – 387.

[Ha]Harrell, E.M. *The Band Structure of a One- Dimensional Periodic Problem in the Scaling Limit.* Ann. Phys., **119** (1979), pp. 351 – 369.

[HemHerb1]Hempel, R., Herbst, I., *Bands and Gaps for Periodic Magnetic Hamiltonians.* Preprint ESI, **162** (1994).

[HemHerb2]Hempel, R., Herbst, I., *Bands and Gaps for Periodic Magnetic Hamiltonians.* in Operator Theory: Advances and Applications, **78**, Birkhäuser, 1995, pp. 175 – 184..

[HemHerb3]Hempel, R., Herbst, I., *Strong Magnetic Fields, Dirichlet Boundaries and Spectral Gaps.* Comm. Math. Phys., **169** (1995), pp. 237 – 259.

[HøHoMa]Høegh-Krohn, R., Holden, H., Martinelli, F. *The Spectrum of Defect Periodic Point Interactions* Lett. Math. Phy., **7** (1983), 3, pp. 221 – 228.

[JaMolPas]Jakšić V., Molchanov S., Pastur L. *On the Propagation Properties of Surface Waves.* IMA preprint **1316** (1995).

[KargKor]Kargaev P., Korotyaev, E. *Effective Masses and Conformal Mapping.* Commun. Math. Phys., **169** (1995), pp. 597-625.

[K1]Karpeshina, Yu.E. *Expansion Theorem in Eigenfunctions for Scattering Problem by Uniform Periodic Carriers of Chain Types in Three-dimensional space.* Problems of Mathematical Physics, **10** (1982) Leningrad, pp .147 – 175; Engl. transl.: Sel. Math. Sov., **4**, 3 (1985), pp. 259 – 276.

[K2]Karpeshina, Yu.E. *Spectrum and Eigenfunctions of Schrödinger Operator with Zero-Range Potential of the Homogeneous Lattice Type in Three Dimensional Space.* Teor. i Mat. Fiz., **57** (1983), 2, pp. 304 – 313; Engl. transl.: Theor. and Math. Phys., **57** (1983), pp. 1156 - 1162.

[K3]Karpeshina, Yu.E. *Spectrum and Eigenfunctions of Schrödinger Operator with the Zero-Range Potential of the Homogeneous Two-Dimensional Lattice Type in Three-Dimensional Space.* Teor. i Mat. Fiz., **57** (1983), 3, pp. 414 – 423; Engl. transl.: Theor. and Math. Phys., **57** (1983), pp. 1231-1237.

[K4]Karpeshina, Yu. E. *Perturbation Theory for a Polyharmonic Operator with a Nonsmooth Periodic Potential.* Notes of LOMI Sci. Seminars, **169(8)** (1988), pp.76-84 (in Russian).

[K5]Karpeshina, Yu. E. *Perturbation Theory for the Schrödinger Operator with a Periodic Potential.* Schrödinger Operator Standard a Non-standard, (Dubna 1988). World-Scientific, 1989, Ed. P. Exner and P. Seba, pp. 131-145.

[K6]Karpeshina, Yu. E. *Geometrical Background for the Perturbation Theory of the Polyharmonic Operator with a Periodic Potential.* Topological Phases in Quantum theory (Dubna 1988). World Scientific, 1989, Ed. Markovski, S. Vinitsky, pp. 251-276.

[K7]Karpeshina, Yu. E. *Analytic Perturbation Theory for a Periodic Potential,* Izv. Akad. Nauk SSSR Ser. Mat., **53** (1989), 1, pp. 45-65; English transl.: Math. USSR Izv., **34** (1990), 1, pp. 43 – 63.

[K8]. Karpeshina, Yu.E. *Perturbation Theory for Polyharmonic Operator with a Nonsmooth Periodic Potential.* Problems of Mathematical Physics, **13** (1991) Izd. Leningr. Univ., pp. 132-153 (in Russian).

[K9]Karpeshina, Yu.E. *Perturbation Theory for the Schrödinger Operator with a Nonsmooth Periodic Potential.*, Dokl. Akad. Nauk SSSR, **309** (1989), 5, pp. 1055 – 1059; Engl. transl.: Soviet Math. Dokl., **40** (1990), 3, pp. 614 – 618.

[K10]Karpeshina, Yu.E. *Perturbation Theory for the Schrödinger Operator with a Periodic Potential.* Trudy Mat. Inst. Steklov, **188** (1990), pp. 88 - 116; Engl. transl.: Proceedings of the Steklov Institute of Mathematics, 1991, Issue 3, pp. 109-145.

[K11]Karpeshina, Yu.E. *Perturbation Theory Formulas for the Schrödinger Equation with a Nonsmooth Periodic Potential.* Matem. Sbornik, **181** (1990), 9; pp. 1256 - 1278; Engl. transl.: Math. USSR Sbornik, **71** (1992), 1, pp. 101-123.

[K12]Karpeshina, Yu.E. *Perturbation Formulae for Polyharmonic Operator with a Periodic Potential near Planes of Diffraction.* Preprint ETH, Zurich, 1992, pp. 1-21.

[K13]Karpeshina, Yu.E. *The Formulae for Eigenfunctions of the Periodic Schrödinger Operator on the Isoenergetic Surface.* Preprint University of Augsburg, **312**, 1994.

[K14]Karpeshina, Yu.E. *Perturbation Formulae for the Schrödinger Operator with a Periodic Potential near Planes of Diffraction.* Mathematical Research Letters, **2** (1995), pp. 59-74.

[K15]Karpeshina, Yu.E. *Perturbation Series for the Schrödinger Operator with a Periodic Potential near Planes of Diffraction.* Communications in Analysis and Geometry, **4**, (1996), 3, pp. 339 - 413.

[K16]Karpeshina, Yu.E. *Interaction of a Free Wave with a Semicrystal.* Zapiski Nauchnich Seminarov LOMI, **186** (1990), pp. 107-114; Engl. transl.: Journal of Mathematical Sciences, **73** (1995), 3, pp. 366 - 369.

[K17]Karpeshina, Yu.E. *Interaction of a Free Wave with a Semicrystal.* Equations aux Derivees Partielles (Proceedings of the International Conference on Partial Differential Equations. Saint-Jean-de-Monts, 29 Mai au 2 June 1995), pp. XIX.1-XIX.7.

[Kato]Kato, T., *Perturbation Theory for Linear Operators.* Springer-Verlag, 1966.

[Ki]Kittel, Ch., *Introduction to Solid State Physics.* New-York, Wiley, c1976.

[KirSi]Kirsch, W., Simon, B. *Comparison Theorems for the Gap of Schrödinger Operator.* J. of Functional Analysis, **75** (1987), 2, pp. 396 -410.

[KnTr]Knörrer, H., Trubowitz, E. *A Directional Compactification of the Complex Bloch Variety.* Comment. Math. Helvetici, **65** (1990), pp. 114 - 149.

[Kob]Kobayashi, T., Ono, K., Sunada T. *Periodic Schrödinger Operators on a Manifold.* Forum Math., **1** (1989), pp. 69 - 79.

[Kohn]Kohn, W. *Analytic Properties of Bloch Waves and Wannier Functions.* Phys. Rev., **115** (1959), pp. 809 - 821.

[Kor1]Korotyaev, E.L. *On some Properties of Quasimomentum of One Dimensional Hill Operator.* Zap. Nauchn. Sem. Leningrad Oldel. Mat. Inst. Steklov (LOMI), **195** (1991), pp. 48-57; English transl.: J.Soviet Math., **62** (1992), 6, pp. 3081 - 3087.

[Kor2]Korotyaev, E.L. *The Propagation of Waves in Periodic Media at Large Time.* Preprint ESI, Vienna, **152**, 1994.

[Kr]Krichever, I.M. *Spectral Theory of Two-Dimensional Periodic Operators and its Applications.* Russian Math. Surveys, **44** (1989), pp. 145 - 225.

[Kuch1]Kuchment, P. *Floquet Theory for Partial Differential Equations.* Russian Math. Surveys, **37** (1982), 4, pp. 1 - 60.

[Kuch2]Kuchment, P. *Floquet Theory for Partial Differential Equations.* Birkhäuser, Basel, 1993.

[KuchVa]Kuchment, P., Vainberg, B. *On Absence of Embedded Eigenvalues for Schrödinger Operators with Perturbed Periodic Potentials.* Preprint 1996.

[KupMakPav]Kuperin, Yu.A., Makarov K.A., Pavlov B.S. *An Exactly Solvable Model of a Crystal with Non-Point Atoms.* In the book : Lecture Notes in Physics, **324**, Springer-Verlag 1989.

[KurPa1]Kurasov P.B., Pavlov B.S. *Electron in a Homogeneous Crystal of Point Atoms with Internal Structure II.*, Teor. i Mat. Fiz., **74** (1988), 1, pp. 82 – 93; Engl. transl.: Theor.and Math. Phys., **74** (1988), 1, pp. 58 -66.

[KurPa2]Kurasov P.B., Pavlov B.S. *Surfaces with Internal Structure.*, in the book: Lecture Notes in Physics, **324**, Springer-Verlag 1989.

[KurPa3]Kurasov P.B., Pavlov B.S. *Micro Spectral Properties of Crystals and their Band Structure.* Proc. Int. Conf. "Mathematical Physics: On Three Levels", Leuven, Belgium, 1993.

[LaPan]Lazutkin, V.F., Pankratova T.F., *Asymptotics of the Width of Gaps in the Spectrum of the Sturm-Liouville Operator with a Periodic Potential.* Dokl. Akad. Nauk SSSR, **215** (1974) ,5, pp. 1048-1051; Engl. transl.: Soviet Math. Dokl., **15** (1974), 2, pp. 649 – 653.

[Le]Lenahan, T.A. *On Quantum Mechanical Scattering on Periodic Media* Ph.D. Thesis, University of Pennsylvania, Philadelphia, PA, 1970.

[Lo]Love, A.E. *A Treatise on the Mathematical Theory of Elasticity.* Dover, N.Y. 1944.

[Mad]Madelung, O., *Introduction to Solid State Theory.* Berlin, New-York, Springer-Verlag, 1978.

[Mag]Magnus, W. *Monodromy Groups and Hill's Equation.* Comm. Pure Appl. Math. **29** (1976), pp. 701 – 716.

[MagWin]Magnus, W., Winkler, S. *Hill's Equation.* New-York, Dover, 1979.

[Mar1]Marchenko, V.A., *Sturm-Liouville Operators and their Applications.* Basel: Birkhäuser, 1986.

[Mar2]Marchenko, V.A., Ostrovski, I.V., *A Characterization of the Spectrum of the Hill Operator.* Mat.Sb., **97(139)** (1975), 4(8), pp. 540 – 606; Engl. transl.: Mathematics of the USSR – Sbornik, **26**, 4, pp. 493 – 554.

[McKTr1]McKean, H.P., Trubowitz, E. *Hill's Equation and Hyperelliptic Function Theory in the Presence of Infinitely Many Branch Points.* Comm. Pure Appl. Math., **29** (1976), pp. 143 -226.

[McKTr2]McKean, H.P., Trubowitz, E. *Hill's surfaces and their Theta Functions.* Bull. Amer. Math. Soc. **84** (1978), 6, pp. 1042 – 1085.

[Ne.G]Nenciu, G., *Existence of Exponentially Localised Wannier Functions.* Comm. Math. Phys., **91** (1983), 1, pp. 81 – 85.

[No]Novikov, S.P. *Two-Dimensional Scrödinger Operators in Periodic Fields.* In Itogi Nauki i Techniki, Contemporary Problems of Math., **23**, pp. 3 – 23, VINITI, Moscow, 1983.

[OdKe]Odeh, F., Keller, J.B. *Partial Differential Equations with Periodic Coefficients and Bloch Waves in Crystals.* J. Math. Phys., **5** (1964), pp. 1499 –1504.

[Ol]Oleinik, V.L. *Asymptotic Behaviour of Energy Band Associated with Negative Energy Level.* Journal of Statistical Physics, **59** (1990), 3/4, pp. 665 – 678.

[Out]Outassourt, A. *Comportement Semi-Classique Pour l'Opérateur de Schrödinger à Potentiel Périodique.* Journal of Functional Analysis, **72** (1987), 1, pp. 65 – 93.

[Pas]Pastur, L. *Surface Waves: Propagation and Localization.* Journees "Equations aux Derivees Partielles" (Proceedings of the International Conference on Partial Differential Equations. Saint-Jean-de-Monts, 29 Mau au 2 June 1995), pp. IV.1-IV.12.

[Pav]Pavlov B.S. *Electron in the Homogeneous Crystal of Point Atoms with Internal Structure I*, Teor. and Mat. Fiz., **72** (1987), 3, pp. 403 – 415; Engl. Transl.: Theor. and Math. Phys., **72** (1988), 3, pp. 964 – 972.

[PavPol]Pavlov B.S., Popov I.Yu. *Surface States and Extension Theory.* Vestnik Leningr. Univ., 1986, **4**, pp. 105 – 107.

[PavSm1]Pavlov, B.S., Smirnov, N.V. *Resonance Scattering by One-Dimensional Crystal and Thin Film.* Vestnik Leningr. Univ., 1977, **13**, 4.

[PavSm2]Pavlov, B.S., Smirnov, N.V. *Spectral Properties of One-Dimensional Disperse Crystals.* Zapiski Nauchnich Seminarov LOMI, **133** (1984), pp. 197-211; Engl. transl.: Journal of Soviet Mathematics, **31** (1985), 6, pp. 3388 – 3398.

[PavSm3]Pavlov, B.S., Smirnov, N.V. *A Model of a Crystal from Potentials of Zero-Radius with Internal Structure.* Probl. Mat. Fiz., **12**, (1987), Leningrad, pp. 155-164 (in Russian).

[Pinc]Pinchover, Y. *On Positive Solutions of Elliptic Equations with Periodic Coefficients in Unbounded Domains.* Maximum Principles and Eigenvalue Problems in PDE, Pitman Res. Notes in Math., **175**, Longman, London, 1988, pp. 218 – 229.

[Pins]Pinsky, R.G., *Second Order Elliptic Operators with Periodic Coefficients; Criticality Theory, Perturbations and Positive Harmonic Functions.* J. of Functional Analysis, **129** (1995), 1, pp. 80 – 107.

[PoV,Sk]Popov, V.N., Skriganov, M.M., *Remark on the Structure of the Spectrum of a Two-Dimensional Schrödinger Operator with Periodic Potential.* Zap. Nauchn. Sem. Leningrad. Otdel. Mat. Inst. Steklov. (LOMI), **109** (1981), pp. 131-133; English transl.:J. Soviet Math., **24** (1984), 2, pp. 239 – 240.

[Rai]Raikov, G.D. *Eigenvalue Asymptotics for the Schrödinger Operator with a Perturbed Periodic Potential.* Invent. Math., **110** (1995), 1, pp. 75 –93.

[Ray]Rayleigh, J.W.S. *The Theory of Sound.* Dover, N.Y. 1945.

[ReSi1]Reed, M., Simon, B. *Methods of Modern Mathematical Physics.*, Vol I, Academic Press, 3rd ed., New York – San Francisco – London (1987).

[ReSi4]Reed, M., Simon, B. *Methods of Modern Mathematical Physics.*, Vol IV, Academic Press, 3rd ed., New York – San Francisco – London (1987).

[Scha]Scharf, G. *Das Blochsche Theorem für Unendliche Systeme.* Helv. Phys. Acta, **39** (1966), pp. 556 – 560.

[Schr]Schroeder, C. *Green's Functions for the Schrödinger Operator with a Periodic Potential.* J. Funct. Anal., **77** (1988), 1, pp. 60 – 87.

[SheShu]Shenk, D., Shubin, M. *Asymptotic Expansion of the State Density and the Spectral Density and the Spectral Function of a Hill Operator.* Math. USSR Sbornik, **56** (1987), 2, pp. 473 – 490.

[Si1]Simon, B. *Trace Ideals.* London Math. Soc. Lect. Notes, **35**, Cambridge Univ. Press, 1979.

[Si2]Simon, B. *Semiclassical Analysis of Low Lying Eigenvalues III – Width of the Ground State Band in Strongly Coupled Solids.* Ann. Phys., **158** (1984), 2, pp. 415 – 420.

[Sj] Sjöstrand, J. *Microlocal Analysis for the Periodic Magnetic Schrödinger Equation and Related Questions.* Microlocal Analysis and Applications, Lecture Notes in Physics, **1495**, Springer-Verlag, Berlin, e.a., 1991, pp. 237 – 332.

[Sk1]Skriganov, M.M. *Proof of the Bethe-Sommerfeld Conjecture in Dimension Two.* Dokl.Akad. Nauk SSSR, **248** (1979), 1, pp. 49-52; English transl. in Soviet Math. Dokl., **20** (1979), 5, pp. 956 – 959.

[Sk2]Skriganov, M.M. *On the Bethe-Sommerfeld Conjecture,* Dokl.Akad. Nauk SSSR, **244** (1979), 3, pp. 533-534; English transl.: Soviet Math. Dokl., **20** (1979), 1, pp. 89 – 90.

[Sk3]Skriganov, M.M. *Finiteness of the Number of Lacunae in the Spectrum of the Multidimensional Polyharmonic Operator with a Periodic Potential.* Mat. Sb., **113(155)** (1980), pp. 133-145; Engl. transl.: Math. USSR. Sb., **41** (1982), pp. 115 – 125.

[Sk4]Skriganov, M.M. *Geometric and Arithmetic Methods in the Spectral Theory of Multidimensional Periodic Operators.* Trudy Mat. Inst. Steklov., **171**, pp. 1 – 121. (1985); Engl. transl.: Proc. Steklov Inst. Math., **171**, 2 (1987).

[Sk5]Skriganov, M.M. *The Multidimensional Schrödinger Operator with a Periodic Potential.* Izv. Akad. Nauk SSSR Ser. Mat., **47** (1983), pp. 659-687; Engl. transl.: Math USSR Izv., **22** (1984), pp. 619 – 645.

[Sk6]Skriganov, M.M. *Proof of the Bethe-Sommerfeld Conjecture in Dimension Three.* Preprint R-6-84, Leningrad. Otdel. Mat. Inst. Steklov., Leningrad, 1984 (in Russian).

[Sk7]Skriganov, M.M. *The Spectrum Band Structure of the Three-Dimensional Schrödinger Operator with a Periodic Potential.* Invent. Math., **80** (1985) pp. 107 –121.

[Su1]Sunada, T. *A Periodic Schrödinger Operator on an Abelian Cover.* J. Faculty of Sci., Univ. Tokyo, Sec IA, **37** (1990), 3, pp. 575 – 583.

[Su2]Sunada, T. *Group C^*-Algebras and the Spectrum of a Periodic Schödinger Operator on a Manifold.* Canadian J. of Math. **44** (1992), 1, pp. 180 – 193.

[Th]Thomas, L.E. *Time Dependent Approach to Scattering from Impurities in Crystal.* Comm. Math. Phys., **33** (1973), pp. 335 – 343.

[ThWa1]Thomas, L.E., Wassell, S.R. *Semiclassical Approximation for Schrödinger Operators at High Energy.* Lecture Notes in Physics, **403**, Schrödinger operator, Proceedings, Aarhus, Denmark, 1991, edited by Balslev E. (Springer-Verlag, New-York, 1992), pp. 194 – 233.

[ThWa2]Thomas, L.E., Wassell, S.R. *Stability of Hamiltonian Systems at High Energy,* J. Math. Phys., **33**, 10 (1992), pp. 3367 – 3373.

[Ti]Titchmarsh, E.Ch. *Eigenfunction Expansions Associated with Second Order Differential Equations.* Oxford, Clarendon Press, 1962.

[Tr]Troianiello, G.M. *Scattering Theory for Schrödinger Operators with L_∞ Potentials and Distorted Bloch Waves.* J. Math. Phys., **15** (1974), 2048 – 2052.

[Ve1]Veliev, O.A. *On the Spectrum of the Schrödinger Operator with a Periodic Potential,* Dokl. Akad. Nauk SSSR, **268** (1983), 6, pp. 1289-1292; Engl. transl.: Soviet Math. Dokl., **27** (1983), pp. 234 – 237.

[VeMo1]Veliev, O.A., Molchanov, S.A. *Structure of the Spectrum of the Periodic Schrödinger Operator on an a Euclidean Torus,* Functional. Anal. i Prilozhen., **19** (1985), 3, pp. 86-87; Engl. transl.: Functional Anal. Appl., **19** (1985), pp. 238 – 240.

[Ve2]Veliev, O.A. *Asymptotic Formulas for the Eigenvalues of the Multidimensional Schrödinger operator and Periodic Differential Operators,* Preprint, **157**, Inst. Fiz. Akad. Nauk Azerbaidzan. SSR, Baku, 1985 (in Russian).

[Ve3]Veliev, O.A. *Asymptotic Formulae for Eigenvalues of a Periodic Schrödinger Operator and Bethe-Sommerfeld Conjecture.* Functional. Anal. i Prilozhen., **21** (1987), no. 2, 1–15; Engl. transl.: Functional Anal. Appl., **21** (1987), pp. 87 –99.

[Ve4]Veliev, O.A. *Asymptotic Formulae for Eigenvalues of the Schrödinger Operator.* Spectral Theory of operators and its applications, **8**, Baku, 1987, pp. 41 – 71 (in Russian).

[Ve5]Veliev, O.A. *On the Spectrum of Manydimensional Periodic Operator.* Functions Theory, Functional Analysis and their Applications, **49** (1988), Kharkov University (in Russian).

[Ve6]Veliev, O.A. *Author's Review of the Doctoral Thesis.* 1989, Tbilisi (in Russian).

[Ve7]Veliev, O.A. *Asymptotic Formulae for Bloch Functions of Multidimensional Periodic Schrödinger operator and Some of Their Applications.* Spectral Theory of Operators and its Applications, **9**, Baku, 1989, pp. 59 –76 (in Russian).

[WeKe1]Weinstein, M.I., Keller, J.B. *Hill's Equation with a Large Potential.* SIAM J. Appl. Math. **45** (1985), pp. 200 - 214.

[WeKe2]Weinstein, M.I., Keller, J.B. *Asymptotic Behavior of Stability Regions for Hill's Equation.* SIAM J. Appl. Math. **45** (1987), pp. 941 - 958.

[Wil]Wilcox E. *Theory of Bloch Waves.* J. d'Analyse Mathematique, **33** (1978), pp. 146-167.

[Ya]Yakovlev, N.N. *Asymptotic Estimates of the Densities of Lattice k-Packings and k-Coverings and the Structure of the Spectrum of the Schrödinger Operator with a Periodic Potential.* Dokl. Akad. Nauk SSSR, **276** (1984), 1, pp. 54 - 57; Engl. Transl.: Soviet Math. Doklady, **29** (1984), 3, pp. 457 - 460.

[Zi]Ziman, J.M. *Principles of the Theory of Solids.* Cambridge, University Press, 1965.

Index